Practical Electronic
Reliability Engineering

PRACTICAL ELECTRONIC RELIABILITY ENGINEERING

Getting the Job Done from Requirement through Acceptance

Jerome Klion

VNR VAN NOSTRAND REINHOLD
_____ New York

Library of Congress Catalog Card Number 91-39601
ISBN 0-442-00502-4

Manufactured in the United States of America

Published by Van Nostrand Reinhold
115 Fifth Avenue
New York, New York 10003

Chapman and Hall
2–6 Boundary Row
London, SE1 8HN, England

Thomas Nelson Australia
102 Dodds Street
South Melbourne 3205
Victoria, Australia

Nelson Canada
1120 Birchmount Road
Scarborough, Ontario M1K 5G4, Canada

16 15 14 13 12 11 10 9 8 7 6 5 4 3 2 1

Library of Congress Cataloging-in-Publication Data

Klion, Jerome, 1935–
    Practical electronic reliability engineering : getting the job
done from requirement through acceptance / Jerome Klion.
        p.   cm.
    Includes index.
    ISBN 0-442-00502-4
    1. Electronic apparatus and appliances—Reliability.   I. Title.
TK7870.K554   1992
621.381'54—dc20   91-39601
                                                            CIP

This book is dedicated to my wife Lila. Her encouragement, effort, and suggestions made its completion possible.

Acknowledgement is also given to Mr Mark McCallum for his technical aid in the preparation of the manuscript.

Contents

Preface

This book is intended for the engineer or engineering student with little or no prior background in reliability. Its purpose is to provide the background material and guidance necessary to comprehend and carry out all the tasks associated with a reliability program from specification generation to final demonstration of reliability achieved.

Most available texts on reliability concentrate on the mathematics and statistics used for reliability analysis, evaluation, and demonstration. They are more often suited more for the professional with a heavier mathematical background that most engineers have, and more often than not, ignore or pay short-shrift to basic engineering design and organizational efforts associated with a reliability program.

A reliability engineer must be familiar with both the mathematics and engineering aspects of a reliability program. This text:

1. Describes the mathematics needed for reliability analysis, evaluation, and demonstration commensurate with an engineer's background.
2. Provides background material, guidance, and references necessary to the structure and implementation of a reliability program including:
   - identification of the reliability standards in most common use
   - how to generate and respond to a reliability specification
   - how reliability can be increased
   - the tasks which make up a reliability program and how to judge the need and scope of each; how each is commonly performed; caution and comments about their application.

Jerome Klion

Practical Electronic
Reliability Engineering

1

Reliability: An Introduction to Concepts and Terms

Welcome to the world of reliability. By this time you may have heard and read a number of things about reliability and you may feel confused; you or your boss may feel that you need to be more reliability-literate in order to better perform your duties; you may be an experienced reliability hardware design engineer but feel a need for more insight into the mathematical/statistical aspects of reliability; you may be a reliability engineer experienced in mathematical/statistical modeling and analysis and feel a need for more insight into reliability engineering design-development aspects; or you may have been newly assigned to a reliability organization and have a degree in engineering but the barest acquaintance with probability and statistics. If so, this book is for you.

There are a number of reliability texts in print that are excellent in quantitative and mathematical detail but fail to provide the average engineer with the "feel" he needs to understand why he's doing what he's doing. This text is designed to provide both qualitative and quantitative insights into reliability theory and its statistical and engineering applications. In addition, the quantitative material included is aimed at the engineer with a minimum background in mathematics and statistics. The material that follows is ordered to describe the reliability concept, discuss and provide insight into reliability terms, and provide information necessary for application of various reliability analyses, tools, and tasks.

1.1 RELIABILITY AND FAILURE: THE CONCEPT

In order to get an appreciation of reliability, you first must get an appreciation of failure. Philosophers will tell you that everything must die at some time,

1

including the component parts making up an equipment. Such deaths or failures can be attributed to:

1. Defects in the basic material or parts
2. Defects, accidents, and stresses introduced through manufacturing and use environments
3. Defects induced by application or environmental conditions operating on a part
4. Deficiencies or omissions in the design/development process that allow otherwise avoidable stresses or inappropriate parts to be introduced into an item or equipment design
5. Basic physical and chemical processes inherent in the material that precipitate wearout

The wearout phenomenon, while applicable to mechanical items and equipment, is not generally considered for electronic equipment. This may well be attributable to the fact that a typical service life of less than 10 years is common to electronic equipments. Data shows that this period of time, and even times substantially longer, does not result in an accelerated frequency of failure due to wearout phenomena in the basic material used. (For equipments which contain components subject to wearout, the general policy is to replace such components in accord with a preplanned schedule based on wearout start times. Such replacements are usually integrated with other scheduled maintenance tasks.)

Over service life, each part in either an electronic or mechanical equipment is subjected to a use environment. Such environments can be construed as stresses, temperature, vibration, pressure, etc. Each such stress acts on the component parts of the equipment, much like the way a game fish stresses a fishing line and affects reliability:

1. If there are different types of defects in different places in the line, the line is more prone to breakage under any given weight of fish than one having fewer defects of the same types.
2. Assume that there are defects in the line which are adversely affected by degree of immersion time in salt water and line tension. The line is more prone to fail if it is immersed in salt water for 20 hours than if it is immersed in salt water for 10 hours, and if a fish weighs 20 lb as opposed to 10 lb.

Each equipment is also subjected to a "condition of use" environment. While not always an outright stress on the components of the system, it affects the scenarios and rules associated with care and operation. It includes

TABLE 1-1

Equipment	Number of parts	Mean Time Between Failure
Radar	13 500	400
Communications	6000	1000
Computer	15 000	900

such elements as mission/use profile, mission/use time, duty cycle, maintenance concept, and logistics plan. Such "conditions of use" also affect reliability. (These will be covered in more detail in later chapters.)

If you get the idea by now that I'm telling you to base your concept of reliability on defects and deficiencies in parts and processes that result in failures under service/life stresses, you're right. It's an easy concept to grasp; it will provide a logical feel about what failure rate is all about; and it will make sense out of screening, reliability growth, and the relationship between reliability and quality.

Defects or deficiencies do not have to be of significant proportions to make themselves felt. See Table 1-1, which includes some typical mean-time failure and complexity data for equipment developed in the 1980s.

Assuming a worst case of 24-hour-a-day operation over a service life of 10 years, the equipments in question would suffer 225, 90, and 100 failures, respectively. These figures amount to failure of only 1.7%, 1.5%, and 0.7% of the parts making up each equipment. In other words, in each case over 98% of the parts would still be expected to be operating at the end of service life.

Those failures which occur over service life will appear to occur *randomly over time*. You may have heard the statement that "failures occur randomly." This is wrong. Failures never did and never will occur randomly. Failures occur because of defects. Saying that the combination of defects and accrued stresses acting on such defects results in failures which appear to occur *randomly in time or over a period of time* presents the idea in a more proper light.

If something occurs randomly, it occurs randomly in accord with some characteristics. For example, it may occur randomly but the rate of occurrence may stay the same, increase, or decrease with time. Consider as an illustration insurance mortality tables, which tell us that probability of death for an individual over a 12 month period of time can be treated as random and that the older the individual is, the greater the probability of occurrence of death in that period.

A given item or equipment, depending on the characteristics of its environmental and use stresses and the quantities of its defects, can experience rates of failure which stay the same, increase, or decrease with time or with item age. We describe changes in rate of occurrence of failures over time through the concept of time to failure distributions. To be perfectly accurate we should rather say assumptions of, or approximations to, distributions. This will be discussed in more detail in Chapter 2. The most commonly used distribution in reliability is the exponential, which has as its basis the assumption that failures throughout service life occur at a constant rate. Other distributions are applied to describe the assumption that failures throughout service life occur at either decreasing or increasing rates.

In the majority of cases, a constant or decreasing rate of failure with time is more commonly found for equipments and complex components than increasing failure rates with time. Incidences of increasing failure rates with time are limited to equipment with wearout phenomena or significant quality problems.

Additionally, it is common for decreasing failure rates to be associated with the first few years of field usage or periods of test.

It is the quantity and types of defects within a population of parts (a part lot or an equipment) that will determine the failure distribution associated with the part lot or equipment under a given set of operating stresses. As a consequence, it is possible because of defects introduced through material, design, and manufacturing processes, to observe increasing, decreasing, or constant rates of failure over different phases of operational life. The well-known bathtub curve of reliability exemplifies this well (see Fig. 1-1).

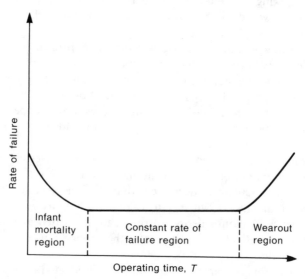

FIGURE 1-1.   The bathtub curve.

1.2   RELIABILITY: ITS ORIGIN

In order to see why and where reliability began, we have to travel back in time to just prior to the United States' involvement in World War II. At that time, modern-day electronics applications such as radar, sonar, and improved communications concepts were in their infancy. At that time, the total electronics in the most sophisticated aircraft cost on the order of a few thousand dollars. It was soon realized that war-fighting capability on all fronts could be significantly enhanced through the application of electronics. As a consequence, electronics use increased, as did its complexity. As the importance of electronics to the war effort increased, and electronics complexity increased, the importance of reliability also increased. This served as an impetus to research and development that continued after the war.

During World War II, what is considered by some to be the first application of reliability engineering was undertaken. In Germany, Robert Lusser, working on the V-2 rocket, showed that the multiplication law of probability (i.e., that the probability that an overall event will occur is equal to the product of the probabilities of the occurrences of a number of other independent events; discussed further in Chapter 5) applied to the reliability of a series-connected system.[1] As electronics design techniques evolved, a trend towards electronic function size "shrinkage" became evident. This occurred first in the development of miniature and subminiature vacuum tubes and other components and then in the introduction of the transistor in the early 1950s. The capability to reduce volume and power needs in conjunction with the need to perform more functions and to perform the functions better and faster led to a rapid increase in electronic equipment/ system complexity by the early 1950s. By that time, failures in electronic equipment were becoming recognized problems for both mission achievement and maintenance and logistics costs.

The need to do something about reliability was first recognized in 1950 when the Ad-Hoc Group on Reliability of Electronics Equipment was formed under the auspices of the Department of Defense (DOD). (The name of the group was officially changed to the Advisory Group on Reliability of Electronic Equipment—AGREE—in 1952.) The group had as its objectives: (1) to monitor and stimulate interest in the art of reliability; (2) to provide recommendations concerning steps that the services should take in this area. Its efforts served as the impetus for additional efforts by the services in this area (see Ref. 1 for details of such efforts). By 1955 both interest and knowledge had reached a point enabling at least rudimentary recommendations to be made in this area. A special AGREE committee made up of selected individuals from both industry and government was established to formulate guidance in reliability application for DOD. The report generated,

"Reliability of Military Electronic Equipment", was published in 1957. In 1956 and 1958 the Air Force's Rome Air Development Center (RADC)* developed, respectively, "Reliability Guidelines of Ground Electronic Equipment,"[2] considered by most to be the first overall reliability design guide published for general use, and the first standard means for reliability prediction of electronic equipment. The procedure formed the basis for Mil-Hdbk-217, "Reliability Prediction of Electronic Equipment."

In the late 1950s and 1960s the reliability standards and handbooks with which we are so familiar today were generated. Reliability as a consideration in design and development was off and running.

1.3   RELIABILITY: A QUANTITATIVE DEFINITION

The quantitative definition of reliability is based on the characterization of three related parameters: hazard rate, failure rate, and mean time to failure. The relationships among these parameters are sometimes confused.

Hazard rate, $h(t)$, can be defined as a rate of failure in terms of percentage or proportion of failures during a particular arbitrarily small interval of time that starts at a specific point in time. It can be interpreted *for our purposes*, as the average rate of failure (failure rate) per hour experienced during a *specific* one-hour interval: as the proportion or number of failures experienced during a *specific* one-hour interval, taking as a base the number of items functioning at the beginning of that time interval. If the hazard rate $h(t)$ increases or decreases with time t, it is indicative of a time to failure distribution that is not exponential. If $h(t)$ remains the same over all time t, then it is a constant that is indicative of an exponential distribution of failure time. The particular trend of $h(t)$ is uniquely linked to the particular failure distribution associated with the part or equipment.

Failure rate, $\lambda(T)$, on the other hand, can be thought of as the *average* hazard rate, per hour, *calculated over any given period of operating time (of many hours)*. Military Handbook-338 (Mil-Hdbk-338, Reliability Design Handbook) likens failure rate to the average speed made over a trip of several hundred miles (e.g., 50 miles per hour) and hazard rate to the speedometer reading at any given point in the trip (e.g., 35 miles per hour). Military Standard-721 (Mil-Std-721, Definitions of Terms for Reliability and Maintainability) defines failure rate as the total number of failures within an item population, divided by the total number of life units (total operating time) expended by that population, during a particular measurement interval

---

*In 1991 the name of the Rome Air Development Center (RADC) was changed to Rome Laboratory.

under stated conditions. Hence, both documents clearly treat failure rate as an average measure calculated over a given period of time. We can express failure rate $\lambda(T)$ as the average value of $h(t)$ over a period of time through the following relationship:

$$\lambda(T) = \frac{\displaystyle\int_0^T h(t)\, dt}{T} \tag{1-1}$$

where

$h(t)$ $\quad$ = hazard rate

$h(t)\, dt$ $\quad$ = the expected number of failures in time dt

$\int_0^T h(t)\, dt$ = the summation (the integral) of the expected number of failures over a period of time $t = 0$ to $t = T =$ the total number of failures expected between $t = 0$ and $t = T$

T $\quad$ = the period of time in question from $t = 0$ to $t = T$

$\lambda(T)$ $\quad$ = the average failure rate per operating hour experienced over T hours of operation

More generally, if failure rate is calculated over a period of time T from $t = T_1$ to T_2, i.e., $T = (T_2 - T_1)$, then

$$\lambda(T_2 - T_1) = \lambda(T) = \frac{\displaystyle\int_{T_1}^{T_2} h(t)\, dt}{T_2 - T_1} \tag{1-2}$$

where

T_1 = the number of hours of prior operating service

T_2 = the number of hours of use on the item after operating over the T-hour interval

$T\ = T_2 - T_1$

Mean time to failure, MTTF, can be defined as the average time to failure of a population of items when that population is operated under a specific set of conditions for a specific period of time. Logically speaking, if failure rate represents the number of failures expected over a *specific* time interval, 0 to T, or more generally from T_1 to T_2, then the mean or average time to failure over T is equal to the reciprocal of the failure rate over T.

$$\text{MTTF}(T_2 - T_1) = \frac{1}{\lambda(T_2 - T_1)} \tag{1-3}$$

where

$\text{MTTF}(T_2 - T_1)$ = the mean time to failure as measured between times T_1 to T_2

$\lambda(T_2 - T_1)$ = the failure rate measured between times T_1 to T_2

When $T_1 = 0$, then $T = T_2$, and (1-3) reduces to

$$\text{MTTF}(T) = \frac{1}{\lambda(T)} \qquad (1\text{-}4)$$

The confusion among the three terms rests with the fact that for the exponential distribution—the most commonly used distribution—there is a clearcut relationship among the three parameters:

$$\text{MTTF} = 1/\text{hazard rate} = 1/\text{failure rate}$$

$$\text{Hazard rate} = \text{failure rate}$$

for *any* and *all* intervals of operating time T (i.e., for all values of T_1 and T_2). That is to say,

$$\lambda = h(t) = \lambda(T) \qquad (1\text{-}5)$$

When other common distributions of failure are assumed, such particular relationships will not hold. While the particular relationships depicted above may not hold for distributions other than the exponential, other more general relationships between hazard and failure rates will always hold: *Failure rate, over a specific period of time T, for any distribution can be defined as in (1-2).*

The use of *failure rate* in describing the results of *hazard rate* characteristics, for distributions other than the exponential, often provides design/development engineers with a better visibility and grasp of implications than the use of the term hazard alone.

Hazard and failure rate are direct components of the most general reliability expression. The reliability equation, regardless of whether or not $h(t)$ is constant or changes with time, takes the general form

$$R(T_2 - T_1) = e^{-\int_{T_1}^{T_2} h(t)\, dt} \qquad (1\text{-}6)$$

where

$R(T_2 - T_1)$ = probability that an item can perform its intended function for a specified interval of time $T = (T_2 - T_1)$ under stated conditions. In other words, the probability that the item will not fail during T assuming it was operable at T_2

$h(t)$ = the hazard rate of the item

T = the interval of operating hours specified

T_1 and T_2 are defined as before

When the distribution of failures is exponential, (1-6) reduces to the familiar form

$$R(T) = e^{-\lambda T} \tag{1-7}$$

because of the memoryless characteristic of the exponential distribution, which will be discussed in the next chapter.

Let us see how it is derived. We will derive it in two separate fashions, one based on a calculus derivation and one based on a rather simple probabilistic concept.

Assume that N different items are put on test. As they fail they are not replaced, but the survivors continue on in the test. Let N items be put on test. At some time during the test, we look at the results:

F = the number of units that failed

S = the number of units that survived

and

$$S = N - F \tag{1-8}$$

At any given point in time, T, the ratio of survivors, S, to N, the original number of items starting the test, represents the proportion of units which survived. This also represents the probability that an item representative of the items under test will not fail during T. Reliability, R, may then be represented as

$$R = S/N = 1 - \frac{F}{N} \tag{1-9}$$

The hazard in accord with the definition provided earlier (the rate of failure during a small increment of time) can be expressed in terms of the rate of change of failure rate (F/S) with time where F = the number of failures occurring during a period of time and S = the number of items operating (nonfailed) at the beginning of the period.

From elementary calculus, hazard $h(t)$ takes the form of

$$h(t) = \frac{d(F/S)}{dt} = \frac{dF}{dt}\frac{1}{S} \qquad (1\text{-}10)$$

Substituting from (1-8) and (1-9), (1-10) can be expressed as

$$h(t) = \frac{1}{R}\frac{dR}{dt} \qquad (1\text{-}11)$$

which, when integrated, takes the form

$$R = e^{-\int h(t)\,dt} \qquad (1\text{-}12)$$

If we assume that an exponential failure distribution holds, then the hazard is a constant over every unit of time comprising the interval 0 to T and

$$R(T) = e^{-\lambda T} \qquad (1\text{-}13)$$

where λ = the hazard (failure) rate per operating hour.

The preceding, while applying only rudimentary calculus skills, may seem a bit involved. So let me provide you with an alternate more direct means to make sense out of the reliability equation. For this I will introduce a rudimentary statistical distribution, the Poisson distribution.

One of the problems most engineers have with statistics, probabilities, and mathematical models in general is the fact that it is not realized that they just tell a story. If the story is understood, the mystery is gone. Here in rudimentary form is the Poisson distribution and its story.

The distribution:

$$P(x) = \frac{a^x e^{-x}}{x!} \qquad (1\text{-}14)$$

where x = number of events and a = expected number of events.
The story:

$P(x)$ = the probability of x events when a events are expected.

Let us tailor the story to meet our needs. Let *events* connote failures. In addition, recall our definition of failure rate, $\lambda(T)$, which was the average rate of failure occurrence per hour over a specific interval of time, $T_2 - T_1 = T$.

$\lambda(T)T$ = expected number of failures in T. Therefore,

$$\lambda(T)T = a$$

Next take a value of $x = 0$. Equation (1-14) now takes the form

$$P(0) = e^{-\lambda(T)T} \qquad\qquad (1\text{-}15)$$

or

$$P(0) = e^{-\lambda T} \qquad \text{(assuming an exponential failure distribution)}$$

where $P(0)$ = the probability of no failures in T operating hours when $\lambda(T)T$ failures are expected. In other words, the probability that the item will not fail during T. Therefore,

$$P(0) = R(T)$$

In Chapter 2 we will discuss $R(T)$ and $\lambda(T)$ characteristics for nonexponential distributions in such detail for those who are interested.

1.4 RELIABILITY: COMPREHENSION OF DEFINITIONS, MEASURES, AND SPECIFICATIONS

Communication between the customer and the designer/developer is necessary for meeting program targets and specifications. In the area of reliability, the critical path for such communication is in the proper definition and understanding of reliability terminology. One of the core areas in defining and interpreting reliability terms and measures is realizing that some make more sense in some situations than in others. For example, whether an item can be maintained during use or cannot has a bearing on the reliability parameters which are appropriate. Whether the item or system is redundant (fault-tolerant) or has a simple series configuration can also effect the sense of the terms and measures applied.

The following represent definitions of terms important to understanding reliability. Where appropriate, explanations and sample applications are provided for each term/measure. In some instances the common literature uses more than one term to describe a given measure or concept. For those cases we have identified the terms which are equivalent. The definitions which follow, tailored to enhance comprehension, are consistent with those which appear in Mil-Std-721, "Definitions of Terms for Reliability and

Maintainability" (a DOD standard which will be further described in Chapter 3) and other documents.

Availability   A measure of the degree to which an item is in an operable and commitable state when the mission is called for at a future unknown (random) time. It is usually expressed as a probability or a proportion and applied when corrective maintenance may be performed during item operation or when the item is in a standby status. The term is appropriate for use for both redundant and series-connected systems.

Application sample: A search radar functions 24 hours a day. Its objective is to acquire information on any unfriendly aircraft entering a given sector. We will assume that the radar has no redundant components, and further that when it fails it is inoperable until a repair is made. For a continuous mission, what is the probability that the radar is operable and operating at the time an unfriendly aircraft enters the sector? What proportion of the time will the radar be in an operating state?

Dependability   A measure of the degree to which an item is capable of performing its function, either at any random time during a mission or through the entire mission term, taking into account its *operating state at the beginning of the mission*. It is usually expressed as a probability and generally applied when associated with "system effectiveness" concepts (described in Chapter 9). It takes into account the combined effects of reliability, maintainability, and maintenance and can be applied to both redundant and series-connected systems.

Application sample: Assume that the search radar above is redundant, that is to say, that it can operate satisfactorily when all of its components are operable and/or when certain of its components are failed. Each possible combination of operating and failed components defines a specific system operating state. A system can start a mission in any operating state that can provide at least a minimum degree of successful performance. However, the reliability of the system over the mission term depends on the particular state that the system is in at the beginning of the mission. Given that the system is in one of its operable states at the beginning of the mission, what is the probability that it will function satisfactorily for the duration of the mission?

Inherent Reliability   A measure of the reliability of an item in an ideal operational and support environment.

Application sample: A system is put on test in a controllable factory test environment which simulates a particular field usage. It is maintained by qualified technicians whose actions are monitored; it is not unduly cycled off and on; after each failure it undergoes a complete checkout; failures caused by outside causes, such as mishandling are not counted.

Mean Time to First Failure (MTFF)   A measure of reliability, usually applied only to redundant systems, or to equipment/components which exhibit increasing or decreasing failure rate with time; in general the mean (average) time it takes to experience a system failure, measured from the time that all system/equipment items are brand new and operable; for the case where each system component has an exponential distribution of failure time, the mean (average) time to experience a system failure measured from the time all components are operable. It is appropriate for application for redundant systems composed of components having exponential failure distributions only when the system is shut down after first indication of system failure and not activated again until *all* failed elements are repaired.

Application sample: The redundant system shown in Fig. 1-2 is made up of units *A* and *B*. Both units have exponential distributions of failure time

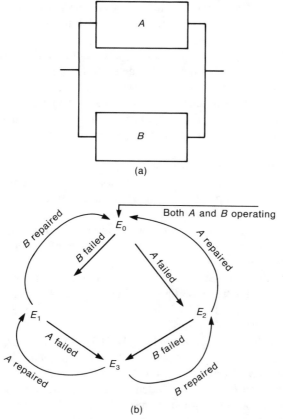

(a)

(b)

FIGURE 1-2.   (a) Two redundant components. (b) Possible state transitions for the subsystem in (a).

and are in operation at the same time. If one fails, the other can carry on the system function. While the surviving unit is carrying on the system function, the failed unit is being repaired. A system failure occurs only if the remaining surviving unit fails before the failed unit is repaired. If that happens, the system is started again only when both units have been repaired. What will be the MTFF for the system (i.e., the mean operating time between system startup and failure)?

Steady-State Mean Time Between Failure (M_{ss})   A system measure of reliability applied to redundant systems where (1) corrective maintenance can be performed during the mission, (2) after the first indication of *system* failure, the system is always returned to use after the *minimum* number of elements necessary to maintain system operation are operable, as opposed to waiting until all failed elements are repaired (as is done when working with MTFF).

It is the mean (average) time it takes to go from such a minimum operating state (defined as a *borderline* operating system state) back to a system failed state.

Application sample: Take again the redundant system shown in Fig. 1-2. All conditions described in the MTFF definition still hold with the exception that in the event of failure in the surviving unit prior to repair of the failed unit, the system enters a failed state but will start to perform its operational function when *either A or B* has been repaired. M_{ss} defines the average time between failures after a system failure occurs and the system is restored to a minimum operating state.

Mean Time Between Failure (MTBF) (see also Mean Time to Failure, MTFF)   A basic measure of reliability for repairable items that have an exponential distribution of failure times; the mean (average) number of hours during which all parts of the item perform within their specified limits, during a particular measurement interval under stated conditions; the reciprocal of failure rate or hazard; the total number of hours of operation of an item divided by the total number of failures.

Application sample: Studies of particular communications equipment all in use in a specific type of fighter aircraft over the past 4 years have accumulated a total of 150 000 hours of operation and experienced 250 failures. Its MTBF = 600 hours.

Mean Time Between Maintenance (MTBM)   A measure of an item's demand for maintenance that combines the effects of reliability, of scheduled maintenance, and of erroneous calls for maintenance; the total number of hours of operation of an item divided by the total number of maintenance events (calls) associated with that item. The term is applicable for use for both redundant and series-connected systems.

Application sample: Over the past 4 years a particular communications equipment in use in a specific type of fighter aircraft accumulated a total of 150 000 operating hours, and accumulated 250 failures, 100 calls for maintenance for which no cause could be found, and 100 calls for scheduled maintenance. MTBM measures the average time between *all* calls for maintenance. Its MTBM = 333 hours.

Mean Time Between Removals (MTBR)   A measure of the system reliability parameter related to demand for logistics support; the total number of system operating hours divided by the total number of items removed from the item during a stated period of time. The measure excludes removals performed to facilitate other maintenance and removals associated with product improvements. It can, however, include the removal of more than one removable unit per maintenance event and it can include removal of units that later test good. Appropriate for use for both redundant and series-type systems.

Application sample: Over the past 4 years a particular communications equipment in use in a specific aircraft accumulated a total of 150 000 operating hours, and accumulated 250 failures that necessitated 320 removals of units (some corrective maintenance events resulted in the removal of more than one unit). No other units were removed during the period. MTBR measures the average time between *all removals*. MTBR = 469 hours.

Mean Time to Failure (MTTF) or Mean Residual Life (MRL)   A term used in many texts and papers synonomously with MTBF (when an exponential failure distribution of time to failure is assumed). More generally, the average time to failure measured from a *particular point* in time, or from a *particular system state*. (M_{ss} defined previously is a special category of MTTF.) The term can be used in conjunction with an item or a population of items, for a repairable or a nonrepairable item. Defined as the total number of operating hours of a population of items divided by the total number of failures within that population, measured from a particular point in item (population) operating history. It is equal to the reciprocal of the failure rate over the particular operating interval and is appropriate for application to both redundant and series-connected systems.

Application sample: A population of 1000 items are put on test when all are brand new until all have failed. What is the MTTF of the population? A population of 1000 items are put on test when each has a prior operating history of 300 hours of operation and are tested until all have failed. What is the MTTF of the population? Only if the items all have exponential distributions of failure time will their MTTF or MRL be the same.

Logistics Reliability (LR) or Single-Thread Reliability (STR)   A measure of an item's demand for maintenance and spares attributable to reliability. It is usually only applied to redundant systems. It treats the system as if all its components were connected in series (i.e., the failure rate of the system is computed as the sum of all the failure rates of the system's components). The measures are calculated by dividing total operating hours by the total number of *component* failures.

Application sample: The redundant *system* in Fig. 1-2, over a period of 21 000 hours of operating time, failed only 14 times; however, unit *A* failed 10 times and unit *B* failed 11 times. LR measures the average time between *component* failures. $(LR/STR) = 1000$ hours.

Mission Reliability   The probability that an item can perform its intended function for a specified interval of time under stated conditions. It is applicable to both redundant and nonredundant systems.

Application sample: The terrain-following radar of a bomber has a maximum mission time of 10 hours. The probability of its providing that function for that period of time is 0.999.

Reliability   A general term that can be expressed in a variety of ways: (1) The probability that an item will perform its required function for a specified period of time under stated conditions (note that many documents define reliability uniquely as this). This is applicable to both redundant and nonredundant systems; it is equivalent to the *mission reliability* measure described above. (2) A duration of failure-free performance under stated conditions over a specified interval of time. It can be expressed in terms of MTBF, LR, M_{ss}, MTFF, etc. applicable to both redundant and nonredundant systems. (3) A failure rate, generally applied at the device level rather than at the equipment or system level.

Application sample: The redundant system in Fig. 1-2 is on continuously. Assuming that it has repetitive missions to perform of 10 hours duration, we most commonly describe its reliability in terms of (1) or (2) above.

References
1. Coutinho, J. de S. (1974). *Introduction to Systems Assurance*. U.S. Army Material Systems Analysis Agency. Special Publication SP-9.
2. Henny, K. (1956). *Reliability Factors For Ground Electronic Equipment*. McGraw-Hill, New York.

2

Application of Failure Distributions to Reliability

Reliability prediction, analysis, and demonstration are based on the characteristics of the distributions of times to failure which we associate with component parts, equipments, and systems. In most instances, the failure distribution assumed is based upon data available, historical precedent, or engineering judgment. When a failure distribution is chosen based upon available empirical data, generally speaking, the best that can be said is that the data closely approximates one distribution or another.

It is possible for distributions of times to failure of parts, components, and systems to take on failure rates (hazard rates) which increase, decrease or remain approximately constant over operating time. As a consequence, general distributions have been suggested which fit those needs. The distribution which has been most closely (and classically) associated with reliability has been the exponential. This is the most simple distribution to apply to prediction, evaluation, and demonstration. If you are comfortable with the assumption of a constant failure rate for parts, components, and equipments, you need just read that portion of subsection 2.1 that pertains to the exponential and skim the remainder of the subsection. In the event that you want to learn more about the ramifications of other failure distributions; on the reliability of an item; on reliability prediction, evaluation, or demonstration process, this chapter will aid your comprehension. The distributions that we will discuss will be those most commonly used in reliability engineering.

Before discussing the characteristics of the distributions used for reliability purposes, it is important to first discuss some general factors which are often ignored, misinterpreted, or glossed over when dealing with failure distributions that have increasing or decreasing hazard rates. First of all, note the existence

17

of a general limit theorem,[1] which implies that (1) if an equipment is made up of very many different items, each with a nonconstant hazard, and (2) each item contributes a relatively small part to failure frequency, and enough failures (and replacements) take place over service life, then eventually the equipment will take on an exponential distribution of failure. Prior to using this as sole justification for applying the exponential distribution to your program, it would be well to perform analyses (i.e., Monte Carlo simulations) to verify that it holds for your particular application and the service life of the equipment you're developing. An equipment made up of items having increasing/decreasing hazard with time will, for some time at least, result in some sort of equipment failure distribution having an increasing or decreasing hazard rate. In all likelihood it will be possible to *approximate* the failure distribution by one of the more common distributions that we will discuss.

Whatever the distribution that is approximated, if its hazard rate increases or decreases with time, and if we assume that the characteristics remain fixed, then:

1. The mean time to the first equipment failure MTFF, which we can denote by $(MTTF)_1$, and the mean time to the second equipment failure, $(MTTF)_2$, will be different, and in general

   $$MTTF_i \text{ will always be greater or less than } MTTF_{i+1}$$

   As a consequence, the average time between failures changes from one failure to the next and it is clear that the mean time *between* failures varies. Hence it becomes clear why the term MTBF is inappropriate when hazard rate is not a constant. If the hazard rate is constant, then

   $$MTTF_i = MTTF_{i+1} = MTBF = MTFF$$

2. If the approximation dealt with the failure distribution of a population of parts as opposed to an equipment, then

   $$MTTF \text{ of the population and MTFF of each part can be}$$
   $$\text{considered equivalent}$$

3. For both an equipment or a part, if a nonconstant hazard applies, then the reliability of the item, $R(T_2 - T_1)$, will vary as the values T_1 and T_2 vary, even though $(T_2 - T_1)$ always equals the same value, T, where $R(T_2 - T_1) = $ the probability that the item will operate satisfactorily over an interval of operating time $T = (T_2 - T_1)$, given that it had accrued a

prior operational history of T_1 hours of use and was in a satisfactory operating condition at the start of the mission.

For example, an item has experienced prior use of 200 hours of operation (i.e., $T_1 = 200$) and will be put to use for a *mission term of 100 hours* (i.e., the item, at the end of the mission, will have accrued an operating history of $T_2 = 300$ hours). Assume now the same item has completed the mission and has been sent out on an identical mission of *100 hours* duration. Now, $T_1 = 300$, and at the end of the mission $T_2 = 400$. Even though $T = 100$ in both cases, the probabilities expressed by $R(T_2 - T_1)$ will be different. If the hazard rate was constant, both values would be identical.

In the sections that follow, guidance will be provided on how to characterize and assess the behavior of parts, equipment, and systems that have both constant and nonconstant hazards.

Reliability engineering also deals with the application of statistical distributions to various reliability design, demonstration, and analysis tasks. These too, can be perplexing to the engineer. At times it may be indicated that a given reliability measurement or evaluation can be performed using a chi-square distribution; another time the same type of assessment may be performed using a Poisson distribution; yet another time it may be performed using a gamma distribution. The very concept of a distribution, a probability density function, is often foreign to the engineer's background and technical cultural experience. The very number of different distributions (see Chart 2-1) applied to one facet or another of analysis and depiction is another compounding factor.

Be that as it may, the proper comprehension of reliability needs and the development of an effective approach to meet such needs is dependent on

---

CHART 2-1 A Sampling of Statistical Distributions Applied to Reliability Tasks

1. The exponential distribution
2. The normal distribution
3. The chi-square distribution
4. The Poisson distribution
5. The gamma distribution
6. The log-normal distribution
7. The beta distribution
8. The Weibull distribution
9. The F distribution
10. The binomial distribution
11. The multinomial distribution
12. The hypergeometric distribution

the engineer having a clear insight into the role and applications of distributions to reliability.

In order to simplify the discussion, we will arbitrarily break down the *application of distributions in reliability* into three separate application areas:

- To describe the distribution of failures associated with items and equipments made up of such items
- To aid in the modeling and analysis of items and systems
- To implement reliability demonstration and measurement tests

2.1 DISTRIBUTIONS ASSOCIATED WITH FAILURE OF ITEMS AND EQUIPMENT

In Chapter 1 it was indicated that defects in material or parts and defects introduced during the development/manufacturing process resulted in failures. Each vendor lot of parts produced includes a quantity of parts which have one or more defects which will cause those parts to fail during *service life* when subjected to a specific application/use environment. Each equipment produced has associated with it defective material or defects introduced during manufacturing, which under a specific use environment will cause failures in the equipment over its service life. The quantity of defects contained and the character of each defect, in combination with the application/use stresses acting, will determine the transition time from symptomless defect to failure incidence and the frequency of such transitions. This is in essence what a failure distribution characterizes. Further, if you get the impression that there is no one standard failure distribution which will work for everything, you're right. Conceptually, depending on its defect makeup, almost any failure distribution for any product is possible. Experience and analysis of available information on the product and similar products, in combination with engineering, process, and physics knowledge, will provide some intelligence in this area. These will aid in the identification of the most likely failure distribution (and the parameters of the distribution) associated with your product or a given family of products.

Discussion in this section will address four failure distributions which are most commonly found in the literature which have a wide range of applicability: (1) the exponential, (2) the Weibull, (3) the log-normal, (4) the gamma.

2.1.1 The Exponential Distribution

The exponential distribution is perhaps the most familiar and most frequently used distribution in reliability engineering. Its particular characteristic property is the fact that a constant hazard rate (and, hence, a constant failure rate) is assumed. This means, for example, that an item which has been operating for 1000 hours is just as likely to survive for an additional 1000 hours as an item which is brand new. It means that an equipment, after failing for the first time or the 100th time (and being repaired) is as reliable as an equipment that is brand new. This is commonly referred to as the *memoryless characteristic* of the exponential distribution. It means, for a population of parts put under life test, that for each interval of time, say of length 500 hours, the same proportion of parts (based on the number still functioning at the beginning of the interval) will, on the average, fail.

The exponential failure distribution is widely applied to electronic parts, equipments, and systems. There is less clear evidence for its application to parts than for its applicability to equipments. Data and experience have historically indicated a good fit to equipments and systems in many reported instances, especially in studies of mature equipments (see for example Refs. 1 and 2).

2.1.1.1 Mathematical Characteristics: Exponential Distribution at the Device or Item Level

We define $f(t)$ = a failure probability (density) function:

$$f(t) = \lambda e^{-\lambda t} \qquad \text{(see Fig. 2-1a)} \tag{2-1}$$

where

λ = failure rate, or hazard rate; for this distribution they are equivalent

t = a particular time

Then

$$\int_0^T f(t)\, dt = \text{probability that an item will fail during any interval of time of length } T$$

and we can describe the reliability $R(T)$ of the item as

$R(T)$ = probability that an item will perform satisfactorily (not fail) during period of operating time T

$$R(T) = 1 - \int_0^T f(t)\, dt = e^{-\lambda t} \qquad \text{(see Fig. 2-1b)} \tag{2-2}$$

Hazard rate $= h(t) = \lambda$ (a constant) (see Fig. 2-1c)

Failure rate $= \lambda =$ hazard rate

Mean time to failure or MTBF $= 1/\lambda$

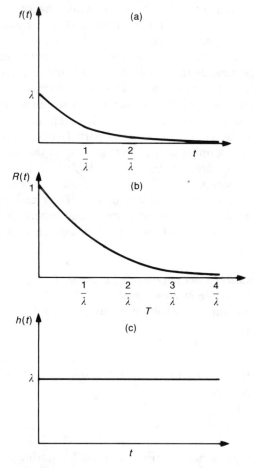

FIGURE 2-1. (a) Exponential density function. (b) Exponential reliability function. (c) Hazard rate associated with an exponential density function.

Special properties Because of the fact that the exponential distribution has a *memoryless characteristic*, T can be interpreted as *any T-hour period* of operating time (see Appendices A0 and A1 for proof).

Example A radar set that has no redundant components has a failure rate that reduces to one failure per 100 hours of operation (we indicate "reduces to" because most failure rates are keyed to percentage of failures per million or billion hours of operation), or an MTBF = 100 hours. It is put into use just after it has been delivered and has had no prior operating history. Its first mission has a duration of 10 hours. What is the probability it will not fail during the mission?

$$R(T) = e^{-\lambda T}$$

$$T = 10, \qquad \lambda = 1/100 = 0.01$$

$$R(10) = e^{-0.1} = 0.905$$

The radar has been in operation for several months and has accumulated 300 hours of use. A new mission of 15 hours duration has been assigned to the radar. (1) What is the reliability of the radar for the new mission?

$$R(15) = e^{-0.15} = 0.86$$

(2) What would the reliability have been for a mission of 10 hours?

$$R(10) = e^{-0.1} = 0.905$$

2.1.1.2   Calculating / Predicting Equipment Reliability When All Items Have Exponential Failure Distributions

Since $R_i(T) = \exp(-\lambda_i T) =$ probability that item i will perform satisfactorily for T hours,

$$R_E(T) = \prod_{i=1}^{n} R_i(T) = \prod_{i=1}^{n} \exp(-\lambda_i T)$$

$$R_E(T) = \exp\left(-T \sum_{i=1}^{n} \lambda_i\right) = \exp(-\lambda_T T) \qquad (2\text{-}3)$$

where
λ_i = hazard rate for any item = failure rate for any item
$\lambda_T = \sum_{i=1}^{n} \lambda_i$ = cumulative hazard rate = total failure rate for equipment
n = Number of items in equipment

Hence, we see that:

1. If all items in the equipment have exponential failure distributions (even if they are different exponential distributions), the equipment will have an exponential failure distribution.
2. The particular exponential failure distribution the equipment will follow will be characterized by a hazard rate which will be equal to the sum of the hazard rates of the items making up the equipment.

Assuming that all parts in an equipment do indeed follow an exponential failure distribution, and assuming that, once a failure occurs in an equipment, the equipment can be repaired, we have set the stage for definition of a most important reliability characteristic, mean time between failure (MTBF).

For an exponential distribution mean time between failure connotes that the *average operating time* to failure between the repair of the kth failure and incidence of the $(k + 1)$th failure will remain the same regardless of the operating age of the item. For any other distribution, a *different average operating time* between *each* and every failure will exist. Also, each average value will be a function of the total operating life history of the item.

The difference in behavior follows from one of the properties cited earlier (in Section 2.1.1) for the exponential distribution—its *memoryless characteristic*, which in short means that a repaired equipment with T hours of operating history is the equivalent of a brand new equipment with no operating history. Further, for the exponential distribution,

$$\text{MTBF} = \text{MTTF} = 1/\lambda_T \qquad (2\text{-}4)$$

The reliability over *any period of equipment operation time, T,* regardless of prior operational history, can be expressed as:

$$R_E(T) = \exp(-\lambda_T T) \qquad (2\text{-}5)$$

Example   An equipment is comprised of 10 000 component parts; each component part follows an exponential failure distribution with a different, but known, hazard (failure) rate. The sum of all the failure rates equals λ_T, i.e.,

$$\lambda_T = \sum_{i=1}^{n} \lambda_i = 0.01$$

What is the MTBF of the equipment? What is its reliability over a 10-hour mission? Over a 25-hour mission? Would there be any difference in mission reliability if the equipment had a prior operating history of 100 hours, or 1000 hours over what it has at present?

$$\text{MTBF} = 1/\lambda_T, \qquad \lambda_T = 0.01$$
$$\text{MTBF} = 100 \text{ hours}, \qquad R(T) = \exp(-\lambda_T T)$$
$$R(10) = e^{-0.1} \sim 0.90, \qquad R(25) = e^{-0.25} \sim 0.78$$

NO!

2.1.2   The Weibull Distribution

If the exponential distribution is the most familiar and frequently used distribution in reliability, the Weibull at least ties for second place. Its characteristic properties are defined by two parameters: a shape parameter β and a scale parameter α, which provide great flexibility. This provides the capability to represent a wide range of different shapes. For $\beta = 0$, it is representative of a constant hazard rate. In fact, it is transformed into the exponential distribution. For values of $\beta < 1$, the hazard rate decreases with time of operation. This will result in a decreasing failure rate with time. It means that for a population of parts put under life test, for each interval of time, say 500 hours, a smaller and smaller proportion of parts (based on the number still functioning at the beginning of the interval) will fail. For values of $\beta > 1$ the hazard rate increases with time of operation, resulting in an increasing failure rate with time and proportionately more failures for each following interval of operating time.

The Weibull failure distribution is widely applied to electronics, semiconductors, electron tubes, optical devices, electromechanical and mechanical devices, and equipments and systems where the shape parameter β can be take on values greater or less than 1.[3,4] Data and experience have historically indicated reasonably good fits to many situations. Within the past 10 years it has been found to be particularly useful for representing reliability growth at the equipment level.

2.1.2.1   *Mathematical Characteristics: The Weibull Distribution at the Item and Device Level*

Any distribution that has an increasing or decreasing hazard with time will have reliability characteristics which are functions of *specific* operating intervals, dependent on prior operating history.

We define $f(t) = $ a failure probability (density) function,

$$f(t) = \frac{\beta}{\alpha}\left(\frac{t}{\alpha}\right)^{\beta - 1} e^{-(t/\alpha)^{\beta}} \qquad \text{(see Fig. 2-2a)} \qquad (2\text{-}6)$$

where
   $\beta = $ shape parameter $\beta > 0$
   $\alpha = $ scale parameter $\alpha > 0$
   $t\ = $ a particular time

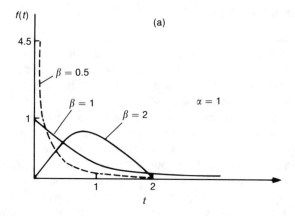

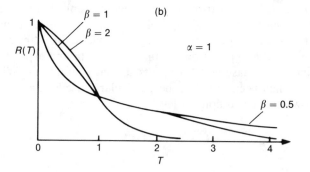

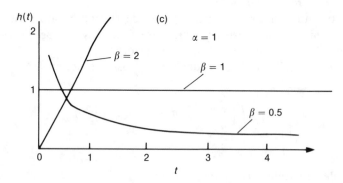

FIGURE 2-2.   (a) Weibull density function. (b) Weibull reliability function. (c) Hazard rate associated with a Weibull distribution.

Then

$$\int_{T_1}^{T_2} f(t)\, dt \bigg/ \int_{T_1}^{\infty} f(t)\, dt = \text{probability that an item will fail in a specific}$$

time interval $T = (T_2 - T_1)$ given the item was operational at time T_1
(see Appendix A0)

where
  T_1 = hours of prior operating time on the item prior to a mission
  T_2 = total hours of operating time on the item after the mission

We can describe the reliability, of the item over a *specific* interval of time as $R(T_2 - T_1)$, where $T_2 > T_1$, as

  $R(T_2 - T_1)$ = probability an item will perform satisfactorily
  (not fail) over the *specific* operating time interval $(T_2 - T_1)$
  given the item was operational at the start of the interval

The reliability $R(T_2 - T_1)$ over $T = T_2 - T_1$ is

$$R(T_2 - T_1) = 1 - \frac{\displaystyle\int_{T_1}^{T_2} f(t)\, dt}{\displaystyle\int_{T_1}^{\infty} f(t)\, dt} = \frac{R(T_2)}{R(T_1)} = \exp[-(1/\alpha)^{\beta}(T_2^{\beta} - T_1^{\beta})]$$

(2-7)

(see Appendix A2 for derivation) where $R(T_i)$ = the reliability over an interval of time T_i when the item has 0 hours of operating history.
  In the special case when $T_1 = 0$ (at the beginning of item operation when the item is brand new),

$$T_2 - T_1 = T_2 = T$$
$$R(T) = e^{-(T/\alpha)^{\beta}} \quad \text{(see Fig. 2.2b)} \tag{2-8}$$

The hazard rate for the Weibull distribution $= h(t) = \dfrac{\beta}{\alpha}\left(\dfrac{t}{\alpha}\right)^{\beta-1}$

(see Fig. 2.2c) (2-9)

Average failure rate per hour, $\lambda(T)$, over a specific interval of time

$T = (T_2 - T_1)$ is

$$\lambda(T_2 - T_1) = \lambda(T) = \frac{\int_{T_1}^{T_2} h(t)\,dt}{T_2 - T_1} \tag{2-10}$$

where $T_1 = 0$ (when the unit has no operational history, i.e., is brand new), $T_2 - T_1 = T_2 = T$, and

$$\lambda(T) = \left(\frac{1}{\alpha}\right)^{\beta} T^{\beta-1} \tag{2-11}$$

(Note that there is a difference between hazard and failure rates.)

$$\text{MTTF} = \alpha\Gamma\left(1 - \frac{1}{\beta}\right) \text{ measured from } T_1 = 0 \tag{2-12}$$

where $\Gamma[1 - (1/\beta)]$ denotes the gamma function of the quantity $[1 - (1/\beta)]$.

The mean time to failure, MTTF, or mean residual life, MRL, for the item or population of items which has an approximate Weibull distribution will vary over different segments of service life (since it will have an increasing or decreasing hazard with time). The following relationship provides a time-based measure of MTTF or MRL based on the operational history, T_1, of the item(s):

$$\text{MTTF} = \text{MRL} = \frac{1}{R(T_1)} \int_{T_1}^{\infty} R(T)\,dT \tag{2-13}$$

(See Appendix A5 for the derivation.)

Special Properties   Because β and α can take on various values, the Weibull distribution can assume a large number of different shapes, and the hazard rate can take on various increasing and decreasing patterns with time. When $\beta = 1$, the Weibull distribution is transformed into the exponential distribution.

Example   The radar set described in the last example has an approximate Weibull distribution of failures with $\alpha = 100$, $\beta = 0.5$. It is put into operation just after it has been developed and has had no prior operating history. Its first mission has a duration of 10 hours. What is the probability that it will not fail during the mission? What is the MTTF (mean time to failure) for a population of such radars assuming that they all start with no prior operational history and operate until they each fail?

$$R(T) = \exp(-(1/\alpha)^\beta (T)^\beta)$$

$$R(10) = \exp(-(0.1)^{0.5}(10)^{0.5}) = 0.73$$

$$MTTF = \alpha\Gamma(1 + (1/\beta))$$

$$MTTF = 100\Gamma(3) = 200 \text{ hours}$$

The radar has been operational for several months and has accumulated 300 hours of use. A new mission of 15 hours duration has been assigned to the radar. What is the reliability of the radar for the first such mission (i.e., $T_1 = 300$, $T_2 = 315$)

$$R(T_2 - T_1) = e^{-(1/\alpha)^\beta (T_2^\beta - T_1^\beta)}$$

$$R(15) = e^{-0.043} = 0.958$$

What would the reliability have been for a mission of 10 hours under the above conditions?

$$R(10) = e^{-0.0286} = 0.97$$

Compare the average failure rate (per hour) over a 10 hour mission with start after 0 hours of previous operating history to the average failure rate per hour with start after 300 hours of previous operating history:
From (2-11):

$$\lambda(T) = (1/\alpha)^\beta (T)^{\beta-1}$$

$$\lambda(10) = 0.1(10)^{-0.5} = 0.0316$$

From (2-10):

$$\lambda(T_2 - T_1) = \frac{(1/\alpha)^\beta (T_2^\beta - T_1^\beta)}{T_2 - T_1}$$

$$\lambda(T_2 - T_1) = 0.1(0.286)/10 = 0.00286$$

2.1.2.2 Calculating/Predicting Equipment Reliability
When All Items Have Weibull Failure Distributions
Both expressions (1) and (2) below connote the probability that an item i will perform satisfactorily for a specific mission time of T hours; the difference

between the two is representative of the operational history of the item, which *will* affect the reliability over T. Note that if $T_1 = 0$, expression (2) transforms to (1) with $T_2 = T$.

1.
$$R_i(T) = \exp(-(T/\alpha)^\beta)$$

assuming mission start after 0 hours of operational history.

2.
$$R_i(T_2 - T_1) = R_i(T_2)/R_i(T_1) = \exp\left[-\left(\frac{1}{\alpha_i}\right)^{\beta_i}(T_2^{\beta_i} - T_1^{\beta_i})\right]$$

assuming mission start after T_1 hours of operational history.

Equipment reliability over a specific interval of time, $R_e(T_2 - T_1)$, can be expressed as

$$R_e(T_2 - T_1) = \prod_{i=1}^{m} R_i(T_2 - T_1) = \prod_{i=1}^{m} \exp\left[-\left(\frac{1}{\alpha_i}\right)^{\beta_i}(T_2^{\beta_i} - T_1^{\beta_i})\right]$$

$$= \exp\left[-\sum_{i=1}^{m}\left(\frac{1}{\alpha_i}\right)^{\beta_i}(T_2^{\beta_i} - T_1^{\beta_i})\right] \qquad (2\text{-}14)$$

where
β_i, α_i = shape and scale parameters of the ith item in the equipment
m = number of items making up the equipment

In the event *that all values of β_i are identical*, then

$$R_e(T_2 - T_1) = \exp\left[-(T_2^\beta - T_1^\beta)\sum_{i=1}^{m}\left(\frac{1}{\alpha_i}\right)^\beta\right] \qquad (2\text{-}15)$$

when $T_1 = 0$, $T_2 - T_1 = T$ and

$$R_e(T) = \exp\left[-T^\beta\sum_{i=1}^{m}\left(\frac{1}{\alpha_i}\right)^\beta\right] \qquad (2\text{-}16)$$

Under the above conditions, the equipment can be construed to have a Weibull distribution of failure time with shape parameter β and a scale parameter α, where

$$\alpha = \left[\sum_{i=1}^{m}\left(\frac{1}{\alpha_i}\right)^\beta\right]^{-(1/\beta)} \qquad (2\text{-}17)$$

For generality and simplicity, assume: (1) that the failures occurring over service life are extremely small compared to m, and (2) that there will be one item replaced per failure. Under those circumstances, the introduction of new items into the equipment will have little significant effect on the value of $R_e(T_2 - T_1)$ during the course of service life. (Note that by repairing the equipment through the introduction of new parts, technically we can no longer represent equipment reliability exactly by (2-14). This will be discussed further in Chapter 9. We, however, can use (2-14) as an approximation in that case.)

The time-dependent hazard rate $h_e(t)$ and failure rate $\lambda_e(T_2 - T_1)$ at the equipment level are expressed as

$$h_e(t) = \sum_{i=1}^{m} \left(\frac{\beta_i}{\alpha_i} \right) \left(\frac{t}{\alpha_i} \right)^{\beta_i - 1} \tag{2-18}$$

$$\lambda_e(T_2 - T_1) = \frac{\sum_{i=1}^{m} \left(\frac{1}{\alpha_i} \right)^{\beta_i} (T_2^{\beta_i} - T_1^{\beta_i})}{T_2 - T_1} \tag{2-19}$$

MTFF for the equipment measured from $T_1 = 0$ can be evaluated as

$$\text{MTFF} = \int_0^\infty R_e(T)\, dT \tag{2-20}$$

This is an important relationship (see Appendix A4 for its derivation).

In the event that all values of β_i are different, the distribution of equipment failures may have an increasing or decreasing hazard, but it will not necessarily be Weibull. In order for an equipment made up of items which have Weibull distributions to have a Weibull failure distribution, $R_e(T)$ must take the form below when $T_1 = 0$:

$$R_e(T) = \exp\left[-T^\beta \sum_{i=1}^{m} \left(\frac{1}{\alpha_i} \right)^\beta \right]$$

$$= \exp\left[-\left(\frac{T}{\alpha_0} \right)^\beta \right] \tag{2-21}$$

where $\alpha_0 = $ a constant, for example, as defined in (2-17).

In order to show that this is not the case when β values for each item differ, take an equipment composed of two items where each item has Weibull parameters β_1 and β_2, respectively, and the same value of α. Applying (2-14) when $T_1 = 0$ yields

$$R_e(T) = \exp\left[-\left(\frac{T}{\alpha}\right)^{\beta_1} - \left(\frac{T}{\alpha}\right)^{\beta_2} \right] \tag{2-22}$$

which cannot be forced into a form like (2-21). Consequently we cannot consider that the two Weibull-distributed items yield an equipment with a Weibull failure distribution. (Note that it is possible for the failure distribution to take on an approximate Weibull form, but this conclusion can only be reached after lengthy analysis usually requiring the use of simulation.)

Example   An equipment is comprised of 10 000 component parts, all with the same value of $\beta = 2$, but different values of α distributed in the following proportions: $10\% = 1000$, $15\% = 2000$, $25\% = 10\,000$, $50\% = 30\,000$. Assume start at $T_1 = 0$ and a mission time of 20 hours. What is the reliability over the mission? What is the failure rate over the mission?

$$R_e(T) = \exp\left[-T^{\beta_i} \sum_{i=1}^{m} \left(\frac{1}{\alpha_i}\right)^{\beta_i} \right]$$

$$\sum_{1}^{4} \left(\frac{1}{\alpha_i}\right)^{\beta_i} T^{\beta_i} = (20)^2 [1000(1/1000)^2 + 1500(1/2000)^2$$

$$+ 2500(1/10\,000)^2 + 5000(1/30\,000)^2]$$

$$= 400[0.001 + 0.000375 + 0.000025 + 0.0000056] = 0.56$$

$$R(20) = 0.57$$

$$\lambda(T) = \frac{\sum_{i=1}^{m} \left(\frac{T}{\alpha_i}\right)^{\beta_i}}{T}$$

$$\lambda(20) = 0.028$$

What would the mean time to failure, MTTF, be if the equipment started operation at $T_1 = 0$?

$$\text{MTTF} = \int_0^\infty R_e(T)\, dT$$

$$R_e(T) = \exp\left[-(T^\beta)\sum \left(\frac{1}{\alpha_i}\right)^\beta \right]$$

assuming all items have the same β characteristics as before.

$$\text{MTTF} = \int_0^\infty \exp[-T^2(0.0014)] \, dT = \frac{1}{2} \frac{\sqrt{\pi}}{0.0374} = 23.7$$

Solve the same problems assuming that the equipment has an operational history of 200 hours (i.e., $T_1 = 200$).

2.1.3   The Log-Normal Distribution

The log-normal distribution is frequently cited with respect to the failure distribution of mechanical/structural items, fatigue analysis, and wearout phenomena, and with accelerated test results of semiconductor devices.[5-7] Its characteristic properties are defined by two parameters, a standard deviation, σ (which can be considered a shape parameter), and a mean, μ (which can be considered a scale parameter). The distribution gets its name from the fact that if we were to take the logarithms of all the times to failure, the distribution of those "log values" would take on a normal distribution. The hazard rate has the characteristic of first increasing with time and then transitioning to a decreasing rate with time.

The log-normal failure distribution has been found to be generally applicable in instances where the degree of damage done due to a given operating stress is proportional to the degree of damage which exists at the time of the stress (e.g., fatigue phenomena) and failure occurs after a given damage threshold is reached.

2.1.3.1   *Mathematical Characteristics: Log-Normal Distribution at the Device or Item Level*

This distribution, like the Weibull, has mission reliability, failure rate, and MTTF characteristics which are keyed to specific operating intervals and dependent on prior operating history. Defining $f(t)$ = failure probability (density) distribution,

$$f(t) = \frac{1}{\sigma t \sqrt{2\pi}} \exp\left(-\frac{1}{2\sigma^2}(\ln t - \mu)^2\right) \qquad \text{(see Fig. 2.3a)} \qquad (2\text{-}23)$$

where
    σ = standard deviation of the logarithms of all times to failure; a shape parameter, $\sigma > 0$
    μ = mean of the logarithms of all times to failure; a scale parameter

For computational purposes it is helpful to represent the density function in terms of "$\ln t$" since "$\ln t$" will be normally distributed.

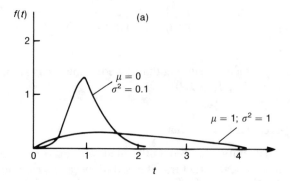

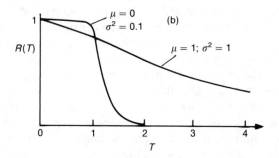

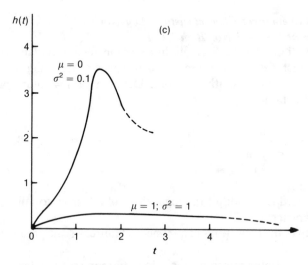

FIGURE 2-3.   (a) Log-normal density function. (b) Log-normal reliability function. (c) Hazard rate associated with a log-normal distribution.

$$f(\ln t) = \frac{1}{\sigma\sqrt{2\pi}}\exp\left(-\frac{1}{2\sigma^2}(\ln t - \mu)^2\right) \tag{2-24}$$

It follows then that the reliability function equals $R(T_2 - T_1)$, where $T_2 > T_1$, and where

$R(T_2 - T_1)$ = probability an item will perform satisfactorily
 (not fail) over a specific operating time interval
 $(T_2 - T_1)$ given that the item was operational at time T_1

  = the reliability over a specific interval

$R(T_2 - T_1)$ can be represented in terms of a standard normal distribution (explained in more detail in Section 2.2.3).

$$R(T_2 - T_1) = \frac{R(T_2)}{R(T_1)} = \frac{(1/\sqrt{2\pi})\displaystyle\int_{(\ln T_2 - \mu)/\sigma}^{\infty} e^{-x^2/2}\,dx}{(1/\sqrt{2\pi})\displaystyle\int_{(\ln T_1 - \mu)/\sigma}^{\infty} e^{-x^2/2}\,dx} \tag{2-25}$$

where $R(T_i)$ = the reliability over a specific time interval T_i when the item has no previous operating history.

This can be evaluated through the use of a table of the standard cumulative normal distribution available in any statistics text or book of tables.

When $T_1 = 0$ (when the unit has no previous operating history), $(T_2 - T_1) = T_2 = T =$ any defined length of operating time, and

$$R(T) = \frac{1}{\sqrt{2\pi}}\int_{(\ln T - \mu)/\sigma}^{\infty} e^{-x^2/2}\,dx \qquad \text{(see Fig. 2.3b)} \tag{2-26}$$

hazard rate $h(t)$ can be represented as

$$h(t) = \frac{(1/\sigma t\sqrt{2\pi})\exp\left[-\dfrac{1}{2}\left(\dfrac{(\ln t - \mu)}{\sigma}\right)^2\right]}{(1/\sqrt{2\pi})\displaystyle\int_{(\ln t - \mu)/\sigma}^{\infty} e^{-x^2/2}\,dx} \qquad \text{(see Fig. 2.3c)} \tag{2-27}$$

which must be evaluated through the use of tables.

Average failure rate per hour over a specific interval of time T_1 to T_2,

$\lambda(T_2 - T_1)$, takes the form

$$\lambda(T_2 - T_1) = \frac{\displaystyle\int_{(\ln T_1 - \mu)/\sigma}^{(\ln T_2 - \mu)/\sigma} h(t)\, dt}{T_2 - T_1} \tag{2-28}$$

which requires numerical solution, as does the expression for average failure rate when $T_1 = 0$, i.e.,

$$\lambda(T) = \frac{\displaystyle\int_{-\infty}^{(\ln T - \mu)/\sigma} h(t)\, dt}{T} \tag{2-29}$$

Mean time to failure, MTTF, or mean time to first failure, MTFF, when measured from time $T_1 = 0$, takes the form

$$\text{MTTF} = \text{MTFF} = \exp(\mu + 0.5\sigma^2) \tag{2-30}$$

The mean time to failure and mean residual life for the item or population of items which have an approximate log-normal distribution will vary over different segments of service life. Time-dependent MTTF/MRL parameters can be calculated for such items using the relationship below:

$$\text{MTTF} = \text{MRL} = \frac{1}{R(T_1)} \int_{(\ln T_1 - \mu)/\sigma}^{\infty} R(T)\, dT \tag{2-31}$$

(See Appendix A5 for the derivation.)

Special Properties   Because μ and σ can take on various values, the log-normal distribution can assume a large number of shapes. The hazard rate can take on increasing values to a specific point in time and decreasing values after that point in time.

Example   The large population of a new type of "dispose-at-failure" equipment was delivered to a particular user. The entire population was put into use at the same time and used continuously. It is known that the items have a log-normal failure distribution with parameters $\mu = 3$, $\sigma = 1.8$ (recall how these parameters were defined earlier). What is the MTTF of the population? If your organization is given one, what is the reliability of your equipment over a first-use mission of 30 hours?

$$\text{MTTF} = \exp(\mu + 0.5\sigma^2)$$

$$\text{MTTF} = \exp[3 + 0.5(3.24)] = 101.5 \text{ hours}$$

$$R(T) = \frac{1}{\sqrt{2\pi}} \int_{(\ln T - \mu)/\sigma}^{\infty} e^{-x^2/2} \, dx$$

$$(\ln T - \mu)/\sigma \sim 0.22$$

From a table of the standard normal distribution, $R(T) = 0.41$.

Given that your equipment has survived 100 hours of operation without failure and you have been assigned a new mission of 30 hours duration, what is the reliability of your equipment over the new mission?

$$R(T_2 - T_1) = \frac{(1/\sqrt{2\pi}) \int_{(\ln T_2 - \mu)/\sigma}^{\infty} e^{-x^2/2} \, dx}{(1/\sqrt{2\pi}) \int_{(\ln T_1 - \mu)/\sigma}^{\infty} e^{-x^2/2} \, dx}$$

For $T_1 = 100$, $T_2 = 130$,

$$(\ln T_1 - \mu)/\sigma \sim 0.89, \qquad (\ln T_2 - \mu)/\sigma \sim 1.04$$

From a table of the standard normal distribution,

$$\frac{1}{\sqrt{2\pi}} \int_{(\ln T_1 - \mu)/\sigma}^{\infty} e^{-x^2/x} \, dx = 0.19, \qquad \frac{1}{\sqrt{2\pi}} \int_{(\ln T_2 - \mu)/\sigma}^{\infty} e^{-x^2/2} \, dx = 0.15$$

$$R(30) = 0.79$$

Does that surprise you? Compute the hazard rate at 15 hours of operation; at 115 hours of operation.

$$h(t) = \frac{(1/\sigma t\sqrt{2\pi}) \exp\left[-\frac{1}{2}\left(\frac{\ln t - \mu}{\sigma}\right)^2 \right]}{\frac{1}{\sqrt{2\pi}} \int_{(\ln t - \mu)/\sigma}^{\infty} e^{-x^2/2} \, dx}$$

For $t = 15$, $(\ln t - \mu)/\sigma \sim -0.16$,

$$h(15) = 0.0146/0.5636 \sim 0.026$$

For $t = 115$, $(\ln t - \mu)/\sigma \sim 0.97$,

$$h(115) = 0.0012/0.166 \sim 0.007$$

2.1.3.2   Calculating/Predicting Equipment Reliability When All Items Have Log-Normal Distributions

Both expressions (1) and (2) below connote the probability that item i will perform satisfactorily for a specific mission time T between T_1 and T_2 hours. The difference between the two is representative of the operational theory of the items which *will* affect the reliability over T. Note that if $T_1 = 0$, the expression (2) transforms to (1) with $T_2 = T$.

1.
$$R_i(T) = \frac{1}{\sqrt{2\pi}} \int_{(\ln T - \mu)/\sigma}^{\infty} e^{-x^2/2} \, dx$$

assuming mission start after 0 hours of operating history, i.e. $T = T_2$

2.
$$R_i(T_2 - T_1) = R_i(T_2)/R_i(T_1)$$

assuming mission start after T_1 hours of operating history

where $R(T_i)$ is defined as previously.

Equipment reliability $R_e(T_2 - T_1)$ can be expressed as

$$R_e(T_2 - T_1) = \prod R_i(T_2 - T_1) = \prod_1^m \left[\frac{(1/\sqrt{2\pi}) \int_{(\ln T_2 - \mu)/\sigma}^{\infty} e^{-x^2/2} \, dx}{(1/\sqrt{2\pi}) \int_{(\ln T_1 - \mu)/\sigma}^{\infty} e^{-x^2/2} \, dx} \right]$$

(2-32)*

When $T_1 = 0$, $T_2 = T$ and a less complicated relationship results:

$$R_e(T) = \prod R_i(T) = \left(\frac{1}{\sqrt{2\pi}} \right) \prod_{i=1}^{m} \left[\int_{(\ln T - \mu)/\sigma}^{\infty} e^{-x^2/2} \, dx \right]_i$$

(2-33)

Both forms require evaluation through a table of the standard normal distribution for each value of T.

---

*Assuming that each component has been failure free for the T_1 hour period. If failures have occurred and have been replaced by new parts, but the quantity of new parts introduced is much less than the total parts making up the equipment, then (2-32) and (2-35) can be used as approximations.

We see also from (2-33) that if all items in the equipment had log-normal failure distributions the equipment failure distribution would not be log-normal (it would have what is termed by some a "mixed distribution," see Ref. 10 for a discussion of mixed distributions) because

$$\prod_i^m \frac{1}{\sqrt{2\pi}} \int_{(\ln T - \mu_i)/\sigma_i}^{\infty} e^{-x^2/2} \, dx$$

cannot be transformed into

$$\frac{1}{\sqrt{2\pi}} \int_{(\ln T - \mu_e)/\sigma_e}^{\infty} e^{-x^2/2} \, dx$$

where

u_i, σ_i = mean and standard deviation of the ith item
μ_e, σ_e = mean and standard deviation of the system

The hazard rate $h_e(t)$ and failure rate $\lambda_e(T_2 - T_1)$ at equipment level will take the forms

$$h_e(t) = \sum_1^m h_i(t) \qquad (2\text{-}34)$$

$$\lambda_e(T_2 - T_1) = \frac{\int_{T_1}^{T_2} \sum h_i(t) \, dt}{T_2 - T_1} \qquad (2\text{-}35)$$

MTTF or MTFF for the equipment, assuming operational start at $T_1 = 0$, can be calculated as

$$\text{MTTF} = \text{MTFF} = \int_0^{\infty} R_e(T) \, dt \qquad (2\text{-}36)$$

Solution of (2-32)–(2-36) must be performed numerically.

Example An equipment is composed of three items each with a log-normal failure distribution and values of $\mu_1 = 4.5$, $\sigma_1 = 1$; $\mu_2 = 6$, $\sigma_2 = 1.5$; $\mu_3 = 6.5$, $\sigma_3 = 1.6$. (Recall that μ and σ are, respectively, the mean value of the *logarithm* of time to failure and the standard deviation of *logarithm* of time to failure.)

Mission time is 40 hours and the equipment has 0 hours of previous operational history. What is its hazard rate at $t = 10$? What is its reliability for the 40-hour mission? What is its failure rate over a 40-hour mission after acquiring an operational history of 400 hours (assume failure-free operation to T_1)?

$$h(t) = \frac{(1/t\sigma\sqrt{2\pi})\exp\left[-\frac{1}{2}\left(\frac{\ln t - \mu}{\sigma}\right)^2\right]}{(1/\sqrt{2\pi})\displaystyle\int_{(\ln t - \mu)/\sigma}^{\infty} e^{-x^2/2}\,dx}$$

$h_1(t) = 0.0036, \qquad h_2(t) = 0.0013, \qquad h_3(t) = 0.0008$

$h_e(t) = h_1(t) + h_2(t) + h_3(t) = 0.0057$

$$R_e(T) = \prod_{i=1}^{m} R_i(T) = \prod_{i=1}^{m}\left[\frac{1}{\sqrt{2\pi}}\int_{(\ln T - \mu)/\sigma}^{\infty} \exp(-x^2/2)\,dx\right]_i$$

$R_1(40) = 0.79, \qquad R_2(40) = 0.94, \qquad R_3(40) = 0.96, \qquad R_e(40) = 0.71$

$$R_e(T_2 - T_1) = \prod_{i=1}^{m}\left[\frac{1}{\sqrt{2\pi}}\int_{(\ln T_2 - \mu_i)/\sigma_i}^{\infty} e^{-x^2/2}\,dx \middle/ \frac{1}{\sqrt{2\pi}}\int_{(\ln T_1 - \mu_i)/\sigma_i}^{\infty} e^{-x^2/2}\,dx\right]_i$$

where

$T_1 = 400, \qquad T_2 = 440$

$R_1(440 - 400) = 0.82, \qquad R_2(440 - 400) = 0.96,$

$R_3(440 - 400) = 0.96, \qquad R_e(440 - 400) = 0.76$

2.1.4 The Gamma Distribution

The gamma distribution is mentioned in much of the literature as a likely distribution which is capable of providing an empirical fit to many distributions of failure. Reports of its application for the depiction of component part failure distributions include failures of vacuum tubes and ball bearings.[8,9] However, this distribution is much less reported in the literature than are applications of the other three distributions. Its characteristic properties are defined by two parameters—a shape parameter β and a scale parameter α, which provide it with great flexibility. For a shape parameter value of $\beta = 1$, it is transformed into the exponential distribution. When $\beta > 1$, the hazard increases with time. When $\beta < 1$, the hazard decreases with time. While both the Weibull and the gamma distribution can take on the shape of the exponential, the remainder of the shapes they are capable of assuming are different. Therefore, the gamma augments the capability to more accurately fit empirical data to part failure distributions.

The gamma distribution too (or its transforms) is in use as a descriptor of failure distributions of fault-tolerant systems which have standby units. Failure distributions for fault-tolerant systems is, however, a subject in itself and will be discussed more fully in separate sections. It has also been found to be generally applicable in instances where a damage threshold exists which will cause a part to fail when reached. The threshold is considered finite in the sense that it will be reached after the part is subjected to a given number of insults (shocks, partial failures, etc.).

2.1.4.1 Mathematical Characteristics: Gamma Distribution At the Device or Item Level

Like the Weibull and log-normal distributions, this distribution also produces reliability characteristics which are dependent upon the specific time interval of operation and on the previous operational history of the item. If $f(t)$ = failure probability (density) distribution,

$$f(t) = \frac{\alpha}{\Gamma(\beta)}(\alpha t)^{\beta-1}e^{-\alpha t} \qquad \text{(see Fig. 2.4a)} \qquad (2\text{-}37)$$

where
  α = scale parameter, $\alpha > 0$
  β = shape parameter, $\beta > 0$
  t = operating time

and

$$\int_{T_1}^{T_2} f(t)\,dt \Big/ \int_{T_1}^{\infty} f(t)\,dt = \text{probability that an item will fail in}$$
  a specific time interval $(T_2 - T_1)$ given that the
  item was operational at time T_1 (see Appendix A0)

Reliability $R(T_2 - T_1)$, where $T_2 > T_1$, is defined as

  $R(T_2 - T_1)$ = probability an item will perform satisfactorily
    (not fail) over the specific operating time interval
  $(T_2 - T_1)$ given that the item was operational at time T_1

$$R(T_2 - T_1) = 1 - \left[\int_{T_1}^{T_2} f(t)\,dt \Big/ \int_{T_1}^{\infty} f(t)\,dt \right] = R(T_2)/R(T_1)$$

$$= \frac{1 - \dfrac{\alpha^{\beta}}{\Gamma(\beta)} \displaystyle\int_{0}^{T_2} t^{\beta-1} e^{-\alpha t}\,dt}{1 - \dfrac{\alpha^{\beta}}{\Gamma(\beta)} \displaystyle\int_{0}^{T_1} t^{\beta-1} e^{-\alpha t}\,dt} \qquad (2\text{-}38)$$

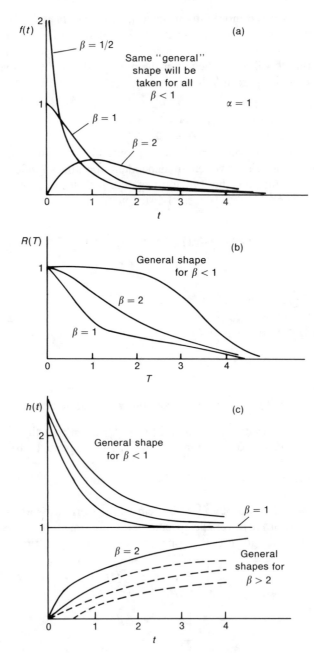

FIGURE 2-4.   (a) Gamma density function. (b) Gamma reliability function. (c) Hazard rate associated with a gamma distribution.

where $[\alpha^\beta/\Gamma(\beta)]\int_0^T t^{\beta-1}e^{-\alpha t}\,dt = $ the incomplete gamma function, which can be evaluated through the use of tables available in any technical library.

Equation (2-38) can also be evaluated through the use of tables of the Poisson distribution when β is an integer. (See Appendix A3 for derivation.) The transformation of the incomplete gamma function to the Poisson takes the form

$$\frac{\alpha^\beta}{\Gamma(\beta)}\int_0^T t^{\beta-1}e^{-\alpha t}\,dt = \sum_{k=\beta}^{\infty}[(\alpha t)^k e^{-\alpha T}]/k! = 1 - \sum_{k=0}^{\beta-1}[(\alpha t)^k e^{-\alpha t}]/k!$$

As a consequence, when $\beta = $ an integer,

$$R(T_2 - T_1) = \exp[-\alpha(T_2 - T_1)]\left[\frac{\displaystyle\sum_{k=0}^{\beta-1}(\alpha T_2)^k/k!}{\displaystyle\sum_{k=0}^{\beta-1}(\alpha T_1)^k/k!}\right] \qquad (2\text{-}39)$$

When $T_1 = 0$ (at the beginning of operation), $T_2 - T_1 = T_2 = T$, the defined length of operating time, and

$$R(T) = 1 - \frac{\alpha^\beta}{\Gamma(\beta)}\int_0^T t^{\beta-1}e^{-\alpha t}\,dt \qquad (2\text{-}40)$$

which can be evaluated through the use of the tables discussed above or, provided β is an integer, as

$$R(T) = 1 - \sum_{i=\beta}^{\infty}[(\alpha T)^i e^{-\alpha T}]/i! = \sum_{i=0}^{\beta-1}[(\alpha T)^i e^{-\alpha T}]/i! \qquad \text{(see Fig. 2.4b)}$$

$$(2\text{-}41)$$

The hazard and the failure rates can be represented as

$$\text{Hazard rate} = h(t) = \frac{\alpha^\beta}{R(t)\Gamma(\beta)}t^{\beta-1}e^{-\alpha t} \qquad (2\text{-}42)$$

where $R(t) = R(T)$ (see Fig. 2.4c).

When β is an integer we can express $h(t)$ as

$$h(t) = \frac{\alpha^\beta t^{\beta-1}}{(\beta-1)!\displaystyle\sum_{i=0}^{\beta-1}(\alpha t)^i/i!} \qquad (2\text{-}43)$$

Failure rate over a specific interval T_1 to T_2 can be expressed as

$$\lambda(T_2 - T_1) = \frac{\int_{T_1}^{T_2} h(t)\,dt}{T_2 - T_1} \tag{2-44}$$

and when $T_1 = 0$ as

$$\lambda(T) = \int_0^T \frac{h(t)\,dt}{T} \tag{2-45}$$

Both of the above require numerical solutions.

Mean time to failure, MTTF, can be expressed as

$$\text{MTTF} = \frac{\beta}{\alpha}, \qquad \text{measured from } T_1 = 0 \tag{2-46}$$

Since the item will have a nonconstant hazard with time, the mean time to failure for the item will vary over different segments of service life. To evaluate mean time to failure or mean residual life over different segments of service life, the following relationship based on prior operational history, T_1, may be used:

$$\text{MTTF} = \text{MRL} = \frac{1}{R(T_1)} \int_{T_1}^{\infty} R(T)\,dT \qquad \text{(see Appendix A5)} \tag{2-47}$$

Special Properties   Because α and β can take on various values, the gamma distribution can assume a large number of different shapes, and the hazard rate a number of increasing and decreasing patterns with time.

Example   The equipment described in the last example has an approximate gamma distribution of time to failure with $\beta = 2$, $\alpha = 0.01$. It is put into operation just after it has been delivered and has no prior operating history. Its first mission has a duration of 10 hours. What is the probability that it will perform satisfactorily during the mission? What is the MTTF for a population of such equipments assuming they all start with no prior operational history, and operate until they each fail?

$$R(T) = 1 - \sum_{i=\beta}^{\infty} [(\alpha T)^i e^{-\alpha T}]/i! = \sum_{i=0}^{\beta-1} [(\alpha T)^i e^{-\alpha T}]/i!$$

From a table of the Poisson distribution,

$$R(10) = 0.995, \qquad \text{MTTF} = \beta/\alpha, \qquad \text{MTTF} = 200$$

The equipment has been in operation for several months and has accumulated 300 hours of use. A mission of 15 hours duration has been assigned to the equipment. What is the reliability over the first such mission (i.e., $T_1 = 300$; $T_2 = 315$)?

$$R(T_2 - T_1) = \exp[-\alpha(T_2 - T_1)] \left[\dfrac{\sum\limits_{k=0}^{\beta-1} (\alpha T_2)^k/k!}{\sum\limits_{k=0}^{\beta-1} (\alpha T_1)^k/k!} \right]$$

$$R(15) = 0.856$$

What would the reliability have been for a mission of 10 hours under the above conditions?

$$R(10) = 0.903$$

Compare this with results from the first example. Using the results of the exercise, compare the failure rate (per hour) over the 10-hour mission with start after 0 hours of prior operating history to the failure rate (per hour) over the 10-hour mission with start after 300 hours of prior operating history.

2.1.4.2   Calculating/Predicting Equipment Reliability
When All Items Have Gamma Distributions

Both expressions (1) and (2) below connote the probability that an item i will perform satisfactorily over a specific interval of time. The difference between the two is representative of the prior operational history of the item, which will affect the reliability over the interval. Note that the expressions (1) and (2) are equivalent if $T_1 = 0$.

1.
$$R_i(T) = 1 - \frac{\alpha_i^{\beta_i}}{\Gamma(\beta_i)} \int_0^T t^{\beta_i - 1} e^{-\alpha_i t}\, dt$$

assuming a mission start after 0 hours of operating history.

2.
$$R_i(T_2 - T_1) = \frac{R_i(T_2)}{R_i(T_1)}$$

assuming a mission start after T_1 hours of prior operating history, where $T = T_2 - T_1$ and $R_i(T_i)$ is defined as before.

Equipment level reliability $R_e(T_2 - T_1)$, when the equipment is made up of m items, is expressed as

$$R_e(T_2 - T_1) = \prod_{i=1}^{m} R_i(T_2 - T_1) = \prod_{i=1}^{m} \frac{R_i(T_2)}{R_i(T_1)} \qquad (2\text{-}48)$$

which requires an iterative solution using a table to discern the values of each $R_i(T)$.

Even when $T_1 = 0$, the simpler form of the relationship is still relatively complex:

$$R_e(T) = \prod_{i=1}^{m} R_i(T) = \prod_{i=1}^{m} \left[1 - \frac{\alpha_i^{\beta_i}}{\Gamma(\beta_i)} \int_0^T t^{\beta_1 - 1} e^{-\alpha_i t} \, dt \right]_i \qquad (2\text{-}49)^*$$

However, when each β takes on integer values (see 2-39),

$$R_e(T) = \prod_{i=1}^{m} \sum_{k=0}^{\beta_i - 1} \frac{(\alpha_i T)^k}{k!} e^{-\alpha_i T} \qquad (2\text{-}50)$$

The relationship is still somewhat cumbersome and requires iterative solution for each particular value of T.

The failure distribution of an equipment made up of items having gamma failure distributions will not be gamma. This can be shown from (2-50). If the equipment failure distribution were gamma, $R_e(T)$ would take the form

$$R_e(T) = \sum_{k=0}^{\beta_e - 1} \left(\frac{(\alpha_e T)^k}{k!} \right) \exp(-\alpha_e T)$$

But (2-50) cannot be forced into that form for values of $\beta > 1$. Try it yourself for $m = 2$; $\beta_1, \beta_2 = 2$; $\alpha_1, \alpha_2 = A$.

Hazard and failure rates at the equipment level can be respectively defined as

$$h_e(t) = \sum_{i=1}^{m} h(t)_i \qquad (2\text{-}51)$$

$$\lambda_e(T_2 - T_1) = \int_{T_1}^{T_2} \frac{\sum_{i=1}^{m} h(t)_i \, dt}{T_2 - T_1} \qquad (2\text{-}52)$$

---

*Assuming that each component has been failure free for the T_1 hour period. If failures have occurred and have been replaced by new parts, but the quantity of new parts introduced is much less than the total parts making up the equipment, then (2-49) and (2-52) can be used as approximations.

When $T_1 = 0$,

$$\lambda_e(T) = \int_0^T \frac{\sum\limits_{i=1}^m h(t)_i \, dt}{T} \tag{2-53}$$

MTTF for the equipment, assuming operational start at $T_1 = 0$, can be calculated as

$$\text{MTTF} = \text{MTFF} = \int_0^\infty R_e(T) \, dT \tag{2-54}$$

Analytical solution of equations (2-50)–(2-54) is laborious and difficult when β can take on noninteger values. When β is restricted to integer values, solution is only laborious.

Example   An equipment is made up of three identical items with $\beta_i = 2$, $\alpha_i = 0.003$. What is its reliability for a mission of 30 hours, assuming each item has a past operational history of zero hours ($T_1 = 0$)? After the equipment (and each item) has acquired 400 operating hours, what is the reliability for the same mission (assume no failures over T_1)? What is the mean time to failure (MTTF) when $T_1 = 0$?

$$R_e(T) = \prod_{i=1}^3 R_i(T)$$

$$R_e(30) = \left[(e^{-0.09}) \sum_{k=0}^1 \left(\frac{(0.09)^k}{k!} \right) \right]^3 \approx 0.989$$

$$R_e(T_2 - T_1) = \prod_{i=1}^3 R_i(T_2 - T_1) = \prod_{i=1}^3 \frac{R_i(T_2)}{R_i(T_1)} \qquad T_1 = 400, \qquad T_2 = 430$$

$$R_e(430 - 400) = \left[\frac{(e^{-0.003T_2}) \sum\limits_{k=0}^1 \frac{(0.003T_2)^k}{k!}}{(e^{-0.003T_1}) \sum\limits_{k=0}^1 \frac{(0.003T_1)^k}{k!}} \right]^3 \approx 0.861$$

$$\text{MTTF} = \text{MTFF} = \int_0^\infty R_e(T) \, dT$$

$$\text{MTTF} = \text{MTFF} = \int_0^\infty \left[e^{-0.003t} \sum_{k=0}^1 \frac{(0.003t)^k}{k!} \right]^3 dt \approx 321$$

2.1.5   Equipment-Level Prediction Using a Hybrid Approach

If you have followed the discussion so far and tried the exercises, it is clear that:

1. We can characterize the behavior of an item with a given failure distribution more precisely than we can characterize the behavior of an equipment made up of many items (except for the exponential failure distribution).
2. Even using a calculator or a computer, defining equipment level characteristics can be laborious for equipments composed of parts having nonconstant hazard rates.
3. From an ease-of-use standpoint, if one could assume that all items exhibited an exponential distribution of failure times, prediction and calculation of equipment reliability would be facilitated.

Most of the data available and used as standards for reliability prediction in electronics (e.g., Mil-Hdbk-217, "Reliability Prediction for Electronic Equipment") indeed assume a constant hazard/failure rate. There are also failure rate sources for nonelectronic parts which conclude that for a majority of parts considered an exponential failure distribution assumption is adequate (see Ref. 11) to characterize failure rate.

Nonconstant hazard/failure rate behavior at both the part and the equipment level is, however, far from a rare occurrence. Hence, there is an apparent need to depict such characteristics.

The ideal solution would be one which would allow the application of existing failure rate data (constant or nonconstant hazard/failure rates) in a framework which provides the computational advantages attributed to the assumption of a constant failure rate. The following describes such an approach.

Assume that the total failure rate $\lambda(T_p)$ for the equipment (see Section 2.1.1.2) actually represents the average failure rate per hour over a given period of time T_p (see Section 1.3), where T_p = a period of operating time to which $\lambda(T_p)$ applies. Total average failure rate per hour, $\lambda_e(T_p)$, for the equipment over the first T_p operating hours can then be calculated as

$$\lambda_e(T_p) = \sum_{i=1}^{m} \lambda_i(T_p) \qquad (2\text{-}55)$$

where $\lambda_i(T_p)$ = average failure rate per hour over the first T_p hours of operation for the ith item/part, estimated from hazard rate information and empirical data on parts and components. T_p is assigned as a standard time period for prediction purposes (i.e., the first 1000, 2000, 3000, etc., hours of use). See Refs. 12 and 13 for information on similar approaches to the problem.

For example, let us assume that based on past experience a particular item developed under a company's standard reliability program has a Weibull distribution of failures over its service life characterized by a given β parameter. If we define a period of service life as T_p, from (2-11) we can define the average failure rate per hour over T_p as

$$\lambda(T_p) = \left(\frac{1}{\alpha}\right)^{\beta} (T_p)^{\beta - 1} \tag{2-56}$$

For any value of $T > T_p$ or $T < T_p$, average failure rate per hour

$$\lambda(T) = \left(\frac{1}{\alpha}\right)^{\beta} T^{\beta - 1} \tag{2-57}$$

Taking the ratio of (2-57) to (2-56)

$$\frac{\lambda(T)}{\lambda(T_p)} = \left(\frac{T}{T_p}\right)^{\beta - 1}$$

and

$$\lambda(T) = \lambda(T_p)\left(\frac{T}{T_p}\right)^{\beta - 1} \tag{2-58}$$

and $\lambda(T_p)$ becomes a surrogate for α, a parameter with more direct relevance to reliability assessment.

Reliability assessment for the equipment for any period of time T can then be performed on the basis of the failure rate of each part over the first T_p hours of operation. In other words, in terms of the $\lambda(T_p)$ value of each part and the value of β associated with each part.

Item reliability characteristics can now be defined in terms of T_p:

$$h(T_p) = \frac{\beta}{\alpha}\left(\frac{T_p}{\alpha}\right)^{\beta - 1} ; \qquad h(t) = h(T_p)\left(\frac{t}{T_p}\right)^{\beta - 1} \tag{2-59}$$

$$\lambda(T_2 - T_1) = \frac{\lambda(T_p)(T_2^{\beta} - T_1^{\beta})}{T_p^{\beta - 1}(T_2 - T_1)} \tag{2-60}$$

(For $\lambda(T)$, see (2-58))

$$R(T) = \exp\left(-\frac{\lambda(T_p)}{T_p^{\beta-1}} T^\beta\right) \qquad (2\text{-}61)$$

$$R(T_2 - T_1) \sim \frac{R(T_2)}{R(T_1)} = \exp\left[-\frac{\lambda(T_p)}{T_p^{\beta-1}}(T_2^\beta - T_1^\beta)\right] \qquad (2\text{-}62)$$

$$\text{MTTF} = \text{MTFF} = \frac{(T_p)^{1-1/\beta}\,\Gamma\left(1 + \dfrac{1}{\beta}\right)}{\lambda(T_p)^{1/\beta}} \qquad (2\text{-}63)$$

measured from time $T_1 = 0$. (See Appendix A6 for derivation.)
When $T_1 \neq 0$, MTTF measured from T_1 is

$$\text{MTTF} = \text{MRL} = \frac{1}{R(T_1)}\int_{T_1}^{\infty} R(T)\,dT \qquad (2\text{-}64)$$

Total average equipment failure rate over the first T_p hours of operation, $\lambda_e(T_p)$, made up of m items, can be expressed as

$$\lambda_e(T_p) = \sum_{i=1}^{m} \lambda_i(T_p) \qquad (2\text{-}65)$$

Equipment reliability over time T_p can be expressed as

$$R_e(T_p) = \exp\{-[\lambda_e(T_p)]T_p\} \qquad (2\text{-}66)$$

Equations (2-60) and (2-62) can be employed to assess reliability characteristics over longer periods of time.

Note that while the illustration above was in terms of the Weibull distribution, the same process can be applied to almost any other distribution equally as well.

2.2 DISTRIBUTIONS USED IN EQUIPMENT AND SYSTEM MODELING

A number of distributions could be included under this category. We will discuss the three most commonly used for this purpose in reliability engineering: the Poisson distribution, the normal distribution, and the binomial distribution.

Distributions used in modeling are used to tell a story. Once you are familiar with the types of stories that each can tell, they tend to lose their mysteries. You were introduced to the story that the Poisson distribution tells in Chapter 1 when we discussed a logical means to derive the basic reliability relationship. Let us examine other stories that distributions tell in a little more detail.

2.2.1 The Poisson Distribution as a Modeling Tool

$$P(x) = \frac{a^x}{x!} e^{-a} \qquad (2\text{-}67)$$

defines the form of the Poisson distribution where $P(x) =$ probability of x events, and $a =$ number of events expected.

In Chapter 1 we defined $\lambda(T)\,T$ as the expected number of failures over a time interval T when the average failure rate per operating hour is $\lambda(T)$. Hence,

$$a = \lambda(T)T \qquad (2\text{-}68)$$

and (2-67) can be expressed as

$$P(x) = (\lambda(T)T)^x \frac{e^{-\lambda(T)T}}{x!} \qquad (2\text{-}69)$$

This will hold regardless of the failure distribution of the item under a generally accepted set of assumptions (failures occur independently; probability of more than one failure in any arbitrarily small period of time Δt is negligible; probability of failure during time Δt approximately proportional to the hazard rate during that time). For simplicity let us assume, however, that the item follows an exponential failure distribution such that

$$P(x) = (\lambda T)^x \frac{e^{-\lambda T}}{x!} \qquad (2\text{-}70)$$

The story associated with (2-67) can be interpreted as follows. An event can be interpreted as a failure; therefore (2-70) can be restated as *the probability that x failures will occur when λT are expected.*
There is a cumulative form of the Poisson distribution,

$$P(x \leqslant A) = \sum_{x=0}^{A} a^x \frac{e^{-a}}{x!} \qquad (2\text{-}71)$$

which expresses *the probability that A or fewer events will occur, if a events are expected.*

This, for example, translates to *the probability that no more than A failures will occur over a total operating time of T hours if the failure rate is λ.*

You will find that the Poisson, if not *the most frequently used distribution* in reliability engineering, is *one* of the most frequently used distributions. One of its most useful applications is in the area of standby fault-tolerant/redundant components. We will discuss fault tolerance in greater detail in Chapter 10. For now, let us just outline the scenario in terms of an example.

There is a critical mission to be performed, so we have elected to employ a type of redundant system. The system chosen is comprised of three units. Only one is required to perform the mission. Hence one is operating the remainder in unpowered standby modes of operation. We have at our disposal a perfect means of detecting failure in any unit, and if failure occurs there is a perfect instantaneous switching apparatus available to switch in the standby unit and switch out the failed unit. We have also come to the conclusion that a unit in unpowered (standby) condition has a failure rate of zero when it is unpowered. The only time any unit has a failure rate is when it is performing the required mission.

Recall that if the mission time of concern for a single powered unit (all that is required for the system to perform its function) is T, and the rate of failure for a powered unit is λ, then λT is the expected number of failures for a powered unit in T hours of operation. We have two spare units, each with no failure rate when unpowered, and perfect detection/switching. Our system is *equivalent to a single unit* which can fail and can be perfectly and instantly resurrected two times. For the third failure, *there is no resurrection.* This means that:

If no failure occurs in time T, the mission is successful.
If 1 failure occurs in time T, the mission is successful.
If 2 failures occur in time T, the mission is successful.

The reliability $R(T)$ of the system is then equal to the probability that no failures, one failure, or only two failures occur in time T. Applying (2-71),

$$R(T) = P(0) + P(1) + P(2) = \sum_{i=0}^{2} (\lambda T)^i \frac{e^{-\lambda T}}{i!} \qquad (2\text{-}72)$$

where
  $\lambda T = a$
  $P(0) =$ probability of 0 failure when λT are expected
  $P(1) =$ probability of 1 failure when λT are expected
  $P(2) =$ probability of 2 failures when λT are expected

and (2-72) describes the probability of two failures or less in time T and expresses all conditions/assumptions of the example problem.

Another application of the Poisson distribution is in the area of logistics. Let us also outline this scenario in terms of an example.

One hundred equipments are in the field; each will operate for T hours per month. Assume that there are more than sufficient spares at each equipment site for a month's operation, but the depot under contract must be in a position to fully replenish each site at the end of each month. In order to do this the depot must be able to ship one replacement for each failure that has occurred. Assume that each equipment is composed of identical modules such that any failure can be replaced with spares from a common pool. What is the probability that a pool size of N will be sufficient for the depot to meet site needs for a given month?

λT = expected number of failures per month from one equipment
$100\lambda T$ = expected number of failures per month from 100 equipments

We know that in order to successfully supply the sites there have to be *no more* than N failures in the month's time. The probability of N or fewer failures over one month of operation is

$$\sum_{i=0}^{N} (100\lambda T)^i \frac{e^{-100\lambda T}}{i!}$$

Tables of the Poisson distribution can be found in most standard statistical texts or in books of tables available in almost any technical library.

2.2.2   The Binomial Distribution as a Modeling Tool

$$P(x) = \frac{N!}{(N-x)!x!} p^x (1-p)^{N-x} \qquad (2\text{-}73)$$

defines the binomial distribution where $P(x)$ = probability of x events, p = probability that an event will occur when it is attempted, or the proportion of occurrences of an event that can be anticipated, and N = number of attempts to achieve an event or number of components involved in one attempt.

The story that (2-73) states is *the probability that x events will occur in N trials when the probability that the event will occur during any single trial is p.*

An event can be interpreted as a successful mission for a component, equipment, or system, where success may be defined as performance of its required function over a given period of time. The parameter p can then be interpreted as the probability of satisfactorily performing the required function over a given mission or time. Hence,

$$p = R(T)$$

Equation (2-73) can then be restated as

$$P(x) = \frac{N!}{(N - x)!x!}[R(T)]^x[1 - R(T)]^{N-x} \tag{2-74}$$

        = the probability that x out of N items will survive a
           mission of duration T when the reliability of each item
           is $R(T)$

        = the probability that x out of N missions of a certain
           type will be successful when the probability of mission success
           is $R(T)$ (i.e., when success of the mission depends on
           the satisfactory operation of a given item, whose reliability is
           $R(T)$)

In the case of single-shot devices such as ordinance, the parameter p is simply defined as the probability that the device performs as desired. Then (2-73) can be used to state *the probability that x out of N devices will perform as required when the probability of a single device performing as required is p.*

There is a cumulative form of the binomial distribution which expresses the probability that at least M events $P(x \geqslant M)$ will occur in N attempts or trials:

$$P(x \geqslant M) = \sum_{x=M}^{N} \frac{N!}{(N - x)!x!}p^x(1 - p)^{N-x} \tag{2-75}$$

This is *the probability that more than M events will occur in N trials when the probability that the event will occur during any single trial is p.* For example, this can translate to *the probability that at least (M/N)* (i.e., a proportion of single shot devices) *will perform satisfactorily when the reliability of a single shot device is p.*

The binomial distribution is one of the most frequently used distributions in reliability engineering. It has applications ranging from reliability assessments of phased-array antennas, to "full-on" fault-tolerant systems, to single-shot devices. Let us look at these in terms of a few more examples.

Example 1 There is a critical mission to be performed and a type of redundant system is selected for use. The system chosen is comprised of three units. Only one is required to perform the mission. Further, we have at our disposal a perfect means of detecting failure in any unit, and if a failure occurs we have a 100%-reliable switching apparatus available to perform the switching function. The conclusion has been reached that in order to perform an instantaneous switchover, the unit to be switched into operation *must be in a powered state* prior to the switchover. Hence, all redundant units must be in the same powered state as the unit which is performing the system function. This means that the redundant units each have the same failure rate, λ, as the operating unit. From the above it is clear that a mission failure will occur only if all three units fail during mission time T. (Conversely, the mission will be a success if one, two, or three units do not fail over the mission.)

The probability p that a unit will perform satisfactorily over mission time T is

$$p = e^{-\lambda T}$$

Therefore, the probability that one out of three, two out of three, or three out of three units survive the mission of length T is

$$R(T) = P(x \geqslant 1) = \sum_{x=1}^{3} \frac{3!}{(3-x)!x!}(e^{-\lambda T})^x(1 - e^{-\lambda T})^{3-x}$$

Example 2   A quantity of ordnance devices are to be exercised against a given target:

$$p = \text{probability of successful performance of each device}$$

$$N = \text{number of devices to be exercised}$$

$$P(x \geqslant M) = \text{probability that at least } M \text{ devices will perform as required}$$

$$= \text{probability that a proportion } M/N \text{ or greater of the devices will perform as required}$$

$$P(x \geqslant M) = \sum_{x=M}^{N} \frac{N!}{(N-x)!x!}p^x(1-p)^{N-x} \tag{2-76}$$

Example 3   A phased-array antenna is composed of N independent elements. The antenna will perform satisfactorily as long as 80% of the elements (in any pattern) are still functioning. The antenna is in a remote area which can be reached only every 30 days (at which time all failed elements are replaced). If less than 80% of the elements are operating, the system is considered failed. The antenna must operate a total of T hours per month. Each element has a failure rate of λ. The reliability for the antenna can be expressed as the probability that more than 80% of its elements survive a 30-day operational period, $P(x \geqslant 0.8N)$.

$$R(T) = P(x \geqslant 0.8N) = \sum_{x=0.8N}^{N} \frac{N!}{(N-x)!x!}(e^{-\lambda T})^x(1 - e^{-\lambda T})^{N-x} \quad (2\text{-}77)$$

A question can be raised about the ease of evaluation of relationships of the type of (2-76) and (2-77) if N is very large. Tables of the binomial distribution are usually common up to values of $N = 50$. Hence, for large values of N, either a computer must be used or an approximation must be applied. It happens that for large values of N a binomial distribution can be approximated through use of the normal distribution provided $Np(1 - p) \geqslant 10$. In that instance, the mean μ of the normal distribution $= Np$ and the standard deviation σ of the normal distribution is $(NP[1 - p])^{1/2}$ (Ref. 14).

We will discuss the normal distribution next.

2.2.3   The Normal Distribution as a Modeling Tool

The probability relationship descriptive of the normal distribution takes the form

$$p(y) = \frac{1}{\sigma\sqrt{2\pi}}e^{-(y-\mu)^2/2\sigma^2}$$

where

  $y =$ the variable in question, for example time to failure or a part characteristic

  $\mu =$ the mean of the variable

  $\sigma =$ the standard deviation of the distribution of the variable's value.

The normal distribution is a continuous distribution, in contrast to the Poisson and binomial, which are discrete. $p(y)$ provides an expression for the frequency of occurrence of an event (a failure) over an arbitrarily small interval of time dy. The cumulative form of the normal distribution will be the focus of our discussion:

$$P(y \leqslant y_p) = \int_{-\infty}^{y_p} p(y)\,dy \quad (2\text{-}78)$$

where $y_p =$ a particular value of y.

For ease of computation, (2-78) is usually evaluated by transforming it into what is termed the *standardized* normal distribution. It is the *standardized* normal distribution that is depicted in most tables of the normal distribution. It takes the form

$$P(x \leqslant K_p) = \int_{-\infty}^{K_p} \frac{1}{\sqrt{2\pi}} e^{-x^2/2} \, dx \tag{2-79}$$

where

$x = (y - \mu)/\sigma$
y, σ, μ are as defined before
$K_p = (y_p - \mu)/\sigma$ is known as the coefficient of the normal distribution
y_p is as defined before

Tables are available for the evaluation of (2-79) in almost any elementary statistics text or book of tables.

For reliability purposes, the basic story that (2-79) states is that *if y (time to failure) is a normally distributed variate with mean μ and standard deviation σ, then the probability of experiencing a failure over operating time y_p is equivalent to the probability that the value of $(y - \mu)/\sigma$, will be less than $K_p = (y_p - \mu)/\sigma$.*

The normal distribution is a very useful tool for defining the performance tolerances of units based on the known performance tolerances of their components and for use as an approximation tool to assess values from other distributions.

One of the most powerful characteristics of the normal distribution is its family of addition theorems. Simply stated, if you have m normal distributions each with a unique value of the mean, μ, and standard deviation, σ, and take a sample value y_i from each and add such values together, so that their sum is

$$Y_t = \sum_{i=1}^{m} y_i \tag{2-80}$$

and the values of Y_t will be normally distributed with a mean μ_t and a standard deviation σ_t given by

$$\mu_t = \sum_{i=1}^{m} \mu_i \tag{2-81}$$

$$\sigma_t = \left(\sum_{i=1}^{m} \sigma_i^2 \right)^{1/2} \tag{2-82}$$

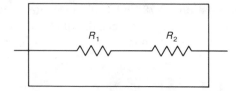

FIGURE 2-5.   Unit comprised of two resistors.

This describes the so-called additive capability of the normal distribution and provides for the analysis of tolerances of the whole by considering the tolerances of its individual parts. We will discuss this application in more depth in a later section. In order to provide an initial feel for what is meant by analysis of tolerance, let us look at a simple example of a unit comprised of two resistors connected in series (Fig. 2-5). The unit is so designed that of the two resistors one has a nominal value of 500 ohms and the other a nominal value of 1000 ohms. The nominal value for the resistance of the unit composed of two resistors is 1500 ohms. However, the unit performs a sensitive function in an equipment such that

- If its resistance is initially greater than 1575 ohms, resistor drift which can be expected to occur as a consequence of aging, and the effects of the system environment will cause premature system failure.
- If its resistance is less than 1400 ohms, the system will not work initially and a costly rework will be required.

Thousands of these units are produced, so that the individual selection and measurement of resistors is not economically feasible. Instead each part is selected from large populations of parts which are ordered. Both populations are normally distributed about the target nominal resistance values. The 500-ohm resistor has a mean of 500 ohms and a standard deviation of 50 ohms. The 1000-ohm resistor has a mean of 1000 ohms and a standard deviation of 100 ohms.

What is the probability that a unit made up of components from such populations will have a resistance value of less than 1400 ohms and hence require rework? (This is equivalent to asking what proportion of units will require rework.) What will be the probability that a unit made up of components from such populations will have a resistance value of greater than 1575 ohms, and hence be victim of premature failure?

Since the resistance values of each of the two populations of devices are normally distributed with mean $\mu_1 = 500$, $\sigma_1 = 50$ and $\mu_2 = 1000$, $\sigma_2 = 75$, respectively a resistor chosen randomly from the first population and connected in series with a resistor chosen randomly from the second population will have a series resistance

$$r_s = r_1 + r_2$$

where

r_1 = value of the resistor taken from the first population
r_2 = value of the resistor taken from the second population

r_s will be normally distributed with mean μ_s and standard deviation σ_s where

$$\mu_s = \mu_1 + \mu_2 = 500 + 1000 = 1500$$

$$\sigma_s = (\sigma_1^2 + \sigma_2^2)^{1/2} = [(50)^2 + (75)^2]^{1/2} \approx 90$$

For the former situation (i.e., the proportion of items produced having resistance values equal to or greater than 1575 ohms

$$\frac{1}{\sqrt{2\pi}} \int_{-\infty}^{K_p} e^{-x^2/2} \, dx = \frac{1}{\sqrt{2\pi}} \int_{-\infty}^{0.83} e^{-x^2/2} \, dx \approx 0.80$$

where $K_p = (y_p - \mu)/\sigma = (r_s - \mu_s)/\sigma_s = (1575 - 1500)/90 = 0.83$.

Looking up the probability associated with this value of K_p in a table of the normal distribution, we find that it is associated with a value of 0.8. This is indicative that the probability of realizing a unit resistance value of *less* than 1575 is 80%, or conversely, the probability is 20% that a unit will be produced which has a resistance greater than 1575 ohms. Therefore, 20% of the production lot will fail prematurely. For the latter situation (i.e., the proportion of items produced having resistance values less than or equal to 1400 ohms

$$\frac{1}{\sqrt{2\pi}} \int_{-\infty}^{K_p} e^{-x^2/2} \, dx = \frac{1}{\sqrt{2\pi}} \int_{-\infty}^{-1.1} e^{-x^2/2} \, dx \approx 0.13$$

where $K_p = (y_p - \mu)/\sigma = (r_s - \mu_s)/\sigma_s = (1400 - 1500)/90 = -1.11$.

Looking up the probability associated with this value of K_p in a table of the normal distribution, we find that 13% of the units produced will have a resistance value less than 1400 ohms and will require rework.

As indicated previously, another use of the normal distribution is as an approximation for other distributions, when direct application of such distributions would be difficult. The instance of normal approximation to the binomial distribution was discussed in the last section. The normal distribution can also be used to approximate a Poisson distribution when the value of the parameter a (in the Poisson distribution) is large. In that instance the number of failures (x) is considered normally distributed with $\mu = a$ and $\sigma = (a)^{1/2}$. Since the gamma distribution can be expressed as a Poisson distribution, the normal distribution also approximates the gamma when the shape parameter β is an integer (see Sections 2.1.4.1 and 2.1.4.2). The normal can also be used to approximate the chi-square distribution, which will be discussed in the next section.

As an illustration of use of the normal as an approximation, take Example (3) from the previous section. Assume that the antenna contains 1000 elements and that 800 is the minimum required for operation. Further, let

$$R_i(T) = e^{-\lambda T} = p = 0.82$$

Equation (2-77) will take the form

$$R(T) = P(x \geqslant 800) = \sum_{x=800}^{1000} \frac{1000!}{(1000 - x)!x!}[0.82]^x[0.18]^{1000-x}$$

Binomial tables not being generally available to evaluate such values, exact solution would require resort to a computer solution. In order to quickly evaluate the reliability, the binomial can be approximated by the normal as $Np(1 - p) = 147.6 > 10$, and hence conditions for the normal approximation discussed previous have been met. Therefore, the binomial distribution can be approximated by a normal distribution with

$$\mu = Np = 820$$

$$\sigma = [Np(1 - p)]^{1/2} \approx 12.15$$

$$x = (y - \mu)/\sigma = (800 - 820)/12.15 \approx -1.65$$

$$R(T) = P(x \geqslant 800) \approx 1 - \frac{1}{\sqrt{2\pi}} \int_{-\infty}^{-1.65} e^{-x^2/2} \, dx = 0.95$$

2.3 DISTRIBUTIONS USED FOR RELIABILITY DEMONSTRATION AND MEASUREMENT

Reliability demonstration and measurement are often an area of confusion for the unwary. Their objectives are basically different; however, the terminology and mathematical descriptors used to describe each are similar. The latter can lead to misunderstandings by the engineer because the mathematical logic used to describe each appears only subtly different.

To start, let us describe what is meant by *measurement*. An item is under test, or in operational use, and a number of failures occur over a given period of time. We use the data on failures occurring over the known period of time to make an *estimate* of the reliability of the item. We can even put a *confidence* interval around the *estimate* as a measure of the degree of certainty of the *estimate*. Such measurements have one objective—*to tell you where you are*.

Demonstration or qualification tests have a completely different objective— they determine *whether or not an item has met a given set of requirements* and the item is either accepted or rejected on that basis. Most demonstration tests have associated with them consumer risks and producer risks which are directly related to the actual reliability of the item.

Both estimation and demonstration employ the same type of statistics and distributions but in different ways (in order to meet their particular objectives). Chapter 12 will provide more detail on each. The distributions most commonly used in demonstration and measurements are the Poisson, the Weibull, the gamma, the chi-square, and the binomial.

2.3.1 The Poisson Distribution in Demonstration and Measurement

The Poisson distribution, for reliability and measurement purposes, may be represented as

$$P(x \leq N) = \sum_{i=0}^{N} \frac{[\lambda(T)T]^i}{i!} e^{-\lambda(T)T} \qquad (2\text{-}83)$$

$P(x \leq N)$ = probability that N or fewer failures occur over a
time period T, when $\lambda(T)T$ are expected

where $\lambda(T)$ = average rate of failure per hour over a specific period of time T. When failures follow an exponential failure distribution, (2-83) reduces to

$$P(x \leq N) = \sum_{i=0}^{N} (\lambda T)^i e^{-\lambda T} / i! \qquad (2\text{-}84)$$

where λ = failure rate per hour over *any* interval of T.

The Poisson distribution is the basis for most of the reliability estimation and demonstration tests in use today. Fixed-time tests, fixed-number-of-failure tests, and sequential test plans have been derived and are available. There are direct, relatively straightforward relationships among the Poisson, chi-square and gamma distributions that allow them to be used interchangeably to solve the same problem. One can even "transform" times to failure from a Weibull distribution (provided the shape parameter β is known) such that the Poisson may be used to demonstrate/estimate lifetime characteristics for that distribution as well.

This distribution is utilized primarily to provide estimates and demonstrations of failure rate λ or its reciprocal MTBF (in many documents MTBF is denoted as θ where MTBF $= \theta = 1/\lambda$) for equipments and systems which have assumed exponential failure distributions.

2.3.2   The Weibull Distribution in Demonstration and Measurement

The Weibull distribution for reliability demonstration may be represented as

$$P(t > T) = \exp[-(T/\alpha)^{\beta}] \qquad (2\text{-}85)$$

where

$P(t > T)$ = the probability that the time to failure of an item will be greater than T

β = a shape parameter indicative of whether or not hazard is increasing ($\beta > 1$), is constant ($\beta = 1$), or is decreasing with time ($\beta < 1$)

α = a scale parameter, sometimes referred to as characteristic life. It is representative of a measure of mean time to failure

This distribution has been utilized directly and through transformation for both measurement and demonstration purposes. It has been applied extensively in the areas of reliability growth (which is discussed in some depth in Chapter 11). Graph paper tailored to the Weibull is available to provide *estimates* of β and α given data on failures and time to failure, and handbooks have been developed for the purposes of Weibull distribution analysis.[15] Sampling plans for Weibull attributes and sequential test plans have been developed for demonstration purposes.[16]

The Weibull distribution is applied to components (from devices to jet engines), to equipments, and to systems.

2.3.3   The Gamma Distribution in Demonstration and Measurement

The gamma distribution is shown below in the form discussed previously:

$$f(t) = \frac{\alpha}{\Gamma(\beta)}[\alpha t]^{\beta-1}e^{-\alpha t} \qquad (2\text{-}86)$$

In the event that the failure times for the items in a population of N items follow an *exponential distribution*, and if one were to sum up the total operating time T, for the population over which the first r failures occur, then

$$T = \sum_{i=1}^{N} t_{i/r} \qquad (2\text{-}87)$$

where

N = number of items in population under test

$t_{i/r}$ = total operating hours on an item, i, at the time of the rth failure. If unit i failed at time t_i prior to the rth failure, then $t_{i/r}$ will equal t_i, the time to failure of the item

Under these conditions, T will follow a gamma distribution and the average item operating hours to failure that will be observed is equal to $\hat{\theta}$, where

$$\hat{\theta} = \frac{T}{r} = \frac{\sum_{i=1}^{N} (t_{i/r})}{r} \qquad (2\text{-}88)$$

It can be shown (see Appendix A7) that $\hat{\theta}$ will also take on a gamma distribution,

$$f(\hat{\theta}) = \frac{a}{\Gamma(r)} r(ra\hat{\theta})^{r-1} e^{-ra\hat{\theta}} \qquad (2\text{-}89)$$

which means that given α and r the probability of realizing a value of $\hat{\theta}$ less than some value A can be ascertained as below:

$$P(\hat{\theta} \leqslant A) = \int_{0}^{A} f(\hat{\theta}) \, d\hat{\theta} \qquad (2\text{-}90)$$

Equation (2-90) can be evaluated provided tables of the gamma distribution are readily accessible. However, we really do not need tables of the gamma distribution to evaluate the function, because (2-90) can be evaluated using either the tables of the chi-square distribution or the tables of the Poisson distribution. Why and how will be discussed in some detail in Section 2.4. Hence, the gamma distribution forms a basis for reliability measurement and demonstration, but its application can be transparent. It is included in this discussion to provide an insight and the logic of why, for example, a chi-square distribution is used as a basis for developing reliability measurement and demonstration procedures (the gamma distribution can be transformed to a chi-square distribution).

2.3.4   The Chi-Square Distribution in Demonstration and Measurement

The chi-square is one of the most common distributions in general use for statistical purposes. The distribution takes the form

$$f(z) = \left[2^{N/2} \Gamma\left(\frac{N}{2}\right) \right]^{-1} e^{-z/2} z^{N/2-1} \qquad (2\text{-}91)$$

where z = the chi-square variable and N = number of degrees of freedom.

Tables for the chi-square distribution are readily available for assessing the probability $P(z_n \leqslant A)$ that any value z of chi-square with a given number n of degrees of freedom will be less than a given value A:

$$P(z_n \leqslant A) = \int_0^A f(z) \, dz \qquad (2\text{-}92)$$

Let us next show how (2-89) evolves into a chi-square distribution. Again, use a simple transformation: let $2r\alpha\hat{\theta} = z$. Following again the general transformation procedure described in equation (A7-2) of Appendix A7, it can be shown that

$$\left(\frac{1}{2r\alpha}\right)\left(\frac{\alpha}{\Gamma(r)}\right) r \left(\frac{z}{2}\right)^{r-1} e^{-z/2}$$
$$= [2^r \Gamma(r)]^{-1} z^{r-1} e^{-z/2} = f(z) \qquad (2\text{-}93)$$

is a chi-square distribution with $2r$ degrees of freedom. Hence, if failures occur in an item or a population of items which exhibit an exponential distribution of times to failure, we can perform measurements and demonstrations with respect to MTTF/MTBF and failure rate of the item or item population through application of the chi-square distribution. As we will see later, through application of a simple time transformation, we can apply the chi-square distribution to the measurement and demonstration of items which follow a Weibull distribution.

2.3.5   The Binomial Distribution in Demonstration and Measurement

The binomial distribution is one of the most commonly used distributions for the development of sampling plans and for item acceptance. The distribution takes the form

$$P(x \leqslant M) = \sum_{x=0}^{M} \frac{N!}{(N - x)!x!} p^x (1 - p)^{N - x} \qquad (2\text{-}94)$$

where

$P(x \leqslant M)$ = probability of M or fewer occurrences when the probability of each occurrence equals p

N = number of components tested

For moderate-sized values of pN, and relatively small values of p or $(1 - p)$, say $0.9 \leqslant p \leqslant 0.1$, the binomial distribution can be approximated by the Poisson distribution. For large values of N such that $Np(1 - p) > 10$, the binomial distribution can be approximated by a normal distribution.

2.4   THE RELATIONSHIPS AMONG DISTRIBUTIONS

Now that you have an acquaintance with some of the major distributions used in reliability and how they are used, some things should be apparent:

- Some distributions have more applications than others. The same distribution may be used differently in different applications, for example, for measurement or demonstration.
- There are equivalences among distributions as discussed in some of the preceding sections. One analyst may apply the Poisson distribution, another the gamma, and a third the chi-square distribution to the same set of data or problem and all three would arrive at the same answer.

Summarizing the relationships among distributions that you should be aware of:

1. *The Weibull distribution* transforms into the exponential distribution when the shape parameter $\beta = 1$.
2. *The gamma distribution*
   a. Transforms into the exponential when shape parameter $\beta = 1$.
   b. Can be expressed in terms of the Poisson distribution when the shape parameter β is an integer. Since failures always take on integer values,

i.e., 0, 1, 2, 3, etc., for reliability purposes, the transformation is always possible. The transformation takes the form

$$1 - \sum_{x=0}^{r-1} \frac{(\lambda T)^x e^{-\lambda T}}{x!} = \int_0^T \frac{\lambda}{\Gamma(r)} (\lambda t)^{r-1} e^{-\lambda t} \, dt$$

(see Ref. 17 for a relatively simple derivation).

   c. Can be expressed in terms of the chi-square distribution when $2r\alpha\hat{\theta}$ is considered.

   d. Observed values of MTBF from an exponential distribution take on a gamma distribution.

   e. Can be approximated by the normal distribution for large sample sizes.

3. *The binomial distribution*

   a. Can be approximated by the Poisson distribution when Np is of moderate size and $0.9 \leqslant p \leqslant 0.1$ (see Section 2.4).

   b. Can be approximated by the normal distribution when $Np(1-p) \geqslant 10$ (see Section 2.2.2).

4. *The Poisson distribution*

   a. Transforms into the exponential when probability of zero occurrences is calculated.

   b. Transforms into the gamma distribution (in cases where the gamma shape parameter β is an integer). See 2.b above.

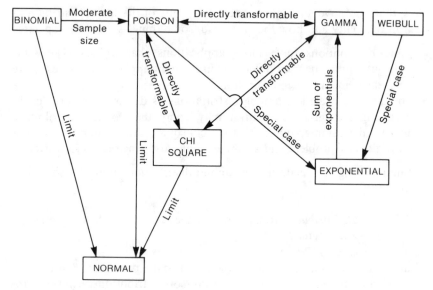

FIGURE 2-6.   Relationship among distributions.

    c. Transforms into the chi-square distribution when $2\lambda T = z$ is considered a chi-square variable, distributed with $2r$ degrees of freedom, i.e., the value of Poisson distribution $\Sigma(\lambda T)^x e^{-\lambda T}/x!$ can be evaluated through use of a chi-square distribution with $\chi^2 = 2\lambda T$, with $2r$ degrees of freedom (we will discuss the chi-square distribution in greater depth in Chapter 12).

    d. Can be approximated by a normal distribution with mean λT and $\sigma = (\lambda T)^{1/2}$, when λT is large.

5. *The chi-square distribution*

    a. Transforms to the Poisson and the gamma as shown above.

    b. Can be approximated by a normal distribution when degrees of freedom N is large.

6. *The normal distribution* serves as an approximation to virtually any distribution when sample size is large enough (due to the central limit theorem, which has not been addressed as yet) but can serve as an approximation to the Poisson, gamma, binomial and chi-square distributions with moderate to large sample sizes. (Note, too, that there are Weibull distributions which can also be approximated by the normal distribution.)

Figure 2-6 shows a depiction of the distributional relationships.

References

1. Drenick, R. F. (1960). "The Failure Law of Complex Electronic Equipment." *J. Soc. Ind. Appl. Math.* **8**, p. 680.

2. Nenoff, L. (1988). "Field Operational R & M Warranty." Proceedings, 1988 Annual Reliability and Maintainability Symposium, IEEE, Piscataway, NJ.

3. Kao, J. (1960). "A Summary of Some New Techniques on Failure Analysis." Proceedings, 6th National Symposium on Reliability and Quality Control, IEEE, Piscataway, NJ.

4. Kerscher, J. (1988). "Failure-Time Distribution of Electronic Components." Proceedings, 1988 Annual Reliability and Maintainability Symposium, IEEE, Piscataway, NJ.

5. Ratnaparchi and Park (1986). "Log Normal Distribution—Model for Fatigue Life and Residual Strength of Composite Materials." *IEEE Transactions on Reliability*, August.

6. Howard, B. and Dodson, G. (1961). "High Stress Aging of Failure of Semiconductor Devices." Proceedings, 7th National Symposium on Reliability and Quality Control, IEEE, Piscataway, NJ.

7. Goldwaithe, L. (1961). "Failure Rate Study for the Lognormal Lifetime Model." Proceedings, 7th National Symposium on Reliability and Quality Control, IEEE, Piscataway, NJ.

8. Singpurwalla, N. (1971). "Statistical Fatigue Models: A Survey." *IEEE Transactions on Reliability*, August.

9. Gupta, S. and Groll, P. (1961). "Gamma Distribution in Acceptance Sampling Based On Life Tests." *Journal of the American Statistical Association*, Dec.
10. Mann, N., Schafer, R. and Singpurwalla, N. (1974). *Methods for Statistical Analysis of Reliability and Life Test Data*. Wiley, New York.
11. Hughes Aircraft Co. (1985). *RADC Nonelectronic Reliability Notebook*. RADC TR-85-194.
12. Musolino, L. and Conrad, T. (1988). "Simplified Weibull Modeling of Component Early Life Reliability." Proceedings, 1988 Annual Reliability and Maintainability Symposium, IEEE, Piscataway, NJ.
13. Tiernan, P. (1988). "Failure Time Distribution of Electronic Components." Proceddings, 1988 Annual Reliability and Maintainability Symposium, IEEE, Piscataway, NJ.
14. Parzen, E. (1960). *Modern Probability Theory and Its Applications*. Wiley, New York.
15. Pratt & Whitney Aircraft (1983). *Weibull Analysis Handbook*. AFWAL-TR-83-2079.
16. Good, H. and Kao, J. (1962). "Sampling Procedures and Tables for Life and Reliability Testing Based on the Weibull Distribution." Proceedings, 8th National Symposium on Reliability and Quality Control, IEEE, Piscataway, NJ.
17. Herd, G. (1959). "Some Statistical Concepts and Techniques for Reliability Analysis and Prediction." Proceedings, 5th National Symposium on Reliability and Quality Control, IEEE, Piscataway, NJ.

3

The Customer's Role in Reliability Programs

Definition of the role and responsibilities of the customer and the contractor has been a matter of evolution over the years. The ideas and concepts discussed in this chapter represent a compendium of conclusions reached by many authors, reports, manuals, and papers. A sampling of documents which have contributed to the ideas expressed in this chapter (and Chapter 4) is contained in a bibliography at the end of the chapter.

Reliability specification and development for a given product can be likened to a series of actions and reactions, first among members of a customer group, then between customer and contractor(s). The customer need not always be the government, an airline describing its needs to an avionics supplier, a utility defining a new power generation system, or a direct consumer. It can be a prime contractor interfacing with a subcontractor. An organization can, at different times, act as both a customer and a contractor.

A reliability engineer, then, must be prepared to wear either hat, and even when wearing a contractor hat it is important that he have an appreciation of what the responsibilities of the customer are and how they can be implemented.

As a customer the two primary hypotheses that we must start with are:

1. Reliability is an item (part, component, equipment, or system) parameter that is controllable by the contractor and measurable by the customer.
2. The contractor is under no obligation to provide the customer with more reliability than is specified by the customer.

3.1   HOW MUCH RELIABILITY DO WE NEED?

Reliability is not an end in itself. Its use is advocated only to allow an item to successfully perform its mission or to perform its mission more effectively or with less cost. In other words, it must have a definable payoff in mission terms. The place to start the reliability definition process, then, is by examination of the item proposed and its intended mission or function. The discussion which follows is couched loosely in terms of a military acquisition, but the same principles hold for any type of an acquisition program.

3.1.1   Examining the Need for the Item

For virtually any new item, from component or equipment to major system, each military service requires some degree of documentation, which will be referred to here as a "Statement of Need" (SON). This is prepared prior to initiation of effort. The SON is prepared by the organization that will ultimately be the item user. The document does the following.

1. Describes the mission or a need
2. Describes the design development alternatives that will be considered
3. Identifies programs that exist or are under development to meet similar needs
4. Describes the maturity of technology needed for each design/development/ manufacturing alternative and risks for each alternative
5. Discusses funding required
6. Describes key criteria or conditions necessary to satisfy the need (i.e., reliability, logistics, support, standardization, etc.)
7. Describes acquisition strategy

   Note that these considerations are not too dissimilar to those which should be considered for any new product, from automobile to television set, if you substitute the words "market place" for "mission".

3.1.2   General Nature of a Requirement

From data contained in the SON, and with continued contacts with user personnel and organizations charged with logistics and support responsibilities for the item, the process of translating operational need to contractual requirement begins. The resultant reliability requirement presented to the contractor can take one of two forms.

   1. A requirement containing specific reliability numbers at the system,

subsystem and/or equipment levels. It is normally levied at the highest item level to allow the contractor the flexibility needed to provide the most cost-effective design structure. The requirement can take on a number of patterns depending on the nature of the item and its mission. One attribute that all such requirements must have in common, however, is that each must describe an item's ability to function satisfactorily under some time constraint unless dealing with a single-shot item such as ordnance. The time constraint can take the form of:

- a specific period of time/cycles relating to a single task or mission
- a specific period of time/cycles relating to a combination of tasks/missions
- a proportion of operating time (for continuously operating systems)
- a mean (mean time between failure, mean time to failure, mean time to first failure, etc.)

Examples

- The probability shall be at least 0.90 that the item will perform satisfactorily during a mission of 3 hours under a specified set of conditions.
- The probability shall be at least 0.90 that the item will survive 20 cycles of operation over a given time period under a specified set of conditions.
- The probability shall be at least 0.90 that the item will perform satisfactorily during a mission of 3 hours, when called on to function at a random point in time, under a specified set of conditions (usually applied to a redundant system).
- The probability shall be at least 0.85 that the item will perform the prescribed three mission functions during a mission of 3 hours under a specified set of conditions.
- The item shall maintain satisfactory performance at least 90% of the time, under a specified set of conditions (usually applied to continuously operating maintained systems).
- The probability shall be at least 90% that the item (under a specific set of conditions) shall be capable of responding satisfactorily when called on to operate at a random point in time (usually applied to systems which are prone to failure even when not actively performing their function).
- The item will have a minimum MTBF, (MTTF), (MTBR), (MTBM), etc., of 400 hours under a specified set of operating conditions.

For single shot items such as ordnance, the requirement might take the form:

- The probability shall be at least 0.95 that the item shall perform its function when called on to do so, at a random point in time under a specified set of conditions.

In addition to the above, in some instances, the requirement may stipulate a reliability growth rate (we will discuss reliability growth rate in more detail in Chapter 11).

2. An operationally based reliability requirement, for example: "The system will be able to operate 30 days without a system-inhibiting failure." This type of requirement is most appropriate for application to a large complex system. It provides a prime contractor with *maximum* flexibility to structure specific higher and lower level requirements. In this case the prime contractor levies more specific requirements to his subcontractors.

Whatever the form of requirement, the contractor will generally be required to demonstrate that the requirement has been attained (we will discuss demonstration in more detail in Chapter 12).

The contractual requirement and its demonstration, as a rule, reflect only on the item's material, design and manufacturing characteristics—those items under contractor control. Negative effects on reliability due to logistics, support, handling, or other factors under the control of, or the responsibility of the customer, are not chargeable to the contractor's design or considered during demonstration. They must, of course, be taken into consideration when the requirement is developed.

A clear distinction should be made between a requirement and a goal. A requirement is an objective that is subject to demonstration. A goal is a desirable objective but is not subject to demonstration. Both requirements and goals can be levied on a program at the same time. For example, an equipment might require a minimum value of MTBF = 500 hours, but the design goal for the item might be 800 hours. In this case the contractor may be asked to demonstrate that the minimum value has been attained but show by virtue of design efforts (e.g., prediction results, allocations, target goals established) that design was to the higher level.

Too stringent a requirement can have costly effects for both the customer and contractor. It can provide misleading premises to logistics and support analyses and it can even lead to litigation.

A requirement less than that which the state of the art can economically provide will result in a higher support cost than necessary. At best, it will result in a slightly lower initial acquisition cost.[1,2]

3.1.3 Generation of Reliability Requirements

Translating a mission need into a contractual requirement requires:

1. Examination and analysis of the mission with respect to how reliability affects its success or effectiveness: with respect to how mission and reliability may be affected by operational, logistics, and support constraints.

2. Acquisition of data or generation of estimates of what the reliability state of the art can provide with one or more design alternatives.
3. Identification and assessment of risks associated with meeting the requirements.

Each of the above is addressed in the following subsections.

3.1.3.1 *Relating Reliability to Mission Needs*

Relating reliability to mission needs is relatively straightforward for components and equipments destined for placement in planned or established systems or subsystems. In such instances the primary linkage between item, system, and mission has already been established. As a result, this particular task is more appropriate for the definition of reliability needs and requirements at the system or major subsystem levels for *new* acquisitions.

Its purpose is to identify the relative importance of each major subsystem with respect to the system and its mission and to provide insight with respect to interfaces and conditions of use. It also provides the information necessary to develop a black box or mathematical model (mathematical models will be covered in more detail in Chapters 9 and 10) of design alternatives.

In all likelihood, the SON and/or prime contract document has described a mission which can be related to reliability needs. Terms and expressions such as:

Probability of mission success
Kill probability
Availability or operational readiness
Equipment must maintain operation for 30 days without a mission inhibition

all provide a linkage between reliability and mission needs and provide information on the types of reliability requirements (i.e., the time constraint measures, previously described) that are most appropriate.

After that general linkage has been established, the role of each design entity (in this context, component or subsystem) in the system must be evaluated for each system mission.

If the system has more than one function (mission), each must be prioritized.

For each mission performed by the system define:

1. A time duration.
2. A frequency of occurrence.
3. A breakdown of each mission into phases to identify the usage profile for each component and/or subsystem:

    a. Identification of those subsystems and equipments which must be operational during each phase.

    b. Proportion of time in each phase.

4. What constitutes a failure for each mission (phase) in terms of the subsystem and equipment functions which have to be working satisfactorily in order to perform the given mission or mission phase.

5. Critical failures (failures which result in total loss of the mission function or the component, subsystem, or system itself).

6. The environments under which the missions will be carried out (temperature, shock, vibration, humidity, etc.).

Both the success of the mission and its reliability needs will be affected by operational, logistics, and support constraints. As a consequence, inputs are acquired from user and logistics organizations to assess potential effects on reliability resulting as a consequence of (1) operational concept, i.e., attended, unattended, degree of access; (2) maintenance plan/concept; (3) logistics concept and plan.

3.1.3.2 Reliability Data Acquisition

The previous section defined the equipment and subsystem functions that had to be operating in order for a given mission to be performed. Therefore, if we were able to ascribe a reliability value which was achievable to each subsystem or equipment making up the item, we would be able to specify an achievable item reliability with respect to its mission(s). In order to accomplish that task, state-of-the-art reliability data on equipments or components of the same type as in our item should be acquired. (Some may argue that "less than current state-of-the-art" estimates would be a more accurate statement. This follows from the fact that data can be acquired and assessments made on units only after they have been in operation perhaps for one or more years and that these actually represent the state-of-the-art that existed a number of years prior to actual field use. Hence, what is measured probably represents less than current state-of-the-art capability.) The data should then be conservatively extrapolated to where the reliability state-of-the-art should be at the time of actual item development.

Most desirably, such data should be acquired from (1) equipment/subsystems similar to those making up the item in question or from similar items, (2) equipment/subsystems having a similar magnitude and scope of associated reliability programs and tasks.

Be aware, however, that a contractual specified (and even demonstrated) value of MTTF or MTBF can be, and in the vast majority of cases is, different from its apparent observed operational value. This differential has been attributed to a number of reasons:

1. Field operational values of reliability are really measured in terms of MTBM or MTBR rather than values of MTTF or MTBF, as that is all that most data systems are capable of providing accurately.
2. Design to meet the specified value is guided by results of reliability prediction. Inaccuracies either in the reliability prediction technique itself or in the way the prediction was performed can result in deviations from the intended specified value which are not caught by the accept/reject criteria of the demonstration.
3. The operational environment introduces happenstances, policies, human errors, and applications of the item not taken into account during prediction.
4. A combination of the above.

To aid in correlating such values of reliability, the Rome Air Development Center developed a series of guides depicting statistical relationships between MTBM and predicted/specified MTBF for a number of different types of electronic equipments. These guides are available to U.S. government agencies and their contractors.[3]

In the event that data on equipments are not available, general relationships developed from either empirical or engineering judgment, or a combination of the two, are available in the literature. As an example, see Chart 3-1 (extracted from the RADC Reliability Engineers Toolkit) which provides "typical" ranges of MTBF for a number of different types of equipments/systems circa 1988.

3.1.3.3 Identifying Risks/Opportunities
As indicated previously, an extrapolation or adjustment for technical progress between current and future state of the art reliability is a part of the requirement-generation process. It is obvious that reliability improvement as a consequence of such maturing/emerging technologies as VLSI, VHSIC, and photonics have to be considered. However, the item design in question will in all likelihood be a true hybrid. It will be made up of a combination of (1) existing design (where the components used are equivalent to those used today but are more reliability mature), (2) modified design (which contains not only more mature versions of today's components but some new components and component technology as well), (3) new design (which has the greatest potential for the use of new components and new technology). The mix estimated to be most appropriate to the development of the item is the logical baseline to start extrapolation.

Application of brand new technology does not always equate to immediate

CHART 3-1 Typical Ranges of MTBF for a Range of Different Types of Equipment

Item	MTBF (hours)
Radar Systems	
Ground rotating search radar	75–175
Large fixed phase array radar	3–6
Tactical ground mobile radar	25–75
Airborne fighter fire control radar	50–200
Airborne search radar	300–500
Airborne identification radar	200–2000
Airborne navigation radar	300–4500
Communications Equipment	
Ground radio	5000–20 000
Portable ground radio	1000–3000
Airborne radio	500–10 000
Ground Jammer	500–2000
Computer Equipment	
Ground computer	1000–5000
Ground monochrome display	15 000–25 000
Ground color display	2500–7500
Ground hard disk drive	5000–20 000
Ground tape storage unit	2500–5000
Ground printer	2000–8000
Ground modem	20 000–50 000
Miscellaneous Equipment	
Airborne countermeasures system	50–300
Airborne power supply	2000–20 000
Ground power supply	10 000–50 000

Source: RADC Reliability Engineer's Toolkit.

significant reliability gains over an existing mature technology. New model versions of an automobile generally have more defects and poorer reliability in the year of their introduction than in ensuing years.

Estimates of the rate of maturation of a component type, a device, or a technology can be assessed in several different ways (discussed in more detail in Chapter 8):

1. *Available data.* If failure rate information relating to a given device, combination of devices, a given device technology, etc., contemplated for use in the item to be developed, is included in Mil-Hdbk-217 (Reliability Prediction for Electronic Equipment), note the item failure rate in

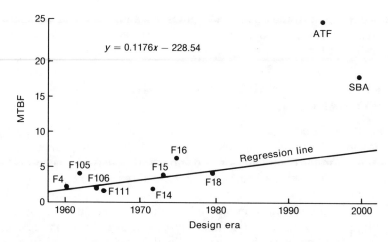

FIGURE 3-1.   Reliability trends over time with respect to fighter aircraft electrical/electronic functional groups.   [From "DOD/Industry—R & M Technology Study Analysis," Lyons and LaSala, Proceedings, 1984 Annual Reliability and Maintainability Symposium. © 1984 IEEE.]

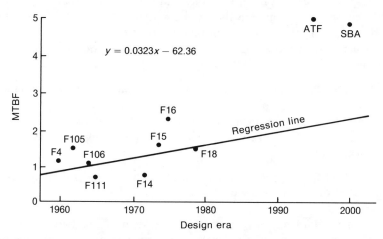

FIGURE 3-2.   Reliability trends over time with respect to total aircraft systems.   [From "DOD/Industry—R & M Technology Study Analysis," Lyons and LaSala, Proceedings, 1984 Annual Reliability and Maintainability Symposium. © 1984 IEEE.]

Mil-Hdbk-217 over a number of revisions of the document (spanning a number of years), to see whether a trend line is apparent. Use that trend line as a basis for extrapolation for those devices.

2. *At the item (equipment/subsystem) level itself.* Track operational or specification requirements levied for similar items in like environments over a number of years. Fit a trend line to the data collected and use that

as a basis for extrapolation. See Figures 3-1 and 3-2 which summarize reliability operational data for electrical/electronic item groups and total aircraft as a function of design era for eight fighter aircraft.[4]

3.2 RELIABILITY WARRANTIES IN LIEU OF REQUIREMENTS

A reliability requirement levied in a contract places a responsibility on the customer for developing a minimum achievable reliability value, and a responsibility on the contractor for its demonstration. Passing the test is the reward and there is no motivation to exceed that which is required.

Contractual provisions dealing with reliability-based incentives, guarantees, and warranties are used as a means to cost-effectively enhance operational reliability and reduce life-cycle cost to the benefit of the customer. At the same time, these provide the contractor with incentives to enhance the reliability of the product. There are various types of incentives, guarantees, and warranties which can be employed. Whether or not it would be advantageous to apply one or the other and what the terms and conditions should consist of depends on the candidate item, its intended application, and the scope of the acquisition program.

A general distinction can be made among warranties, guarantees and incentives:

- A warranty extends the contractor's responsibility to repair, overhaul, or replace items which fail in use.
- A guarantee is a commitment by the contractor to meet operational targets.
- An incentive is a reward or penalty for providing a given level of performance.

Almost 20 different types of warranties, guarantees, and incentives exist.[5] Examples of the most widely used are:

1. Reliability Improvement Warranty with Mean Time Between Failure Guarantee. A series of target values are specified over consecutive time periods of the guarantee. Contractor corrective action is triggered when a target is not reached; the contractor could be rewarded if the contract value is exceeded.
2. Ultimate Life Guarantee. A value of service life is established for an item and remedial action is taken upon failure.
3. Commercial Service Life Guarantee. Extension of warranty provision over service life.
4. Warranty for Contractor Repair/Replacement: appropriate for intermediate or depot levels. Covers repair, overhaul, replacement of items for a given period of time or under a given usage rate.

5. Reliability Improvement Warranty (RIW). Contractor will repair all failures covered by the warranty for a specified period of time, for an agreed single payment.
6. Logistics Support Cost Guarantee. Employs a mathematical model to relate system operational characteristics to support costs. A target value of support cost is chosen and support cost is calculated based on observed operational characteristics. Differences between calculated and target support cost are the basis for reward and penalty.

The airlines were among the first users of warranty to provide for the replacement, repair, maintenance, or cost of maintenance for many different items. The United States Congress in the mid-1980s took a series of actions which resulted in a statute relating to the obligation of DOD prime contractors to provide their products with guarantees with respect to quality.[6] Warranties are presently in place on many items procured by the three services.

3.2.1 When a Warranty/Guarantee Should be Considered

Not every product is a candidate for warranty application, and some products make more effective candidates than others. In order to make a proper judgment, parameters relating to the planned procurement strategy, the characteristics of the product, operational and support factors, and economic factors must all be weighed. Chart 3-2 depicts such parameters, broken down to individual warranty/guarantee application factors which were developed in an Air Force study[7,8] for RIW or RIW/MTBF guarantee procurements, which can be used for that purpose.

A Yes with respect to any factor is a plus with respect to warranty guarantee application. The relative importance of each factor is, of course, dependent on the particular nature of the planned warranty.

3.2.2 Composition of a Warranty/Guarantee

A warranty/guarantee must contain clear and unambiguous statements concerning conditions, customer and contractor obligations and administration detail. For example,

1. The portions of the item excluded from the warranty, such as components subject to scheduled replacement.
2. Term of warranty (function of time and usage rate).

CHART 3-2 Factors That Impact the Practicality of Warranty Application

Procurement factors

Procurement on a fixed-price basis.
Multiyear funding for warranty-guarantee available.
Warranty administration can be efficiently accomplished.
Procurement is competitive.
Potential contractors have proven capability, experience, and cooperative
    attitude in providing warranty-guarantee commitments.
Escalation clause included in contract applicable to warranty-guarantee costs.
Equipment will be in production over substantial portion of
    warranty-guarantee period.

Equipment factors

Equipment maturity at an appropriate level.
Unit can be properly marked or labeled to signify existence of
    warranty-guarantee coverage.
Unit operates independently of other subsystems.
Unit has high level of ruggedization when shipment is required.
An elapsed-time-indicator (ETI) can be installed on the equipment.
Unit has no failure mode that would lead to additional damage to itself or other
    units if not corrected.

Operational factors

Use environment is known and predictable.
Equipment operational reliability and maintainability are predictable.
Equipment wartime or peacetime mission criticality is not of the highest level.
Operational failure and usage information can be supplied to the contractor.
Backup warranty repair facilities are available.
Provision can be made for computing the equipment's MTBF.

Support factors

Control of unauthorized maintenance can be exercised.
Unit is field-testable.
Number of site maintenance personnel tends to be independent of equipment
    performance.
Rapid restoral time is required.
Proper operation can be remotely sensed.

Economic factors

The combination of item reliability, maintainability, deployment schedule, and
    using rate information is sufficient to determine compliance with the contract
    warranty-guarantee provisions and to suggest improvements if necessary.

3. Thresholds—the number of claims, failures or returns that must be exceeded before warranty comes into effect.
4. Exclusions to warranty action; for example, failures and/or damage induced by mishandling damage, act of God.
5. Nature and extent of action required:
   a. Repair, replace, overhaul.
   b. Design and material changes, and if so, to what degree they should be implemented throughout the inventory.
   c. Configuration management.
6. Contractor turnaround time limitations.
7. Disposition of items returned that test good.
8. Data requirements for both contractor and customer.
9. Warranty marking seals.
10. Maintenance facilities required.
11. Assessment criteria (i.e., timekeeping, usage rate calculation, failure reporting).

For more information on warranty application and structure, see Refs. 9–11.

3.3 STRUCTURING PROGRAM REQUIREMENTS: THE FIVE FACTORS

If you have followed all the steps previously outlined, you have a grasp for the steps to be taken by the user in structuring reliability requirements or warranty provisions. The next consideration for the customer is to determine the basic needs of the type of reliability program needed such that those requirements can be practically attained.

Mil-Std-785 "Reliability Program Plan for Systems and Equipment, Development and Production," in existence since 1965 and discussed in Section 3.5, defines the makeup of a *reliability program* in terms of a series of administrative and technical tasks generally suited to the needs of the armed services. While originally structured to meet DOD needs, and subject to periodic revision, the tasks can be logically applied to *any reliability program, military or civilian*. The tasks represent a codification of experience gained and engineering management rationale developed over a broad range of programs. However, over time there is a tendency (1) to treat codification results (the tasks) as if they were in themselves principles or objectives in their own right and (2) to lose sight of the logical principles which resulted in the codification in the first place. To that purpose, let us go back to basics and describe what *any* reliability program must provide for in terms of five generic factors:

1. *Visibility,* i.e., to provide visibility with respect to progress and problems during design to both the contractor and his customer.
2. *Controls,* i.e., to provide for controls and guides to insure that adequate standards are being applied in design, quality, and manufacturing as they affect reliability.
3. *Correction,* i.e., to provide means to find and correct problems after design.
4. *Communication,* i.e., to provide for means to communicate information concerning all the above within the contractor organization and to the customer.
5. *Demonstration,* i.e., to provide for a demonstration showing that the requirement has been achieved.

The makeup of these factors will be further discussed in Section 3.4, for now become comfortable with the concepts.

All are necessary to provide needed direction, control, and incentive to the contractor, and to provide necessary feedback and assurance to the customer. It would be hard to dispute the need of the contractor to apply the five factors or the right of the customer to expect their application.

Each program, the particular acquisition phase of the program, its characteristics, and its procurement strategies mandate differences in how each of the factors must be addressed. Such differences will be discussed next.

3.3.1   The Five Factors and the Acquisition Process

The acquisition process for any item can be divided into four separate phases—concept formulation, validation/demonstration, full-scale engineering development, and production. (See Chart 3-3 for a general definition of each phase.) While the five factors generally must be provided for the acquisition to be successful, each of the factors is not necessarily applicable to every phase of the acquisition process. (Four separate phases of acquisition are far from sacrosanct. In many instances it makes sense to integrate two or more into a single phase. In that case, the five factors must be applied in a manner consistent with such an integration.) As can be seen, each of the phases has a different objective. Generally speaking, we determine what reliability is needed during concept formulation; we see what has to be done to get it during validation/demonstration; we design for it and perform qualification tests to show that we can achieve it during full-scale engineering development; during production we seek to maintain the reliability to which this item was qualified.

Chart 3-4 shows general applicability of the five factors during the four

---

CHART 3-3 The Four Phases of Acquisition

Conceptual phase—Establishment of the technical, military and economic bases for the development program. Outputs include alternate concepts and their characteristics, estimates of schedule, acquisition costs and support parameters.

Validation phase/Demonstration phase—Refinement of major program characteristics through extensive study and hardware development test and evaluation. Purpose is to validate/demonstrate given design alternatives; to provide a basis for proceeding into full-scale engineering development.

Full-scale engineering development phase—Design, prototype production, demonstration, test, and evaluation of the item. Forms the basis of entry to production phase.

Production—Production and delivery of the item and its necessary support equipment and documentation.

CHART 3-4 General Applicability of the Five Factors in Each Phase of the Acquisition Program

	Concept	Validation	FSED	Production
Visibility progress/problems	×	×	×	×
Controls			×	×
Find/correct problems	×	×	×	×
Communication	×	×	×	×
Demonstration			×	×

phases of the acquisition cycle. For example, visibility with respect to progress and problems is of critical concern during all phases of acquisition. Controls and direction for design, quality, and manufacturing may be developed and structured during concept and validation phases, but they are not generally implemented until full-scale engineering development and production phases. Finding and correcting problems of various types goes on during all phases of acquisition, as does communication. Demonstration/qualification/

maintenance of qualification generally take place at the end of full-scale engineering development and during production.

While some factors are appropriate for application in more than one phase of acquisition, the nature (purpose) of the factor is modified by the phase in which it is applied. For example, during full scale engineering development we are concerned with *visibility of progress/problems associated with* design. That translates into use of reliability predictions, logistics support analyses, application of reliability evaluation models, results of failure modes and effects analyses, interpretation of environmental and other test results, future availability of components, etc. During production we are concerned with visibility of progress and problems in maintenance of reliability during production. That translates into use of tests, screens, and quality control management procedures at parts and equipment indenture levels, reliability growth tests, failure analyses, etc.

3.3.2   The Five Factors and Equipment System Characteristics

As the acquisition phase defines the role played by the five factors, the characteristics and parameters of the item under development define the scope and magnitude of the role played by each factor. For each factor in each phase, ask yourself what situations would require an increase (decrease) in activities associated with each factor. For example:

- Complexity of the item (hardware complexity, number of separate subsystems). Factors influenced: visibility, controls, and demonstration.
- Application of new technology. Factors influenced: controls and correction.
- Degree of new design incorporated into item. Factors influenced: controls and correction.
- Number of failure prone or reliability problem devices and components in the design. Factors influenced: controls, correction, and communication.
- Number of nonstandard parts in design. Factors influenced: visibility, controls, and correction.
- Ratio of available qualification test time to MTTF small. Factor influenced: demonstration.
- Significantly higher values of requirements/more stringent tolerances for components. Factors influenced: visibility, controls, correction, and communication.

Most of the above will either trigger actions within the reliability program which otherwise would not be performed, or will increase the magnitude and/or man-hours associated with a given reliability program action. An effort completed in 1987 (Ref. 12) is illustrative of relationships found between levels of complexity, other program characteristic factors, and combined reliability/maintainability (R & M) program scope expressed in terms of engineering/technician man-hours. The following represent samples of the empirical results from that effort.

Average cost of nonstandard parts, including justification/follow up/ specification preparation = 45 labor hours per part.

Average labor hours required for R & M visibility (modeling) and allocation increases as system design complexity increases.

$$H_M = (MC)^{2.03}(U)^{1.7}(RC)^{1.34}$$

where
  H_M = average labor hours required for R & M modeling and allocation
  MC = Modeling complexity
    MC = 1: Simple series configuration
    MC = 2: Simple fault tolerance
    MC = 3: Complex fault tolerance
  U = Number of unique items; each unit is a procurable unit with stipulated values of R & M
  RC = Requirement complexity, the number of distinct subsystem models which have to be integrated into the model

3.3.3   The Five Factors and Procurement Strategy

Procurement strategy can have more profound an influence on the five factors than any other effect discussed to this point, because it can nullify the need for application of one or more of the factors. We will discuss two such procurement strategies—one concerning warranties/incentives/guarantees and the other the purchase of off-the-shelf equipment.

3.3.3.1   *Applications of Warranties and Associated Clauses*

We discussed warranties, guarantees and incentives purposes, applications, and composition in Section 3.2. Warranties, guarantees and incentives may be applied for a number of reliability reasons: to motivate a contractor to provide greater reliability; to provide second-party assurance of spare units; to reduce, at least short term, maintenance burden; to reduce visible cost of the reliability program. (However, it also requires more upfront spending for the warranty.)

Their use can be predicted under the hypothesis that a user is interested in operational field performance, not test or analysis results, because the latter do not fully reflect the efforts of operational and environmental conditions. In the event that the item does not meet warranty provisions, extra spares can be included to maintain field use capability while changes or repairs are being made. Warranties and incentives can also be used in combination, where warranties are imposed to make sure that incentives work.[2] They can also be considered for application when the item to be acquired contains ragged-edge state-of-the-art technology with attendant advantages and risks to both customer and contractor. In particular, with respect to the five factors, warranty-type contracts may well relax the need for demonstration (reliability qualification and production acceptance tests). They will also, in all likelihood, lead to negotiation with respect to providing the contractor with increased levels of flexibility as to customer approval of controls and standards to be applied.

3.3.3.2   Use of Off-the-Shelf and Commercial Items

No one subject can create as much discussion or be deluged with as divergent a set of opinions as the use of existing items (equipments and subsystems) in the configuration of a new system or subsystem. This is particularly true in the case of application of a commercially produced item in a military system.

There are undeniable reliability related payoffs from using off-the-shelf equipment in general: reduced reliability program costs and existence of a proven reliability record with verified logistics and maintenance support needs. (Other reasons, of course, exist that make off-the-shelf acquisitions desirable: reduced development cost; low risks with respect to schedule and cost; potential for reduced unit production costs over those for a new production item.)

There are, however, various types of off-the-shelf items, each of which can have a different effect on the characteristics of the reliability program:

1. An item developed under military specifications for application in another system where:
   a. No modifications are required for performance or quality.
   b. Modifications are required for performance or quality.
2. An item developed commercially where:
   a. No modifications are required for performance or quality.
   b. Modifications are required for performance or quality.

Items Developed Under Military Specifications
Obviously, if the item has already been produced to a military specification, has a satisfactory operational reliability and logistics history, and if necessary modification to the item is slight, then a minimum reliability program would suffice.

However, if the item has both an unknown operational reliability history and an uncertain qualification test/reliability development background, then the *reliability program* should require an analysis of the design and a review of the standards used during development and parts selection, to determine whether any changes are required. The reliability program should require that inadequacies in item reliability be corrected. At minimum a demonstration/qualification test should be performed on the item.

If the item must be modified to meet either performance or reliability levels, then the reliability program must be structured to the five factors with respect to those modifications. The scope of the reliability program will then be proportional to the degree and nature of the modifications made.

If the item is to be used in an environment different from that in which it was previously used, the reliability program should include tests for the effects on reliability of the new environment; engineering design changes, if necessary to provide needed reliability; and appropriate demonstration to show its achievement.

Items Developed Commercially

Escalating costs for development and production of military equipments and systems have led to increasing interest in the use of commercial products in military systems.

Whether or not to include commercially available items in a system acquisition, and what if any modifications to the items are required, depends on at least three different considerations, only one of which deals with reliability and the reliability program.

1. Military demands on the operation of the items, such as security features or antijamming capabilities. This rules out some but not all types of applications. For example, commercial items in the areas of navigation, nonsecure communications, and data processing are presently in use by the military.
2. Administrative/support considerations, which may include the need for special or nonstandard test and support equipment at intermediate and/or depot levels of maintenance, documentation and test manual acquisition, sparing, resupply and nonstandard part needs, configuration management, and data and reprocurement rights.
3. Adequacy of the item to meet reliability needs in the anticipated mission environment. Whether or not the item can be modified to meet reliability needs. The nature of the modification program, its cost and practicality.

While the first two are of at least equal importance to the third, our attention is most directed to the last. To frame a base for adequacy, first consider the particular controls, guides, and standards associated with reliability design typically levied on an equipment acquisition:

Part quality
Packaging
Thermal design
Derating practices
Shock and vibration standards
Temperature standards
Humidity standards
Quality assurance standards

Such activities (included under the *Controls* factor of the Five Factors) are present in various degrees in every program. It is the stringency of such activities that differentiates a military program from a commercial one.

Reference 13 evaluated possible cost/benefits of using commercial off-the-shelf equipment for Air Force applications. The data and charts which follow are based on the results of that effort.

Chart 3-5A defines typical military requirements levied on such elements for acquisitions of ground-fixed and airborne transport equipment. Commercial requirements for these eight program standards and controls vary widely. Chart 3-5B defines minimum commercial requirements which can be encountered. In some instances the environment that an item is exposed to in a given commercial environment can be no less severe than that applied in a typical military environment; for example, a communications set used both in a commercial airliner and in a military transport. As a consequence, some commercial products can be expected to perform as well in military environments, some will not.

In some instances the reliability of commercial items in military application environments can be enhanced by design adaptation:

1. Mil-qualified parts can be substituted for commercial parts, but exact substitutes do not always exist, which would necessitate a redesign effort.
2. Spot bonding of wires can be used to improve vibration resistance; openings can be sealed or modified for dust and moisture protection.
3. Modifications of a degree can be made to improve package ruggedness.

There will be some features, however, that cannot be adapted. The materials and processes used in the fabrication of the item, such as soldering, dissimilar materials, or adhesives, cannot be spot-changed to meet a given acquisition. These affect reliability through their environmental sensitivity.

If the development plan associated with the commercial item is as stringent

CHART 3-5A Typical Military Program Requirements

Parts quality
- Program parts Selection list (PPSL) (a list of all parts approved for use in the program)
- Qualified products list
- Screened/established reliability parts

Packaging
- Ground based
    Enclosures defined
    Cooling methods standardized
- Airborne
    Packaging standards
    Unique for high-performance aircraft
    High density
    More EMI/RFI shielding

Thermal design
- Formal testing
- Critical part identified with adequate margins provided
- Tasks particular to thermal analysis performed (particularly for high-density aircraft packages)

Derating practices
- Contractually required
- Typically more stringent
- Required for all parts

Shock and vibration standards
- Ground based
    30 g peak operational shock
    No crash safety limit
    Sinusoidal vibration
        2.5 g peak (5–2000 Hz)
- Airborne
    Shock
        6 g operational shock
        15 g crash safety shock
    Vibration
        Random per Mil-Std-810C
            Performance levels (2 hr per axis)
                Cockpit (0.7 g RMS/15–2000 Hz)
                Fuselage (8 g RMS/15–2000 Hz)
            Endurance levels (3 hr per axis)
                Cockpit (2.0 g RMS/15–2000 Hz)
                Fuselage (17.0 g RMS/15–2000 Hz)

Continued

Temperature standards
- Ground based
    - Operating (0 to 50°C)
    - Nonoperating (−62 to +71°C)
- Airborne
    - Cockpit (50 000 ft)
        - Operating (−54 to +55°C), (71 C DASH)
        - Nonoperating (−57 to +85°C)
    - Fuselage (70 000 ft)
        - Operating (−54 to +71°C), (95 C DASH)
        - Nonoperating (−57 to +95°C)

Humidity standards
- Ground based
    - 90 to 95% at +50°C (2 days)
- Airborne (transport)
    - 95 to 100% at 50°C (2 days)

Quality assurance standards
- Final test (all equipment)
- Temperature cycling burn-in
- Discretionary reliability/longevity testing
- Production qualification tests
- Medium-quality audit level of effort
- Limited failure reporting to customer

CHART 3-5B Minimum Requirements for Commercial Programs

Parts quality
- Limited vendor control
- Vendor standard parts
- No change control authority
- Limited receiving inspection
- Minimal specification definition

Packaging
- Not compact
- Lightweight construction
- Plastic rather than metal structure

Thermal design
- Typically convection cooled
- Limited thermal testing
- No special heat exchanger
- Critical parts not identified

Derating practices
- Applications with vendors maximum rating
- No derating policy

Shock and vibration standards
- Ground based
   No specification
- Airborne
   6 g operational shock
   15 g crash safety shock
   1.5 g peak (5–55 Hz)
   0.25 g peak (55–2000 Hz)

Temperature standards
- Ground based
   Operating (0 to 30°C)
   Nonoperating (no limit)
- Airborne
   Operating (-15 to $+71$°C)
   Nonoperating (-54 to $+85$°C)

Humidity standards
- Ground based
   80 to 90% at 30°C (8 hr)
- Airborne (transport)
   95 to 100% at 50°C (2 days)

Quality assurance standards
- Final test (all equipment)
- Limited burn-in
- No sampling plans
- No failure reporting to customer
- Low-quality audit level of effort

as one that would be applied to a military procurement, and if modification to the item is slight, or if a satisfactory reliability operational history on the item exists (in an environment equivalent to that of the contemplated mission), then a minimum reliability program will suffice.

If the item has a reliability history in a known commercial environment less severe than the one contemplated for your application, the reliability program should, as a minimum, provide for the following before a decision is made:

1. Reliability predictions taking into account the contemplated system environment.
2. Analyses of the reliability controls and standards under which the item was designed to determine risk of operation in contemplated environment.

3. Verification that the item can function successfully in the planned environment.

If the item has a reliability history in a known commercial environment similar to the one contemplated but requires modification in order to meet system needs, the reliability program must be structured to the five factors with respect to the particular changes made.

3.4 COMMUNICATING WITH THE CONTRACTOR: SPECIFICATIONS AND STANDARDS

Up to this point we have described the development of the requirement, the general make-up of the reliability program (the five factors) and the options and provisions which directly impact reliability. All of these require communication. The customer must make his needs known to the contractor and the contractor must convey his understanding of those needs back to the customer. Communication between customer and contractor is carried out through use of contractual statements of work (SOW), through specifications, and through standards. Virtually all significant hardware acquisitions where reliability is an issue, whether of a commercial or military nature, have these in common.

- A specification defines the performance or other attributes of a product. (It may describe the reliability of the product.)
- A SOW is a portion of the contract. It will define the reliability requirements of the contract and is the governing document. (In a typical military procurement, a SOW takes precedence over a specification.)
- A standard promotes commonality with respect to processes and practices within a given organization or industry. It defines rules and criteria that have been established for the conduct of processes and practices. Reliability program standards may be unique to a given company or a DOD service organization or global to an industry or to DOD in general.

Because of their prevalent use within the industrial complex and because of the frequent misunderstandings which prevail about their application, we will discuss in some detail the use of current DOD standards (Military Standards) as an approach to the treatment of the five factors.

Military Standards, or Mil-Stds, are not requirement documents. They are used to provide commonality within the services with respect to processes and practices that are performed by contractors. Their purpose is to alleviate, if not eliminate, the need for a contractor to establish three different sets of procedures (one for each service) to perform a given process or practice. A

Mil-Std also tells the contractor what the military "standards" are concerning the performance of given practices and processes.

Military Handbooks or Mil-Hdbks provide guidance and information concerning how certain tasks included in a Mil-Std may be performed.

Neither a Mil-Std or a Mil-Hdbk is a static document. Both are subject to change, integration with other standards and handbooks, outright replacement, or elimination.

In order to further clarify the role of Mil-Stds, let us go back to our earlier discussion (Section 3.3) of the five factors: i.e., provide visibility with respect to design progress/problems; provide for implementation of proper controls; provide means to find and correct problems; provide means of communication; provide for demonstration. How does the customer know that any given contractor has a program in place which satisfactorily covers the five factors? How does a contractor know that his program is consistent with the needs of the customer? The answers to both questions are *through the use of Mil-Stds*.

The customer uses a Mil-Std tailored to the needs of a particular acquisition as a means of communicating information about the perceived general and specific needs of a given reliability program to the contractor. The contractor then communicates actions taken and results obtained to the customer via the *Data Items* (DI) associated with each Mil-Std. A DI is a specific report associated with the performance of a task. Its purpose is to describe the makeup, detail, and results of the contractor's approach to performing a particular task. It is usually subject to approval by the customer.

In the area of reliability, Mil-Stds are commonly broken down into a series of general and specific tasks or procedures which represent a codification of the standard's objectives. Not all tasks in a given standard will be applicable to all acquisitions, nor will the scope of each task be identical.

Let us look at the codification process in a little more detail, for example in terms of translating the five factors into the reliability engineering tasks of the current version of Mil-Std-785, "Reliability Program for Systems and Equipment Development and Production."

3.4.1 Transforming General Reliability Program Factors to Engineering Tasks

As indicated in Section 3.3, Mil-Std-785 defines the makeup of a reliability program generally acceptable for military acquisitions. It is broken down into the following administrative and technical tasks:

Defining the program plan makeup   (Task 101)
Monitor/control of subcontractors and suppliers   (Task 102)

Performance of program reviews	(Task 103)
Instituting a failure reporting, analysis, and correction action system (FRACAS)	(Task 104)
Establishing a failure review board (FRB)	(Task 105)
Providing for reliability modeling	(Task 201)
Providing for reliability allocation	(Task 202)
Providing reliability predictions	(Task 203)
Performing failure modes effects and criticality analysis (FMECA)	(Task 204)
Providing for sneak circuit analysis	(Task 205)
Providing for electronics parts/circuits tolerance analysis	(Task 206)
Providing for a parts program	(Task 207)
Identifying reliability critical items	(Task 208)
Assessing the effects of testing, storage, handling, packaging, transport, and maintenance	(Task 209)
Providing for environmental stress screening (ESS)	(Task 301)
Providing for reliability growth testing	(Task 302)
Providing for reliability qualification test program	(Task 303)
Providing for a reliability acceptance test program	(Task 304)

Each defines an action to be performed by the contractor and most require information be specified by the customer as an input to the task performed.

Chart 3-6 shows a matrix relating the reliability tasks of Mil-Std-785 to the five factors.

3.5 RELIABILITY STANDARDS AND HANDBOOKS THAT YOU SHOULD KNOW ABOUT

As indicated previously, Mil-Stds and Mil-Hdbks are subject to revision and update, to combination, and to elimination. (Integration of such documents, however, is historically rare and total elimination rarer still.) The following, based on circa 1991 versions of such documents, should be viewed in that light. It should as a minimum acquaint the reader with the basic purpose of each document, which should be revision-independent; with samplings of their tasks; and with their rationale and other factors deemed appropriate. For more detail the current version of document itself should be reviewed. (Complete and current copies of each document available from the Naval Publications and Forms Center, Philadelphia, PA.) Detailed summaries of each current standard and handbook are available.[14] Our discussion will be limited to the following:

CHART 3-6 Relating the Five Factors to the Reliability Tasks of Mil-Std-785

	Provide Visibility, Progress During Design	Provide Controls and Standards	Provide Means to Find/Correct Problems after Design[a]	Provide Means to Communicate Progress and Problems	Provide Means to Demonstrate that Requirements are Achieved
Program plan makeup		×			
Monitor/control venders	×	×			
Program reviews	×			×	
FRACAS			×	×	
FRB		×	×	×	
Reliability modeling	×			×	
Reliability allocation	×	×			
Reliability prediction	×	×		×	
FMECA	×		×		
Sneack circuit analysis			×		
Tolerance analysis	×	×			
Parts program	×	×			
Reliability critical items	×	×			
Testing, storage, handling	×	×			
ESS		×	×		
Reliability growth			×		×
Reliability qualification test					×
Reliability acceptance test					×

[a] May trigger a redesign effort.

Mil-Std-785, "Reliability Program and Equipment Development and Productions"
Mil-M-38510,* "General Specifications for Microcircuits"

* Other standards and specifications dealing with parts will be discussed in Section 6.2.

Mil-Std-883, "Test Methods and Procedures for Micro Electronics"

Mil-Std-756, "Reliability Modeling and Prediction"

Mil-Hdbk-217, "Reliability of Electronic Equipment"

Mil-Hdbk-338, "Electronic Reliability Design Handbook" (Volume I)

Mil-Std-2164 (Navy), "Environmental Stress Screening Process for Electronic Equipment"

DOD-Hdbk-344 (USAF), "Environmental Stress Screening of Electronic Equipment"

Mil-Std-2155 (Navy), "Failure Reporting Analysis and Corrective Action System"

Mil-Std-1629, "Procedures for Performing a Failure Modes Effects and Criticality Analysis"

Mil-Hdbk-251, "Reliability/Design Thermal Applications"

Mil-Std-781, "Reliability Testing for Engineering Development Qualification and Production"

Mil-Hdbk-781, "Reliability Test Methods, Plans and Environments for Engineering Development, Qualifications and Production"

Mil-Hdbk-189, "Reliability Growth Management"

Mil-Std-1521, "Design Reviews"

Mil-Std-721, "Definitions of Items for Reliability and Maintainability"

3.5.1   Mil-Std-785: "*Reliability Program for Systems and Equipment Development/Production*"

The application of Mil-Std-785 defines the structure of the reliability program in terms of tasks to be performed. The basis for the tasks chosen and their scope is a function of the mission need, the acquisition phase of the program, the characteristics of the item itself (for example, the degree of new, modified, or existing design/development required), and the acquisition strategy chosen (e.g., off-the-shelf buy, warranty, etc.). One purpose of the program plan is to provide guidance to the contractor in planning and scoping the needed reliability program for a particular acquisition. Tailoring of the tasks by the customer provides that guidance. A second purpose of the program plan is to evaluate the contractor's approach to understanding, execution, planning, and control of the tasks. The evaluation is performed in two ways: (1) the review of "data items" and other reports the contractor must provide, indicating his response/efforts taken with respect to each task, and (2) design reviews.

The standard contains 18 tasks (listed in Section 3.4.1). Guidance in performing the vast majority of the tasks is contained in other ad-hoc standards, handbooks, or both. (Standards/handbooks which will be discussed were listed above: note the relationship among these and the 18 tasks of Mil-Std-785.)

Associated with each task in the standard is a section entitled "Details to be Specified by the Procuring Activity." This contains information for the contractor from the customer which is necessary to scope, plan, and perform the tasks satisfactorily. Examples of the type of information which could be contained in such "Details" are identification of mission or use parameters and operational constraints; levels to which analyses and predictions are to be performed (i.e., piece part, printed circuit board, module, equipment, etc.); definition of terms (e.g., definition of critical failure); particular design guides to be used; part selection policies; and environmental stress screening levels.

The relationship of each task in Mil-Std-785 to other standards will be described in some detail and an overview of "Details to be Specified" will be provided in the discussions which follow.

3.5.2   Mil-M-38510: "General Specifications for Micro Electronics"

(1) Mil-M-38510 establishes requirements and procedures that must be followed in order to qualify and maintain qualification of microcircuits for general DOD use and hence provide a JAN (a part which has successfully passed a series of defined tests and screens) designation for the products. It covers required quality assurance activities in terms of materials, tests processes, screens, inspection, failure analyses, and corrective action.

(2) It requires extensive characterization of the functional, packaging, electrical, and mechanical characteristics of each device qualified through the standard.

(3) It is used in conjunction with Mil-Std-883 ("Test Methods and Procedures for Micro Electronics"), which provides the characteristics of the various tests and screens which can be performed.

3.5.3   Mil-Std-883: "Test Methods and Procedures for Micro Electronics"

Mil-Std-883 defines screens, testing procedures, and standards for the test of microelectronic devices in the military and space environments. It is used for device qualification, screening, and quality conformance inspection. (It supports Task 207 of Mil-Std-785.)

Test methods in the standard cover environmental, mechanical, and electrical stresses. Procedures are defined for part approval, screening, failure analysis, and other aspects of microcircuit acceptance and control.

3.5.4   Mil-Std-756: "*Reliability Modeling and Prediction*"

Mil-Std-756 defines uniform general and specific guidelines, rationale, and procedures to be followed to identify and implement a reliability modeling and prediction approach. (It supports Tasks 201, 202, and 203 of Mil-Std-785.) Prediction and modeling results provide visibility with respect to progress and problems in meeting reliability targets. The customer uses the standard to define the level of detail required in prediction and modeling tasks. Also defined is information relating to the level of indenture of the model (i.e., printed circuit board, module, equipment, subsystem); the type of (or combinations of types) prediction appropriate (i.e., similar item method, similar circuit method, active element group method, parts count method, or parts stress analysis method); the environment(s) of operation; etc.

The standard is broken down into two sections, one dealing with modeling and one with prediction.

3.5.4.1   *Mil-Std-756 Modeling Considerations for Logistics/Support and Mission Needs*

Logistics Reliability/MTBR/MTBM   Development of a reliability model of the unit (equipment, subsystem, or system) to the level of indenture specified, assuming all items making up the unit were connected in a series and any failure of any item would cause an item removal/replacement.

Mission Reliability, R(T), MTTF, MTFF, Steady-State MTTF Development of a reliability model of the unit taking into account alternate modes of operation; degraded but satisfactory performance; fault tolerance to the level of indenture specified taking into account the fact that an item failure will not necessarily cause unit failure.

3.5.4.2   *Reliability Prediction: Estimating the Reliability of Each Item in the Above Model*

Different prediction means are available for use. The means most appropriate for application in a given program depend on the phase of acquisition, characteristics of the unit design (degree of new, old, existing), and acquisition strategy (off-the-shelf, warranty, etc.):

1. Prediction by comparison with similar items (contained in Mil-Std-756).
2. Prediction by comparison with similar circuits (contained in Mil-Std-756).
3. Prediction by active element group method (contained in Mil-Std-756).
4. Prediction by parts count (contained in Mil-Hdbk-217).
5. Prediction by part stress analysis (contained in Mil-Hdbk-217).

(1), (2), and (3) are more appropriate for early acquisition phases; (4) and (5) are more appropriate for later acquisition phases. More detail will be provided on prediction methods in Chapter 8.

3.5.5 Mil-Hdbk-217: *"Reliability Prediction of Electronic Equipment"*

Mil-Hdbk-217 provides information and data necessary to perform a reliability prediction by parts count or part stress analysis. (It supports reliability prediction Task 203 of Mil-Std-785 and reliability prediction Task section 200 of Mil-Std-756.) Reliability prediction at this level provides contractor and customer visibility with respect to risks of not meeting requirements and serves as a basis for design and material change. The handbook is broken down into two sections: one dealing with part stress reliability prediction, which takes into account the effects of environment, application, power, voltage and thermal stresses, quality, construction, and other factors on each component part making up the item; and one dealing with parts count reliability prediction, which takes into account part environment (ground, airborne, inhabited, space, etc.) and quality level and which assumes a moderate overall level of stress.

3.5.5.1 *Part Stress Reliability Prediction: Requiring the Greatest Amount of Detail*
This type of prediction is applicable during later design phases where actual hardware and circuits are being designed. It assigns failure rates to virtually every component part category. These include:

Microelectronics	Connectors
Discrete semiconductors	Printed wire boards
Tubes	Connections
Lasers	Meters
Resistors	Quarts crystals
Capacitors	Incandescent lamps
Inductive devices	Electronic filters
Switches	Fuses

The prediction technique takes into account the following factors (where appropriate) for the device:

Quality level and maximum rating of each part
Environment (e.g., ground, airborne, naval, space, missile launch, missile flight, manpack environment)

Application stress (e.g., power, voltage, thermal, construction, function, duty cycle, etc.)

3.5.6 Mil-Hdbk-338: "*Electronic Reliability Design Handbook*" (Volume I)

Mil-Hdbk-338 provides theory, background philosophy, and guidelines for virtually every aspect of reliability design and analysis. Its purpose is to support virtually every task in Mil-Std-785 and other reliability Mil-Stds. The document at present covers the following subject areas

Definitions
Reliability/maintainability/availability theory and modeling
Reliability specification and apportionment
Reliability measurement
Reliability engineering guidelines
Reliability data collection and analysis
Software reliability
System reliability engineering
R & M management considerations

3.5.7 Mil-Std-2164 (EC): "*Environmental Stress Screening Process for Electronic Equipment*"

Mil-Std-2164 (EC) is a military standard approved for use within the Navy. While "limited" documents of this nature have not been coordinated and approved for tri-service use, it can, nevertheless, be specifically cited for use as a standard for any particular acquisition program regardless of service. (However, modifications may be required to fit the standard to the policies, definitions, etc., in use by the particular service.)

Environmental stress screening (ESS) is a process whereby environmental stresses are applied to subassemblies, assemblies, equipments, etc. Such stresses act on existing material and workmanship defects to precipitate failures which otherwise would occur during field usage. The standard identifies specific screens, screen durations, test conditions, and general administrative procedures under which screens should be run. (It supports Task 301 of Mil-Std-785.) The screens usually involve the application of vibration and thermal cycling.

3.5.8   DOD-Hdbk-344 (USAF): "*Environmental Stress Screening of Electronic Equipment*"

DOD-Hdbk-344 (USAF) is a military handbook approved for use within the Air Force. While "limited" documents of this nature have not been coordinated and approved for tri-service use, it can, nevertheless, be specifically cited for use as a handbook for any particular acquisition program regardless of service. (Modifications, however, may be required to fit the handbook to the policies, definitions, etc., in use by the particular service.) Reference the discussion relating to Mil-Std-2164 for an explanation of environmental stress screening (ESS). The handbook defines a sequence of procedures to be taken by a contractor to structure, plan, monitor, and control an ESS program tailored to a particular hardware acquisition program. (It supports Task 301 of Mil-Std-785.) Specific guidance covered by the handbook includes:

- *Development of planning estimates of incoming defect density.* Procedures and tables are included to aid in estimating incoming defect density as a function of unit complexity, the quality of the items making up the unit, the operational environment of the unit, and other characteristics.
- *Screen selection and placement.* Procedures and tables are included to aid in assessing the strength and effectiveness of various environmental screens against the unit to which they are applied.
- *Structuring screening acceptance tests.* Procedures and tables are included to aid in defining the length of a test which provides a given statistical confidence that screening objectives have been achieved.
- *Cost-effectiveness analysis.* Procedures are provided to identify the most cost-effective screen selection and placement sequences.
- *Monitoring, evaluation, and control.* Procedures are provided to aid in determining whether or not observed results are within predetermined control limits, to determine what corrective actions are required, and to prioritize corrective actions needed.

3.5.9   Mil-Std-2155 (AS): "*Failure Reporting Analysis and Corrective Action System*"

Mil-Std-2155 is a military standard approved for Navy use. (It supports Tasks 104 and 105 of Mil-Std-785.) While limited documents of this type have not been coordinated and approved for tri-service use, they can, nevertheless, be specifically cited for use as a standard for any particular acquisition program regardless of service. (Modifications, however, may be required to fit the standard to the policies, definitions, etc., in use by the particular service.) The standard describes the makeup and function of a failure reporting, analysis, and corrective action system (FRACAS), and a Failure Review Board (FRB). FRACAS and FRB have as their respective objectives:

- The establishment of a closed loop failure reporting system; of procedures for analysis of failures to determine cause; and of procedures for documentation and recording of corrective actions taken.
- To review failure trends, significant failures, corrective action status, and to insure that adequate corrective actions are taken in a timely manner and recorded during the development/production phases of the program.

The general concept of the composition of FRACAS is defined. General procedures (tasks) which make up FRACAS are described. The tasks take the form of a closed loop going from failure *observation* to *documentation* to *verification* to *failure analyses* to *corrective action* with the FRB as a hub of the process.

3.5.10   Mil-Std-1629: "*Procedures for Performing a Failure Mode Effects and Criticality Analysis*"

Mil-Std-1629 defines the tasks and procedures necessary to the performance of a failure modes/effects and criticality analysis (FMECA). (It supports Tasks 204 and 208 of Mil-Std-785.) The document establishes standards for:

1. The evaluation and documentation of the potential impacts of each functional or hardware failure on mission success, safety, performance, maintainability, and maintenance requirements. This is accomplished through a failure mode and effects analysis (FMEA).
2. Criticality analysis (CA) of each of such potential impact.
3. Ranking of each potential failure by severity of its effect.

With respect to (1), guidelines are provided as to when a *functional* or a *hardware* based FMEA is appropriate. With respect to (2), both a qualitative and a quantitative approach to assess criticality are defined. Information and data components necessary to the performance of a FMECA are identified. Formats for documentation of results in the form of worksheets are provided. Chapter 11 (Section 11.1) provides descriptions and examples of (1)–(3). Interfaces of FMECA with maintainability, safety, and survivability are defined.

3.5.11   Mil-Hdbk-251: "*Reliability/Design Thermal Applications*"

Mil-Hdbk-251 provides thermal design guidelines for the purpose of enhancing reliability. It provides guidance for the definition of thermal requirements, for the selection of cooling methods, and for thermal design approaches in general.

3.5.12   Mil-Std-781: "*Reliability Testing for Engineering Development, Qualifications and Production*"

Mil-Std-781 establishes standards for the performance of reliability testing performed for the integrated test programs specified in Mil-Std-785. Task descriptions for the reliability development/growth testing (RD/GT), reliability qualification testing (RQT), production reliability acceptance tests (PRAT), and environmental stress screening (ESS) are defined. (It supports Tasks 104, 301, 302, 303, and 304 of Mil-Std-785.) The standard is applicable to six broad categories of military equipment/systems: fixed ground, mobile ground, shipboard, jet aircraft, turbo prop aircraft, and helicopter equipment and missiles.

The standard contains four classes of tasks, all to be tailored to the particular needs of the acquisition.

1. Test planning and control, which covers the *development of the integrated reliability test plan; development of test procedures keyed to each test included in the test plan; provision for test program reviews;* and *establishment of a joint test group to provide coordination of the test program.*
2. Development testing, which covers *thermal and vibration survey testing to detect potential design weakness* and *reliability development/growth tests.*
3. Reliability accounting tests, which cover *demonstration that the unit conforms to reliability requirements* and *production reliability acceptance tests.*
4. Environmental stress screening (ESS), which covers the formulation and implementation of an ESS program.

3.5.13   Mil-Hdbk-781: "*Reliability Test Methods, Plans and Environments for Engineering Development, Qualification and Production*"

Mil-Hdbk-781 provides and explains the test methods and test environmental profiles used to implement the integrated test programs defined in Mil-Std-781. (It supports Tasks 104, 301, 302, 303, and 304 of Mil-Std-785.)

Statistical test methods for evaluating reliability growth and for evaluating environmental stress screening programs are provided. Reliability assurance tests covering a wide range of consumer and producer risk are provided through a variety of test plan options: failure-free test plans; sequential test plans; fixed-duration test plans; all equipment production reliability test plans.

Guidelines for choice of test methods and test plans are provided. Guidelines for choice of test environments and profiles are provided.

3.5.14   Mil-Hdbk-189: *"Reliability Growth Management"*

Mil-Hdbk-189 defines guidance and procedures for conducting and managing a reliability growth program. Reliability growth involves finding and removing material and design deficiencies from a product to a point where reliability needs are satisfied. A reliability growth program provides visibility and control to that process. (It supports Tasks 104, 302, and 303 of Mil-Std-785.) The handbook contains guidance with respect to: the reliability growth concept; structure of reliability growth curves; evaluation of reliability growth.

For the last, two methods of evaluation (assessment) are provided. One is of a qualitative nature based on monitoring reliability program tasks such as prediction, choice of parts, FMECA, FRACAS, ESS, etc., and gauging whether or not the results/outputs are consistent with reliability needs. The second is of a more active, quantitative nature based on the application of defined growth standards and the application of mathematical/plotting techniques to assess progress in meeting standards.

3.5.15   Mil-Std-1521 (USAF): *"Program Reviews"*

Mil-Std-1521 is a military standard approved for Air Force use. (It supports Task 103 of Mil-Std-785.) While limited documents of this type have not been coordinated and approved for tri-service use, they can, nevertheless, be specifically cited for use as a standard for any particular acquisition program regardless of service. (Modifications, however, may be required to fit the standard to the policies, definitions, etc., in use by the particular service.) This standard defines and describes the types, makeup and timing of program reviews during the acquisition process.

3.5.16   Mil-Std-721: "*Definition of Terms for Reliability and Maintainability*"

Mil-Std-721 defines the most commonly used words and terms associated with reliability and maintainability. It is intended to be used as a common basis for reliability and maintainability term definitions for the services. However, if a specific definition is provided for a given term in a particular development program in a request for proposal, statement of work, or specification, the definition provided in that document generally prevails.

3.6   CUSTOMER RESPONSIBILITIES

Customer responsibilities range from preparing for the acquisition program, to determining and communicating requirements and other information essential to the program, to assuring that all requirements have been satisfied.

3.6.1   Customer Responsibilities: Preparing for the Program

The responsibilities of the customer start in the conceptual phase of development and extend to the end of the production phase. Before the beginning of full-scale engineering development, the customer should have the answers to such considerations as the following:

What reliability requirements are needed?
System complexities that might mandate need for extreme parts control, or a need for unique design tolerance.
Design concepts that might result in need for application of new or immature technology.
Parts control policies necessary.
Logistics and support policies/plans contemplated.
Design guidelines, if any, to be imposed.
Special requirements, if any, for tests.
Special facilities needed to perform tests.
Applicability of warranties, guarantees, incentives.
Potential reliability problems based on past experiences.

All of the above represent information that is needed by the customer: to scope the shape of the reliability requirements; to address any special needs that must be associated with the reliability program; and to focus on and provide foresight relating to potential problems associated with meeting reliability needs.

3.6.2 Customer Responsibilities in the Application of Mil-Stds

In the event that the customer is one of the services, the vehicles for program plan development may be the standards and handbooks described earlier. In that case, the tasks of the standards are tailored to the needs of the particular acquisition.

Only those tasks are selected which make sense to the nature of the acquisition. Usually a tailoring appendix will be available to provide guidance in that area.

For each task called out, necessary guidance and constraint information (detail) must be provided. This data is consistent with the information that should be available to the customer at the time (see Section 3.6.1). Most modern standards make provision for such guidance under sections titled "Details to Be Specified," associated with each task. An example summary of the information included in such sections in Mil-Std-785 is illustrated in Chart 3-7.

The Data Items (DI) (see Section 3.4) associated with each task provide a means for the contractor to describe the approach taken, and the scope and results of each task performed. DIs, then, provide a means of surveillance of contractor activity on each task. In order to insure that such reports are

---

CHART 3-7 Summary of "Details to be Specified" for Tasks

- Identification of the reliability tasks relevant
- Identification of the contractual status of the program plan
- Identification of additional tasks to be performed or guidelines to be followed
- Identification of any DIs required and delivery dates
- Identification of level of assembly for failure reporting
- Identification of requirements for logistics support analyses
- Identification of mission parameters and operations constraints
- Identification of applicable alternative modeling techniques
- Identification of the level to which allocations shall be made
- Identification of the item life profile, to include one or more mission profiles
- Identification of alternate reliability prediction methods
- Identification of part approval procedures
- Identification of detailed design guidelines for parts (screening, derating, etc.)
- Description of environmental stress screens to be applied
- Identification of level (board, subassembly, assembly) at which testing will be accomplished
- Identification of equipment and quantity to be used for reliability development/ growth testing
- Identification of Mil-Std-781 or alternative procedures to be used for reliability qualification testing

CHART 3-8 Examples of Data Items Called Out in Mil-Std 785, 781, and 756.

Mil-Std-785B (Selected Data Items)
DI-R-7079 Reliability Program Plan
DI-R-7080 Reliability Status Report
DI-R-7041 Failure Summary and Analysis Report
DI-R-7081 Reliability Mathematical Models
DI-R-2114 Reliability Allocation Report
DI-R-7082 Reliability Predictions Report
DI-R-1734 Failure Modes, Effects, and Criticality Analysis Report
DI-R-7083 Sneak Circuit Analysis Report
DI-R-7084 Electronic Parts/Circuits Tolerance Analysis Report
DI-R-35011 Critical Item Control Plan

Mil-Std-781D (Selected Data Items)
DI-RELI-80250 Reliability Test Plan
DI-RELI-80251 Reliability Test Procedures
DI-RELI-80252 Reliability Test Report
DI-RELI-80253 Failed Item Analysis Report
DI-RELI-80254 Corrective Action Plan
DI-RELI-80255 Failure Summary and Analysis Report

Mil-Std-756B (Selected Data Items)
DI-R-7094 Reliability Block Diagrams and Mathematical Models Report
DI-R-7095 Reliability Prediction and Documentation of Supporting Data

provided, the DIs must be specified as deliverable items in the contract. Usually this is done through a vehicle known as the Contract Data Requirements List (CDRL). Chart 3-8 lists examples of DIs that are currently called out in Mil-Std-785, Mil-Std-756 and Mil-Std-781.

Applications of Mil-Std-785 or other Mil-Stds are not, however, panaceas for meeting either customer or contractor responsibilities. Proper tailoring often requires a degree of knowledge and experience unavailable to the customer. It has also been argued by some that:

- Tasks which are unnecessary, not cost-effective, or duplicative of other tasks can be called out by the unwary.
- Information unavailable to the customer during the tailoring process can result in the choice of inappropriate tasks or task scope.
- Application of Mil-Stds can provide for too much technical micromanagement of, and too little flexibility for, the contractor.

As a consequence, even within the services, different approaches to the application and use of Mil-Stds can be found.

3.6.3 Customer Responsibilities Irrespective of Standards Used

It does not matter whether the standards that are used are unique to company or military service organization or global to the industry or to DOD in general. It does not matter whether the customer is the government, a utility, or a prime contractor. The customer—every customer—assumes much the same responsibilities. Even if the authority to structure a program plan is delegated to the contractor, the customer still retains ultimate responsibility. The responsibilities include:

1. Defining the requirement to be met in clear, unambiguous terms.
2. Making sure that the reliability program established is adequate.
3. Providing the contractor with the information and detail necessary to implement the reliability program.
4. Monitoring progress/problems and providing for independent assessment and evaluation.
5. Providing coordination with other engineering disciplines.

The first element has been addressed in some detail in preceeding sections. Let us briefly discuss some approaches that can be taken for the remainder.

3.6.4 Structuring the Reliability Program

As indicated previously, the structure of the reliability program plan is a function of the acquisition phase, the characteristics of the item to be developed, and the acquisition strategy. Some subjective rules of thumb have been suggested with respect to task criticality and priority:

1. A series of Institute for Defense Analyses (IDA) studies[15] indicated that the following reliability tasks are major payoff items:
   a. Failure reporting and corrective action systems (FRACAS)
   b. Test, analyze, and fix (TAAF)
   c. Worst-case analysis
   d. Thermal analysis
   e. Stress screens
   f. Reliability growth
2. The Rome Air Development Center (RADC) in the "RADC Reliability Engineers Toolkit"[16] indicates a task priority/importance matrix basically in accord with Chart 3-9.
3. Mil-Std-785 (dated 1980) provides a reliability task importance prioritization, abstracted in Chart 3-10.

CHART 3-9 Task Priority Matrix

Task	Full-scale Development			Production			Off-the-Shelf Hardware		
	Ground	Airborne	Space	Ground	Airborne	Space	Ground	Airborn	Space
Design review	R	R	R	E	E	E	R	R	R
FRACAS	E	E	E	E	E	R	E	E	E
Subcontractor control	R	R	R	R	R	R	E	E	E
Part selection	E	E	E						
Part derating	E	E	E						
FMECA	O	R	E						
Prediction and allocation	E	E	E				E	E	E
Sneak circuit analysis		O	R						
Critical items	O	O	E	O	O	O			
Thermal analysis	R	E	E	E	E	E	E	E	E
Effectiveness of storage and handling	R	R	R	O	O	O			
Reliability qualification test	E	E	R				O	O	
ESS	P	P	P	E	E	E	E	E	E
Production reliability test				E	E	E			
Relilability growth	O	E	E				O	O	O

E = essential; R = recommended; P = plan for; O = optional.

Source: RADC Reliability Engineer's Toolkit

CHART 3-10 Task Applicability Matrix

Task	Title	Task Type	Concept	Valid	FSED	PROD
					Program Phase	
101	Reliability program plan	MGT	S	S	G	G
102	Monitor/control of subcontractors and suppliers	MGT	S	S	G	G
103	Program reviews	MGT	S	S(2)	G(2)	G(2)
104	Failure reporting, analysis, and corrective action system (FRACAS)	ENG	NA	S	G	G
105	Failure review board (FRB)	MGT	NA	S(2)	G	G
106	Thermal management control (TMC) program	MGT	G	G	G	G
201	Reliability modeling	ENG	S	S(2)	C(2)	GC(2)
202	Reliability allocations	ACC	S	G	C	GC
203	Reliability predictions	ACC	S	S(2)	C(2)	GC(2)
204	Failure modes, effects, and criticality analysis (FMECA)	ENG	S	S(1)(2)	C(1)(2)	GC(1)(2)
205	Sneak circuit analysis (SCA)	ENG	NA	NA	C(1)	GC(1)
206	Electronic parts/circuits tolerance analysis	ENG	NA	NA	C	GC
207	Parts control/application program	ENG	S	S(2)(3)	G(2)	G(2)
208	Reliability critical items	MGT	S(1)	S(1)	G	G
209	Effects of functional testing, storage, handling, packaging, transportation, and maintenance	ENG	NA	S(1)	G	GC

CHART 3-10 *Continued*

Task	Title	Task Type	Concept	Valid	FSED	PROD
210	Thermal/reliability design trade studies	ENG	G	G	GC	GC
211	Thermal/reliability design analysis	ENG	NA	G	G	GC
301	Environmental stress screening (ESS)	ENG	NA	S	G	G
302	Reliability development/growth testing	ENG	NA	S(2)	G(2)	NA
303	Reliability qualification test (RQT) program	ACC	NA	S(2)	G(2)	G(2)
304	Production reliability acceptance test (PRAT) program	ACC	NA	NA	S	G(2)(3)
305	Thermal design validation test (TDVT) program	ENG	NA	S	G	GC

S = Selectively applicable; G = Generally applicable; GC = Generally applicable to design changes only; NA = Not applicable
(1) Requires considerable interpretation of intent to be cost effective.
(2) Mil-Std-785 is not the primary implementation requirement. Other Mil-Stds or statement of work requirements must be included to define the requirements.
Source: Mil-Std-785.

The choice of any particular task should be based on need, payoff, cost, schedule, and applicability to a given acquisition phase and procurement strategy. The problems with many customer generated program plans are:

- Reliability program tasks are sometimes called out under the impression that resulting reliability is proportional to the number and scope of the tasks called out.
- Tasks are often called out just because they were called out in other programs in the past.
- Series of tasks are often called out as "boilerplates" without any effort to tailor them to the needs of the program.

The structure of an appropriate reliability program plan is a logical process that cannot be put in terms of a mathematical formula.

Subjective program structures, on the other hand, as described in the preceeding charts have their place in default decisions if no better structure can be defined. Before resorting to a default decision, however, the acquisition program should be reviewed to see whether a better course of action can be generated. This is accomplished by considering the reliability program tasks which are candidates for inclusion in an acquisition program, by asking some logical questions and by considering various facts. As an illustration, let us consider the types of tasks identified in Mil-Std-785. Regardless of which different standards and policies may be used to structure a reliability program, the types of tasks described in Mil-Std-785 will be common to most.

The general features which dictate the makeup and magnitude of the reliability program include:

1. The particular phase of acquisition. Ask the question whether the information and data necessary to perform a particular task will be available. If so, to what degree of accuracy can the task be performed? What would be the task payoff(s) for the particular acquisition phase?

2. The degree of new, modified, or existing design contemplated. Generally speaking, the more new or modified the design, the greater the required magnitude of the reliability program. An off-the-shelf item will, as a whole, require a reliability program of less magnitude than one requiring new or modified design.

3. The complexity of the item. Generally speaking, the more complex an item is, the greater the magnitude in terms of engineering hours required for the reliability program. This may be best exemplified in reliability prediction, in parts control, and in the FMECA tasks, all of which require a level of effort proportional to the complexity of the item to which they are applied.

4. The utility of the task. Does the task provide just information or information that can be used constructively to improve item reliability? A task has minimum utility if used to evaluate a design or a certain aspect of design performed or completed after the time has passed to take practical corrective action.

5. The cost of the task. See Ref. 17, which provides empirical information on reliability program costs associated with missile and space systems, and Ref. 12, which provides empirical information on reliability program costs associated with ground electronics systems.

6. The impact on schedule. This is especially relevant to testing tasks. Consideration can be given to trading off number of units under test, allowable test time, and consumer and producer risks.

Specific considerations as to applicability and scope of individual tasks for a particular acquisition include:

1. Degree of subcontractor control needed. Are subcontractors contemplated? If so, to what degree? Will subcontractors be supplying off-the-shelf designs or components? If so, will modifications have to be made to meet the needs of the program? How will the customer and his prime contractor know that the component or design will meet reliability needs? Will the subcontractor be supplying new designs or components? If so, how is consistency between the prime contractor and subcontractor program responsibilities, requirements, and tasks maintained?

2. Necessity for design reviews to provide the customer with the visibility needed. The level of need increases with degree of new or modified design, number of different item interfaces, and number of subcontractors. Reviews should be keyed to specific development milestones. Generally speaking, there is more need to hold reviews more frequently during the early phases of a design/development program than in the later phases. Scope of the review includes all reliability analysis, modeling, prediction, failure analysis details and, results; problems related to vendors and subcontractors, design, and parts; program progress; and adherence to schedule.

3. Necessity for a FRACAS activity to identify failure causes and implement corrective action. What should be its scope? Consider the complexity of the item to which it pertains; the maturity of the design; the manufacturing process; the components making up the item; the rigor of the environmental envelope of operation; the number and types of tests to be performed; the level(s) of tests and failures at which FRACAS will be applied (component, subassembly, assembly, item); whether FRACAS activity will be limited to certain types of failures; whether FRACAS activity should be open-ended or limited to critical failures. Failure analysis, reporting, and corrective actions associated with ESS and reliability growth should be linked to the overall FRACAS.

4. Necessity for the reliability modeling task which is used for reliability allocation and evaluation. What should be its scope? The following should be taken into account: the nature and complexity of the item; whether or not fault-tolerant design is contemplated; whether or not the mission is composed of multiple phases, each requiring different hardware capabilities or different modes of hardware operation; whether or not different mission phases are subject to different levels of environmental stress; whether or not repair is possible during the mission; whether or not repair can be effected periodically. An affirmative response to any one of the preceeding is indicative of a need for a model to provide needed visibility and evaluation capability. The model will usually be based on a block diagram representation of the system. Each block will be representative of a hardware component or function and have associated with it a reliability estimate, goal or prediction. As phases of design progress, blocks will represent lower and lower orders of hardware or item indenture (subsystems, equipment, modules, assemblies, subassemblies, etc.). As a consequence, the model will grow in complexity and detail as design progresses. A reliability prediction on a single item is the simplest form of a mathematical model. It is appropriate for use when the item in question has no redundant modes of operation or hardware, when it operates in what may be considered a single environment, and when no repair during mission is performed—for example, a radio receiver in a home.

5. Necessity for a reliability prediction to provide visibility of basic item and mission reliability and to compare the reliability of alternate designs. What should be its scope? Consider the amount of information available on the item (other than from predictions) such as from operational experience. Such information can be used in place of predictions. Predictions performed in early phases of acquisition are less refined and detailed than predictions performed in later phases of the acquisition process. There are various types of predictions that can be made, each with a different level of accuracy and each requiring different levels of information and effort. The types of predictions that can be applied include: estimates at the item level based on experience with similar items; parts count reliability based on estimates of complexity of the item; reliability predictions based on the stresses acting on each component part in the item. The type and detail of data available during a given phase of acquisition determines which types of prediction are practical. Consider, too, logistics needs for reliability prediction data and the need for reliability predictions for evaluating design alternatives.

6. Necessity for a FMECA to identify potential design weaknesses, to identify critical parts. To what level should it be performed? Consider the complexity of the item, the number of component interfaces which the item contains and the number of new components making up the item. As each of these increases, so does the probability of unforseen failure modes which can result in incidence of secondary and critical failures. Consider carefully the level at which FMECA is to be applied and whether or not sufficient information is available to perform it effectively. Failure mode definition (representative of an operational failure) at the component part level can be difficult to express, especially for more sophisticated and complex devices such as VLSI and VHSIC. Furthermore, information of the type needed may not be easily available. FMECA can be performed functionally or parametrically (see Section 11.1 for details). Functional FMECA is the simpler of the two but is less comprehensive than a parametric FMECA. Take into account, also, the needs for FMECA for maintainability, testability, logistics, and safety analyses and consider integrated FMECA programs with such disciplines.

7. **Necessity for a sneak circuit analysis to identify latent paths that cause occurrence of unwanted functions or inhibit desired functions.** The need for the task increases: with criticality of the item, especially when active control functions are performed by the item; as item interface with other items increases; if the baseline design is subject to frequent modification; if new design has to be performed frequently. The availability of design information and the time of availability of such information is critical to the practicality of the task. (See Ref. 18 for additional information about sneak circuit analysis.) Consider integrating the data needs of sneak circuit analysis with the data needs of FMECA.

8. **Necessity for components tolerance analysis.** The purpose of this task is to determine whether or not an item or the components making up the item remain within specified tolerances during service life. The need for the task increases as the tolerance bands for inputs/outputs tighten, as system complexity increases, and as interfaces among components within the item and with other items increase. The need for tolerance level analysis might be at the component part level, the printed circuit board or component level, or at higher levels of indenture. Tolerance analysis procedures applied vary in level of sophistication from worst-case analysis to the application of computer-driven circuit analysis programs. Data needs for this task vary with its level of application and the complexity of the item to which it is applied. The degree and level to which it can be applied also depends on the particular design/development phase.

9. **Necessity for parts selection/application.** Take into account: effects on reliability; the costs associated with procurement; the costs associated with reprocurement and logistics which result from reducing the number of types of parts and the number of nonstandard parts. The effort generally starts during the validation-of-concept phase of development. The overall task performed should include consideration of such elements as part application,

selection, control, production quantities needed and their availability, the need for testing/screening special quality parts that may be required, and the use of standard and nonstandard parts. Interfaces and coordination with customer logistics organizations should be effected.

10. Necessity for environmental stress screening (ESS) to screen out failures due to defects in material and workmanship. This task has been shown to improve the reliability of fielded systems. See, for example, Ref. 19, which indicated a 2:1 reduction in field removal/maintenance action rates for hardware subjected to ESS when compared to identical equipment not subjected to ESS. ESS is applied during the production phase of development, but planning for ESS takes place most effectively during full-scale engineering development. It can be applied at all levels of assembly, even part level. Generally speaking, the higher the level of application, the higher the cost of application. If successfully applied at lower levels, it can lead to reduced rework costs at higher levels. The less mature the manufacturing process, the more need for ESS. Take into account linkages of ESS to FRACAS activities. Recall that ESS is concerned primarily with defects in workmanship (production) and material, not design.

11. Necessity for reliability growth tests to increase reliability through the identification, analysis, and correction of design-linked failures and verification of the corrective action. A reliability growth test allows inferences to be drawn concerning the reliability of an item as tests and corrective actions continue. It is primarily concerned with correction/improvement of reliability design and of material specifications which impact reliability. It is a planned prequalification test, analyze, and fix (TAAF) process in which items are tested under actual, simulated, or accelerated environments to disclose design deficiencies and defects. Reliability growth is generally applied during full-scale engineering development and selectively applied during production. It may be applied to an entire item or to selected "most-critical" components of the item. The schedule of acquisition program must be compatible for growth testing. Provisions must be made for corrections and modifications of item configuration management as a consequence of design component changes in the item.

12. Necessity for reliability qualification tests to provide the customer reasonable assurance that requirements have been met. The task is generally performed at the end of the full-scale engineering phase of acquisition to provide assurance that requirements have been met prior to start of production. A requirement which has been levied requires a demonstration that the requirement has been met. The task is appropriate for all new designs. For modified designs, a degree of judgment must be exercised to determine whether the entire item or just portions of the item should undergo qualification test. The decision is dependent on the degree of modified design and the nature of the modifications. Off-the-shelf items with an historical record of operation in the intended operational environment may be accepted on the basis of that record. Those off-the-shelf items with previous operational histories in environments less severe than that intended should be subjected to qualification test under the intended environment. There is a statistical probability, *producer risk*, that a product which has a reliability as good as or better than that specified will fail a qualification test. There is a statistical probability, *consumer risk*, that a product which has a reliability worse than that specified will pass a qualification test. (These will be discussed in more detail in Chapter 12.) Consider the fact that, for any qualification test, both consumer and producer risks are inversely proportional to needed test times. For items that have reliabilities so large that an inordinate and impractical amount of test time would be required, alternate means of providing the customer with the assurance he needs must be provided. This may mean more severe test environments; more stringent selection and tests of components; use of warranties/guarantees; tests which provide assurance of a minimum value of reliability (less than that required), consistent with maximum test time available and coupled with one or more of the preceeding.

13. Necessity for production reliability acceptance tests (PRAT) to provide assurance that the reliability of the item qualified is not degraded as a result of production, parts selection, workmanship, or processes. It is usually performed as a continuing process throughout production. It can involve test of every production item or samples from each production lot. At first, testing is more extensive (to the level of consumer risk associated with the qualification test), until evidence is apparent that production is under control. At that time, testing may be performed to higher statistical levels of risk (fewer samples or shorter test duration) until evidence of declining quality becomes apparent. Testing should be performed in an environment which simulates the operational environment of the item. It is appropriate for production runs of new, old, or existing designs or hardware. ESS tests should be performed to screen out workmanship/production and material defects prior to the start of PRAT tests.

Note: Each task above will be described in more detail in separate sections of the chapters which follow.

3.6.5   Customer Responsibilities: Providing Information and Detail

In order for a contractor to properly structure a reliability program to meet acquisition and mission needs, information vital to reliability program structure and to contractor tasks must be provided, or provisions must be made for its acquisition. Such information which must be provided to the contractor by the customer includes:

1. Definition and constraints of the environment in which the item will be operated. The operational use environment, physical/location environment, maintenance environment (will repair/replacement be possible during, after, before mission?) and their constraints, logistics, and support constraints.
2. Definition of the mission to be performed. Function(s) to be performed by the item or its major components; criticality of each function to the mission; duration of the mission; feasibility of maintenance during the mission; allowable downtime for system during the mission.
3. Identification of demonstration tests to be performed, e.g., growth, qualification, production, along with test environments and information as to when and where performed.
4. Identification of special factors, as applicable, e.g., screens, part control, design guidelines to be applied.
5. Logistics and support interface needs.

We discussed earlier that one means through which such information, or at least significant portions of such information, may be disseminated is through the "Details to be Specified" sections of Mil-Stds. Mil-Stds, however, are not applied to every acquisition. Be that as it may, many of the "details to be specified" provide useful *prompts* to the customer as to what types of information a contractor needs to receive or *consider* in structuring a reliability program plan. Chart 3-7 provides a sampling of such "Details" taken from Mil-Std-785.

3.6.6   Customer Responsibilities to the Customer Organization

The following represent responsibilities the customer owes not to the contractor but to the customer organization to which the customer reliability engineer (CRE) belongs.

3.6.6.1   *Monitoring Progress/Problems*
The purpose is to review progress and identify potential problem areas which would impact the attainment of reliability objectives. While the CRE may receive information from the contractor, for this purpose it is the CRE's interpretation/assessment of this which forms the brunt of CRE responsibilities. Monitoring takes various forms. Review of the technical reports submitted by the contractor (or the data items associated with Mil-Stds which were

discussed earlier) and informal technical communication with the contractor can be loosely grouped as a type of monitoring called *surveillance*. Activities or actions which occur periodically, to assess overall progress and targets, is a second type of monitoring that may be termed *review*. A third type of monitoring is *independent assessment and evaluation*. The latter take various forms, and is thought by some to be one of the most valuable of the monitoring functions. Take, for example, a reliability prediction performed on an item by the contractor for CRE review, approach, or information. The independent assessment and evaluation could taken one of two forms:

1. Performance of a complete, independent prediction by the customer using the part stress data supplied by the contractor. This would most likely be performed using a reliability prediction software program in the possession of the customer, or by a second contractor.
2. Same as the above, but performed on a randomly selected portion of the item, assuming that the accuracy of the prediction performed on the selected portion of the item is consistent with the accuracy of the prediction as a whole.

3.6.6.2 *Providing Coordination and Interface*

We have previously discussed warranties and incentives, which must be addressed directly through contractual provisions. Other aspects of the reliability program must also be accounted for in the contract. These include, for example: the minimum reliability requirements, the major milestones associated with the reliability program plan schedule, the provisions associated with the demonstration test, design reviews, product inspection, subcontract delegation, and corrective actions. As a consequence, the customer reliability organization must foster close working links with contracting.

The customer reliability organization must also interface with the user and with other system engineering elements which require reliability inputs. These include logistics, maintainability, safety, and test system organizations.

References

1. Coppola, A. and Sukert, A. (1979). *Reliability and Maintainability Management Manual*. RADC-TR-79 200. Available from The Defense Technical Information Center. Report AD-1073299.
2. USAF (1988). *USAF R&M 2000 Process*. First Edition.
3. Boeing Aerospace Co. (1984). *Reliability/Maintainability Operational Parameter Translation*. RADC-TR-84-25. Available from The Defense Technical Information Center. Reports AD-B087426, AD-B087507.
4. Lyon, H. and LaSala, K. (1984). "DOD/Industry—R&M Technology Study

Analysis." Proceedings, 1984 Annual Reliability and Maintainability Symposium, IEEE, Piscataway, NJ.

5. Fleig, N. (1986). "The United States Air Force Product Performance Agreement Center." Proceedings, 1986 Annual Reliability and Maintainability Symposium, IEEE, Piscataway, NJ.

6. Yuspeh, A. et al. (1986). "Legislation on Weapon System Warranties." Proceedings, 1986 Annual Reliability and Maintainability Symposium, IEEE, Piscataway, NJ.

7. ARINC Research Corporation (1979). *Warranty—Guarantee Application Guidelines for Air Force Ground Electronic Equipment.* RADC-TR-79-287. Available through Defense Technical Information Center. Report AD-A082318.

8. ARINC Research Corporation (1976). *Guidelines for Application of Warranties to Air Force Electronic Systems.* RADC-TR-76-32. Available from The Defense Documentation Center. Report AD-A023956.

9. Rypka, E. and Kujawski, G. (1983). "Repair/Reliability Guarantee Programs Can Work." Proceedings, 1983 Annual Reliability and Maintainability Symposium, IEEE, Piscataway, NJ.

10. Nenoff, L. (1988). "Field Operational R&M Warranty." Proceedings, 1988 Annual Reliability and Maintainability Symposium, IEEE, Piscataway, NJ.

11. Kruvand, D. (1987). "Army Aviation Warranty Concepts." Proceedings, 1987 Annual Reliability and Maintainability Symposium, IEEE, Piscataway, NJ.

12. Hughes Aircraft Company (1987). *R&M Program Cost Drivers.* RADC-TR-87-50. Available from the Defense Documentation Center. Report AD-A182773.

13. Rockwell International Corporation (1983). *Reliability, Maintainability and Life Cycle Costs Effects of Using Commercial Off the Shelf Equipment.* RADC-TR-83-29. Available from the Defense Technical Information Center. Reports AD-A129596, AD-129597.

14. Reliability Analysis Center (1988). *R, M & S Standards Primer.* RAC publication PRIM-1.

15. Institute for Defense Analysis (1973). *IDA/OSD Reliability and Maintainability Study.* Report R-272, Vols. 1, 2, 3, 4.

16. Morris, S. et al. (1988). "RADC Reliability Engineer's Toolkit." Available through The Reliability Analysis Center.

17. Boeing Aerospace Co. (1983) *Missile and Space Systems Reliability vs. Cost Tradeoffs Study.* RADC-TR-83-13. Available from the Defense Documentation Center. Report AD-A129328.

18. Boeing Aerospace Co. (1982). *Sneak Circuit Application Guidelines.* RADC-TR-82-179. Available from the Defense Documentation Center Report AD-A118479L.

19. Grumman Corp. (1987). *Improved Readiness Through Environmental Stress Screening.* RADC-TR-87-225. Available from the Defense Documentation Center Report AD-A193788.

Bibliography

Babin, R. (1974). "DC-10 Avionics Parts Reliability in Review." Proceedings, 1974 Annual Reliability and Maintainability Symposium, IEEE, Piscataway, NJ.

Bajakian, A. et al. (1981). "Reliability Optimization of a Surveillance Radar." Proceedings, 1981 Annual Reliability and Maintainability Symposium, IEEE, Piscataway, NJ.

Ball, L. (1970). "1970 Annual Symposium on Reliability Tutorial Sessions/Program Planning and Organization." Proceedings, 1970 Annual Reliability and Maintainability Symposium, IEEE, Piscataway, NJ.

Belev, C. (1989). "Guidelines for Specification Development." Proceedings, 1989 Annual Reliability and Maintainability Symposium, IEEE, Piscataway, NJ.

Coppola, A. (1972). "A Program Guide to R/M Requirements." Proceedings, 1972 Annual Reliability and Maintainability Symposium, IEEE, Piscataway, NJ.

Coppola, A. (1979). *Reliability and Maintainability Manual.* RADC Report RADC-TR-79-200. Defense Documentation Center Report AD-A073299.

Coutinho, J. de S. (1974). *Introduction to Systems Assurance.* U.S. Army Material System Analysis Agency Publication SP-9.

Cox, T. and Bohan, E. (1987). "High Reliability by Automation and Design Quality." Proceedings, 1987 Annual Reliability and Maintainability Symposium, IEEE, Piscataway, NJ.

Coy, R. (1980). "Reliability Program Planning and Avionics Systems." Proceedings, 1980 Annual Reliability and Maintainability Symposium, IEEE, Piscataway, NJ.

Curtiss, D. and Ouellette, T. (1971). "Shillelagh Reliability Program Development to Deployment." Proceedings, 1971 Annual Reliability and Maintainability Symposium, IEEE, Piscataway, NJ.

Department of the Army (1979). *Systems Engineering, Field Manual.* Headquarters, Department of the Army Report FM770-78.

Electronic Systems Division U.S.A.F. (circa 1987). *Readiness Improvement Through Systems Engineering (RISE).* Publication of Headquarters, Electronic Systems Division, Hanscom AFB, Mass.

Feigenbaum, I. (1980). "INTELSAT System Reliability." Proceedings, 1980 Annual Reliability and Maintainability Symposium, IEEE, Piscataway, NJ.

Fowler, P. and Bailey, H. (1970). "Some Practical Parts Criteria." Proceedings, 1970 Annual Reliability and Maintainability Symposium, IEEE, Piscataway, NJ.

Giordano Associates (1988). *Testability/Diagnostics Encyclopedia Program Part 1.* RADC Report RADC-TR-88-211. Defense Documentation Center Report AD-A205346.

Giordano Associates (1990). *Testability/Diagnostics Encyclopedia.* RADC Report RADC-TR-90-239. Defense Documentation Center Report AD-A230067.

Karmiol, E. and Greenberg, S. (1970). "Reliability Factors in the Design process." Proceedings, 1970 Annual Reliability and Maintainability Symposium, IEEE, Piscataway, NJ.

Liberman, D. (1971). "A Facelifting for NASA's Reliability Requirements." Proceedings, 1971 Annual Reliability and Maintainability Symposium, IEEE, Piscataway, NJ.

Liebesman, B. (1985). "Reliability Requirements and Contractual Provisions." Proceedings, 1985 Annual Reliability and Maintainability Symposium, IEEE, Piscataway, NJ.

Morris, S. et al. (1989). *RADC Reliability Engineer's Toolkit.* RADC Report

RADC-TR-89-171. Defense Documentation Center Report AD-A215977.

Muller, F. (1970). "Subsystem Requirements." Proceedings, 1970 Annual Reliability and Maintainability Symposium, IEEE, Piscataway, NJ.

Munech, J. (1972). "A Complete Reliability Program." Proceedings, 1972 Annual Reliability and Maintainability Symposium, IEEE, Piscataway, NJ.

Pilny, M. (1978). "Viking Lander Reliability Program." Proceedings, 1978 Annual Reliability and Maintainability Symposium.

Provett, W. and Ullman, R. (1976). "Effective Reliability Planning and Implementation." Proceedings, 1976 Annual Reliability and Maintainability Symposium, IEEE, Piscataway, NJ.

Rickets, M. (1982). "F/A-18 Hornet Reliability Challenge: Status Report." Proceedings, 1982 Annual Reliability and Maintainability Symposium, IEEE, Piscataway, NJ.

Thomas, E. (1975). "Pitfalls in Contracting for Reliability." Proceedings, 1975 Annual Reliability and Maintainability Symposium, IEEE Piscataway, NJ.

Thomas, E. (1976). "Pitfalls in Reliability Program Management." Proceedings, 1976 Annual Reliability and Maintainability Symposium, IEEE, Piscataway, NJ.

Tri-Service Report (1975). Final Report of the Joint Logistics Commanders Electronic Systems Reliability Workshop.

U.S.A.F. Headquarters (1987). *USAF R&M 2000 Process.* Publication of Headquarters, U.S.A.F.

Zebick, H. and Hoerster, M. (1986). "R&M in a Systematic Trade Study Process." Proceedings, 1986 Annual Reliability and Maintainability Symposium, IEEE, Piscataway, NJ.

4

The Role of the Contractor/Developer in Formulating Reliability Approaches and Needs

Definition of the role and responsibilities of the customer and the contractor has been a matter of evolution over the years. The ideas and concepts discussed in this chapter represent a compendium of conclusions reached by many authors, reports, manuals, and papers. A sampling of such documents which have contributed to the ideas expressed in this and the last chapter is contained in a bibliography at the end of Chapter 3.

The previous chapter discussed the customer's role in the reliability development process. This chapter will mark the beginning of discussion on the role played by the contractor/developer during that process. Its particular objective will be to scope and outline the different reliability activities to be performed/generated by the developer during the various phases of the program keyed to information available and system engineering process constraints. The developer's reliability effort takes on a slightly different perspective from that of the customer in the fact that it must be consistent with the particular design approach espoused by the developer as well as with the mission needs. In addition, it must be consistent with the system engineering program structure associated with the acquisition program.

Reliability must be a concern in system planning, engineering, and management, starting in the very early phases of the design development program. Reliability engineering must be an integrated working part of the development organization; it must play a role in technical planning and concept development. Its approach must correlate with the particular technical approach strategy chosen and with the general systems engineering approach that implements the design/development process.

4.1   THE RELIABILITY ORGANIZATION: FORM AND RESPONSIBILITIES

The developer's reliability organization can take many different forms and shapes. It can be centralized or ad-hoc (a separate reliability organization dedicated to a single acquisition); it can be situated in a system engineering organization, a design engineering organization, or a quality assurance organization. Its nature is largely a function of the policies or "traditions" of the development organization and the nature of the product that is being developed. Each type of reliability organization structure has associated with it certain inherent advantages and disadvantages. The bottom line is, however, that each can get the job done, provided that developer management is convinced that a significant reward is associated with achieving or exceeding reliability requirements, or that significant penalties will accrue if the requirements are not met.

The developer's reliability organization plays two types of roles: a general overhead role and an acquisition specific role. The first involves support of the development organization with respect to establishing and generating reliability policies and operating procedures. These may take the form of guidelines for part acquisition and application/derating; for internal reliability design review/coordination practices; for establishment of company standards for generating reliability programs on new items or systems; and for establishment of reliability design practices to be followed. It may provide for reliability indoctrination and education activities throughout the development organization, for establishment of a company product-based failure reporting system, and for providing interfaces with other development and customer organizations.

The second role is the one which will be of most interest to the typical reader. This involves support of a developer's product with respect to:

1. Definition of reliability objectives.
2. Interpretation of reliability requirements and interfaces.
3. Participation in negotiating and reviewing contractual reliability issues.
4. Interface with the customer with respect to reliability.
5. Identification of areas that need improvement and ways that improvement can be achieved.
6. Development and implementation of a reliability program plan designed to meet the needs of the acquisition, structured to the design/development process, its schedule, data availability and its flow. This includes:
   a. General and specific reliability design guidance.
   b. Guidance on part selection and control.
   c. Providing for integration of tests that relate to reliability and other features; failure reporting and analysis.

d. Generation of reliability reports and data items.
e. Providing reliability assessment and evaluation services, i.e., prediction, modeling.
f. Coordination and interface with part, standardization, failure effects, sneak circuit, logistics specialists.
g. Interpretation and analysis of test requirements and qualification, and acceptance test risks.

4.2   RELIABILITY TECHNICAL PLANNING: A DEVELOPER'S VIEW

Almost all contractors maintain contacts with potential customers to evaluate future need and get involved with development efforts in advance of planned contractual activity. By the time of contract award, a company will, in all likelihood, have been working on and establishing boundaries and basic design for the acquisition item for some time. In order to present a single-thread approach to developer activities, let us go back to the time when that company decides to make a concerted effort to go after a given acquisition. The first step involves efforts to dimension the problem, to develop a qualitative assessment relative to potential project difficulty, and to identify potential strategies which can be used.

Initially, the ball will be in the court of systems engineers who will define potential alternate basic means of achieving the performance required for the system (item). This begins with an analysis of mission objectives, constraints, and functions to be performed. Alternate technical approaches for meeting product objectives are examined. Existing, modified, or updated versions of company products, new designs, subcontracted products, or a combination of these will be considered. The role of the reliability engineer in this phase is to:

1. Provide estimates of the reliability associated with each basic means of meeting performance needs
2. Assess upper and lower reliability bounds for each
3. Provide information on general means through which reliability can be enhanced
4. Identify potential reliability impediments and problems
5. Acquire familiarization with those reliability standards, handbooks, and policies associated with the acquisitions of the particular customer

The information with which the reliability engineer works to provide estimates of the reliability for each initial alternate design will be dependent on the degree of detail that can be provided for each of those alternate approaches. Usually, even an alternate design concept can be expressed in terms of a functional block diagram where inputs and outputs can be grossly defined but not characterized. At this point, system engineering should be able to provide at least qualitative estimates of the complexity of each functional block (i.e., less complex than existing product A, more complex than existing product B). It is from such estimates of complexity that the first reliability approximations are made. The form of such approximations will be discussed in more detail in Chapter 8.

The reliability engineer uses available information on mission objectives, constraints, and functions performed. Tracing how each of these may influence reliability positively or negatively. The engineer identifies potential strategies which can be used as *reliability multipliers*. Such multipliers will be discussed in more detail in later sections. A brief description of each multiplier concept is provided here as a precursor.

4.2.1 Cycling Constraints: Reliability Multipliers

Do all components have to be in a full operating mode during all of the mission? If not, operation in less than a 100% operating mode is usually equated with a lower failure rate. As an example, let us take a transceiver with independent transmitter and receiver functions.

Performance of the transmit function in this particular instance is not necessary when in the receive mode. As a consequence, the transmitter does not have to be "fully operational" at all times the transceiver is functioning.

4.2.2   Allowable Downtime: Reliability Multipliers

Can the mission be satisfactorily performed even if the system is inoperative for a given proportion or a given period of mission time? If the answer is yes, and if corrective maintenance can be performed during the mission, then mission reliability can be enhanced (see Section 4.2.4).

4.2.3   Fault Tolerance: Reliability Multipliers
(Functional or Replicative)

Functional fault tolerance results when a failure can occur in one component or a system but the remaining operational components (not necessarily identical in form or function to the failed component) can sufficiently take over the function of the failed component to carry out the mission. An example of this is the presence of VHF and UHF radios in an aircraft. Both have different basic functions, but upon failure of one, basic communications can still be carried out.

Replicative fault tolerance can be defined as two or more identical components which perform, or are capable of performing, the same function, where the failure of one or more still allows for performance of mission functions. An example is that, while humans have two kidneys, only one is needed to satisfactorily maintain life.

4.2.4 Maintenance Allowed: Reliability Multipliers

Provisions for maintenance can take various forms, from corrective maintenance which can be performed during missions, to periodic or scheduled maintenance, to periodic checkout, to repair only after system/mission failure. Each can have a different impact on reliability.

4.2.4.1   Corrective Maintenance During a Mission

If maintenance is allowable or feasible during a mission, it acts as a reliability multiplier for all fault-tolerant systems, and is a reliability multiplier for some series-connected (nonfault-tolerant) systems.

For the former, if corrective maintenance can be performed on failed components during the mission, the basic concern after occurrence of a component failure is that the remaining components continue to operate until the failed component can be repaired (or the mission ends). Since the time needed to effect repair is usually very small compared to remaining mission time, reliability is enhanced.

For the latter, if a given proportion of downtime is acceptable, either for a continuously operating system or for a mission, failures are allowable if they can be corrected within a given period of time. As a result, reliability is enhanced.

4.2.4.2   Periodic Corrective Maintenance

If corrective maintenance on a fault-tolerant system can be performed only periodically to repair any components which have failed over a period of time, the more frequent the maintenance, the more reliability is enhanced. This assumes that a system can satisfactorily perform its function with fewer than all its components operational. During each periodic maintenance visit, all the failed system components are replaced or repaired. This is applicable to systems operating at remote sites, for unattended operation, and for deferred maintenance scenarios.

4.2.4.3   Periodic Checkout (and Repair if Necessary)

If (1) a system either redundant or series-connected is in a standby state of operation (a time between when it was last used and its next mission) and subject to be called to perform a mission at a random point in time, and (2) a possibility of failure exists during the standby phase, the system's next mission reliability will be enhanced by "checking" the status of the system periodically during its standby state. The more frequent the checks, the greater the enhancement.

4.2.5 Technology Advances: Reliability Multipliers

Applications of new technology may lead to reductions in complexity of systems and/or reduced thermal and power stresses in components. An example of this is the application of large-scale VLSI or VHSIC technology where a *single* device can perform a function previously performed by many. In addition, existing technology and device types can be expected to improve in quality as they mature. See Section 3.1.3.3 for similar projections a customer must make, and Section 8.1.1.1 for further discussion of the subject.

4.2.6 Integration and Sharing of Resources

A design configuration composed of several subsystems can present opportunities for combination/integration/sharing of components to reduce system complexity (and reduced complexity enhances reliability), weight, and volume. Examples can include shared use of processors, communication paths, or power supplies. In many instances, a component designed to perform a joint purpose will be significantly less complex than two separate components. For such designs, fault tolerance can be selectively applied to further enhance the reliability of key item functions.

4.2.7 Use of Higher-Reliability Parts

In many instances, baseline estimates of component and subsystem reliability are founded on previously designed hardware which was made up of parts of various quality levels. If so, then reliability can be expected to increase with the application of higher-quality parts.

4.3 FORMULATING THE RELIABILITY DESIGN/DEVELOPMENT APPROACH

In addition to performing a basic feasibility study relative to the accomplishment of reliability objectives, the first steps in formulating the needs of the reliability program are taken. Information and lessons learned from previous acquisitions are major tools for this purpose. These include similarity of system function, potential similarity of system components, similarity of environment, similarity of reliability requirements, similarity of reliability program magnitude and scope, and, last but not least, problems encountered and successes achieved.

Near the end of this early phase in the development process there should be at least qualitative estimates of:

1. The degree of new, modified, or existing design that will be required. This will provide at least rudimentary insights into the magnitude and degree of effort that will be required for:
    a. Reliability prediction.
    b. Consideration of part quality and part tolerances required for the new design and efforts for their control and direction.
    c. Analyses of the new and modified components to ensure that they are being operated within their environmental limits.
    d. Reliability allocation of the new products from the subsystem to the module level.
    e. FMECA and sneak circuit analysis for the new and modified products.
    f. Qualification and reliability growth testing of the new products.
2. The relative complexity of the system (item) developed. This will effect the effort required for prediction, modeling, parts quality, allocation,

---

CHART 4-1 Linkages Between Early Development Determinations and Reliability Program Attributes

Degree of new, modified, existing design	Affects magnitude of effort required for reliability allocation, qualification testing, reliability analysis tasks (sneak circuit analysis, thermal analysis, FMECA, tolerance analysis, etc.), reliability growth tests
New technology application	Increases difficulty in reliability prediction (absence of failure-rate models and data); uncertainty regarding design/characteristics results in ambiguities with respect to FMECA performance; unavailability of reliability design application guidelines; environmental application uncertainties
Fault tolerance	Affects magnitude of reliability modeling and analysis effort, FMECA effort, qualification test structure

analyses (thermal, shock, vibration), FMECA, sneak circuit analysis, FRACAS,ESS, testing, etc.

3. The new technologies that are candidates for application. These introduce levels of uncertainty and increase effort required for performing reliability prediction, for definition of design guidelines, for determining operational and environmental operating characteristics (thermal, shock vibration), for FMECA, and for sneak circuit analysis.

4. The likelihood of application of fault tolerance. This will affect the magnitude of the reliability analysis effort needed to provide visibility, the magnitude of the FMECA at the level at which fault tolerance is applied, and the structure of the reliability qualification test.

Chart 4-1 shows the linkages of such determinations to the various reliability engineering program tasks.

Combining technical planning results and lessons learned with the general acquisition policies and procedures inherent to the development organization, a preliminary working document reliability program plan is developed. It will identify the reliability engineering tasks to be performed, tests to be performed, minimum and maximum time and facility needs, and manpower loading. The information presented will serve as a basis for developing an initial budget proposal for reliability activities.

4.4 CORRELATING THE RELIABILITY PROGRAM WITH THE SYSTEM ENGINEERING PROCESS

In Section 3.6 we discussed the reliability program in terms of both the general necessity of individual reliability tasks and the relevance to particular phases of the acquisition process. Those criteria and guidelines are as appropriate for the developer as for the customer in structuring a reliability program plan. As indicated in Section 3.6.4, one of the criteria that must be considered when structuring a reliability program is the quantity and type of data and information necessary to support the performance of a given task. In some instances, tasks may be performable during a number of different phases, but to different levels of accuracy, to different levels of completeness and detail, and to different levels of system indenture. The bottom line is that the type and quantity of data and information available for reliability program task performance is nothing more or less than the information available during the system engineering process.

We will consider three phases of the acquisition process discussed previously: demonstration/validation, full-scale engineering, and production, which were defined in Chapter 3. Bear in mind, however, that each program is not necessarily structured to follow such phases. Some programs can collapse the three phases into two or even a single contractual effort.

4.4.1 Demonstration/Validation Phase of Development

Chart 4-2 depicts a simplified version of the information usually generated during the demonstration/validation phase. This information flow can be used as inputs for reliability engineering activities during this phase.

Information is provided by the customer on characteristics covering a wide range of system parameters and needs, including reliability (see Section 3.6). The preacquisition studies performed by the developer (see Sections 4.2 and 4.3) also serve as inputs. The objectives of this phase are to establish a sound technical approach, to investigate the advantages and payoffs associated with different technical approaches and their risks.

In order to put what follows in proper perspective, let us very generally and briefly outline what can be expected in this phase.

During the phase, design architectures are examined and validated. A relatively accurate estimate of the degree of new, modified, or existing design

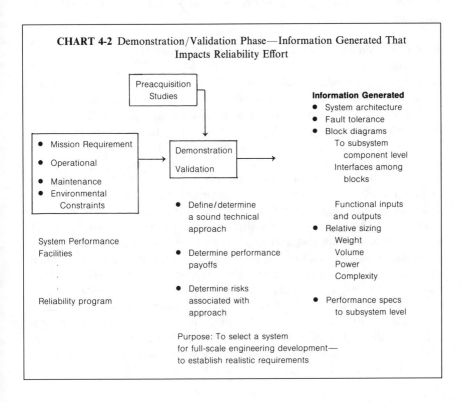

CHART 4-2 Demonstration/Validation Phase—Information Generated That Impacts Reliability Effort

will be determined as will (a) the type of new design and where in the system it will be applied, and (b) the types and nature of modified design to be implemented.

The system is broken down into block diagrams. Blocks are generally representative of subsystem components. Rough sizes of estimates of weight, volume, power requirement, and complexity can be provided. Performance specifications to at least the subsystem level are generally available and functional inputs and outputs to each subsystem component can be defined. Segments of the architecture which are capable of providing replicative or functional fault tolerance are identified. The various system and subsystem operational performance states are related to satisfactory or unsatisfactory (failed) levels of performance. (i.e., the system will perform its function satisfactorily if subsystems A and B are in a failed state and subsystem C is operational. The system will not be capable of performing its function if subsystem C is in a failed state regardless of the condition of subsystems A and B).

Experiments and tests are performed to validate the practicality and effectiveness of new design concepts. Application studies of new technologies are performed. Linkages with needed subcontractors and vendors are established.

4.4.1.1 Establishing Reliability Targets and a Technical Foundation

At this point, reliability predictions can be performed based on estimates of complexity or experience with similar component designs. Based on the predictions, consideration of possible reliability multipliers (see Section 4.2), and contemplated design architectures, a *reliability target* for the system can be defined. A *reliability target* is an estimate of the maximum reliability possible provided a given course of design action is followed.

The reliability target or goal is usually greater than the contemplated minimum requirement levied by the customer for the following reasons. (1) The greater the target value (provided the target value is achieved) with respect to the required value, the greater is the probability that the system will pass qualification test and product reliability acceptance tests. This is discussed in further detail in Chapter 12. (2) Target values for each major subsystem or component are usually assigned as part of the reliability allocation process. That would mean that if each target were established to satisfy the minimum value of reliability for the system and if only one failed to meet its *target*, the system would fail to meet its minimum requirement. A target value higher than a minimum (but within current state-of-the-art limits) provides room for tradeoffs in the event that, due to unforeseen circumstances, a given target cannot be achieved.

Rudimentary reliability allocation based on function, design approach espoused, and estimates of equipment and subsystem complexity can be performed. This basic allocation can determine which subsystems must employ fault tolerance; which must have higher-quality parts in its composition; and which must use new technology to reduce complexity. The effects of fault tolerance on system reliability will be visible during this phase. FMECA can be performed on a functional basis to the subsystem level.

Other contributions of this phase include: (1) providing information and strategies to be used in the reliability program during the following acquisition phase. It provides information necessary to scope various individual reliability program tasks, their schedule, and cost: for example, the parts program, specification, and requirements to be levied on the subcontractors and the identification of various tests available. (2) Identifying reliability risks and fallback positions which can be adopted. The need for the last is most likely to occur in areas which involve application of new technology, components designed/delivered through other parties, problems associated with test schedules, and potential for higher than anticipated costs.

4.4.2   The Full-Scale Engineering Phase Development

Chart 4-3 depicts a simplified representation of the objectives of and the information available from the full-scale engineering phase of development and outputs which are used for reliability activities. The customer provides system requirements structured to either the system as a whole or to particular subsystems. The previous phase's development results are available to the developer as an input.

The approaches and rudimentary designs of the previous phase are transformed into a detailed design capable of meeting system performance requirements during this phase. The functional analyses performed during the previous phase are expanded to lower and lower levels of indenture. Functional and parameter values for input and output are associated with each level of indenture. Input/output goals are transformed into hardware and software designs. Tests are performed at each level of indenture to satisfy performance and environmental operational requirements, ending with formal qualification tests.

During this phase, the following are among data and information generated: (a) The system hierarchy is defined from system to subsystem, to equipment and component levels. (b) Protocols for system operation. (c) Sizing characteristics (weight, volume, packaging, costs) for each element of the system hierarchy. (d) Detailed design specifications, requirements and schematics for each element of the hierarchy. (e) Stresses on each element identified and analyzed. (f) Circuit analyses performed on each circuit down

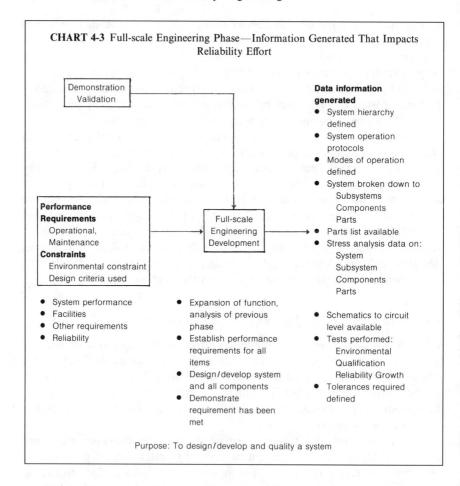

CHART 4-3 Full-scale Engineering Phase—Information Generated That Impacts Reliability Effort

Demonstration / Validation

Full-scale Engineering Development

Performance Requirements
Operational, Maintenance
Constraints
Environmental constraint
Design criteria used

- System performance
- Facilities
- Other requirements
- Reliability

- Expansion of function, analysis of previous phase
- Establish performance requirements for all items
- Design/develop system and all components
- Demonstrate requirement has been met

Data information generated
- System hierarchy defined
- System operation protocols
- Modes of operation defined
- System broken down to
  Subsystems
  Components
  Parts
- Parts list available
- Stress analysis data on:
  System
  Subsystem
  Components
  Parts

- Schematics to circuit level available
- Tests performed:
  Environmental
  Qualification
  Reliability Growth
- Tolerances required defined

Purpose: To design/develop and quality a system

to stresses encountered to the part level. (g) Parts lists established (and part quality levels established). (h) Procurement policies for parts and components are established. (i) Tolerance levels necessary for inputs and outputs of each element of hierarchy. (j) Design criteria to be used are established (at beginning of phase). (k) Schedule and milestones are established (at beginning of phase). (l) Performance test results used to validate performance in operating environment. (m) Qualification test results used to verify that performance requirements have been met.

With information such as the above available, reliability predictions can be performed on an evolutionary basis. First on the basis of complexity, then as the engineering estimates of system, subsystem, equipment, and component makeup mature, predictions are made to finer grains of complexity. Finally, predictions are performed taking into account application and operational stresses on each part (when schematics to that level are developed).

When, in addition, (1) the different modes of system operation which can affect the failure rates of various portions of the system are defined, (2) the maintenance and repair policies and constraints are established, and (3) fault tolerance considerations are taken into account, then an updated reliability mathematical model can be developed that takes into account all of the above. This model is used for final reliability allocation, and tradeoff decisions, and for providing visibility into the relative effects on reliability of various fault-tolerance and maintenance/repair policies.

Reliability allocations are extended to the lower tiers of the system during this phase. Initial allocations performed during previous phases were employed primarily to grade and scope the relative reliability needs of system and subsystems. Reliability allocation in this phase takes on a management function and must take cognizance of the various *ways* the subsystem reliability requirements can be met. This involves taking into account weight, power needs, volume, and cost of each means, and making the most cost-effective choice.

Failure modes, effects, and criticality analyses can be applied to equipments, components, printed circuit cards, and individual circuits. Failure mode information of a functional nature can be generated at any level. Information necessary to a parametric FMECA (open, short, etc.) will generally be available to the part and circuit level.

Sneak circuit analysis can be employed using the schematics and interface design specifications established.

Electronics parts/circuits tolerance analysis can be performed to subsystem, equipment, component, printed circuit board, or part levels through analyses of schematics and available performance specifications.

A plan for environmental stress screens (ESS) can be developed for future application in the production phase. Units to which ESS can be applied can be chosen, screen types and durations determined from information available on assembly/equipment makeup, parts employed, and production process to be implemented.

Thermal analyses from the part to the equipment level can be performed to determine hot spots and problem areas, and to reduce thermal stresses.

Tests can be performed on breadboard, brass board, prototype versions of printed circuit boards, components, equipments, and subsystems to verify environmental robustness and performance in given environments, to provide for reliability growth, and for reliability qualification.

Failure reporting and corrective action can be keyed to the results of every test.

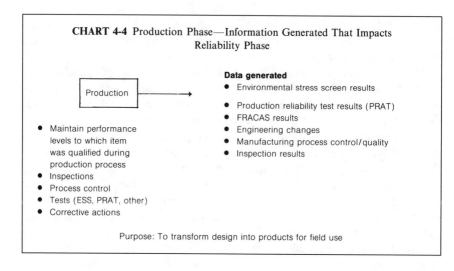

CHART 4-4 Production Phase—Information Generated That Impacts Reliability Phase

Production

- Maintain performance levels to which item was qualified during production process
- Inspections
- Process control
- Tests (ESS, PRAT, other)
- Corrective actions

Data generated
- Environmental stress screen results
- Production reliability test results (PRAT)
- FRACAS results
- Engineering changes
- Manufacturing process control/quality
- Inspection results

Purpose: To transform design into products for field use

4.4.3   The Production Phase of a Development Program

Chart 4-4 depicts a simplified representation of the objectives of, and information resulting from, the production process which are used by reliability activities.

The objective of the production phase is to deliver quantities of the item developed during the previous phase while maintaining the performance and quality levels to which it qualified. An item may be produced by a number of different means (e.g., different methods of soldering) through a number of different processes. Each combination has an impact on manufacturing quality and, hence, reliability. The quality of materials used in the manufacturing process and their variability also impacts the quality and reliability of the produced items. During production, quality/reliability is controlled/maintained through inspections (at part, assembly, and higher levels), process control at all levels of system indenture, and tests. If defects are found in process, parts, or design, rework and corrective actions are taken. The information available for reliability purposes during this phase is the results from ESS and other screens, production reliability test results, FRACAS results, data from inspections (incoming inspection results on parts, assembly, subassembly, equipment inspection results), and statistical process control data generated.

4.5 THE ROLE OF THE DEVELOPER/PRODUCER AFTER CONTRACT AWARD

After a developer/producer has been selected, the details of a contract are negotiated. The nature of the responsibilities of the contractor's reliability organization are, of course, dependent on the particular acquisition phase for which the contract was awarded and the characteristics of the program. Regardless of program phase, the following general tasks are universal:

- Review and interpretation of all reliability issues in the contract and their potential impacts.
- Negotiating possible difference between the customer-defined reliability program plan and the plan tailored to follow the contractor's particular design/production approach.
- Resolving ambiguities with respect to requirement failure definitions and reliability engineering tasks to be performed.

Such tasks do not only involve communication/negotiation with the customer; they also involve communication within the contracting organization as well.

Reliability issues to be reviewed or interpreted range from choice of a suitable reliability figure of merit; to growth test assumptions, to consumer and producer risks associated with qualification or production reliability tests; to guidelines and standards to be used during design; to interfaces between reliability and logistics, maintainability, and safety analysis; to parts control and selection policies; to possible incentives and warranties; to integration of tests; to definition of failure.

Particularly stringent requirements or design complexities may force the need for extremely conservative designs, the use of high-quality parts or tight-tolerance parts, and special design analyses and special or extended tests.

Differences between the reliability program tasks initially outlined by the customer and the tasks mandated by the particular approach taken by the winning contractor must be adjusted. A given development/production approach may result in some reliability engineering tasks becoming more or less important. In addition, since constraints will in all likelihood be levied on the cost of the reliability program, these too, must be addressed at this time.

Resolution of ambiguities with respect to reliability requirements and definitions is not a matter to be addressed just between the contractor and the customer. While some of the contractor program team will have participated in preproposal and proposal activities, most will not have participated. Hence, there is a potential for differences in interpretation between working team members. And there is nothing more dangerous to a program than having members of the working acquisition team holding different interpretations of reliability definitions and requirements.

5

Rudimentary Probabilistic/ Statistical Concepts Used in Performance of Reliability Program Tasks

Before describing the reliability program tasks which must be carried out by the developer, this is an appropriate time to introduce a few simple probabilistic/statistical principles and concepts which are needed to understand and implement certain facets of those tasks. They may not necessarily be addressed in the way a mathematician would choose, but hopefully they are expressed in a way simple enough to provide some insight to the non-mathematician and yet not be inconsistent with basic mathematical definitions.

5.1 THE MEAN

A mean is the average value of something. In this particular text, we will be interested in the average value of such things as time to failure or hazard over a given interval.

The mean is a measure of central tendency. The fact that an equipment is said to exhibit a mean time to failure of, say, 500 hours, does not mean that each equipment of that type will have a measured value of mean time to failure of 500 hours. Some equipments will exhibit a mean time to failure greater (sometimes very much greater) and some will exhibit means far less. When the times to failure for all equipments are combined, the overall mean calculated is the mean time to failure for the equipment.

The mean is calculated as

$$\bar{x} = \frac{\sum\limits_{i=1}^{N} x}{N} \tag{5-1}$$

where
  x_i = a measure of a parameter, time to failure, power output, voltage, etc.
  N = the number of measures taken

If x_i = time to failure, t_i, then the mean time to failure (MTTF) is

$$\text{MTTF} = \frac{\sum_{i=1}^{N} t_i}{N} \qquad (5\text{-}2)$$

5.2   A PROBABILITY

Probability is the chance that something will happen or will not happen. A coin has two sides. If a coin is flipped, each side has an equal chance of coming up. Hence, the probability of a head is $1/2$; the probability of a tail is $1/2$. A die has six sides. After a single toss, the probability of showing a "1" is $1/6$, of a "2" is $1/6$, of a "3" is $1/6$, and so on.

Let us take the coin example a bit further. If the coin were flipped 1000 times, how many heads would you *expect* to see? (500, of course.) What proportion of flips would you expect to be heads? (1/2.) Note that the terms of probability of an event and the proportion of occurrence of an event really say the same thing. In virtually every instance where you have a statement in terms of probability, you can transform it into a statement of proportion.

Reliability, $R(t)$, can be stated in terms of a probability: the probability that an item will perform its required function for a specified period of time under stated conditions. This means, for example, that if we choose an equipment at random which has an exponential distribution of failure times and operate it for t hours, the probability will be $R(t)$ that it will not fail during that period of time. It also means that if we chose a very large number of equipments of that same type and started operating them at the same time, by the time t hours had passed the proportion of them that would not have failed would be $R(t)$.

5.3   AN EXPECTED VALUE

An expected value is a projected mean value which takes into account an event or a set of events, where (a) each event has a given probability of occurrence (or occurs in a given proportion) and a probability of not occurring, (b) each event has payoffs associated with its occurrence or nonoccurrence; and the products of each probability and its associated payoff are summed.

For example, take a game where a coin is tossed. When a head shows, I will pay you $1.00. When a tail shows, you pay me $1.50. On the average, how much will I make per toss? Two events are possible:

- A head, which will occur with probability $P_h = 0.5$.
- A tail, which will occur with probability $P_t = 0.5$.

If a head shows, the payoff associated with it for me is $-$1.00$. (I have to pay a dollar to you.) If a tail shows, the payoff associated with it for me is $+$1.50$. (You pay me this sum.) The expected earning, E, for me is

$$E = P_h(-1) + P_t(1.50)$$
$$E = 0.5(-1) + 0.5(1.50) = 0.25$$

so that, on the average, I can project a winning for myself of $0.25 per toss. This obviously will not happen on any one toss. On each toss I will either lose $1.00 or win $1.50, but over a long series of tosses, I will *average* a winning of $0.25 per toss.

Let us take an example closer to a reliability situation. Assume that failure rate is constant and that we are analyzing the reliability of a component. During operation, the component goes through three separate operational states: standby (S_s), regular operation (S_r), full operation (S_f). Each state has associated with it a different set of stresses such that a different failure rate is associated with each:

The component will be in state S_s for $P_s = 0.5$ proportion of the time and, in S_s, will have a failure rate (λ_s) of 0.01.
The component will be in state S_r for $P_r = 0.25$ proportion of the time and, in S_r, will have a failure rate (λ_r) of 0.02.
The component will be in state S_f for $P_f = 0.25$ proportion of the time and, in S_f, will have a failure rate (λ_f) of 0.06.

The expected or mean value of failure rate $E(\lambda)$ experienced over operating time for that component is

$$E(\lambda) = P_s\lambda_s + P_r\lambda_r + P_f\lambda_f$$
$$E(\lambda) = 0.5(0.01) + 0.25(0.02) + 0.25(0.06) = 0.025$$

which means that the component exhibits an average failure rate over time equal to $E(\lambda)$.

5.4 AN INDEPENDENT EVENT

When the occurrence or nonoccurrence of an event or more than one event does not affect the probability of occurrence of another, the events are independent. For example, take the case of a sequential toss of a die and a coin. The probability of getting a head (tail) is not effected by the face of the die showing after its toss.

In reliability testing or analysis, *generally* speaking, independence among failures is assumed. That is to say, given two identical units are under test, a failure in the first does not effect the probability of failure of the second.

5.5 A CONDITIONAL EVENT

When the occurrence or nonoccurrence of one event affects the probability of occurrence of another, the events are conditional. For example, there are two jars; one contains 1 black ball and 2 white balls. The second contains 2 black balls and 3 white balls. One jar is chosen randomly and a ball is picked. What is the probability that the ball chosen was black? That probability depends on which of the jars was chosen. If the first jar was chosen, the probability is 1/3 (since 1 out of the total of the 3 balls is black). If the second jar was chosen, the probability is 2/5 (since 2 out of the total of 5 balls are black). Hence the probability of picking a black ball is *conditional* on the choice of the jar.

5.6 MUTUALLY EXCLUSIVE EVENTS

When the occurrence of one event makes the occurrence of one or more other events impossible, the events are mutually exclusive. For example, in the single toss of a coin, of the two possibilities—heads or tails—only one can result. Another example is the toss of a die, where only one face can show at a time.

When two or more units are connected in series to form a system as in Fig. 5.1, where any failure in any unit can cause a system failure and the system to be deactivated for repair, and no additional failures can occur once the system is deactivated (failure rate = 0 for deactivated systems), then a

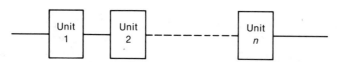

FIGURE 5-1.   A system made up of *n* units in series.

single failure in a unit excludes others from happening in the system (except when the failure in one unit causes a secondary failure elsewhere in the system) and such failures are mutually exclusive events.

5.7 THE ADDITION LAW FOR PROBABILITIES

Probabilities can often be a function of alternate events. One example of this is when any one of two or more independent events is capable of providing a given result or objective. If we want to determine the probability of achieving that objective, taking into account the probability associated with those events, we apply the addition law. Basically, the addition law acts in much the same way as an OR gate. That is to say, you will get a "1" at the output if one or more inputs are "1."

We will consider first the case when the contributing events are *not mutually exclusive* and we will illustrate the principle by way of an example.

In order to go on a fishing trip, you need a boat. Two of your friends, A and B who act independently, both have boats and know that you plan to go fishing. Let A represent the event that your friend A will offer you his boat. You know from past history that the probability P_a that A will offer his boat to you is 0.8. Let B represent the event that your friend B will offer you his boat. You know from past history that the probability P_b that B will offer his boat to you is 0.5. Similarly to Boolean notation, let

$$P(A + B) = \text{probability that } A \text{ or } B \text{ or both will occur}$$

According to the addition law for probabilities, if two events are independent and not mutually exclusive (while you might not be able to use two boats at a time, both *can* be *offered* to you) then

$$P(A + B) = P_a + P_b - P_a P_b \tag{5-3}$$

or, in the case of our example,

$$P(A + B) = 0.8 + 0.5 - 0.4 = 0.9$$

Hence, the probability that either A or B or both will offer you a boat is 0.9.

Equation (5-3) is the general form for the addition law for two events. It can easily be expanded to include more than two events. For three events A, B, C, let the event $A + B$ be denoted by Z_1 such that $P_{z_1} = P(A + B)$.

$$P(Z_1 + C) = P(A + B + C) = P_a + P_b + P_c - P_aP_b - P_bP_c$$
$$- P_aP_c + P_aP_bP_c \qquad (5\text{-}4)$$

For four events A, B, C, and D, let the event $A + B + C$ be denoted by Z_1 such that $P_{z_1} = P(A + B + C)$:

$$P(Z_1 + D) = P(A + B + C + D) = P_a + P_b + P_c + P_d - P_aP_b - P_bP_c$$
$$- P_aP_c - P_aP_d - P_bP_d - P_cP_d + P_aP_bP_c + P_aP_bP_d$$
$$+ P_aP_cP_d - P_aP_bP_cP_d \qquad (5\text{-}5)$$

and so on.

Now consider the case where the alternate events are mutually exclusive. Again we will illustrate the principle by way of an example.

You have one die and your objective is to throw either a 1 or a 5. Let A represent the event you throw a 1. You know that the probability P_a of throwing a 1 is 1/6. Let B represent the event you throw a 5. You know that the probability P_b of throwing a 5 is 1/6. Let

$P(A + B) =$ probability that A or B or both will occur, as before

Hence, as before,

$$P(A + B) = P_a + P_b - P_aP_b$$

However, A and B are mutually exclusive, which means they cannot both happen at the same time; since it is impossible for both to happen at the same time.

$$P_aP_b = 0$$

and for A and B mutually exclusive,

$$P(A + B) = P_a + P_b \qquad (5\text{-}6)$$

For our example,

$$P(A + B) = 1/6 + 1/6 = 2/3$$

Equation (5-6) is the general form for the addition law for two mutually exclusive events. It can easily be expanded to cover more than two events:

$$P(A + B + \cdots + R) = P_8 + P_b + \cdots + P_r \qquad (5\text{-}7)$$

for any number of mutually exclusive events.

5.8 THE MULTIPLICATION LAW FOR PROBABILITIES

While the addition law is applicable to calculating the probability of achieving a given objective when *any single event out of several can achieve* that objective, the multiplication law is applicable to the calculation of the probability of achieving a given objective when *all of a given series of events is required.* Basically speaking, the multiplication law acts in much the same way as an AND gate. That is to say, a 1 will appear at the output only when all inputs are 1.

We will consider first the case when all contributing events act independently. That is to say, the occurrence or nonoccurrence of one event will not affect the probability of occurrence of another event.

Let our objective be tossing two heads in a row. Let us denote the event of getting a head on the first toss as A; the probability P_a of getting a head on this toss is $1/2$. Similarly denote the event of getting a head on the second toss (the coin was retrieved after the first toss and flipped again) as B; $P_b = 1/2$.

Similarly to Boolean notation, let

$$P(A \cdot B) = \text{probability that } A \text{ and } B \text{ will both occur}$$

According to the multiplication law, for two independent events,

$$P(A \cdot B) = P_a \cdot P_b \tag{5-8}$$

or in the case of our example,

$$P(A \cdot B) = (1/2) \cdot (1/2) = 1/4$$

Hence, the probability that a head will appear in two consecutive tosses of a coin is $1/4$.

Another example of the multiplication law is of special significance to reliability engineering. Take two resistors as in Figure 2.5. Each unit must operate satisfactorily in order to provide the proper output for the system made up of the two units. Suppose that our objective is that the system operate satisfactorily for a particular span of t hours. In order for this objective to occur, it is obvious that both unit A and unit B must operate satisfactorily for that same period of time t.

Let the probability that unit A will operate successfully for t hours be P_a. P_a, however, is in actuality the reliability of unit A over t hours of operation. Hence, $P_a = R_a(t)$. Let the probability that unit B will operate successfully for that same t-hour span be $P_b = R_b(t)$. Let A = the event that unit A will operate successfully for t hours and B = the event that unit B will operate successfully for t hours. Then,

$$P(A \cdot B) = P_a \cdot P_b = R_a(t) \cdot R_b(t) \tag{5-9}$$

Hence, the reliability of a series system is equal to the product of the reliabilities of the units making up the system.

The multiplication law (5-8) can be easily expanded to include more than two events:

$$\prod_{i=1}^{r} P_i = P_t \tag{5-10}$$

where
P_i = the probability of the occurrence of event i
P_t = the probability that all r events will occur
r = the number of events required in order to achieve a given objective

We will consider next the case when contributing events act conditionally. We have 100 balls in a hat, made up of 25 white balls and 75 black balls. Our objective is to pick three consecutive black balls in the first three picks. After each pick the ball chosen will not be returned to the hat. What is the probability of achieving our objective?

Let A = the event of picking a black ball on the first pick; $P_a = 75/100 = 0.75$.
Let B = the event of picking a black ball on the second pick, given that a black ball was chosen on the first pick. The probability of that event is denoted as P_b/A and is $74/99 = 0.7475$ (since, if one black ball was chosen on the first pick, only 74 remain out of a population that now totals 99).
Let C = the event of picking a black ball on the third pick, given that black balls were chosen on the first and second picks. The probability of that event is denoted as P_c/AB and is $73/98 = 0.745$ (since two black balls were chosen in prior picks, leaving 73, and the population now totals 98).

The probability of picking three consecutive black balls can then be expressed as

$$P(A \cdot B/A \cdot C/AB) = P_a \cdot P_b/A \cdot P_c/AB \tag{5-11}$$

This can be read as the probability of picking a black ball on the first try, times the probability of picking a black ball on the second pick given that we picked one on the first, times the probability of picking a black ball on the third try given we have already picked one on the first and second tries.

5.9 ADDITION AND MULTIPLICATION GUIDES FOR DISTRIBUTIONS

If $x_1, x_2, \ldots, x_n$ are independent variates with known means, $\mu_1, \mu_2, \ldots, \mu_n$ and known variances $\sigma_1^2, \sigma_2^2, \ldots, \sigma_n^2$, and if

$$f(x_1, x_2, \ldots, x_n) = a_1 x_1 + a_2 x_2 + \cdots + a_n x_n$$

is a linear function of the variates, then the mean, μ_s and variance σ_s^2 of $f(x_1, x_2, \ldots, x_n)$ is

$$\mu_s = a_1 \mu_1 + a_2 \mu_2 + \cdots + a_n \mu_n \tag{5-12}$$

$$\sigma_s^2 = (a_1 \sigma_1)^2 + (a_2 \sigma_2)^2 + \cdots + (a_n \sigma_n)^2 \tag{5-13}$$

regardless of what the distribution of the x_i is.

If, however, the distributions of $x_1, x_2, \ldots$ are all normal, then $f(x_1, x_2, \ldots, x_n)$ is a normally distributed quantity with mean μ_s, and variance σ_s^2.

If the distributions of $x_1, x_2, \ldots, x_n$ are not normal, but if n is large, then, by use of what is called the *central limit theorem*, $f(x_1, x_2, \ldots, x_n)$ will be approximately normally distributed with mean μ_s and variance σ_s^2.

If $x_1, x_2, \ldots, x_n$ are all normal variates with known means $\mu_1, \mu_2, \ldots, \mu_n$ and known variances $\sigma_1^2, \sigma_2^2, \ldots, \sigma_n^2$, and if $f(x_1, x_2, \ldots, x_n)$ is a nonlinear function, then the mean μ_s and the variance of $f(x_1, x_2, \ldots, x_n)$ can be expressed as

$$\mu_s \sim f(u_1, u_2, \ldots, u_n) \tag{5-14}$$

$$(\sigma_s)^2 \sim \sum_{i=1}^{n} \left(\frac{\partial f(x_1, x_2, \ldots, x_n)}{\partial x_i} \sigma_{xi} \right)^2 \tag{5-15}$$

and $f(x_1, x_2, \ldots, x_n)$ can be considered approximately normally distributed.

Products and quotients of normally distributed variates can also be shown to assume approximately normal distributions under certain assumptions. The mathematics and details are, however, beyond the scope of this text. The important thing to remember is that many functions of normal distributions tend to approximate normality. See Refs. 1 and 2 for more detail.

5.10   THE TCHEBYCHEV INEQUALITY

In the event that a variate x or a function $f(x_1, x_2, \ldots, x_n)$ takes on a distribution which is not normal or, more importantly to our purposes, *is unknown* but has a known mean μ and a standard deviation σ, we can apply the Tchebychev inequality. This will always provide *conservative* estimates of proportions of a population within or outside given limits, regardless of the distribution type. On that basis if it was applied to a function which *had* a normal distribution, the resulting estimates would of course be conservative. The Tchebychev inequality states that the probability is at least $[1 - (1/K)^2]$ that a value will fall inside an interval, $\mu \pm K\sigma$, or

$$P[(\mu + K\sigma) > x > (\mu - K\sigma)] \geqslant 1 - \frac{1}{K^2} \qquad (5\text{-}16)$$

where

μ = mean of any distribution
σ = standard deviation of any distribution

Then in accord with the properties of the above, for values of $K > 1$, equal to 1.5, 2, 2.6 and 4.47, respectively

- At least 56% of all values will be between $\mu \pm 1.5\sigma$.
- At least 75% of all values will be between $\mu \pm 2\sigma$.
- At least 85% of all values will be between $\mu \pm 2.6\sigma$.
- At least 95% of all values will be between $\mu \pm 4.47\sigma$.

We can get a feel for the degree of conservativism of the results by contrasting the above with the same percentages, calculated assuming the distribution was normal:

- 56% of all values will be between $\mu \pm 0.77\sigma$.
- 75% of all values will be between $\mu \pm 1.2\sigma$.
- 85% of all values will be between $\mu \pm 1.44\sigma$.
- 95% of all values will be between $\mu \pm 1.96\sigma$.

If the value of μ is much greater than σ, say $\mu/\sigma = 10$, then the percentage of values within a given range estimated by each procedure when $\sigma = \mu/10$ can be shown to be:

	Normal Estimate	*Tchebychev Conservative Estimate*
56%	$\mu(1 \pm 0.077)$	$\mu(1 \pm 0.15)$
75%	$\mu(1 \pm 0.12)$	$\mu(1 \pm 0.2)$
85%	$\mu(1 \pm 0.144)$	$\mu(1 \pm 0.16)$
95%	$\mu(1 \pm 0.196)$	$\mu(1 \pm 0.447)$

which are reasonably close within the ranges of 56% to 85% indicating only a moderate loss in precision by using the conservative approach.

References
1. DeZur, R. and Donahue, J. (1965). *On Products and Quotients of Random Variables.* Aerospace Research Laboratories Report ARL-65-71.
2. Severo, N., Montzingo, L. and Schillo, P. (1963). *Asymptotic Distribution of Functions of a Normal Random Variable.* Aerospace Research Laboratories Report ARL-62-468.

6

The Parts Program—What You Should Know, What You Can Do

The parts program is comprised of not only the reliability task associated with parts control, but the reliability task associated with analysis and control of part tolerances as well. The parts program, as we will see, impacts not only reliability, but also the degree of quality (and rework required) achieved in the production process. This chapter will deal with those things a reliability engineer should know about a parts program.

Discussion will include: (a) The impact and role of the parts program in other system development activities than reliability. (b) The relationship between part quality and reliability. (c) The identification and description of the most frequently used part standards and specifications and the rationale for their use. (d) The effects of part latent and patent defects on operational reliability and production rework costs and how such effects can be assessed and controlled. (e) The effects of part tolerances on production rework costs and operational reliability and how such effects can be assessed and controlled.

While the reliability of an item is dependent on the choice of parts used, the types of parts used are not solely dependent on the reliability required nor on decisions made by the reliability engineer. The parts program associated with any development effort is influenced by:

1. The design engineer, whose tendency is to select parts with respect to the performance they can provide, even if it means operation of the part up to its design limit; the power necessary to make the part operate; the weight and volume burdens of the part—the lighter and smaller, the better.
2. The standardization engineer, whose tendency is to minimize the usage of the new or nonstandard parts regardless of their potential to increase

performance. Use of nonstandard parts is a cost driver for both the developer and the customer. It requires the development of specifications and drawings by the developer. It can require special ordering and stocking procedures by both customer and developer and may represent a single-source delivery potential. Use of standard parts, parts with a history of use by the developer, results in the use of existing drawings, and reduction in the number of purchase orders and of part types necessary.

3. The parts specialist, whose tendency is to advocate the use of the most cost-efficient mix of parts possible. Parts that have been frequently used—have been proven—have well-defined performance characteristics. New or nonstandard parts may provide degrees of uncertainty with respect to performance characteristics and application/use criteria, and the part specialist is charged with the responsibility for the documentation of both.

4. The reliability engineer, whose tendency is to advocate a parts program containing parts of higher reliability levels, parts capable of reducing overall system complexity, designed in accord with good derating practices. This can require the consideration of additional screens or burn-in on existing parts or the application of higher-reliability parts, which will cost more and increase the concern of both the part specialist and the standardization engineer. It can mean design restrictions on the design engineer with respect to how hard a part can be driven.

As a consequence, the parts program which results reflects a balance of the concerns and compromises made among a number of different organizations. Be that as it may, the actions of each organization have an impact on the reliability of the parts and the reliability design of the system which the parts form. It will be from that focus that we will proceed.

6.1 RELIABILITY AND QUALITY INTERFACES AND EFFECTS

Part reliability and quality are both affected by defects in materials and variations (and defects) in the manufacturing process. Problems in reliability and quality manifest themselves in:

- A large number of products initially defective.
- Significant variations in the performance characteristics of parts coming from the same production line. Consequences include: (1) high rates of rework for circuits and assemblies made up of such parts; (2) premature field operational failure of equipment caused by degrees of change in part parameters which over time cause the product to perform outside of specification limits.

- Premature field operational failures caused by defects in materials and mistakes associated with manufacturing and assembly processes. Such defects can be transparent to a typical performance measurement, but under the stresses of operation, may trigger or greatly accelerate a mechanical, physical, or chemical failure mechanism and result in premature failure.

All three are functions of part process and material control and are influenced by the character of the qualification process and the inspections, tests, and screens called out in the standards and specifications discussed previously.

6.2 RATIONALE AND USE OF PART SPECIFICATIONS AND STANDARDS

Even parts that provide the same function can be manufactured using different types of machines, through processes which are not identical, using material of different qualities, under various strictness of controls, and subject to different test and screening regimes after they are produced. As a consequence, two units from two different vendors which look the same and have the same nominal performance characteristics can be worlds apart with respect to defects contained or to reliability. (Recall that in Chapter 1 we defined reliability in terms of defects in a material, design, or process.) Almost from the time that the Department of Defense realized that reliability was a military concern, efforts were undertaken to establish standards for the manufacture, qualification, and test of parts to insure that minimal levels of part quality and reliability necessary for use in military systems were available; and to provide information on products which met such levels. *Information contained in such documents can provide useful data for use in a parts program whether or not the standards are specifically invoked.* The following standards have been developed and are now currently in use for that purpose.

Mil-Std-965, "Parts Control Program"
This document provides general guidance for the establishment of a parts control program. It provides information for both the customer and the contractor. Its basic objective is the development of the part selection list to be used during the development program.

For Microcircuits
1. Mil-M-38510, "General Specification for Microcircuits," (see Section 3.5.2) provides criteria for qualifying a microcircuit for general tri-service use. It covers qualification of the manufacturing process for the part and necessary documentation of part characteristics. A device which is qualified

and documented in accordance with Mil-M-38510 procedures is provided a JAN designation. Of particular usefulness is the fact that Mil-Std-38510 contains a description of each part type qualified under the standard and a Qualified Producer List (QPL—a list of producers which have met all Mil-Std-38510 requirements for a given part) associated with each part. Currently the list is updated several times a year.

Almost all military programs involving the use of microcircuits require that such parts used be "qualified" under Mil-Std-38510 and procured from a Qualified Producers List (QPL). This requires that qualification be performed on a specific product-to-product basis. Technology today, however, is moving at a pace which is faster than the qualification procedure for given device lines. The result can be a less cost-effective useful life for the product after qualification is complete. For that reason, movement is apparent to institute what may be considered "generic" qualification procedures. Following this concept, if the same process is capable of producing a variety of different specific items (e.g., hybrid packages, microcircuits, etc.), the *process itself* is qualified with respect to such elements as:

- Technology applied
- Materials used
- Packaging employed
- Design data necessary
- Inspection screens and tests
- Manufacturing process employed
- Manufacturing process controls employed

In the late 1980s and early 1990s efforts in this area resulted in the development of two military standardization documents aimed at qualification of manufacturing processes associated with the manufacture of devices belonging to a given family rather than individual members of that family: Mil-H-38534 "General Specification for Hybrid Microcircuits"; and Mil-I-38535 "General Specification for Integrated Circuits (Microcircuits) Manufacturing".

2. Mil-Std-883, "Test Methods and Procedures for Microelectronics," (see Section 3.5.3) establishes screens and test methods for use in microelectronic qualification. It covers tests against environmental, mechanical, and electrical stresses, and screens to detect quality defects introduced by the manufacturing process or through material used in the process. It can be applied as a component of Mil-M-38510 and it can be applied independently to microcircuits not qualified as JAN products (e.g., not qualified with respect to Mil-M-38510).

The failure rate which can be associated with a device is dependent on

CHART 6-1 Microcircuit Quality Levels and Standards Employed

Quality Level	Procedure	Π_q
S	Procured in full accord with Mil-M-38510, and Class S requirements of Mil-Std-883	0.25
S-1	Procured in full compliance with requirements of Mil-Std-975 "NASA Standard Electrical, Electronic, and Electromechanical Parts List" or Mil-Std-1547 "Parts, Materials and Processes for Space Launch Vehicles, Technical Requirements for" and have customer specification approval	0.75
B	Procured in full accord with Mil-M-38510, and Class B requirements of Mil-Std-883	1.0
B-1	Compliant with Class B Requirements of Mil-Std-883 and procured to a Mil-Drawing or other approved documentation. (Not Mil-M-38510 qualified.)	2.0
B-2	Not completely in accord with all requirements of Mil-Std-883 Class B requirements, but tested in accord to Government-approved documentation. (Not Mil-M-38510 qualified.)	5.0
D	Hermetically sealed parts with normal reliability screening and manufacturer's quality assurance practices	10.0
D-1	Commercial part	20.0

whether or not a device is JAN-designated and the particular series of Mil-Std-883 tests and screens associated with the device. For microcircuits Mil-Hdbk-217E lists seven failure-rate quality levels depending on the degree of qualification/test/screens provided for a product line. These are shown and described in summary form in Chart 6-1. The Π_q factor shown on the right is a failure rate multiplier to be used to modify the "base" failure rate provided in the handbook. That is to say, the nominal failure rate for the device is equal to $\Pi_q \lambda$.

3. Mil-Std-1562, "Lists of Standard Microcircuits," provides lists of microcircuits which are acceptable, potentially acceptable, and not recommended for application in military programs. Devices covered in Mil-M-38510 having a Qualified Producer List are identified.

For semiconductors

1. Mil-S-19500, "General Specification for Semiconductor Devices," provides

CHART 6-2 Semiconductor Quality Levels and Standards Employed

Quality level		Π_q
JANTXV	Procedures as in Mil-S-19500	0.12
JANTX	Procedures as in Mil-S-19500	0.24
JAN	Procedures as in Mil-S-19500	1.2
Plastic		12.0

criteria for qualifying a semiconductor device for general tri-service use. It provides information on product assurance, documentation, information, and tests and screens necessary to qualify a device to one of several reliability quality levels. Mil-Hdbk-217E defines three such quality levels—JAN, JANTX, JANTXV—in ascending order of quality as shown in Chart 6-2. Of note is the fact that Mil-S-19500 contains a description of each part type qualified under the standard and a Qualified Producer List associated with each part. Currently the list is updated several times a year.

2. Mil-Std-701, "Lists of Standard Semiconductor Devices," provides lists of semiconductors which are approved for application in military programs. Cross references to appropriate Mil-S-19500 documentation are provided.

For Passive Electronic Parts
Established reliability (ER) specifications have been established covering resistors, capacitors, inductors, connectors, etc. These provide for various reliability levels depending on device type, makeup, and tests performed. Chart 6-3, based on Mil-Hdbk-217E, shows the number of quality levels associated with various different part types. An example of the quality levels associated with composition resistors is shown in Chart 6-4.

1. Mil-Std-199, "Resistors, Selection and Use of," provides information on resistors covered by military specifications and identifies whether or not higher-reliability versions are available.

2. Mil-Std-198, "Capacitors, Selection and Use of," provides information on capacitors covered by military specifications and identifies whether or not higher-reliability versions are available.

Parts not having ER tags but still included in military specifications (Mil-Specs) are referred to as Mil-Spec parts (the higher reliability parts are a special class of Mil-Spec part). In general, these include specifications which were organized prior to the ER specification era. Mil-Hbdk-217 treats these

---

CHART 6-3 Failure-Rate Quality Levels Associated with Passive Components

Established Reliability	Number of Different Quality Levels
Capacitors	8
Resistors	4
Coils	4
Relays	4

CHART 6-4 Failure-Rate Quality Levels of Composition Resistors

Failure-Rate Level Indicator Based on Material, Process, Tests Performed	Π_q
S	0.03
R	0.1
P	0.3
M	1

---

as a distinct quality level. If a commercial part is utilized, the parts used are considered by the prediction procedure to be of the "Commercial" quality level.

6.3 APPLYING CONTROLS TO ENHANCE PART RELIABILITY AND QUALITY

Tri-service standards and specifications have been adopted by industry as a point of reference representing the highest levels of reliability and quality available. They are applied not only to military programs but to commercial operations, nuclear energy, and space applications as well. Also they are applied in general where a failure may be costly in terms of safety or expense of repair.

From the preceeding sections it is clear that reliability and quality levels are dependent on the part manufacturing process; the degree of control associated with such processes; and the inspections, tests, and screens associated with the product before and after such processes. Reliability and quality achieved are directly related to the extent that such actions are exercised. However, at present, the higher the levels of control, the more electrical and environmental tests and screens applied, the higher is the resultant cost per unit produced. In some cases the result will be a low yield ratio for the item in question.

For some applications the most reliable part available will be too costly to acquire or be unavailable in the quantities needed to meet program schedule. In other instances the part needed may not be available in a high-reliability version or may even be available only in a commercial version. In an extreme example, the part needed may be a brand new or custom device.

The following sections will discuss courses of action which can be adopted in such situations.

6.3.1 When Tri-service Standards and Specifications Are Not Applicable as Controls

If, for one reason or another, DOD standards and specifications are not or cannot be applied, other steps or criteria aimed at providing the needed result must be employed. These can include the following.

1. Vendor surveys and analyses to determine vendor/part past performance, and claims made versus products delivered. While this is no assurance of future quality, it identifies a source with a better potential for providing reliability needed than a source with a worse track record. Consideration should be made of past cooperative efforts of the vendor in working with the developer to achieve the reliability needed.
2. Product and product process analyses, which involve:
   a. Acquiring information on the design, the stress limits, and the environmental operating performance of the part. It can include acquisition of sample devices to verify operation at the required stress levels in the planned operational environment.
   b. Acquiring data on materials used in production, in the manufacturing process employed, and procedures followed to maintain control. This information can provide the developer's part and materials specialists with insights into potential failure mechanisms.
   c. Acquiring data on the field reliability of the parts, from long-term operating tests and failure analysis. Such information can be used as evidence that the basic process is capable of providing reliability of the magnitude required.
   d. Acquiring data on process yield. A low process yield after short-term tests and screens can sometimes signify a large number of defects in a product lot. Tests and screens may not be sensitive to all defects. The fact that they detect defects in a large proportion of the population can signify that there is a large proportion of defects remaining which

the tests/screens have not detected. Generally, when given the choice of two vendors, A and B, and it is known that

- Both applied identical short term tests and screens
- Vendor A's available part population was the result of rejecting 50% of a manufacturing lot
- Ventor B's available part population was the result of rejecting 10% of a manufacturing lot

which vendor would you choose? If you were in the market for a used car and you had a choice of two, both with documented maintenance histories, would you buy one with a record of a lot of breakdowns or one with a record of no breakdowns?

3. Performing tests and screens to detect and eliminate both latent and patent defects (see Section 6.6.1). Testing of a part to its specification requirements will suffice for the detection of patent defects. Tests for latent defects may include high-temperature operating burn-in, step stress tests, leakage, thermal shock, moisture resistance, etc. These may take on any of the forms included in Mil-Std-883 and, indeed, are one of the means used to upgrade the reliability of parts. Each test proposed should be referenced to the failure mechanisms in the part to which the test/screen is sensitive.

4. Improving the reliability of a given part by such techniques as failure analysis to determine failure mechanisms. Implementing changes in material or process characteristics to eliminate failure mechanisms found or known to exist.

Depending on the reliability needs and the exact nature of the vendor process and tests already associated with the part product, the effectiveness of each of the above courses of actions can vary from product to product. Two factors, however, are a necessity for almost any part procurement:

1. Preparation of a part specification based on the developer's *interpretation* of the performance that the part must provide; the part's form, fit and function; tolerances; tests and screens to be performed.

2. Requirements associated with any proposed changes in product design, material, production processes:
   a. Documentation
   b. Sample of new products provided for test
   c. Special tests/failure analyses, etc.
   d. Approval requirements

Hence, given inapplicability of DOD standards and specifications, or standards and specifications which are unique to the development organization (individual companies have generated company-specific policies, standards, and specifications dealing with part acquisition), the means to effect part quality and reliability are available.

6.4 IMPROVING THE RELIABILITY AND QUALITY OF AN EXISTING PART

The reliability or quality of an existing device can sometimes be improved through the application of special burn-in procedures. Such procedures may be applied under a variety of environments ranging from less than full-power operation in a benign environment to over full-power operation in a stressful environment. Additional screens, inspections, and tests can also be applied in order to eliminate those parts containing defects which are sensitive to such actions. Application of Mil-Std-883-type screens to microcircuits that are not JAN products is an example of this.

Tests of that nature can be applied to commercial and nonstandard parts to increase their quality. However, the degree of extra reliability and quality improvement achieved due to additional tests and screens is more of a qualitative than a quantitative judgment. An increase certainly occurs, but the nature of its magnitude cannot be quantitatively defined. For example, a JANTX (tested extra) microcircuit has a greater reliability than a JAN microcircuit, but *a burned-in* JAN microcircuit may well be more reliable than a JANTX device; the degree of potential improvement is a decision based on experience and judgment.

Attempting to approach the reliability and quality level of a part which is both qualified under Mil-M-38510 and subjected to Mil-Std-883 tests by merely subjecting commercial or nonstandard parts to Mil-Std-883 tests, by burn-in, or by cycling (on–off, temperature, vibration) can have undesirable, unanticipated results. Qualification of a part under Mil-M-38510 or other qualification procedure implies that the manufacturing process has been described, has met the requirements for qualification, and that controls are in place to monitor the integrity and stability of the process. A part not qualified under Mil-Std-38510 has less assurance of continuing process integrity. The potential results are as follows.

1. More defects can be introduced into the part population due to process change and variation. The tests and screens will detect defective parts having the type of defects to which the tests and screens are sensitive. However, not all defects are detectable by the tests and screens which are presently within the state of the art. For every defect detectable through a part, screen, or series of screens, there are others that will not be detectable. The larger the quantity of parts found defective by a series of tests and screens, the larger the potential population of parts containing defects which are not detectable by tests and screens.

As a consequence, application of tests and screens will improve the quality of a given lot of parts, but will not necessarily provide reliability consistency on a lot-to-lot basis.

2. Burn-in and other tests and screens applied to a population of parts can change the performance characteristics of the parts such that they do not meet the design center values of the original part specification. When such parts are inserted into circuits and modules, they may not perform as anticipated and as a consequence will result in rework at the circuit or higher levels of system indenture.

6.5 DEALING WITH NEW OR UNIQUE PARTS

Reliability and quality actions with respect to system-unique parts or custom devices necessary to provide needed performance present special challenges. One example of such a part is a hybrid circuit. These are generally custom designed and, as a consequence, lack a history of quality achieved. In addition, they often require the application of more than one technology, increasing the risk of unforseen synergistic effects. Due to their "newness," such devices are more prone to have inadequacies and omissions in test characteristics and in their reactions to particular environmental stresses. As a consequence, a reliability program specific to such devices must be generated that assures process validity and control. Tolerance analyses at the lowest levels, thermal analyses, and other types of engineering analyses must be used to the maximum extent possible to assure proper initial design. Suitable in-process controls and tests to provide the required quality and performance are required.

Reliability estimation procedures for hybrid microelectronic circuits are provided in Mil-Hdbk-217. Reliability is calculated by summation of the failure rates of all hybrid components, taking into account construction, interconnection, and other factors.

6.6 EFFECTS OF PART DEFECTS AND TOLERANCE LEVELS ON ASSEMBLY, QUALITY, AND EQUIPMENT RELIABILITY

Part defects and tolerances affect both reliability and incidence of rework required during equipment manufacture. For reasons of convenience, we will address each separately.

6.6.1   Part Defects and the Quality and Reliability of Assemblies

Consider two types of defects:

1. A patent defect is a defect which prevents a part from performing its function or meeting its specification. A patent defect is obvious if the part is tested. It will cause a circuit or assembly that contains the part to test "bad." The more controlled the part manufacturing process, the fewer the number of patent defects that occur.
2. A latent defect is a defect in the part makeup which is not detected by part tests and screens and does not initially prevent a part from performing its function or meeting its specification. Circuits and assemblies which contain the part will test "good." It will precipitate failure during early life under a combination of application and operational stresses. The more controlled the part manufacturing process, the more sensitive the types of defects are to the screens and tests applied, the fewer the number of latent defects that will be contained in the shipped product.

Patent defects manifest themselves in the degree of rework and retest required at the printed circuit board and assembly level (assuming test is performed at the printed circuit board and assembly levels). To be sure, board and assembly interaction and manufacturing/fabrication processes also impact rework and will be considered when ESS (environmental stress screening) is discussed in Chapter 11. At this time, however, we will consider only the effects of part patent defects on rework.

Latent defects manifest themselves in the form of *premature* failures during operational life caused by defects introduced either at the parts level or during the manufacturing process. At this time we will consider only the effects of part latent defects on reliability.

Patent and latent defects, as well as part tolerances, are controllable during manufacture through proper structure and control of production processes coupled with inspections, tests, and screens on the parts produced. The objectives of the part-standards and specifications discussed earlier can be looked at as a means to: (1) Provide a stable process which produces a population of parts with predictable characteristics. Included in that population *may* be parts with defects, but since the process is stable, the types of defects tend to be known. (2) Provide a series of tests and screens capable of identifying parts with defects and eliminating such parts from the population. The more tests and screens or the more stringent the tests and screens performed, the more defective parts that are identified and eliminated.

6.6.1.1   Patent Defects: Effect on Quality

It is obvious that if the population of parts that you are using to fabricate printed circuit boards (PCBs) and assemblies contains defective parts, some of the products fabricated from such parts will also be defective and require rework. If populations of parts from which devices were selected to populate an assembly contained $P\%$ defective devices (where the incidence of one or more defective parts in an assembly would cause the assembly to be defective), it would appear intuitive, to most, to expect approximately $P\%$ of the assemblies to be defective. Intuition, however, can err. In actuality, a part population with a rather small proportion of defectives can result in a much larger proportion of defective assemblies. Let us illustrate this with a simple example.

Assume that a PCB is composed of n identical parts. The parts that we will use to populate the board are chosen from an infinite population of parts which contain a small but finite proportion of defective parts p. The defectives are there either because the cost of individual test of the parts would be excessive, or because the parts were tested but the test regimen was less than 100% effective. Parts are chosen at random to populate each PCB. (Remember that one or more defective parts in a PCB will cause the PCB to be defective.)

Recall our discussion of the binomial distribution in Section 2.2, where we defined the expression

$$P(x) = \frac{n!}{(n-x)!x!}p^x(1-p)^{n-x} \qquad (6\text{-}1)$$

where

$P(x)$ = probability of x events, or proportion of occurrence of x events

p = probability that an event will occur when it is attempted, or anticipated proportion of occurrences of an event

n = number of attempts or components involved in one attempt

We define an event as choosing a defective part from the part population for insertion in a PCB, $P(x)$ can now be interpreted as

$P(x)$ = probability that x defective parts will be inserted into a PCB

We take $x = 0$, which indicates the occurrence that no defective parts are in a PCB, and $P(x)$ can now be stated as

$P(0)$ = probability that no defective parts from the part population will be inserted into a PCB, or the probability that a PCB will not be defective, or the proportion of PCBs produced which will contain no defective parts from the part population

p = probability that a defective part will be chosen from the part population, which equals the proportion of the part population which is defective

(Note that the implicit assumption here is that we are choosing *each* part as needed from an infinite population, or at least an extremely large population of parts, very much larger than the number of parts that are needed. Hence p is considered constant for *every* part.)

n = number of parts from the population chosen/inserted in a single PCB

Relationship (6-1) can now be expressed as

$$P(0) = (1 - p)^n \qquad (6\text{-}2)$$

This relationship assumes either that the PCB contains only a single part type from a population with a fraction defective p, or that it contains more than one part type but all lots of each part type contain the same fraction defective p. Nevertheless, it provides a general insight into the effects of proportion of lot defective on the PCBs and assemblies which are populated from those lots.

Figure 6-1 shows the proportion of PCBs which are defective and require rework as a function of fraction of part population defective and PCB complexity. As can be seen, generally speaking, very small values of part population defective yields very large values of fraction of PCBs defective and the disparity increases with PCB complexity.

When there is more than one part type making up a PCB or assembly, and each part type comes from a part population with a different fraction part population defective, the effect of that too can be evaluated.

Assume that a PCB is composed of n parts, n_1 parts of type 1, n_2 parts of type 2, ..., n_r parts of type r, with

$$n = \sum_{i=1}^{r} n_i \qquad (6\text{-}3)$$

where

n = number of parts per PCB

n_i = number of parts of type i in each PCB

r = number of different type parts making up each PCB

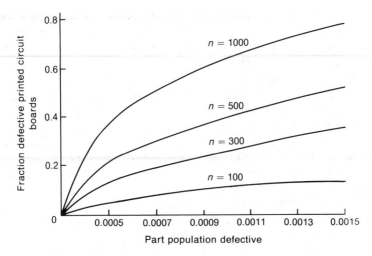

FIGURE 6-1.   Proportion of defective printed circuit boards as a function of board complexity and proportion of part population defective.

Let

p_i = the proportion of the type i part population that is defective.

Then in the same way that (6-2) was derived,

$$P_i(0) = (1 - p_i)^{n_i} \tag{6-4}$$

where
  p_i, n_i are defined as before
  $P_i(0)$ = the probability that no defective parts of type i will be inserted in a PCB,
      the proportion of PCBs produced that will contain no type i defective parts

Since the selection of parts of one type is independent of the parts selected from other part types, the probability $P(0)$ of no *defective parts at all* being in an assembled PCB is equal to

$$P(0) = \prod_{i=1}^{r} P_i(0) \tag{6-5}$$

where $P(0)$, $P_i(0)$, and r are defined as before.

A rough but simple approximation to (6-5) using the concept of average value of proportion of total parts defective, $\bar{p}$, for the overall population of part types can also be developed. Using that option, $\bar{p}$ is calculated and relationship (6-2) is applied substituting $\bar{p}$ for p. We will explain the logic used through use of a simple example.

We have plans to fabricate 100 PCBs. Each PCB is to be made up of a total of 100 parts: 50 parts of type A, 50 parts of type B. The proportion of the type A part population which is defective is 0.01. The proportion of the type B part population that is defective is 0.05. Assuming extremely large populations of both A and B from which to choose, 5000 parts are chosen from population A, and 5000 from population B. The expected number of defectives in the 5000 A-type parts is 50. The expected number of defectives in the 5000 B-type parts is 250. The total expected number of defectives in the two groups is 300, and the total number of parts in the two groups is 100000. Hence, the average proportion of defectives $\bar{p}$, associated with all the parts making up the PCB is 300/10000:

$$\bar{p} = \frac{300}{10\,000} = 0.03$$

Generally speaking, $\bar{p}$ can be calculated as

$$\bar{p} = \frac{\sum\limits_{i=1}^{r} sp_i n_i}{sn} = \frac{\sum\limits_{i=1}^{r} p_i n_i}{n} \qquad (6\text{-}6)$$

where

p_i, n_i, and n are as defined previously
s = number of PCBs to be fabricated

As indicated previously, the relationship (6-1) is based on the assumption that there is available an extremely large (infinite) population of parts from which to choose. Adherence to such an assumption is approached when parts for a given assembly are ordered from a vendor who has a very large population of homogeneous parts from which to fill your order, or from a vendor whose process is in control so that parts with the same characteristics are always produced. In the event of relatively small orders of parts and the PCB requiring a number of different types of parts with different associated levels of quality, another procedure may be applied to estimate the relationship between different part quality levels and fraction of PCBs defective. We will illustrate this technique, too, with the aid of a similar example.

Assume that a PCB is composed of n parts, n_1 parts of type 1, n_2 parts of type $2, \ldots, n_r$ parts of type r, with

$$n = \sum_{i=1}^{r} n_i \tag{6-7}$$

where n, n_i, r are defined as previously

Let

s = the number of PCBs to be fabricated

m = the total number of parts ordered to fabricate s PCBs. (In actuality, a quantity of additional parts will have to be included in the order as replacements for defectives. For simplicity these will not figure into our calculations.)

m_i = number of parts of type i ordered as part of m

p_i = the proportion of type i parts that are defective

$$m = \sum_{i=1}^{r} m_i \tag{6-8}$$

and

$$\frac{m}{n} = s$$

The probability that a PCB will contain no defective parts of type i is $P_i(0)$:

$$P_i(0) = \frac{(m_i - p_i m_i)!}{(m_i - p_i m_i - n_i)! \, n_i!} \bigg/ \frac{(m_i)!}{(m_i - n_i)! \, n_i!} \tag{6-9}$$

(see appendix A8 for derivation).

Assuming all part types are independently and randomly chosen and no testing is done until the PCB is completely fabricated,

$$P(0) = \prod_{i=1}^{r} P_i(0) \tag{6-10}$$

Note: In comparing the application of (6-2) to the application of (6-9), the difference is in one case an infinite population (or a population so large that it will be far from exhausted by the part needs of the fabrication) and in the second a finite population which will be exhausted by the part needs of the fabrication.

When m_i and n_i are large, however, (6-9) becomes unwieldy to apply. In that instance (6-2) and (6-5) may be used for approximation purposes.

6.6.1.2   Latent Defects: Effect on Reliability

Simply defined, a latent defect in a part is a defect which escapes detection by screen or test at the part, assembly, and equipment level. When the unit is put into operational use, the defect precipitates a part failure which results in failure of an assembly, equipment, or system. One could logically argue that all failures over service life not caused by misuse or wearout are caused by such defects. Philosophically that may be so. However, conventional reliability theory generally accepts the occurrence of an average number of failures (caused by an average occurrence rate and magnitude of defects; see Section 1.1) over a period of time as the norm. It defines a latent defect as a special class of defect, one having a special degree or magnitude of inherent weakness that has a high probability of resulting in an *early* life failure under operating (field) life stress conditions (see Ref. 2 for an almost identical definition). It can manifest itself as a hard failure (i.e., open, short, etc) or by a drift out of tolerance for critical part parameters. The proportion of such defects traceable to part defects has been estimated by several sources to be of the order of 1/3 or more. (Ref. 1 analyzed a number of failures during exploratory reliability testing and came up with a proportion of over 0.4.)

To give you a feel about potential impact of latent defects on reliability, let us *arbitrarily* define a premature failure as one which occurs in under 1000 hours of operation in a given environment. Let us further assume that defects capable of causing premature failure are present in a proportion of parts making up the part population, introduced through material and/or part manufacturing process defects or variation. We will further assume that no test or screen available is capable of detecting such defects.

Let us assume that the proportion of devices having this defect is p_d. Use equation (6-1) to determine the proportion of assemblies manufactured which have one or more such defective parts. As an example, assume $p_d = 0.005$ and $n = 100$ and a production quantity of 1000 assemblies. Chart 6-5 shows the proportion of assemblies produced which contain 0 to 4 of these defective parts.

Hence, approximately 40%, or 400, of the produced assemblies will fail prematurely. Out of those that fail prematurely, 304, or 77% (304 out of 394), will fail only once. Nineteen percent (76 out of 394) will fail two separate times in the first 1000 hours of operation. Three percent (12 out of 394) will fail three separate times in the first 1000 hours of operation, thus providing the impression of a "problem" assembly. As this analysis shows, a small proportion of "problem" assemblies is to be expected when part populations with even small proportions of latent defects are used.

The bottom-line effect is that each defect will result in an additional failure for an assembly. In the case of the example, $p_d n$ times 1000 assemblies

CHART 6-5 Effect of 0.005 Latent Defect Proportion on a 100-Part Assembly

Number of Defects / Assembly	Approximate Probability of Occurrence	Number Having Defects in Production Lot of 1000
0	0.606	606
1	0.304	304
2	0.076	76
3	0.012	12
4	0.002	2

produced results in 500 assembly failures being introduced through the use of latent defects. As n increases, the number of failures caused by latent defects also increases.

Such failures result in observed values of MTTF during early life which are lower than they will be during later portions of service life and provide the appearance of a decreasing hazard rate over time. In the extreme, latent defects may be so perfuse that MTTF is adversely effected over much of service life.

6.6.2 Part Tolerance vs. Assembly Quality and Reliability

A given part (a microcircuit, semiconductor, capacitor, etc.) can have a number of different parameters which are critical to the proper operation of a circuit or assembly design. Examples of such parameters include gain, transconductance, frequency response, timing, impedance, etc. The part production process produces parts which are not identical. Given a part with, say, three different measurable parameters A, B, and C, there will be variations of such parameters from part to part even from the same production lot. In order to cope with such differences, allowable performance tolerance envelopes are developed for each parameter, associated with a given part. Parts whose parameters test within those envelopes are considered within specification. Those that test outside one or more such envelopes may be discarded or included in part lots which allow wider ranges of variation. The number of parameters tested per part and the size of the envelope range per parameter can vary among vendors even for the same type of part. Both affect the price of the part. The need for a particular envelope is a function of the design generated by an equipment developer and is generally represented in the part specification prepared by the developer.

Circuit and assembly performance is a function of the exact performance parameters of the parts making up the circuit/assembly and the mathematical relationship(s) that describes the function to be performed in terms of the part parameters. This relationship(s) is defined as the circuit/assembly transfer function.

The transfer function usually developed as part of the design effort is a valuable tool in assessing part tolerance needs and determining impacts of tolerance envelopes. Most of the procedures outlined in the sections that follow are based on knowledge of the circuit/assembly transfer function. The question might logically be posed "What can we do if such a relationship is not available?" The answer is that there are other fallback positions which may be applied, and these, too, will be addressed. For the moment, however, it is important to recognize that the performance of a circuit/assembly, measured in such terms as gain, bandwidth, noise, rise time, impedance, etc., will vary with the variation of each part parameter. As a result, the population of circuits/assemblies produced will have a distribution of different performance outputs. As a consequence:

- A quantity of circuits and assemblies produced might be outside their performance characteristics and require rework.
- A quantity of the circuits and assemblies produced might be so close to the limits of their requirements that a little drift or change in one or more parameters over operating time will produce a failure of the circuit or assembly (it will not function as specified).

In order to minimize the effects of such part-related problems, two basic philosophies have evolved, which can be applied singly or in combination:

1. Reducing, or exerting control over, the variability of part parameter values by part specifications, which directly or indirectly involves:
   a. The choice of a vendor product which already meets parameter variability needs.
   b. The choice of a vendor product which requires modification to its manufacturing process so as to provide parts (i.e., populations of parts) which have small variance in values.
   c. Culling from production lots of parts produced in accord with existing processes those items which have values above or below prescribed values. This usually requires 100% measurement/test of the entire population.
2. Designing the circuit/assembly such that it can tolerate worst-case variation conditions.

The choice of whether to improve the quality by tests/screens and elimination of high-variability parts, to reduce general variability by changing the production process, to practise worst-case design or provide designs which are maximally tolerant of part variation, or to do nothing, depends on the costs vs. payoff of the action.

Against the above-mentioned costs must be weighed the substantial costs of rework at printed circuit board, assembly, and higher levels due to failure of such units to meet specification requirements; and the future effects on the reliability over service life must be considered.

In the sections which follow we will discuss steps common to each philosophy; procedures and considerations associated with such philosophies; and means to determine the effects of particular sets of part tolerances on product rework and quality/reliability, and the fallback positions which may be taken when not all of the information required to exercise a given procedure is available.

6.6.2.1 Performing Sensitivity Analysis

Every part in a circuit/assembly can have a different effect on the output characteristics of that circuit/assembly. As a consequence, variability of a part value may have a minimal effect on circuit/assembly output or a critical one. Since, in general, the cost of a part is related to the degree of variation in the population of that part, it is important to determine which parts are more in need of control than others. It is just as wrong to "gold plate" a design by using expensive material when there is no need for such material as it is to try to justify the use of parts that are unlikely to do the job. As a consequence, part sensitivity analysis must be associated with each strategy considered.

Given the transfer function of a circuit/assembly as $f(x_1, x_2, \ldots, x_n)$, take the partial derivative of the transfer function with respect to each parameter (i.e., $x_1, x_2, \ldots, x_n$). Recall from elementary calculus that the partial derivative of a function $f(x_1, x_2, x_3, \ldots, x_n)$ with respect to any x_i, i.e., $\partial f(x_1, x_2, x_3, \ldots, x_n)/\partial x_i$, can be loosely interpreted as the approximate increment of change, Δx_i, induced in $f(x_1, x_2, \ldots, x_n)$ due to a *unit* increment of change in x_1:

$$\Delta x_i = \frac{\partial f(x_1, x_2, \ldots, x_n)}{\partial x_i} \tag{6-11}$$

But, if (6-11) represents the increment of change induced in $f(x_1, x_2, \ldots, x_n)$ due to a unit increment of change in x_i, then we can define Δx_{pi} as the ratio of the proportion of change induced in $f(x_1, x_2, \ldots, x_n)$ to the proportion of change occurring in x_i:

$$\Delta x_{pi} = \frac{\Delta x_i}{f(\mu_1, \mu_2, \ldots, \mu_n)} \Bigg/ \frac{1}{\mu_i} = \frac{\mu_i \Delta x_i}{f(\mu_1, \mu_2, \ldots, \mu_n)} \qquad (6\text{-}12)$$

where

$i = 1, 2, \ldots, n$, and each parameter other than the one that is the subject of the partial derivative is considered a constant

μ = nominal/mean value of x_i

$f(\mu_1, \mu_2, \ldots, \mu_n)$ = nominal/mean value of $f(x_1, x_2, \ldots, x_n)$

Hence, Δx_{pi} is a measure of the relative sensitivity of $f(x_1, x_2, \ldots, x_n)$ to changes in x_i. The larger the value of Δx_{pi}, the greater the sensitivity to changes in x_i when all other parts are at their nominal value.

Calculating Δx_{pi} for each part and ranking the result with respect to the *absolute* value of its effect will yield visibility on the relative impact of variation in each part to system performance.

If, for example, the circuit/assembly is a series–parallel combination of resistors, as shown in Fig. 6-2, the resistance of the circuit (its transfer function) R_s is

$$R_s = R_1 + \frac{R_2}{2} = 150 \text{ ohms}$$

$$\Delta R_1 = \frac{\partial R_s}{\partial R_1} = 1; \qquad \Delta R_{p1} = \frac{R_1 \Delta R_1}{R_s} = \frac{2}{3}$$

$$\Delta R_2 = \frac{\partial R_s}{\partial R_2} = \frac{1}{2}; \qquad \Delta R_{p2} = \frac{R_2 \Delta R_2}{R_s} = \frac{1}{3}$$

Hence, the sensitivity of R_s to a change in the nominal value of R_1 is

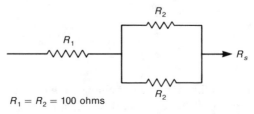

$R_1 = R_2 = 100 \text{ ohms}$

FIGURE 6-2.   A circuit made up of resistors in series and parallel.

greater than the sensitivity of R_s to a change in the nominal value of R_2, indicating that if tighter tolerance levels on R_s were desired, narrower tolerances on R_1 resistors would provide more impact than narrower tolerances on R_2 resistors.

If, however, the design nominal values of resistance in the figure were $R_1 = 1$, $R_2 = 100$, application of (6-12) would yield the resistance of the circuit (its transfer function) R_s as

$$R_s = R_1 + \frac{R_2}{2} = 51 \text{ ohms}$$

$$\Delta R_1 = \frac{\partial R_s}{\partial R_1} = 1; \qquad \Delta R_{p1} = \frac{R_1 \Delta R_1}{R_s} = \frac{1}{51}$$

$$\Delta R_2 = \frac{\partial R_s}{\partial R_2} = \frac{1}{2}; \qquad \Delta R_{p2} = \frac{R_2 \Delta R_2}{R_s} = \frac{50}{51}$$

Hence, the sensitivity of R_s to a proportional change to the nominal value of R_2 is greater than the sensitivity of R_s to a proportional change in the nominal value of R_1, indicating that if tighter tolerance levels on R_s were desired, narrower tolerances on R_2 resistors would provide more impact than narrower tolerances on R_1 resistors.

When a Circuit/Assembly Transfer Function
Is Not Available

There are instances when a sensitivity analysis should be performed and a transfer function is not available. Two general means are available in this situation, an experimental means and an engineering means.

The experimental means requires a breadboard of the circuit/assembly and means to vary the parameter values of each part in the circuit/assembly. In some cases, it can take the form of a test jig capable of switching in and switching out different-valued parts at the same schematic location. In other cases, it can take the form of modifications to the circuit/assembly such that characteristics of given parts can be changed at will. By noting the change in the desired circuit/assembly output(s) with each change of part value, a sensitivity analysis can be performed.

The engineering means requires the availability of a computer program which is capable of performing circuit analyses or of device emulators. There are a number of such programs available commercially or within the public domain. Most developers use or have available one or more of such tools. While such tools will not be able to develop a transfer function, they will be capable of determining the effects on an output of variations in one or more part parameters. Using such tools, the engineer can empirically compute a Δx_{pi} for each part parameter by increasing/decreasing each parameter value by a reasonable increment (say, 10%) and calculating the effects on the output(s) desired.

6.6.2.2 Strategies for Reducing the Variability of Part Populations

There are only two means available for reducing the variability (i.e., the variance) of part populations.

1. Structure and control of a part manufacturing process that is capable of producing parts with the variance required. This allows use of the entire population of parts produced (almost 100%) with a minimum of ad-hoc tests.

A product population (from which each part is randomly chosen for insertion in a circuit/assembly) is made up of a distribution of different values (for each part parameter). The distribution of values for each parameter has a given mean μ and standard deviation σ. The distribution can take a number of different forms, such as those discussed in Chapter 2. Figure 6-3a illustrates a product parameter population in the form of a normal distribution. As we recall from our earlier discussion of the normal distribution in Chapter 2,

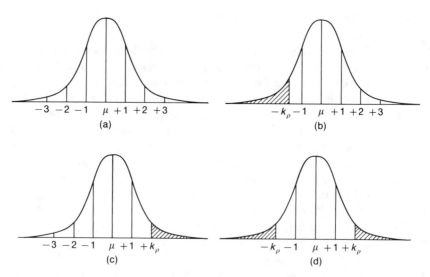

FIGURE 6-3.   (a) Normal distribution of part values. (b) Normal distribution of part values truncated below the mean. (c) Normal distribution of part values truncated above the mean. (d) Normal distribution of part values truncated above and below the mean.

approximately

38% of the values in the population will be within $\mu \pm 0.5\sigma$
68% of the values in the population will be within $\mu \pm \sigma$
87% of the values in the population will be within $\mu \pm 1.5\sigma$
95% of the values in the population will be within $\mu \pm 2\sigma$
97.7% of the values in the population will be within $\mu \pm 3\sigma$

Hence, if the part in question was a resistor coming from a production line producing resistors with mean value $\mu = 100$ ohms with a standard deviation $\sigma = 10$ ohms, virtually 87% of the parts chosen for insertion will be between 85 and 115 ohms. If the design application was such that a resistor between 70 and 130 ohms would always provide the result required, then the entire production lot would, for all practical purposes, be acceptable, since 99.7% of the product would suit our needs.

While matching production line characteristics to particular design needs is highly desirable and advantageous, attempts to implement changes in established product/process lines to produce reduced levels of variance can be difficult.

2. Culling from the part population those parts with values above a given range, below a given range, or both above and below a given range. This results in smaller yields with respect to part production populations and requires the use of tests/measurements/inspections to cull the population. Culling of one type or another is performed on a production lot when design needs dictate the use of parts with one or more parameters within a narrower range than the production process allows. It results in what are called *truncated* distributions, shown for example purposes in Figs. 6-3b to 6-3d. Such truncation, as is intuitive, reduces the standard deviation of the population of part values and can change the mean of the population as well. The subject of the treatment of truncated distributions is beyond the scope of this text. For more information on the subject, in particular when assuming normality of the parent distribution, see Refs. 3 and 4.

For other distributions, the procedures involved in defining the characteristics of the "truncated" distribution and analyzing its effects, as we did for the normal distribution in the last section, become even more mathematically complex and involved. In most cases, if such information is required, the simplest recourse is to determine the parent distribution and perform a computer simulation to determine the characteristics of the resulting "truncated" distribution.

The important thing to remember is that culling the production lot with respect to parameter values can often occur without the customer realizing it. (If asked, the vendor can usually provide the culling parameters employed.) It results in a distribution with different characteristics and, in almost all cases, a different distribution from the parent production distribution. The cost of the part will be a function of the tests and measures performed, the resultant yield, and the possible disposition of the parts rejected.

6.6.3 Using Statistical Design Techniques to Reduce Rework

This section addresses what can best be termed statistical circuit design techniques. Their application first attracted wide attention in the reliability community 30 years ago and has in the interim become a mainstay more for designers than reliability engineers. Discussion is included here because of its relationship to the parts program and parts specification and because of the fact that marginal performance of an item due to poor part choice can result in poor reliability. For that reason, the fundamental concepts of its application will be discussed. Since most of the papers and reports in the reliability literature outlining its application appeared during that era, these will remain the focus of references.

6.6.3.1 Assessing Part Tolerance Effects on Rework by the Distribution Mean/Variance Approach

This approach is based on only general knowledge or assumptions about the distributions taken by the part values making up the item (and the item's transfer functions). It relies heavily either on the assumption of normality or on less precise nonparametric measures. In the event that complete distribution information is available on population part values *and distributions of part values are not normal* and *a more exact assessment of the performance characteristics* of a circuit/assembly (described by a transfer function) is required, the Monte Carlo method of Section 6.6.3.2 should be applied. The mean/variance method can nevertheless be applied simply and quickly and will provide sufficiently accurate results for engineering purposes.

For circuits and assemblies with linear transfer functions: (1) when the populations of parts making up the circuit/assembly have normal distributions of values, it provides exact results. (2) When they have nonnormal distributions of values, but the ratio of the standard deviation to the mean is relatively small and the number of parts making up the assembly is at least moderately large, it provides approximate results. (3) When they have nonnormal distributions of values, but the number of parts making up the assembly is very large, it provides approximate results.[5,6]

In general, if an assembly is made up of many parts and the vast majority act independently (both of which are usually the case), the application of this procedure will provide first-cut planning criteria which are at least reasonable no matter what distribution the individual part values take.

For circuits and assemblies with nonlinear transfer functions, when the population of parts making up the circuit/assembly is large, it provides approximate results when parts have either normal or nonnormal distributions of values.

Linear Transfer Functions
If the transfer function of a component can be expressed in terms of a linear function

$$f(x_1, x_2, \ldots, x_n) \tag{6-13}$$

where x_i are all random normal variates with different means μ_i and standard deviations σ_i, then $f(x_1, x_2, \ldots, x_n)$ will be normally distributed with mean μ_s and standard deviation σ_s,

$$\mu_s = f(\mu_1, \mu_2, \ldots, \mu_n) \tag{6-14}$$

$$\sigma_s = \left[\left(\frac{\partial F}{\partial x_1}\right)^2 (\sigma_1)^2 + \left(\frac{\partial F}{\partial x_2}\right)^2 (\sigma_2)^2 + \cdots + \left(\frac{\partial F}{\partial x_n}\right)^2 (\sigma_n)^2\right]^{1/2} \tag{6-15}$$

where $F = f(x_1, x_2, \ldots, x_n)$.

The distribution of $f(x_1, x_2, \ldots, x_n)$ will then be normally distributed with mean μ_s and standard deviation σ_s and

$$\mu_s \pm k_p \sigma_s \tag{6-16}$$

where k_p = the coefficient of the normal distribution, which defines the proportion of performance values of $f(x_1, x_2, \ldots, x_n)$ *within that range.*

If each production quantity of parts was controlled such that each contained $P_z = (2p - 1)$ proportion of parts inside the range of

$$\mu_i \pm k_p \sigma_i \tag{6-17}$$

where k_p = the coefficient of the normal distribution which defines the proportion P_z of part values *between the limits* described in (6-17), $p = \int_{-\infty}^{k_p} \phi(x)\, dy$, $\phi(x)$ = the normal distribution, and μ_i and σ_i are defined as before, then P_z of all assemblies produced will have performance values $f(x_1, x_2, \ldots, x_n)$ in the range

$$\mu_s \pm k_p \sigma_s \tag{6-18}$$

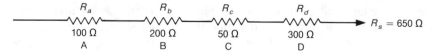

$R_s = 650\ \Omega$

R_a	R_b	R_c	R_d
100 Ω	200 Ω	50 Ω	300 Ω
A	B	C	D

FIGURE 6-4.   An assembly made of four resistors.

This has some interesting ramifications that the engineer should be aware of. *For linear functions, the proportional performance variation of an assembly will be less than that of its parts.* We will illustrate by an example. Suppose that the resistors in Fig. 6.4 have means and standard deviations as follows:

Resistor A mean = 100 ohms; standard deviation = 10
Resistor B mean = 200 ohms; standard deviation = 25
Resistor C mean = 50 ohms; standard deviation = 6
Resistor D mean = 300 ohms; standard deviation = 20

Assume each resistor population takes on a normal distribution and that approximately 90% of the population of each resistor type is in the range $\mu_i = 1.6\sigma_i$ (i.e., $P_z = 0.9$, $p = 0.945$, $k_p = 1.6$).

For resistor A this constitutes a range of $\pm 16\%$ variation from its mean. For resistor B it constitutes a range of $\pm 20\%$ variation from its mean. For resistor C it constitutes a range of $\pm 19\%$ variation from its mean. For resistor D it constitutes a range of $\pm 11\%$ variation from its mean.

From (6-14) and (6-15) the mean resistance of the assembly $\mu_s = 650$ and its standard deviation $\sigma_s = 34$, which means that 90% of the assemblies produced will have a range of variation of slightly over $\pm 8\%$ from its mean.

In order to illustrate (1) what impacts of part tolerance levels are on rework, (2) how and when the sensitivity analysis may be applied, and (3) considerations associated with the application of strategies to reduce part variability, an example simple scenario is provided. We will assume that an assembly is made up of four resistors as in Fig. 6-4, and hence can be represented by a linear transfer function.

Resistor A of 100 ohms
Resistor B of 200 ohms
Resistor C of 50 ohms
Resistor D of 300 ohms

The transfer function to describe the resistance R_s takes the form:

$$R_s = R_a + R_b + R_c + R_d, \qquad R_s = 650 \text{ ohms}$$

which would result if R_a, R_b, R_c, and R_d were exactly 100, 200, 50, and 300 ohms, respectively.

Further, suppose that our system would meet all specifications if $615 \leqslant R_s \leqslant 685$. Resistors A, B, C, and D come from different production lines. Let us for convenience assume that all four come from different normal distributions with means μ_i and standard deviations σ_i (see Section 2.2.3 for discussion of the normal distribution).

$$\mu_a = 100, \sigma_a = 10$$
$$\mu_b = 200, \sigma_b = 25$$
$$\mu_c = 50, \sigma_c = 6$$
$$\mu_d = 300, \sigma_d = 20$$

Recall the addition theorem for the normal distribution discussed in Section 2.2.3 and Section 5.8, where we found that the sum of two or more independent normal variates is normally distributed with mean μ_s given by (6-14) and standard deviation σ_s given by (6-15). In this instance,

$$R_s = f(x_1, x_2, \ldots, x_n) = R_a + R_b + R_c + R_d$$

and relationship (6-14) reduces to

$$\mu_s = \sum_{i=1}^{n} \mu_i = \text{the average value of } R_s$$

or

$$\mu_s = 100 + 200 + 50 + 300 = 650$$

This means that if we randomly choose a resistor of type A from a normal distribution and a resistor of type B from another normal distribution and do the same for resistors C and D, and do this as many times as is necessary to form a large quantity of circuits or assemblies, then the values of R_s of the circuits/assemblies will take on a normal distribution.

Also, due to the addition theorem for the normal distribution (see Section 5.8), R_s will have a standard deviation σ_s expressed by (6-15). In this instance (6-15) reduces to

$$\sigma_s = \left[\sum_{i=1}^{n} (\sigma_i)^2 \right]^{1/2} \tag{6-19}$$

or

$$\sigma_s = [(\sigma_a)^2 + (\sigma_b)^2 + \cdots + (\sigma_n)^2]^{1/2}$$

and

$$\sigma_s = [(10)^2 + (25)^2 + (6)^2 + (20)^2]^{1/2} = 34$$

What percentage of the circuits/assemblies produced will be within the limits prescribed and hence require no rework? Note that the requirement can now be represented in terms of the number of standard deviations above and below the mean:

- The value of 615 ohms, our design lower limit, is equivalent to 35/34 or 1.03 standard deviations below the mean of 650, when $\sigma_s = 34$.
- The value of 685 ohms, our design upper limit, is equivalent to 35/34 or 1.03 standard deviations above the mean of 650 when $\sigma_s = 34$.

In other words, any values of R_s within ± 1.03 standard deviations of the mean will be within the limits prescribed.

Using any readily available table of the normal distribution, it can be found that 70% of the population will be between ± 1.03 standard deviations of the mean. As a consequence, 70% of the circuits/assemblies produced will be within design requirements, and 30% will require rework or to be scrapped.

If we wanted to increase the proportion of circuits/assemblies produced that were within specification requirements while keeping this particular design, we would first perform a sensitivity analysis on the design to determine the relative criticality of each part.

In this case, the sensitivity analysis for the example provides the following. Since

$$R_s = R_a + R_b + R_c + R_d$$

$$\Delta r_a = \frac{\partial R_s}{\partial R_a} = 1; \qquad \Delta R_{pa} = \frac{100 \Delta R_a}{650} = 0.154$$

$$\Delta r_b = \frac{\partial R_s}{\partial R_b} = 1; \qquad \Delta R_{pb} = \frac{200 \Delta R_b}{650} = 0.31$$

$$\Delta r_c = \frac{\partial R_s}{\partial R_c} = 1; \qquad \Delta R_{pc} = \frac{50 \Delta R_c}{650} = 0.08$$

$$\Delta r_d = \frac{\partial R_s}{\partial R_d} = 1; \qquad \Delta R_{pd} = \frac{300 \Delta R_d}{650} = 0.46$$

which indicates that proportionate changes in B and D will affect the value of R_s more strongly than changes in A and C.

After the analysis is complete and decisions have been made relative to the tolerable variability in each part population, determinations of which parts should have narrow tolerances and how such tolerances may best be met must be reached.

In this case assume that we have decided to reduce the variance of all four parts. If we elect either to modify the manufacturing process to reduce the standard deviation of the parts, or to change to a vendor with better controls such that the population is "unculled," we can use the relationships of (6-14) and (6-15) and the procedure illustrated in the example to determine the number of circuits/assemblies produced which will meet specification requirements without rework.

For our example, let us continue to assume that all parts considered take on normal distributions. Let us further say that we have chosen to modify the manufacturing processes for all parts such that the standard deviation of each part population is reduced to 4/5 of what it was previously. Our part characteristics are now:

$\mu_a = 100, \sigma_a = 8$
$\mu_b = 200, \sigma_b = 20$
$\mu_c = 50, \sigma_c = 4.8$
$\mu_d = 300, \sigma_d = 16$
$R_s = 650$, as before

$$\sigma_s = [(8)^2 + (20)^2 + (4.8)^2 + (16)^2]^{1/2} \approx 27.26$$

Representing, as we did earlier in the illustration, the performance requirements for the circuit/assembly, $615 \leqslant R_s \leqslant 685$, expressed in terms of standard deviations above and below the mean: (1) the value of 615 is now equivalent to 35/27.26 or 1.28 standard deviations below the mean, and (2) the value of 685 is now equivalent to 35/27.26 or 1.28 standard deviations above the mean. In other words, any values of R_s within ± 1.28 standard deviations of the mean will be between the performance limits prescribed. From a table of the normal distribution we find that approximately 80% of the population will be within those limits. As a consequence, by modifying the part production processes to reduce variance, 80% of the circuits/assemblies produced will be within design requirements and 20% will require rework or scrap.

If we elect to perform tests/screens to eliminate (cull) from our populations of A, B, C, or D (or any combination of these) the values of parts which are above and below defined limits, we are in fact, *truncating* the distributions.

Each truncated distribution will take on a new set of values of mean and standard deviation. In instances where the distribution of part values is still approximately normal, or our assembly meets a condition where R_s can be considered approximately normally distributed, the same procedure will as before provide reasonably accurate estimates.

For our example let us assume that each population of resistors is approximately normally distributed after the culling process. The parent populations were measured/tested and:

1. For the population of type A resistors all resistors with values greater than 115 ohms and less than 85 ohms were removed. In other words, all parts outside of $\mu_a \pm 1.5\sigma_a$ were removed.
2. For the population of type B resistors all resistors with values greater than 237.5 ohms and less than 162.5 ohms were removed, again removing all parts outside μ_b, $\pm 1.5\sigma_b$.
3. For type C and D resistors, respectively, parts with values outside the ranges 41 to 59 ohms and 270 to 330 ohms were removed, again corresponding to the removal of parts outside $\mu \pm 1.5\sigma$.

Since each parent part population is assumed normally distributed, removal of all parts from all populations outside the $\mu \pm 1.5\sigma$ range is tantamount to discarding approximately 13% of the part population in order to reduce variance (see Fig. 6-2d).

The new mean, μ_{ti}, and standard deviation, σ_{ti}, for each part population is estimated from information available as

$$\mu_{ta} = 100, \qquad \sigma_{ta} = 7$$

$$\mu_{tb} = 200, \qquad \sigma_{tb} = 18$$

$$\mu_{tc} = 50, \qquad \sigma_{tc} = 4$$

$$\mu_{td} = 300, \qquad \sigma_{td} = 14$$

$$\mu_{ts} = 100 + 200 + 50 + 300 = 650$$

$$\sigma_{ts} = [(7)^2 + (18)^2 + (4)]^2 + (14)^2]^{1/2} = 24$$

Assuming R_s takes on an approximately normal distribution, and given the performance requirements for the circuits/assemblies of $615 \leqslant R_s \leqslant 685$, what proportion of circuits/assemblies produced would meet this requirement? Expressing the requirement in terms of the standard deviation of R_s, as before, (1) the value of 615 is now equivalent to 35/24 or 1.46 standard deviations below the mean, and (2) the value of 685 is now equivalent to

35/24 or 1.46 standard deviations below the mean. In other words, any value of R_s within ± 1.35 standard deviations of the mean will be between the performance limits prescribed.

From a table of the normal distribution, approximately 86% of the population will be within these limits. As a consequence, 86% of the circuits/assemblies produced will be within design requirements and 14% will require rework or scrap.

The above total illustration is a simple one, but it serves to bring out three important points:

- Tolerance levels for parts can have a significant impact on whether or not a product can meet requirements.
- There are relatively simple means available to assess the impacts of part tolerance on meeting product requirements.
- There are alternate courses of action that can be taken to improve a given situation.

With respect to the last, each course of action has its advantages, disadvantages, and costs, and each must be weighed when making a decision. To illustrate this, consider again our example where three courses of action were possible: do nothing, modify the part production process, or cull out what we considered the most extreme part values:

1. If we did nothing, 30% of the finished products produced would require rework or scrap due to part tolerance considerations alone. However, the entire population of each part type would have been used; only go–no-go testing of parts would have been required, and hence the price per part would be the minimum.

2. If we chose a vendor having a well-controlled production process or modified an existing part production process to reduce the variance of part values, 20% of the finished products produced would require rework or scrap due to part tolerance considerations alone. However, the entire population of each part type can be used; only go–no-go testing or screening of the parts will be required. However, changes to the part production process (or considerations of a vendor with the type of process results needed) must take place and controls must be implemented to assure stability of the process. Hence, there is potential for increased cost in order to acquire a better product or a greater cost per part (either by the vendor or the developer) until stability of the process has been established.

3. If in our example we culled out the most extreme part values to reduce their variance, 14% of the finished products produced would require rework or scrap due to part tolerance considerations. However, it would

require discarding 13% of the part population and performing tests on the entire population in order to cull out those extremes. Hence, this course of action also has its costs.

As indicated previously, the part populations do not have to take on a normal distribution in order for $f(x_1, x_2, \ldots, x_n)$ to approximate a normal distribution. The more they approximate normality, the more parts making up the assembly, the smaller the value of the part value standard deviation with respect to the value of its mean, the closer the values of $f(x_1, x_2, \ldots, x_n)$ to a normal distribution.

An Alternative When Part Population
Standard Deviation Is Unknown

What happens, however, if we do not know what the standard deviation of the part populations is? How can we arrive at an estimate of product quality?

Many times even if we do not have a quantitative number for the value of a part population standard deviation, we have a feel for the range of values possible in a population. Let us assume that a part population is composed of transistors of a given type with a nominal power output of 250 milliamps. While we might not know the standard deviation of power output for the population, we may very well have reason to believe that virtually all the population is within some range of the nominal value (say $\pm 10\%$ of 250 milliamps). The important thing is the ability, from sampling, from experience, etc., to define what we will call a *virtual range* (VR) for the part values in a population. VR then may be defined as our estimate of the maximum deviation of the part parameter in the given population from its nominal value. VR can be translated into terms of a maximum number of standard deviations. We did something similar previously when we expressed performance requirements for circuits/assemblies in terms of $\pm k\sigma$ from a given mean.

If we view VR as equivalent to the maximum value of $\pm k\sigma$ possible, it represents the number of standard deviations which will include virtually 100% of the part population (for normal distributions a value of $k \geq 3$ meets that criterion).

If we equate

$$\text{VR} = k\sigma \tag{6-20}$$

where k represents the number of standard deviations equal to or *greater than* that required to include virtually 100% of the distribution of part values, we can express

$$\sigma = \frac{\text{VR}}{k} \tag{6-21}$$

If we assume that the distribution of the values of a given parameter is approximately normally distributed, and if we have a good estimate of the virtual range VR (e.g., VR $= \pm 10\%$ of the nominal value), we can approximate the value of σ as

$$\sigma = \frac{VR}{3} \tag{6-22}$$

σ_s for an assembly can be calculated using (6-15), where $(VR/3)_i$ is substituted for σ_i. μ_s for the assembly is approximated using (6-14), treating the nominal value for each part as μ_i. And if we assume a normal distribution of values for each part, we can similarly assume a normal distribution for $f(x_1, x_2, \ldots, x_n)$. Furthermore, the VR for $f(x_1, x_2, \ldots, x_n)$, VR_s, will be

$$VR_s = 3\sigma_s \tag{6-23}$$

so that virtually all values of $f(x_1, x_2, \ldots, x_n)$ will be in the range $\mu_s \pm 3\sigma_s$.

However, applications of the VR technique need not be restricted to the assumption of normality. It can be thought of as a nonparametric procedure capable of handling any distribution or mix of distributions.

Assume that a sufficiently large single value of k exists which is applied to each distribution of parts. Bear in mind that if virtually all values of a part population lie in the range $\mu \pm k_1\sigma$ when $k_1 = 3.6$, virtually all values of that population will lie in the range $\mu \pm k_i\sigma$ when $k_i > 3.6$. Hence we can consider a single value of k which is arbitrarily large. But since the quantity $k\sigma$ is replaced by the value of VR in the calculation, no specific values of k or σ need be considered. Substituting $\sigma = VR/k$ in (6-15),

$$VR_s = \left[\sum_{i=2}^{n} \left(\frac{\partial F}{\partial x_i} \right)^2 (VR_i)^2 \right]^{1/2} \tag{6-24}$$

and the virtual range for the group of assemblies produced can be expressed as

$$\mu_s \pm VR_s \tag{6-25}$$

Equation (6-24) provides a linkage between the range of circuit/assembly performance and the tolerance range of its parts.

Considering the last example, assume each resistor to be as before (note, since normality was assumed originally, we will take $k = 3$),

$$\mu_a = 100, \quad \text{VR}_a = 3\sigma_a = \pm 30 = \pm 30\% \text{ of } \mu_a$$
$$\mu_b = 200, \quad \text{VR}_b = 3\sigma_b = \pm 75 = \pm 37.5\% \text{ of } \mu_b$$
$$\mu_c = 50, \quad \text{VR}_c = 3\sigma_c = \pm 18 = \pm 36\% \text{ of } \mu_c$$
$$\mu_d = 300, \quad \text{VR}_d = 3\sigma_d = \pm 60 = \pm 30\% \text{ of } \mu_d$$

We will assume no knowledge of the values of σ_i, just of VR_i.
The mean, or nominal target of performance expected, μ_s, is

$$\mu_s = \mu_a + \mu_b + \mu_c + \mu_d = 650 \text{ ohms}$$

From (6-24), $\text{VR}_s \approx 102$. Therefore, the maximum range of performance for the assemblies produced is

$$650 - 102 \leqslant R_s \leqslant 650 + 102$$

or

$$R_s = \mu_s \pm 15.7\% \text{ of } \mu_s$$

which again exemplifies the characteristic that for a linear function, proportionately speaking, the *variation of performance at the assembly level will be less than that experienced at the part level.*

Nonlinear Transfer Functions

We will now consider the effects of part value variations on assembly performance when transfer functions are nonlinear. In such cases, where $f(x_1, x_2, \ldots, x_n)$ describes the transfer function, the mean μ_s and standard deviation σ_s of assembly performance take forms similar to those presented previously, except that now they represent approximations:

$$\mu_s \approx f(\mu_1, \mu_2, \ldots, \mu_n) \tag{6-26}$$

$$\sigma_s \approx \left[\Sigma \left(\frac{\partial F}{\partial x_i} \right)^2 (\sigma_i)^2 \right]^{1/2} \tag{6-27}$$

where μ_i, σ_i, and F are defined as before.
The sensitivity analysis remains the same.
The derivation of (6-26) and (6-27) for nonlinear functions can be shown

through application of a multivariate Taylor series (see Refs. 5–8). Extensions of the procedure to cope with nonnormal distributions of the parameters x_i can be provided. However, such procedures are mathematically very complex and beyond the scope of this text. We will concentrate here on the discussion of simpler procedures which are either based on the assumption of a normal distribution or provide for worst-case "and/or" nonparametric estimates of range of circuit/assembly performance based on the tolerance of its component parts.

In Section 5.8 we described some functions of normal variates and their characteristics when the variates were summed. Under some general conditions common to part characteristics and the complexity levels of circuits/assemblies (e.g., standard deviation very small compared to mean, independence among components, large quantity of parts making up each assembly), when independent normal variates are multiplied, taken to negative and positive powers, added, or subtracted, the result for practical purposes approximates or converges to a normal distribution (see Refs. 9 and 10). As a consequence, it is not unreasonable, at least for preliminary engineering decisions, to assume that performance attributes of groups of circuits/assemblies take on approximately normal distributions. This can occur if the component part values making up the transfer function take on approximately normal distributions even when the transfer function is nonlinear.

Under those conditions, μ_s and σ_s are calculated using (6-26) and (6-27) and the performance measure $f(x_1, x_2, \ldots, x_n)$ will be bounded as

$$\mu_s - k_p \sigma_s \leqslant f(x_1, x_2, \ldots, x_n) \leqslant \mu_s + k_p \sigma_s \qquad (6\text{-}28)$$

where k_p = the coefficient of the normal distribution associated with the proportion of assemblies P_z which will have performance values in that range where $P_z = (2p - 1)$; $p = \int_{-\infty}^{k_p} \phi(x)\,dy$, $\phi(x)$ = the normal distribution, and μ_s and σ_s are defined as before.

In some cases, even if the part values making up a transfer function are not normally distributed, it is possible for $f(x_1, x_2, \ldots, x_n)$ to take on a normal distribution, but no general rules have been established to help in making that judgment.

A conservative worst-case answer may be developed by applying a combination of the Tchebychev inequality described in Section 5.9 and the virtual range (VR) procedure discussed above. This estimate will be nonparametric in nature, requiring only judgments on the range of values associated with any distribution of part values.

The expression for the virtual range of the assembly, VR_s, is developed in a similar fashion as previously, except that now it takes the form of an approximation (in the same way that (6-26) provides an approximation to σ_s).

$$\text{VR}_s \approx \left[\sum_{i=1}^{n} \left(\frac{\partial F}{\partial x_i} \right)^2 (\text{VR}_i)^2 \right]^{1/2} \tag{6-29}$$

The virtual range for the group of assemblies produced can be expressed as

$$\mu_s \pm \text{VR}_s \tag{6-30}$$

We can justify the use of the virtual range on the basis of the Tchebychev inequality:

$$P[(\mu + k\sigma) > x > (\mu - k\sigma)] \geq 1 - \left(\frac{1}{k} \right)^2 \tag{6-31}$$

Since $k\sigma = \text{VR}$, (6-31) can be expressed as

$$P[(\mu + \text{VR}) > x > (\mu - \text{VR})] \geq 1 - \left(\frac{1}{k} \right)^2 \tag{6-32}$$

When k is considered arbitrarily large, which it can be when considering virtual range (see discussion following (6-24)), then

$$1 - \left(\frac{1}{k} \right)^2 \to 1$$

Hence, (6-30) results.

We also could apply the Tchebychev inequality directly to the problem if we had information on the standard deviations of the parts making up the assembly. This will be illustrated in an example which follows.

To illustrate the application of the mean/variance approach to a nonlinear transfer function, let us consider a slightly more complex amplifier circuit as shown in Figure 6-5. Nominal value, mean and standard deviation are

$$g_m = \mu_{gm} = 0.02, \qquad \sigma_{gm} = 0.003$$
$$R_c = \mu_c = 600 \text{ ohms}, \qquad \sigma_c = 50$$
$$R_a = \mu_a = 4000 \text{ ohms}, \qquad \sigma_a = 200$$
$$R_b = \mu_b = 3000 \text{ ohms}, \qquad \sigma_b = 300$$

where g_m = gain of the active element and the gain, G, for the circuit can be

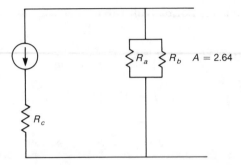

FIGURE 6-5.   An amplifier equivalent circuit.

expressed as

$$G = \frac{g_m(R_a R_b)}{(1 + g_m R_c)(R_a + R_b)} \qquad (6\text{-}33)$$

Assume in this case that the distribution of the values of g_m, R_a, R_b, and R_c are known and that the transfer function (6-33) is nonlinear. Performing a sensitivity analysis:

$$\Delta g_m = \frac{\partial F}{\partial g_m} = 10.1; \qquad \Delta g_{mp} = 0.02(10.1)/2.64 = 0.08$$

$$\Delta R_c = \frac{\partial F}{\partial R_c} = 0.004; \qquad \Delta R_{pc} = 600(-0.004)/2.64 = -0.91$$

$$\Delta R_a = \frac{\partial F}{\partial R_a} = 0.0003; \qquad \Delta R_{pa} = 4000(0.0003)/2.64 = 0.45$$

$$\Delta R_b = \frac{\partial F}{\partial R_b} = 0.0005; \qquad \Delta R_{pb} = 3000(0.0005)/2.64 = 0.57$$

where $F = f(x_1, x_2, \ldots, x_n)$.

The above identifies those parts whose changes will impact $f(x_1, x_2, \ldots, x_n)$ the most. This procedure is required sooner or later for two related steps: (1) determining which part tolerances would impact assembly performance the most; (2) serving as a basis for minimizing part cost, to determine which parts can be purchased with wider tolerance bands without increasing rework.

We next proceed to define approximations for μ_s and σ_s, first assuming that G is approximately normally distributed.

$$\mu_s \approx f(\mu_1, \mu_2, \ldots, \mu_n) = \frac{\mu_{gm}(\mu_a\mu_b)}{(1 + \mu_{gm}\mu_c)(\mu_a + \mu_b)} = 2.64$$

$$(\sigma_s)^2 \approx \sum_{i=1}^{n} \left(\frac{\partial F}{\partial x_i}\right)^2 (\sigma_{xi})^2 = (\Delta g_m \sigma_{gm})^2 + (\Delta R_c \sigma_c)^2 + (\Delta R_a \sigma_a)^2 + (\Delta R_b \sigma_b)^2$$

Note how the results of the sensitivity analysis can be used to calculate σ_s^2:

$$\sigma_s^2 \approx (10.1)^2(0.003)^2 + (0.004)^2(50)^2 + (0.0003)^2(200)^2$$
$$+ (0.0005)^2(300)^2 \approx 0.067$$

or

$$\sigma_s \approx 0.256$$

Assuming normality for G:
75% of all circuits/assemblies produced will have gains

$$2.64 - 1.15(0.256) \leqslant G \leqslant 2.64 + 1.15(0.256)$$

or

$$2.34 \leqslant G \leqslant 2.93$$

85% of all circuits/assemblies produced will have gains

$$2.64 - 1.5(0.256) \leqslant G \leqslant 2.64 + 1.15(0.256)$$

or

$$2.26 \leqslant G \leqslant 3.02$$

95% of all circuits/assemblies produced will have gains

$$2.64 - 2(0.256) \leqslant G \leqslant 2.64 + 2(0.256)$$

or

$$2.13 \leqslant G \leqslant 3.15$$

Applying the Tchebychev inequality to the values of μ_s, σ_s. In instances where the distribution of parts making up the assembly are either not normal or unknown but their standard deviations are known or can be estimated, we can apply the Tchebychev inequality. This will provide us with information concerning the assemblies' performance with respect to G. In this case, >75% of all circuits/assemblies produced will have gains

$$2.64 - 2(0.256) \leqslant G \leqslant 2.64 + 2(0.256)$$

or

$$2.128 \leqslant G \leqslant 3.15$$

>87% of all circuits/assemblies produced will have gains

$$2.64 - 2.77(0.256) \leqslant G \leqslant 2.64 + 2.77(0.256)$$

or

$$1.93 \leqslant G \leqslant 3.35$$

>95% of all circuits/assemblies produced will have gains

$$2.64 - 4.47(0.256) \leqslant G \leqslant 2.64 + 4.47(0.256)$$

or

$$1.50 \leqslant G \leqslant 3.78$$

Note the similarities and differences between the results of the Tchebychev nonparametric estimate and the estimate made assuming a normal distribution of G (see Section 5.9).

Applying the Virtual Range (VR) procedure in the manner described earlier, assuming μ and VR (expressed as a function of μ) as below,

$$g_m = \mu_{gm} = 0.02, \qquad VR_g = \pm 30\% \text{ of } \mu_{gm}$$
$$R_c = \mu_c = 600, \qquad VR_c = \pm 30\% \text{ of } \mu_c$$
$$R_a = \mu_a = 4000, \qquad VR_a = \pm 30\% \text{ of } \mu_a$$
$$R_b = \mu_b = 3000, \qquad VR_b = \pm 30\% \text{ of } \mu_b$$

$$VR_s^2 = (10.1)^2(0.006)^2 + (0.004)^2(180)^2 + (0.0003)^2(1200)^2 + (0.0005)^2(900)^2$$

$$VR_s = 0.92$$

Therefore, the maximum range of performance for the circuits produced is $G = 2.64 \pm 0.92$.

The general statement that was made earlier for the linear functions—that the performance variability for the assembly will be less than that of the parts making up the assembly—does not hold true for a nonlinear transfer function. Our example bears this out.

6.6.3.2   Assessing Part Tolerance Effects on Rework by Monte Carlo Methods

This approach is based upon specific knowledge or assumptions about the distributions of part values making up transfer functions. While it is technically an approximation procedure, it can provide a level of precision of results significantly greater than that of the mean/variance procedure. As in the previous method, the procedure requires a transfer function(s) that relates circuit/assembly performance measures to part values. The distribution of part values associated with each part must also be provided (see Refs. 11 and 12).

The approach is much less mysterious and complex than we may think. Monte Carlo analysis, at least as we will apply it here, is no more than a "test and see" procedure. While we can perform a Monte Carlo analysis manually, it is almost always done with the use of a computer. Let us look at what we are actually doing. Suppose it is our intention to build 1000 circuits/assemblies. Furthermore, suppose that we want *both* to (1) observe the effects of possible variations in the part values that must be used on the distribution of performance characteristics of the circuits/assemblies we will produce, *and* (2) determine the effects of component part drift on the performance or reliability of the item over time in operational use.

Assume that we know the exact transfer function of our circuit and the distribution of part values for each different part in the circuit. Further assume that the part values all take on different distributions; that some populations are culled, some not; that some distributions are normal, some not; and furthermore, that we need as precise an answer as we can get. It would be very helpful if, before we started fabrication, we could *simulate* the fabrication process to provide us with some insight as to what we can expect.

When the fabrication process starts, a part of type i is chosen at random from the type i population and placed in the circuit board; the same is done for part $(i + 1)$, and so on up to part n. (We assume that the circuit/assembly is composed of n parts.) The fabrication process is completed and tests

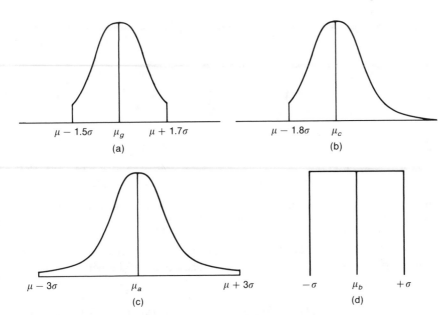

FIGURE 6-6. (a) Distribution of g_m in part population of transistors. (b) Distribution of resistance in part population of resistor C. (c) Distribution of resistance in part population of resistor A. (d) Distribution of resistance in part population of resistor B.

(measurements) are performed on the product to see whether it meets performance objectives. A simulation of this process which will take into account the effects of part variability on the finished product can be performed through the following.

1. Define a transfer function or transfer functions which define the performance requirement of the circuit/assembly in terms of values of the parts making up the circuit. Evaluation of that transfer function for any mix of part values will produce values of performance of the circuit with that particular part value mix.
2. Define the distribution of part values for every population of parts from which a random selection would be made during the fabrication process. For example, taking the illustration in the preceding section, suppose that the distributions of g_m, R_c, R_a, and R_b took on forms such as in Fig. 6-6.
3. Choose one value randomly from Fig. 6-6a, 6-6b, 6-6c and 6-6d.
4. Evaluate the value of the transfer function(s) when the parts take on the values chosen.
5. Repeat (3) and (4) above a large number of times, and group the results. Figure 6-7 shows the results grouped in two ways: as a distribution of

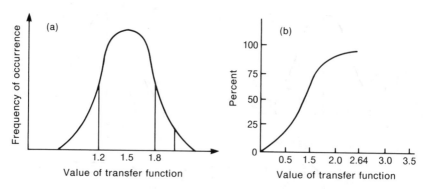

FIGURE 6-7. (a) Distribution of performance values for population of circuit produced. (b) Percentage of circuits that exceed a given transfer function value.

values, or as a cumulative distribution. Either can provide information on portions of circuits/assemblies produced that will be within required performance limits.

After an initial simulation, the distributions in Fig. 6-6 can be changed to represent tighter tolerances or wider tolerances in order to provide the most cost-effective mix possible.

In the event that the parts making up an assembly are subject to drift or other change with time which can be represented statistically (i.e., a distribution is associated with a change in value over a given period of time), a similar process can be employed to predict (1) the proportion of assemblies still providing performance over a certain level after T hours of use; (2) the probability that a given assembly will still provide the performance desired over T hours of use.

6.6.3.3 Worst-Case Design as a Means
to Reduce Rework

Worst-case design is a procedure that can be applied equally well to both linear and nonlinear functions. Its rationale differs markedly from that of the other procedures just discussed. Those are keyed to control and tailoring of part tolerances to enhance the probability of a given circuit/assembly design. Worst-case design, on the other hand, is keyed to circuit/assembly design to accommodate the tolerance limits of parts such that the circuit/ assembly will perform within specification limits even when all parts are at their most extreme values (see Refs. 13, 14).

In order to develop or evaluate a worst case design, (1) a circuit/assembly transfer function, a bread board of the circuit, or capability to perform circuit analysis must be available; (2) information must be available as to the absolute limits of values each part can take (the end points). Each end-point value singly and in combination is inserted into the transfer function, the bread board, or circuit analysis program (any one of such vehicles or a combination can be used for this purpose) and a determination is made as to whether or not the circuit/assembly is within performance limits. For even moderate complexity, the design effort required for such a task can be significant. The number, C, of different combinations eligible for consideration takes the form $C = 2^n$ where n = the number of parts making up this circuit/assembly.

Of course, if a transfer function were available it could provide insight into which part extremes would tend to affect most the value of the transfer function. Alternately, if the circuit/assembly design makeup were such that judgments of that nature could be made, then the number of combinations of end points considered could be reduced.

In the event that one or more extreme value(s) or extreme value combination(s) causes performance considered to be outside the requirement range, the design of the circuit/assembly is modified. In the event that the necessary modification is infeasible, impractical, or would result in other undesirable effects, consideration is given to reduction in the end-point values of the components in question by culling or by going to alternative sources of supply.

The advantage of worst-case design is that, like the other techniques discussed, it minimizes the degree of rework necessary. The analysis involved with worst-case design may be time-consuming but it is relatively simple in nature and does not necessarily require a transfer function.

The disadvantage of worst-case design is that any design so tolerant to change may of necessity, be more complex or require higher power levels.[15] Either exacts a penalty with respect to reliability.

A worst-case design can also provide an overly conservative design. Take, for example, Fig. 6-4. Assume that it makes up a component in a larger assembly. Assume that each population of resistors is approximately normally distributed such that the end points for each resistor are located at $\pm 3\sigma$ from the mean, or

$\mu_a = 100,\ \sigma_a = 10$; end points $\pm 3\sigma = \ \ 70$ and 130, respectively
$\mu_b = 200,\ \sigma_b = 25$; end points $\pm 3\sigma = 125$ and 275, respectively
$\mu_c = \ \ 50,\ \sigma_c = \ \ 6$; end points $\pm 3\sigma = \ \ 32$ and $\ \ 68$, respectively
$\mu_d = 300,\ \sigma_d = 20$; end points $\pm 3\sigma = 240$ and 360, respectively

The worst-case end points for the series combination shown in Fig. 6-4 are $70 + 125 + 32 + 240 = 467$, and $130 + 275 + 68 + 360 = 833$, respectively. It follows under worst case design principles that the assembly must be designed to function properly within such component value limits. However, since the distributions of all part values are normal, the value of R_s will be normally distributed with mean μ_s and standard deviation σ_s given by

$$\mu_s = 650, \qquad \sigma_s = 34 \qquad \text{(see Section 6.6.3.1)}$$

which means that 99.87% of the items (made up of the four resistors) produced will have $548 \leqslant R_s \leqslant 752$, which is significantly inside worst-case limits, and which represents a range that will only be exceeded by 0.0013 of the population despite the fact that the "literal" worst-case design values are $467 \leqslant R_s \leqslant 833$.

We must conclude, therefore, that "literal" worst-case values can often provide overly conservative estimates of ranges of component performance which must be accounted for during design.

If the above then represents an example of a "literal" worst-case analysis, a "figurative" form of worst-case analysis is the VR procedure which was discussed earlier. This can be used as the basis of what might be termed a hybrid worst-case analysis since $\pm \text{VR}_s$ covers the requirement limits of the circuit/assembly.

6.6.4   Part Tolerances and Reliability

If an equipment made up of its associated circuits and assemblies passes all tests and screens and is accepted into field use, its reliability during service life is affected not only by its material defects which precipitate into hard failures, but by the particular value taken by its transfer functions over time as well. Latent and other material defects can cause sudden and abrupt changes in transfer function values caused by what is construed as sudden changes in component part parameters—opens, shorts, etc. Gradual changes or degradation in part parameters also result in equipment failures. Some changes in part values are gradual, with rates of change increasing or decreasing with time and use in an operational environment. Some are prone to stabilize after an initial change. Some occur only for the duration of time that an equipment is within a given operating environment. Such changes in parameter values make up a significant number both of total failures and of calls for maintenance which occur during service life.

The degree to which a part is subject to value change in an environment is a function of time, the degree to which the part is stressed, the environment of operation, and the basic quality of the part. Hence, a higher-power device might be more prone to change in value than a low-power device. A device used at its full power potential might be more prone to change than one operated in a derated condition. A device in an extreme temperature environment might be more prone to change than one operated in a moderate temperature environment. A device which meets combined Mil-M-38510 and Mil-Std-883 requirement might be less prone to change than a device procured commercially.

Immediately after production a circuit or assembly making up an equipment may be close to or a significant distance from its performance requirement end points. Assuming that drift (or change) in a component parameter exists to at least some degree, the circuits most at risk from drift failures are those with transfer function values closest to the requirement end points when first put into service.

In order to estimate proneness to failure simply and quickly, from the sensitivity analysis performed previously select those parts which affect sensitivity the most or consider all parts. Increment (or decrement, or both) the mean μ_i by a value $\Delta\mu_i$,

$$\Delta\mu_i = c_i\mu_i$$

where c_i = proportional change expected in the value of μ_i, in a given environment, over a period of time, etc., and can take on a positive or negative value. Compute an estimate of the mean, μ_{cs} (denoting the effect of c), of the transfer function $f(x_1, x_2, \ldots, x_n)$ if μ_i takes the value ($\mu_i + \Delta\mu_i$):

$$\mu_{cs} \approx f(\mu_1 + \Delta\mu_1, \mu_2 + \Delta\mu_2, \ldots, \Delta\mu_n) \qquad (6\text{-}34)$$

Assume that the standard deviation, σ_s, will not change and calculate σ_s as before in (6-27):

$$\sigma_s \approx \left[\sum_{i=1}^{n} \left(\frac{\partial F}{\partial x_i}\right)^2 (\sigma_i)^2 \right]^{1/2} \qquad (6\text{-}35)$$

where F is defined as before.

From calculations of μ_s and σ_s, estimates of the proportion of circuits/assemblies, P_s, which will be within requirement limits initially after fabrication can be made (computed using the general procedures used earlier in Section 6.6.3.1 for both linear and nonlinear functions).

For the sake of argument, assume that units have been fabricated and without any tests, measurements, or changes are operated for a period of time in a given environment. Further assume that parts can change in value during periods of operation and that the change in value can be represented by a shift in the mean of each part population as illustrated in (6-34). Assume for simplicity that the standard deviation will not be affected by the shift.

From calculations of μ_{cs} and σ_s, the proportion of circuits/assemblies, P_{cs} (again computed using the general procedures discussed earlier in Section 6.6.3.1), which will be within requirement limits after a given period of operating time can be estimated. Then

$P_s - P_{cs} =$ approximately the proportion of units produced that will drift outside performance limits due to test/operation over a given period of time in a given environment (assumes only a very small number of units originally outside specification limits would drift back inside such limits during the operating time interval).

If it is assumed that those assemblies that are not initially within limits are reworked immediately after fabrication and have no better or no worse performance after rework than those assemblies that initially do not require rework, then:

$P_s - P_{cs} =$ approximately the proportion of the assembly population that will fail due to drift/degradation over a period of time T

And

$R_c(T) = 1 - (P_s - P_{cs}) =$ the proportion of units that will go through an interval of operation, T, without experiencing a drift/degradation out of tolerance

For example, take the illustration in Fig. 6-4. Assume that each resistor incurs a positive change in mean value μ_i equivalent to $\Delta\mu_i = +0.05\mu_i$, over 1000 hours of operation in the field such that

$$\mu_a = 105, \qquad \sigma_a = 10$$
$$\mu_b = 210, \qquad \sigma_b = 25$$
$$\mu_c = 52.5, \qquad \sigma_c = 6$$
$$\mu_d = 315, \qquad \sigma_d = 20$$
$$\mu_s = 650, \sigma_s = 34, \qquad \text{as from prior calculations}$$
$$\mu_{cs} = 685.5, \sigma_s = 34$$

Recall the performance requirement $615 \leqslant R_s \leqslant 685$ on R_s. Assuming, as before, that prior to the operational period of 1000 hours R_s is normally distributed with a mean μ_s and standard deviation σ_s, using calculations performed earlier in this section, we know that 70% of the products initially fabricated will be within these limits. Hence, $P_s = 0.70$.

After 1000 hours of operation, the distribution R_s is now normally distributed with a mean μ_{cs} and standard deviation σ_s, where $\mu_{cs} = 682.5$ and $\sigma_s = 34$. A general means to determine P_s and P_{cs} follows.

The proportion of circuits having resistance values R_s between 615 and 685 ohms, $P(615 \leqslant R_s \leqslant 685)$, can be determined, using the principles discussed in Section 2.2.3. Let $R_1 =$ the lower value of R_s, $R_2 =$ the higher value of R_s. Let $\phi(x) =$ the normal probability density function. Then

$$P(R_1 \leqslant R_s \leqslant R_2) = \int_{-\infty}^{K_2} \phi(x)\, dx - \int_{-\infty}^{K_1} \phi(x)\, dx \qquad (6\text{-}36)$$

where

$$K_1 = \frac{R_1 - \mu}{\sigma}$$

$$K_2 = \frac{R_2 - \mu}{\sigma}$$

To evaluate P_s we take $R_1 = 615$, $R_2 = 685$, $\mu = \mu_s = 650$, $\sigma = \sigma_s = 34$. Inserting these values into (6-36) and using a table of the normal distribution,

$$P_s = P(615 \leqslant R_s \leqslant 685) \approx 0.70$$

as we calculated earlier. To evaluate P_{cs} we take $R_1 = 615$, $R_2 = 685$, $\mu = \mu_{cs} = 682.5$, $\sigma_s = 34$. Inserting these values into (6-36) and using a table of the normal distribution,

$$P_{sc} = P(615 \leqslant R_s \leqslant 685) \approx 0.55$$

$$P_s - P_{cs} = 0.70 - 0.55 = 0.15$$

Hence, it can be expected that approximately 15% of the circuits/assemblies produced will incur failures in the field due to part parameter drift.

Bear in mind that our simple example was of a simple resistor circuit with a single transfer function. A typical circuit/assembly is much more complex and may well have associated with it several performance requirements which would have to be treated as a group. In that case, a failure would occur when any one requirement parameter went beyond its limits.

6.7   THE PARTS PROGRAM: THINGS A RELIABILITY ENGINEERING MANAGER SHOULD KNOW

The preceding sections covered primarily the technical issues and concerns of establishing and implementing a parts program. Here the general management/administrative concerns and considerations associated with a parts program will be discussed. In particular, factors involved in meeting the defect, reliability, and tolerance levels discussed previously will be covered. For this purpose we will arbitrarily break down such considerations into two areas—part selection and part acquisition.

6.7.1   Part Selection Criteria

The first step involves determination of what parts are needed to meet *design* needs. Factors which must be considered as a minimum include performance desired, environmental stresses anticipated, and reliability. This task should encompass the evaluation of all parts capable of meeting the need and the basic suitability of each part of the design/environment application.

Once part candidates are chosen, factors associated with cost/risk/experience judgment come into play. Factors which will be considered include the existence and composition of a *preferred parts list*. Such lists are generally particular to a given development, established to foster standardization relative to the use of part types, to influence the use of types of parts which are less costly than their equivalents, and to foster use of parts which provided good performance previously. The fact that a part is included or not included in a preferred parts list does not always bear relationship to its reliability.

Other factors which must also be considered are the cost of the part, its availability in the quantities required with respect to program schedule, long-term availability of the part, and whether the part is available from a single or from multiple vendors. If part performance requirements (e.g., operating characteristics, environmental stress needs) can be best met by a "new" part or by enhancement of an already existing one, the risk associated with meeting not only performance and quality but also cost and schedule objectives must also be considered. For each part chosen, electrical, thermal, and environmental stresses applied as a consequence of design strategy must be related to reliability effects. A part might meet its performance objective more than adequately, but might require operation of the part close to or beyond its stress limit to do so. This would decrease the reliability of the part. Application of nonstandard parts is generally considered a last resort, as these generally have associated with them ambiguity with respect to reliability, tolerances, and quality in general; the potential need for the development of special drawings and particular specifications; the necessity for the application of additional testing to insure a match with environmental requirements; and the potential for complications in long-term logistics needs.

Selection of vendors is the next step. This usually starts with a vendor survey. The magnitude and depth of the survey required for achievement of program objectives depends on the character of the program and its part needs. It can include a formal vendor survey encompassing a general review of vendors' claims, past and current, and the degree to which such claims have merit; and review of past experiences with the vendor. More specifically, it should relate to the vendor's ability to provide the part needed to the performance, tolerance, and reliability levels needed; whether or not existing processes and controls already in place can provide those attributes desired; whether or not additional tests, inspections, and process controls will be required; the risk associated with meeting objectives with such steps; and the costs.

When it is planned to enhance the reliability of an existing part by burn-in other test/screen procedures, such actions are capable of changing the values of part parameters. As a result, the performance characteristics of the circuit/assembly design may change.

For program-critical parts, proof should be available that the manufacturing process is mature; that an adequate process qualification was carried out; that standards, controls and inspection procedures have been established; and that necessary levels of tests have been included (thermal, vibration and shock, humidity, dust, salt, etc.) which are consistent with the application. For these, information as to whether or not the part was qualified to Mil-M-38510, Mil-Std-883, or other standards will provide the information needed. There may be instances where a new part (or a nonstandard part) is required or enhancements to an existing part are required to meet environmental, performance, reliability, or tolerance needs. Then a plan covering qualification procedures, screening, burn-in, environmental tests, and identification of failure modes and mechanisms should be established.

6.7.2   Quality Differences Between Mil-Std and Non-Mil-Std Parts

The differences between parts procured through military standards and those procured through commercial specifications were wide at one time. Recent studies show that a difference still exists with respect to variation in electrical and mechanical characteristics and appearance attributes between Mil-Std standard and commercial parts, but the difference is shrinking. The DESC (Defense Electronics Supply Center, Dayton, Ohio) testing program,

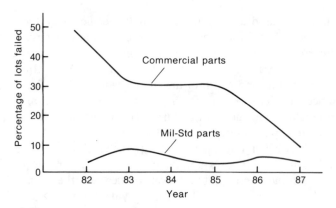

FIGURE 6-8.   Observed quality trends. [From Reference 16.]

established in the early 1980s to detect quality and workmanship problems in parts, published results depicted in Fig. 6.8 in 1989 (see Ref. 16).

The data represented, in large part, the results of tests of semiconductors. As can be seen, a significant improvement occurred between 1982 and 1987 in the quality levels of non-Mil-Std parts. Whether this can be attributed to a general maturation of the part production process, to adaptation or application of certain Mil-Std processes/procedures to commercial parts manufacture, or to a reaction to the DESC testing program cannot be ascertained. While data available is not sufficient to draw inferences about the quality of all possible part types, it does identify possibilities that should not be rejected outright. At any rate, experience data on the particular types that might be included in a design is always of value in making part decisions.

6.7.3   Part Acquisition/Procurement

After selection of the parts and potential part vendors, the requirements necessary for performance, tests, and measurements to be applied to each part type must be established and translated into procurement/specification documents. These include a part drawing in the case of nonstandard or new parts as well as specific information concerning the qualities of the part in question. While a part specification prepared by the vendor might already exist and apparently fit the need, unless the developer/designer is familiar with the part the specification may be ambiguous, contain omissions, or be incomplete. The specification should identify tests and screens to be performed, as well as define the performance, tolerance, and quality requirements needed.

If any specific qualification/control tests or screens are needed, for example to insure that the value of a circuit critical parameter is in a narrow range, they too should be included in the part specification.

The procurement should make provisions, to the degree needed, for vendor and process monitoring, inspections, and timely notice of changes in part manufacturing process or material. In the event of product change, an advance sample of parts should be provided to the developer for test. Results of tests performed by the vendor on the new parts should also be provided.

Beware statements that the parts delivered will come from XYZ lines. You may get the rejects from such lines unless additional safeguards are included.

References

1. Codier, E. (1971). "Reliability in the Third Generation." Proceedings, 1971 Annual Symposium on Reliability, IEEE, Piscataway, NJ.
2. Fiorentino, E. (1986). *RADC Guide to Environmental Screening*, RADC-TR-138. Available from the Defense Documentation Center, Report AD-A174333.
3. Johnson, N. L. and Leone, F. C. (1964). *Statistics and Experimental Design*. Wiley, New York.
4. Hald, A. (1952). *Statistical Theory With Engineering Applications*. Wiley, New York.
5. Dreste, F. (1960). Circuit Design Concepts for High Reliability." Proceedings, 6th National Symposium on Reliability and Quality Control, IEEE, Piscataway, NJ.
6. Hahn, G. and Shapiro, S. (1967). *Statistical Models in Engineering*. Wiley, New York.
7. Hellerman, L. and Racite, M. (1958). "Reliability Techniques for Electronic Circuit Design." *IRE Transactions on Reliability and Quality Control*, September.
8. Morrealle, M. et al. (1958). "Propagation of Error Technique As Applied to Electronic Circuit Design." Proceedings, 6th National Symposium on Reliability and Quality Control, IEEE, Piscataway, NJ.
9. Severo, N. et al. (1963). *Asymptotic Distribution of Functions of a Normal Random Variable*. Aeronautical Research Laboratories U.S.A.F. Report ARL 62-468.
10. DeZur, R. and Donahu, J. (1965). *On Products and Quotients of Random Variables*. Aerospace Research Laboratories U.S.A.F. Report ARL 65-71.
11. Sylvania Electric Products (1961). *Mathematical Simulation for Reliability Prediction*. RADC Report RADC-TR-61-299.
12. Autonetics (1961). *Description and Comparison of Computer Methods of Circuit Analysis*. Autonetics Report EM-6839.
13. Ashcraft, W. and Hochwald, W. (1961). "Design By Worst Case Analysis: A Systematic Method to Approach Specified Reliability Requirements." *IRE Transactions on Reliability and Quality Control*, November.
14. Burns, R. and Lawson, A. (1964). "Quantized Probability Circuit Design Principles Applied to Linear Circuits." *IEEE Transactions on Reliability*, June.
15. Suran, J. (1961). "Effect of Circuit Design on System Reliability." *IRE Transactions on Reliability and Quality Control*, March.
16. Robinson, G. and McNicholl, P. (1989). "DESC Testing Program Is a Frontline Defense for Reliability." Proceedings, 1989 Annual Reliability and Maintainability Symposium, IEEE, Piscataway, NJ.

7

Providing a Basis for Detailed Design. The Reliability Block Diagram and Reliability Apportionment

In the preceeding chapter we discussed the impacts of the parts program and proper design management of part tolerances which affected reliability, part quality, and other design development activities. In this chapter we will discuss those uniquely reliability-oriented tasks which must be performed to provide a foundation for the detailed reliability design/development activities that follow. These include:

1. Establishing a structure for describing system reliability in terms of its component reliabilities, their functions, and relationships. This serves as a management aid which forms the basis for communication within the developer's organization and between the developer and the customer. It serves also as a technical tool which forms a common starting point for the reliability design tasks which must follow. It takes the form of a reliability block diagram.
2. Establishing or allocating a reliability target for each component of the system to provide engineers and designers associated with the design/ development of each component with unambiguous reliability objectives. The allocation also serves as a management aid during the overall program development process. Periodic reports with respect to progress in meeting, exceeding, or not meeting targets provide managers with information needed to determine what areas need more attention and where adjustments to targets can be made most cost-effectively.

7.1 THE RELIABILITY BLOCK DIAGRAM: THE STARTING POINT OF MANY RELIABILITY ACTIVITIES

Reliability allocation, prediction, evaluation, and failure modes and effects analysis (FMEA) all use as a point of origin the idea of a reliability block diagram. The block diagrams take two basic forms. One is descriptive of the *dependency* of required system/equipment level outputs on the satisfactory operation of the system/equipment components. A component can be a subsystem, equipment, assembly, module, or printed circuit board depending on what level of indenture and degree of detail is depicted by the block diagram. Block diagrams of this type are commonly used in the performance of reliability prediction, allocation, and evaluation tasks.

The other form is descriptive of the *interdependency* of the inputs and outputs of system/equipment components and their relationship to necessary system/equipment level outputs. This type of diagram is most commonly used for the purpose of FMEA, maintainability, and testability analyses.

In either instance the system/equipment is broken down into blocks representing a function or set of functions associated with a given hardware component (subsystem, equipment, module, etc.). The block diagram can be made to represent any level of system or equipment indenture. It can represent a system in terms of its subsystems, in terms of all the individual equipment making up each subsystem, or in terms of the lower-level components making up each equipment. Both types of diagrams serve to provide the engineer with the ability to visualize the system/equipment in other than strict mathematical or reliability accounting terms and provide insights not otherwise possible.

The former type of diagram assigns a block to represent each component which must operate satisfactorily in order to meet system/equipment performance requirements. For example, Fig. 7-1a represents a simple radar system in terms of its major component makeup. In order for the radar system to perform its function, the input must be correctly processed through the antenna and receiver/transmitter. The tracking computer must operate on the data satisfactorily and the radar/controller display must acquire an accurate representation of the information received for relay to a communications system. If there is any break in the flow of information, or if erroneous information or signals are introduced due to a failure in any component, then the system can no longer perform its function and is considered failed. Since any failure in any one component can cause the system to fail, the system can be considered to be in simple series.

In the event that one component fails and another is able to perform its function such that at least the bare minimum of performance can be

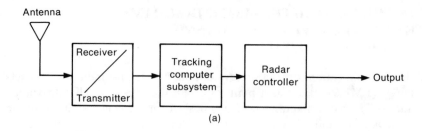

(a)

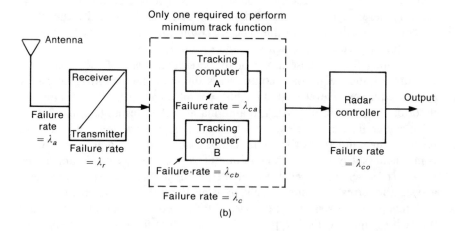

(b)

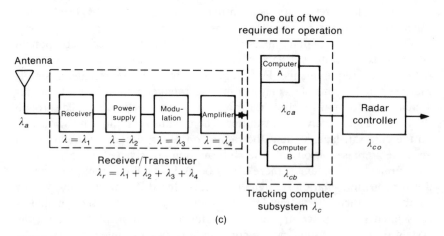

(c)

FIGURE 7-1. (a) Simplified reliability block diagram of a radar system. (b) Simplified reliability block diagrams of a radar system showing redundancy effects. (c) Block diagram of a radar system depicting various levels of indenture.

maintained, a redundant rather than simple series system results. A redundant system can be depicted by placing blocks in a parallel configuration and providing notations for the form and the nature of the redundancy. For example, assume that for our radar system we have two separate computers to perform our tracking function. Two are necessary to handle peak loads, but one computer can perform the function adequately but less efficiently, if required. Figure 7-1b exemplifies that situation.

Information is associated with each block, defining its failure rate or other measure of reliability. Note that for some systems contained in aircraft, aerospace vehicles, or power plants, a given component can exhibit different failure rates for different phases of the mission or use, depending on the stresses on the component or its mode of operation at the time. In that case, a different failure rate must be associated with each component mission/use phase and the proportion of time in that phase must be identified. This will be discussed in more detail in Section 8.5.

Any component can be broken down to lower indenture level components. For example, assume that the transmitter is composed of distinct power supply, modulation, and amplifier functions which can be treated separately in design. Figure 7-1b can then be represented in the form of Fig. 7-1c.

For the type of block diagram used for FMEA, maintainability, and testability analyses, we take into account not only the functions to be performed but the nature of the inputs, outputs, and feedback as well. This is shown in the system-level block diagram in Fig. 7-2, which provides insight as to how a failure in one major component can affect the operation of other components. For example, in a traffic control system a failure in the radar control component can send an unintended control message to the transmitter component which could cause a threat to safety.

This type of block diagram and its generation makes the engineer more aware of the existence of such interdependencies, serves as a basis for their examination, and makes errors of omission less likely. For more information on block diagrams, reference is made to Mil-Std-756 discussed earlier. As

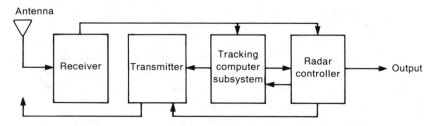

FIGURE 7-2.   A system-level block diagram showing interdependencies of function.

this chapter proceeds, additional detail and applications of block diagrams will be provided.

7.2 RELIABILITY APPORTIONMENT/ ALLOCATION IN THE DEVELOPMENT PROCESS

We have used the words reliability apportionment and allocation more than once so far. You have, in all likelihood, heard the terms in your workplace. What exactly is reliability apportionment and allocation? When is it applied? What forms does it take?

7.2.1 What Reliability Apportionment and Allocation Is

In Section 4.4.1.1 we discussed the establishment of system *reliability design targets* which represented an achievable reliability objective for the system. In Chapters 3 and 4 we discussed reliability keyed to the breakdown of the system to its subsystems and lower-level components. These are all functions of the reliability apportionment process which provides reliability objective values for each system component block and which, if *all* realized, will achieve the system reliability target.

The reliability apportionment process is the means through which reliability requirements are identified and communicated to each design organization having responsibility for the development of a particular subsystem or subsystem component. (In the case of subcontractors, it is the means through which reliability requirements are generated.)

There exist two separate philosophies for reliability apportionment—top-down and bottom-up. *Top-down reliability* apportionment involves a start from the system-level target. Reliability objectives for each subsystem block are established such that, if all objectives for all blocks are achieved, the system level target will be achieved. Next, the reliability objective for each subsystem block is treated as if it were the reliability target, and reliability objectives are generated for each subsystem component. The process is repeated from indenture level to indenture level until the lowest indenture level block has associated with it a reliability objective. The top-down approach is the procedure of necessity in any development phase when the reliability ramifications and detail of *all* possible block level designs are known. Even when such information is not available, it is generally the *first procedure applied to provide first-cut estimates* at apportionment in the early design development phases. When used to provide initial estimates, as design progresses, adjustments to the apportionment may be desirable or even required to meet system needs in the most cost-effective way possible.

Bottom-up reliability apportionment involves the exploitation of specific detail design concepts; specific type(s) of parts, components, hardware/ software combinations, etc., that will provide performance, cost, power, weight, etc., advantages which are desirable or necessary to meet system needs. As a consequence, apportionment starts with a "given" associated with a lower block level design and its associated reliability. It proceeds upward, structuring an overall design (and reliability allocation) which is both consistent with the "given" and meets system objectives.

Most apportionment approaches are hybrids of the two. The top-down approach providing first-cut visibility and guidance, modified later as specific detail design concepts and part types to be used are identified. At that time, a final apportionment is made (usually at the start of full-scale engineering development, see Chapter 4) that takes into account detailed design and part use concepts.

7.2.2   Reliability Apportionment: The First Cut

When reliability apportionment is first applied, its objective is usually to size the problem and furnish estimates of general reliability attainable for each major subsystem/equipment. It is performed using the results of the "concept" reliability prediction, which will be discussed in a later section. The purpose of this apportionment is to provide guidance concerning the reliability necessary and achievable for each subsystem and, in some cases, equipment making up the system.

The apportionment is developed as an input to the general reliability planning effort aimed at defining the overall reliability status of the system. Preliminary allocations are established for the highest-level blocks of the system.

One of the most fundamental types of an apportionment that can be performed in this early phase of development takes a "relative" form. It is based on information available that provides at least *relative* differences among failure rates for different major components, such as can be discerned from Chart 3-1. It is usually used at a time when no data are available on components similar to the ones to be developed. An illustration of this basic apportionment process follows.

Suppose, for example, that from previous development programs or from the literature we can find information on the reliability of radar receivers, transmitters, trackers, and controllers. We find from that information, and from experience in general, that: (a) the ratio of receiver/transmitter failure rate, λ_r, to track computer failure rate, λ_c, is k_r; (b) the ratio of controller failure rate, λ_{co}, to track computer failure rate, λ_c, is k_{co}.

- The failure rate of the receiver/transmitter can be represented as $\lambda_r = k_r \lambda_c$.
- The failure rate of the track computer can be represented as $\lambda_c = 1\lambda_c$.
- The failure rate of the controller can be represented as $\lambda_{co} = k_{co} \lambda_c$.
- The failure rate of the entire system (λ_t) can be represented as

$$\lambda_t = k_r \lambda_c + \lambda_c + k_{co} \lambda_c = (k_r + k_{co} + 1)\lambda_c$$

- The proportion of system failure rate represented by the receiver/transmitter is

$$\frac{k_r \lambda_c}{(k_r + k_{co} + 1)\lambda_c} = \frac{k_r}{k_r + k_{co} + 1}$$

- The proportion of system failure rate represented by the track computer is

$$\frac{\lambda_c}{(k_r + k_{co} + 1)\lambda_c} = \frac{1}{k_r + k_{co} + 1}$$

- The proportion of system failure rate represented by the controller is

$$\frac{k_{co} \lambda_c}{(k_r + k_{co} + 1)\lambda_c} = \frac{k_{co}}{k_r + k_{co} + 1}$$

In general, the proportion of system failure rate represented by any component is equal to

$$\frac{k_j}{\sum\limits_{j=1}^{n} k_j} \tag{7-1}$$

where $k_j = 1$ for the component whose failures rate was used as a base (the track computer in the example).

Hence, from the target failure rate associated with the system, a preliminary allocation to each subsystem can be made by first assuming a series-connected system and determining the failure rates required from each subsystem in order to meet the system target. Information and judgments with respect to impacts of new technology and application of reliability multipliers (see Section 4.2.5) are brought to bear and the first allocation targets are generated.

7.2.3   Reliability Apportionment with Respect to Criticality

In many instances it is possible to improve the reliability of a number of different components. Not all components are as important or critical to system operation as others. Moreover, a system which is comprised of many different components can achieve its target reliability in many different ways. A reliability apportionment process can be applied which takes into account the criticality associated with each component or subsystem. Large, complex systems are usually made up of many independent subsystems and components usually developed by different sources. Not all subsystems are equally critical to mission success or to safety. Take as a simple example your family car. Evey trip to your dealer to repair or correct a failure is a mark against the car's reliability. Not all malfunctions are equivalent in seriousness, however. A failure in your instrumentation is not as critical as failure of an engine, transmission or brakes; a failure in your brake subsystem may be potentially more serious than an engine failure with respect to safety, but less serious with respect to cost. If you knew your car was going to breakdown n times a year, the nature of those breakdowns would be at least as important to you as the magnitude of the number.

For those same reasons, reliability apportionment with respect to criticality must be applied to complex systems developed. To this end, weighting or importance factors must be developed for each subsystem. The development of such factors is at this time more of an art than a science. Customer representatives and other experts may be surveyed; committees of engineers, technicians, operators, and logisticians experienced with systems of the type to be developed may be formed; the statement of need and other customer-related documentation (described in Chapter 3) should be further reviewed to gain insights missed previously. Delphi procedures (defined overly simply as decisions which have 100% accord among a pool of experts, reached after many iterations and discussions) may be applied.

Various criticality weighting schemes both simple and complex may be employed. Some examples follow.

If a subsystem or component provides a single output or function, rank its criticality on a scale between 0 and 1 based on factors established, where a "1" indicates the highest degree of criticality and a "0" the lowest.

If a subsystem or component provides more than one type of output, estimate the relative frequency of failure of each output of the item from the failure rates of the subcomponents used to develop such outputs. For example, a given component provides three different outputs A, B, and C. Based on the types of hardware in the component which produce each output and the quantities of the component's hardware used to produce each, the conclusion can be reached that, given a failure in the component,

$p_1\%$ of the time just A will be affected
$p_2\%$ of the time just B will be affected
$p_3\%$ of the time just C will be affected
$p_4\%$ of the time both A and B will be affected
$p_5\%$ of the time both A and C will be affected
$p_6\%$ of the time both B and C will be affected
$p_7\%$ of the time A, B, and C will be affected

If an output failure (or combination of output failures) will cause a definite threat to mission success or safety, allocate it a weight of 1. If the output failure (or combination) will cause less an effect than that above, give it a value of $0 < c < 1$ depending on the criterion developed for ranking. Applying that criterion, suppose values of p_i and c_i result for our component as in Chart 7-1. Note that if either output A or C is affected, a definite threat to mission success or safety is present. As a result any output failure affecting either or both is critical.

Multiplying each p_i and c_i and summing the result over all outputs, an *average* criticality weight (W_j) results for a component j:

$$W_j = \sum_{i=1}^{7} p_i c_i \qquad (7\text{-}2)$$

where for this case $W_j = 0.98$. W_j can be defined at the system, subsystem, or lower component levels, depending on the level of application of the apportionment and the amount of information available.

The criticality or (cost) of a system failure can be expressed in terms of

---

CHART 7-1 Weighting Components for Definition of Criticality

p	c	pc
$p_1 = 0.10$	$c_1 = 1$	0.10
$p_2 = 0.05$	$c_2 = 0.6$	0.03
$p_3 = 0.10$	$c_3 = 1$	0.10
$p_4 = 0.15$	$c_4 = 1$	0.15
$p_5 = 0.20$	$c_5 = 1$	0.20
$p_6 = 0.15$	$c_6 = 1$	0.15
$p_7 = 0.25$	$c_7 = 1$	0.25
		0.98

average criticality per failure, W_{rf},

$$W_{rf} = \sum_{j=1}^{n} \frac{\lambda_j}{\lambda_{ct}} W_j \qquad (7\text{-}3)$$

where

λ_j = contemplated failure rate of the jth component
λ_{ct} = contemplated failure rate of the system, where $\lambda_{ct} \leqslant \lambda_t$
λ_t = the target failure rate of the system

[Note that it is possible that after design starts it is found that a target originally set for a component was conservative. λ_{ct} makes provision for that contingency. If, on the other hand, it is found that a target value was optimistic, the target values are readjusted.]

λ_j/λ_{ct} = the proportion of contemplated system failures attributed to component j
W_j = the criticality associated with the failure of component j

$$\sum_{j=1}^{n} \frac{\lambda_j}{\lambda_{ct}} = 1$$

and W_{rf} takes the form of an expected value described in Chapter 5, i.e., the average criticality per failure.

For any given value of λ_{ct}, the optimum apportionment is one that minimizes W_{rf}. In practical terms it means that:

1. All other things equal, if we have the choice of increasing the reliability (decreasing the failure rate) of either a component with a high level of criticality or a component with a lower level of criticality, we select the component with the higher criticality index.
2. If we decrease the failure rate of a component with a high criticality index, at a given cost, we can consider making cost-effective reductions to the failure rate of a component with a lower criticality index (say, by changing quality of parts) provided $\lambda_{ct} \leqslant \lambda_t$.

Note that if failure rate were proportionately reduced for all components, W_{rf} would not be affected. That is because it is a measure of criticality of the average failure, not a measure sensitive to the total number of failures.

Take, for example, a complex vehicle comprised of a radar system, a communication system, and a navigation system:

W_r = weight of criticality of the radar system = 0.8

W_c = weight of criticality of the communication system = 0.7

W_n = weight of criticality of the navigation system = 0.98

$$W_{rf} = (\lambda_r/\lambda_{ct})(0.8) + (\lambda_c/\lambda_{ct})(0.7) + (\lambda_n/\lambda_{ct})(0.98)$$

Given that alternative choices of design/hardware are available for each system, we choose the values of λ_r, λ_c, λ_{co} that will minimize the value of W_{rf} subject to the constraint

$$\lambda_{ct} = \lambda_r + \lambda_c + \lambda_{co} \leqslant \lambda_t$$

where λ_t = the reliability target for the system.

We have a reliability target for the vehicle expressed in terms of failure rate λ_t = 0.00370 (MTBF = 270 hours). Our initial design efforts have come up with the conclusions that:

1. Failure rates of λ_{r1} = 0.002, and λ_{r2} = 0.0013, are achievable for the radar system at a cost of 500 and 700 cost units, respectively.
2. Failure rates of λ_{c1} = 0.001 and λ_{c2} = 0.00067 are achievable for the communications system at a cost of 200 and 300 cost units, respectively.
3. Failure rates of λ_{n1} = 0.001 and λ_{n2} = 0.00067 are achievable for the navigation system at a cost of 150 and 250 cost units, respectively.

Chart 7-2 indicates that the average criticality per failure is a minimum when λ_{r1}, λ_{c2}, and λ_{n1} design approaches are taken. Note that other combinations meet the failure rate target at the same cost, but all have higher values of W_{rf}. No other will yield $W_{rf} \leqslant 0.74$ and $\lambda_T \leqslant \lambda_t$ at lesser cost.

CHART 7-2 Apportionment by Criticality Breakdown and Analysis

Combinations	λ_t	W_{rf}	Cost
$\lambda_{r1}, \lambda_{c1}, \lambda_{n1}$	0.00400	0.82	850
$\lambda_{r1}, \lambda_{c1}, \lambda_{n2}$	0.00367	0.81	950
$\lambda_{r1}, \lambda_{c2}, \lambda_{n1}$	0.00367	0.74	950
$\lambda_{r1}, \lambda_{c2}, \lambda_{n2}$	0.00334	0.82	950
$\lambda_{r2}, \lambda_{c1}, \lambda_{n1}$	0.00330	0.82	1050
$\lambda_{r2}, \lambda_{c1}, \lambda_{n2}$	0.00297	0.81	1150
$\lambda_{r2}, \lambda_{c2}, \lambda_{n1}$	0.00297	0.84	1150
$\lambda_{r2}, \lambda_{c2}, \lambda_{n2}$	0.00264	0.82	1250

7.2.4   Reliability Control, A Key Element in Apportionment

Reliability apportionment implies an ability to control the reliability attainable by a given component. Let us consider some means that can be applied to effect control.

1. *Introduction of fault tolerance.* If reliability is a critical concern for the program, then fault tolerance will be a candidate for application as a means to improve reliability. In many system development programs, however, reliability is not as important to designers and managers as other performance measures. Furthermore, fault tolerance levies penalties in terms of weight, volume, power, and acquisition cost. In many situations, however, the basic design approach chosen lends itself to a type of fault-tolerant design with a minimum of additional complexity. Examples of these include diversity techniques in communication systems; gangs of amplifiers where the primary function of a system could continue even if one or more amplifiers failed; two or more computers necessary to cover peak periods of activity where only one would be necessary to provide for minimum needs; or four engines on a transport where only two are necessary to maintain flight. The reliability engineer should be sensitive to such opportunities as a means of reliability control, especially when the payoff is worthwhile in terms of improving the reliability of a component or in relieving the pressure to improve the reliability of another component.

2. *Reduction of system complexity.* There are basically three ways to reduce the complexity of a system: (1) design it in a simpler way with no frills; (2) apply advanced/new technology that can reduce the number of piece parts necessary for the function to be performed; (3) integrate and share system components and functions. The first is a fundamental of reliability design and provides payoffs by reducing unneeded functions. It can, however, sometimes also remove portions of the design needed for component safety and longevity, for example, surge protection circuits in power supplies. Hence, reduction of system complexity by "deletion" should be carefully analyzed prior to its acceptance as a design alternative. The last two means of reducing complexity should always be considered, as these can almost always provide for enhanced performance and reduced power, weight, and size characteristics.

3. *Proper comprehension of the mission and maintenance concept and their bounds.* As indicated in Section 4.2.4, maintenance concepts and mission maintainability allowances can be used as reliability multipliers. (The effects of maintainability and maintenance on reliability are covered in more detail in Chapters 9 and 10.) Reliability enhancements at the system, subsystem, and equipment levels are possible with little effect on design/acquisition cost:

(1) if a relatively small amount of downtime during a given mission time is allowable and corrective maintenance can be performed during the mission; (2) if for systems operating in a standby mode of operation subject to mission calls at random points in time, the system is subjected to periodic performance checks to detect failures and the capability for corrective action exists; (3) if scheduled maintenance is provided for a fault-tolerant system; (4) if corrective maintenance may be performed on fault-tolerant systems while they are in operation.

4. *Application of higher-reliability parts.* As we have discussed in Chapter 6, the use of higher-quality parts can provide payoffs with respect to reducing failures caused by latent defects and drift failures due to part tolerances. The effects of using higher-quality parts are represented by the failure rate quality factors in Mil-Std-217.

7.2.5 Reliability Apportionment with Respect to
Identified Engineering Alternatives

The definitive aspects of apportionment can only be truly established when general estimates of the complexity of the system components can be made; when engineering approaches are being selected; when design constraints and requirements have been fully identified by the customer; when design concept alternatives for each major system component have been established; and when reliability design alternatives such as those discussed in the previous section have been considered.

At that time, lower-level detail designs of the printed circuit board probably have not yet been performed and reliability prediction information is not available. Estimates of reliability based on empirical information, on similar components, and on engineering judgment (see Section 8.3) are used to provide estimates of component reliability.

Associated with each design alternative is a resulting estimated reliability, a set of performance characteristics, and, in some cases, a set of costs for each component in terms of power, weight, volume, acquisition, and life-cycle costs, etc., that will result as a consequence of a particular choice. Reliability apportionment for a system means choosing the combination of design alternatives which will meet or exceed reliability and other system performance characteristics while at the same time keeping particular costs below given constraints and/or minimizing others.

Take as an example the sample system depicted in Fig. 7-1 and structure it as in Fig. 7-3. Suppose that the reliability target for the radar system is taken to be MTTF of 800 hours. For convenience we will assume that failure rate will be approximately constant over service life so that our reliability

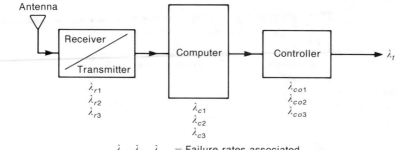

$\lambda_{ri}, \lambda_{ci}, \lambda_{coi}$ = Failure rates associated
with the ith design alternative

FIGURE 7-3.   Apportionment diagram for a radar system showing design alternatives.

CHART 7-3 Failure Rate and Cost Units Associated with Choice of a
Given Design Alternative

Block Design Alternative	Failure Rate	Development Product Costs	LCC	PWR	WT	Volume
R1	0.0090	7	35	40	100	3
R2	0.00080	8	42	42	120	3.2
R3	0.00070	10	44	50	110	2.8
C1	0.00030	3	16	15	70	2
C2	0.00040	2.5	11	10	60	2.5
C3	0.00020	4	8	8	50	1.7
CO1	0.00020	1	5	10	60	2
CO2	0.00010	1.5	6	5	50	1.6
CO3	0.00015	1	5	5	70	2.5

target translates to a failure rate $\lambda_t = 1/800 = 0.00125$. Assume that for each major component depicted there are three design alternatives possible, all of which meet performance requirements for each component but have reliability and cost consequences as shown in Chart 7-3. Assume that we have:

1. A weight (WT) constraint of WT $\leqslant$ 240 units
2. A power (PWR) constraint of PWR $\leqslant$ 65 units
3. A volume (Vol) constraint of Vol $\leqslant$ 7.2 units

CHART 7-4 System Failure Rate and Total System Cost Units Resulting from Choice of a Given Set of Alternatives

Design Combination	Total Failure Rate	Development Product Costs	LCC	PWR	WT	Volume
R1, C1, CO1	0.00140	11.0	56	65	230	7
R1, C1, CO2	0.00130	11.5	57	60	220	6.6
R1, C1, CO3	0.00135	11.0	56	60	240	7.5
R1, C2, CO1	0.00150	10.5	51	60	220	7.5
R1, C2, CO2	0.00140	11	52	55	210	7.1
R1, C2, CO3	0.00145	10.5	51	55	230	8
R1, C3, CO1	0.00130	12	48	58	210	6.7
R1, C3, CO2	0.00120	12.5	49	53	200	6.3
R1, C3, CO3	0.00125	12	48	53	220	7.2
R2, C1, CO1	0.00130	12	63	67	250	7.2
R2, C1, CO2	0.00120	12.5	64	62	240	6.8
R2, C1, CO3	0.00125	12	63	62	260	7.7
R2, C2, CO1	0.00130	11.5	58	62	240	7.7
R2, C2, CO2	0.00130	12	59	57	230	7.3
R2, C2, CO3	0.00135	11.5	58	57	250	8.2
R2, C3, CO1	0.00120	13	55	60	230	6.9
R2, C3, CO2	0.00110	13.5	56	55	220	6.5
R2, C3, CO3	0.00115	13	55	55	240	7.4
R3, C1, CO1	0.00120	14	65	75	240	6.8
R3, C1, CO2	0.00110	14.5	66	70	230	6.4
R3, C1, CO3	0.00115	14	65	70	250	7.3
R3, C2, CO1	0.00130	13.5	60	70	230	7.3
R3, C2, CO2	0.00120	14	61	65	220	6.9
R3, C2, CO3	0.00125	13.5	60	65	240	7.8
R3, C3, CO1	0.00110	15	57	68	220	6.5
R3, C3, CO2	0.00100	15.5	58	63	210	6.1
R3, C3, CO3	0.00105	15	57	63	230	7

and we want to apportion component reliability targets so that the system reliability target is realized and both developer production cost and user life-cycle cost are minimized.

Chart 7-4 represents the system failure rate λ_t and the total system costs which result from all combinations of design alternatives possible. As can be seen, the various constraints rule out various combinations of design alternatives immediately. There are other combinations which may meet such constraints but are not consistent with our reliability target. We are left with eight possible design alternatives which meet our criteria: R1, C3, CO2; R1, C3, CO3; R2, C1, CO2; R2, C3, CO1; R2, C3, CO2; R3, C2, CO2; R3, C3, CO2; and R3, C3, CO3. Of those eight, the one that provides the lowest development production cost and the lowest life-cycle cost is R1, C3, CO3. Hence, the design alternatives that yield a failure rate for the receiver transmitter of 0.00090, a failure rate for the computer tracker of 0.00020, and a failure rate for the controller of 0.00015 would be selected.

As can be seen, cost-effective reliability apportionment can be performed iteratively, as in the example just illustrated. While it can be done manually, as shown for a relatively simple example, for a more complex system usually the use of a computer is required to perform the required calculations efficiently.

Iterative methods are among the most simple to apply to apportionment tasks. There are, however, other, more analytical procedures which may be applied as well. They may be applied in almost all cases where the costs or payoffs associated with a given value of reliability are expressible in terms of a function of reliability or any other system parameter. An example is life-cycle cost, which can be expressed as a function of the reliability of a component. We will discuss such approaches next.

7.2.6 Reliability Cost/Payoff Approach to Apportionment

The apportionment technique is capable of defining the most cost-effective reliability allocation for any given cost. Cost in this sense can be regarded as any parameter that can be considered to be either an analytical or empirical function of the failure rate of a component. Let

$C_f(\lambda_i)$ = the field cost incurred due to component i because the failure rate of i equals λ_i

$C_a(\lambda_i)$ = the cost incurred to achieve a component failure rate of λ_i,

$C_f(\lambda_i)$ and $C_a(\lambda_i)$ can be expressed in terms of different parameters, and there can be different types of costs associated with the achievement of a given value of λ_i: costs in terms of dollars, power required, weight, volume, etc.

First let us consider the example of the simple series system in Fig. 7-1 and a "cost" in question in terms of dollars, where

$C_f(\lambda_i)$ = the support cost associated with component i due to a failure rate of λ_i

$C_a(\lambda_i)$ = production acquisition cost of component i with a failure rate of λ_i

If our objectives were:

1. To provide a system design with a value of failure rate, λ_{ct}, less than or equal to an established target value of failure rate, λ_t; as well as
2. To provide a design that would result in the lowest value of support cost

we would apportion our failure rate objectives for each component such that the value of $C_f(\lambda_{ct})$ is minimized where

$$C_f(\lambda_{ct}) = C_f(\lambda_r)\frac{\lambda_r}{\lambda_{ct}} + C_f(\lambda_c)\frac{\lambda_c}{\lambda_{ct}} + C_f(\lambda_{co})\frac{\lambda_{co}}{\lambda_{ct}} \qquad (7\text{-}4)$$

where

λ_r/λ_{ct} = proportion of system failures occurring over service life attributable to R, the receiver/transmitter. Recall that $\lambda T =$ the number of failures expected over any time period T. As a consequence, $\lambda_r T/\lambda_{ct} T = \lambda_r/\lambda_{ct}$

λ_c/λ_{ct} = proportion of system failures occurring over service life attributable to C, the computer tracker

$\lambda_{co}/\lambda_{ct}$ = proportion of system failures occurring over service life attributable to CO, the controller

$C_f(\lambda_r)$ = the average cost of a failure in component R over service life, taking into account costs of support equipment, material, and manpower required for corrective maintenance

$C_f(\lambda_c)$ = the average cost of failure in component C over service life

$C_f(\lambda_{co})$ = the average cost of a failure in component CO over service life

λ_{ct} = $\lambda_r + \lambda_c + \lambda_{co}$

If the objective of the allocation was to minimize support cost and the cost of achieving any value of λ_i was not a consideration, then the values of λ_i that minimize (7-4), where $\lambda_{ct} \leqslant \lambda_t$, would provide the optimum allocation. At first glance the solution seems to be trivial. Drive the value of each λ_i to its minimum! This improvement might mean more costly parts, modules, or packaging in general; a need for added complexity due to the application of fault tolerance; or a need for additional support equipment for newly introduced technology. All would affect acquisition costs.

If, however, the cost of attaining a given value of λ_{ct} is a concern, then the cost to achieve a particular mix of λ_i must be taken into account:

$$C_a(\lambda_{ct}) = C_a(\lambda_r) + C_a(\lambda_c) + C_a(\lambda_{co}) \qquad (7\text{-}5)$$

Equation (7-5) itself can be used as an allocation objective pertinent to minimizing production costs associated with delivering a system with a failure rate λ_{ct} less than or equal to a failure rate target λ_t. In that case, the mix of λ_i that provides $\lambda_{ct} \leqslant \lambda_t$ at the lowest production cost would provide the optimum allocation.

If the objective of the apportionment is to minimize the sum of production and support costs, $C_s(\lambda_{ct})$,

$$C_s(\lambda_{ct}) = C_a(\lambda_{ct}) + C_f(\lambda_{ct}) \tag{7-6}$$

then the values of λ_i that minimize this sum, where $\lambda_{ct} \leqslant \lambda_t$, would provide the optimum allocation.

Suffice it to say that the quantities to be minimized or maximized by a particular reliability apportionment are dependent on the characteristics of the system or equipment to be developed, the needs of the customer, and the options available to the developer. The apportionment allocation process can be rudimentary or can be extended to any level of detail that is practical consistent with the development phase.

8

Reliability Prediction During the Development Process

Reliability prediction/estimation takes place throughout the development program. Its purpose is to provide insight concerning future reliability performance of an item in use and during reliability test/demonstration. In doing so, it provides the developer with information concerning the ability of a design to meet reliability requirements and to identify those portions of the item which require additional reliability effort. It most often takes the form of an iterative process where previous prediction results are revised as design progresses. Different forms of reliability prediction are associated with each level of design keyed to the information available during that phase of the design development process.

Most commonly used prediction procedures are based on empirical data collected over time from other development programs, data collected to support "standardized" prediction procedures, tests performed on other equipments, components, and devices, and from field failure data. As a consequence, all such prediction procedures are subject to two types of variability or error: (1) One due to the variability of the estimates used, which are averages. Only rarely will an observed value exactly coincide with an average value (e.g., a coin tossed 100 times can be expected, on the average, to display 50 heads. However, we would not be surprised if when we tried it we observed more or less than that number). (2) Another due to the fact that the averages calculated are based on data on items produced in the past. Items over time are likely to undergo reliability change, usually improvement as the processes involved in component and device manufacture improve.

Of the two, as long as the data base is sufficiently large and representative, the latter presents the greater problem. There are, however, two ways to logically cope with the issue: (1) Develop correction factors which account for changes in failure rate of devices/components over time. (2) Accept the empirical values as they exist as minimum standards against which competing design alternatives can be evaluated.

Three such empirical approaches to prediction are currently defined in official military standards and handbooks (one procedure in Mil-Std-756, two in Mil-Hdbk-217). The fact that a procedure is advocated in a military standard does not mean inapplicability for industrial or other commercial use. Over the years, for example, Mil-Hdbk-217 has been applied to a wide variety of nonmilitary items from entertainment unit to nuclear power plant reliability prediction.

A number of other empirical predictions, approaches, and sources of failure rate exist which are organization- or company-particular. (Deterministic approaches to reliability prediction are also possible. In the past the latter have generally been associated with structural integrity. Of late, suggestions with respect to the adoption of such concepts to electronic components, devices, and equipments have been made. This will be addressed further in Section 8.4.2.)

Virtually all forms of empirical reliability prediction are based on the calculation of the total failure rate, λ_e, for a given item. That quantity is then transformed into the terms in which the reliability requirement is couched (i.e., mean time to failure; mean time between failure; $R(T)$, the probability that an item will satisfactorily perform its function for a given period of time; etc.).

Total failure rate, λ_e, is conventionally treated as a sum of all failure rates of all parts making up an equipment or component. How easily this can be done and how straightforward the transformation of that quantity into the terms generally used in reliability specification and evaluation is largely the function of the failure distribution assumed. This was covered in some depth in Chapter 2, where the reliability prediction characteristics of some of the most commonly applied distributions were discussed.

Reliability prediction can be performed at various times during the development cycle to various degrees of accuracy. A number of different prediction procedures are available from which to choose. The choice of which to use is largely a function of the information available on the item which is the subject of the prediction and the amount of effort allocated to the prediction effort. For ease of discussion, reliability prediction procedures will be broken down into three separate classifications: initial, intermediate, and detailed prediction approaches.

8.1   INITIAL RELIABILITY PREDICTION APPROACHES

Initial reliability estimates can be provided by three processes: (1) extrapolation, which is usually the earliest prediction means; (2) function or module count estimation; and (3) gross part count methods.

8.1.1   Extrapolation Approaches to Initial Reliability Estimation

We discussed earlier the need for initial reliability estimates for system development and apportionment purposes. It was indicated then that the initial prediction was most commonly based on the reliability characteristics of similar equipment under development now, or developed in the past and updated to take advantage of both new and more mature technology. A general procedure to achieve the former is outlined as the "Similar Item Method" in Mil-Std-756 mentioned in Section 3.5.4. It involves simply comparing the functional and/or the hardware makeup of the item to be developed to that of items previously developed or in more advanced phases of design. The idea is that a new item similar in function, complexity, and technology to items which have previously been developed will have a similar reliability. Hence analyses and predictions already performed on *similar equipments or components* can serve as a basis for initial reliability estimates, as can data acquired from field use.

Factors which should be considered when comparing similar equipment/ components/functions include: physical performance characteristics, packaging and manufacturing similarities, operational and environmental usage similarities, reliability program similarity, age of the design and technology applied, and similarities in complexity. It is unlikely that close similarities in all of these will be found for the majority of components making up the new item. Hence engineering judgment will generally be required to make adjustments, using those components most similar as a base.

When some of the components of the new item are similar to components developed for other items *under development currently*, then the "Similar Item Method" will provide a first-order estimate of the reliability that is achievable for the components (based on reliability information available on those components). If, however, some of the components only bear such a resemblance to components *developed some years previously*, then extrapolation will be necessary to provide reliability estimates commensurate with the current state of the art.

Extrapolation in this sense involves drawing inferences about current/future reliability values based on information, trends, and/or knowledge of the reliability evolution process over time. The objective is to provide predictions of what can be attained that, if not exact, represent *reasonable* estimates of the maximum or the minimum values attainable. Reliability is subject to evolution in one of three ways:

1. We learn how to make the part better with fewer defects in its makeup. This is best exemplified through use of the so-called learning factor, Π_l in Mil-Hdbk-217, which is applied to microcircuits. Π_l measures impact of product line age on device quality.
2. We learn how to make the part do more things, therefore reducing the number of parts required to perform a given function (see Ref. 1). This also is best exemplified by microcircuits, where by an increase in gate count the device can be used to perform a greater number of functions.
3. We learn how to develop a completely new generation of parts which are either more inherently reliable or capable of functions that until now could only be performed by many more parts. This is best exemplified by the transistor as opposed to the vacuum tube, and by the microcircuit as opposed to the transistor.

The logic behind the effects of all three is evident. It is reasonable to assume that the longer a part is produced with at least moderate pressure to improve/maintain quality, the more stable, controlled, and better understood the process and the material used will become. Perhaps more homogeneous material with fewer defects will slowly evolve, or a superior material class that was once expensive will slowly become inexpensive. As a consequence, over periods of time reliability/quality will change.

The second and third factors of the extrapolation–evolution process deal with use of advances in technology to reduce the number of individual parts used to perform a given equipment or component function. These are reflected both in the improvement of the capabilities of existing technologies and in the introduction of new ones. This is evident in the increase in gate count per device and in the transition from the transistor to the integrated circuit; to small-scale integration (SSI); to medium-scale integration (MSI); to large-scale integration (LSI); to very large-scale integration (VLSI); to very high-speed integration (VHSI) technology.

The results of the three working together can be observed in Figs. 8-1 and 8-2, presented originally as Figs. 3-1 and 3-2. These show a steady growth both in mean time between failure (reduction in failure rate) in aircraft electronic systems and in total overall aircraft reliability with time. The growth occurred despite the fact that there was a general trend to increase performance and functions performed by aircraft subsystems over the 20-odd

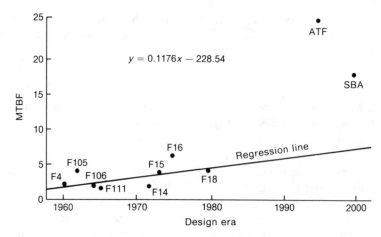

FIGURE 8-1. Reliability trends over time with respect to fighter aircraft electrical/electronic functional groups. [From DOD/Industry—"R & M Technology Study Analysis," Lyon and LaSala, Proceedings, 1984 Annual Reliability and Maintainability Symposium. © 1984 IEEE.]

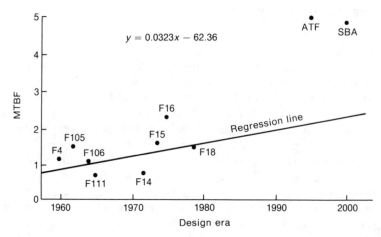

FIGURE 8-2. Reliability trends over time with respect to total aircraft systems. [From DOD/Industry—"R & M Technology Study Analysis," Lyon and LaSala, Proceedings, 1984 Annual Reliability and Maintainability Symposium. © 1984 IEEE.]

years depicted. Hence, even as complexity increased, reliability improved, providing evidence of increasing levels of part reliability with time.

In addition, consider Chart 8-1, which was extracted from the 1957 Advisory Group on Reliability of Electronic Equipment (AGREE) report. It provides a summary of recommended highest minimum acceptable values of MTBF (based on an analysis of the then current state of the art) for

CHART 8-1

Air Force		Navy	
Communications		*Communications*	
UHF	79	UHF transmitter	180
HF	90	MHF	113
Datalink	46	Datalink	17
Fire Control/Defense		*Navigation*	
		TACAN	59
Ranging radar	54	ASTAL compass	61
Bomb/Navigation system	88		
Air data computer	66	*Flight Control*	
Bombing computer	108	Radar altimetry	500
Radar control receiver			
Beacon	144	*Data Handling*	
ECM	216	Intercept computer	11
		Fire control	
		computer	37
		Bombing computer	16
		Detection Tracking	
		Primary Radar	90
		IFF	133
		Bombing navigation	
		(radar + computer)	16

Source: Advisory Group on Reliability of Electronic Equipment, AGREE, Report. 1957.

various types of equipment in the various services. Compare it to the more current state of the art values of MTBF achievable today shown in Chart 3-1.

There is little doubt that reliability evolution is a fact. However, there are only bits of information available which allow inferences to be drawn about what the reliability of a product developed in the past would be if it were presently in development. Let us consider some of these.

8.1.1.1 Dealing with Technology Maturity/Evolution
Since technology maturity and advances in technology have their roots at the device and component level, the discussion will be directed at those points.

Consider the part technical maturity factor exemplified for microcircuits as the Π_l factor. This term is associated with microelectronics reliability prediction in Mil-Hdbk-217E. It is a multiplying factor for device failure rates, the larger the value of Π_l, the larger the failure rate for the part. As the production process associated with the device stabilizes and is better understood and controlled, the quality of the device produced improves.

Both patent and latent defects are reduced. A study performed for RADC, summarized in Ref. 2, gathered and analyzed data from a variety of sources and developed a relationship (under a set of assumptions outlined in the reference) which indicated that the Π_l factor for microcircuits varied as a function of product-line age Y measured in terms of years,

$$\Pi_l = 0.01e^{(5.35 - 0.35Y)} \tag{8-1}$$

As a consequence, all other things the same, the decrease in failure rate (improvement in reliability), $\Delta\lambda$ (i.e. $\Delta\lambda = \Delta\Pi_l$) between a product first used when it was Y_1 years into production and the present, when it has been in production (by the same vendor) for a total of Y_2 years is

$$\Delta\Pi_l = 0.01e^{5.35}(e^{-0.35Y_1} - e^{0.35Y_2}) \tag{8-2}$$

and the proportion decrease in failure rate, $P_{d\lambda m}$, can be expressed as

$$P_{d\lambda m} = \frac{\Delta\Pi_l}{\Pi_l} \tag{8-3}$$

and the extrapolated failure rate for the more mature part

$$\lambda_m = \lambda_1(1 - P_{d\lambda m})$$

where $\lambda_1 =$ the part failure rate at age Y_1.

While discrete electronics devices in Mil-Hdbk-217 are not afforded a Π_l factor, a failure rate improvement factor for a discrete semiconductor device can be estimated. This can be accomplished by comparing the failure rate associated with a given type of semiconductor discrete device in a version of Mil-Hdbk-217 or similar document prepared a number of years in the past with the failure rate provided for the same device type in a later revision of the document. For example, the discrete semiconductor failure rates in Mil-Hdbk-217E were first developed in 1978. A study performed for RADC in 1987 (see Ref. 3) updated failure rates for most devices. Hence, a reliability growth factor over a 9-year period can be estimated. For example, Mil-Hdbk-217E provides an estimate of failure rate λ_{pe} for a silicon NPN general purpose JAN-grade transistor, in linear use, at 40% rated maximum power, at 25°C and in a fixed ground environment application, as

$$\lambda_{pe} = 0.0101$$

The referenced report yields a value of failure rate, λ_{pr}, of

$$\lambda_{pr} = 0.0063$$

for the same device. Hence, the proportional decrease in failure rate, $P_{d\lambda m}$, can be expressed as

$$P_{d\lambda m} = \frac{\lambda_{pe} - \lambda_{pr}}{\lambda_{pe}} \approx 0.38 \qquad (8\text{-}4)$$

over a 10-year period, or an average improvement expressed in terms of $P_{d\lambda m}/\text{year}$ of

$$P_{d\lambda m}/\text{year} \approx 0.038$$

for that device type. The extrapolated failure rate for the more mature part

$$\lambda_m = \lambda_1 (1 - N P_{d\lambda m})$$

where
λ_1 = failure rate last documented for the part
N = number of years since last documentation

Relationships such as (8-1) and (8-4) are usually developed on the basis of a limited data set and associated with sets of assumptions. The relationships above, and all like them developed through such means, cannot be considered as validated or true until or unless an ad-hoc effort is instituted for verification. They do, however, represent the best estimate for the relationship based on existing *available* data and, as such, are usually more helpful and defendable than estimates based on no data.

Advances in technology as a means to reduce complexity and hence increase reliability are best examplified by the area of microelectronics. Microelectronic devices with greater numbers of gates and larger word sizes can perform the same number of functions that could previously be performed only by a larger number of devices. Reliability enhancements accrue due to the fact that fewer devices with fewer connections and fewer solder joints are required. At the same time, recent studies have provided evidence that device complexity does not increase failure rate to the degree that was previously expected.[3,4]

Three different methods to estimate the impact of the application of advances in technology on the reliability of existing "similar" equipment components will be discussed.

1. A method suited to the case where the advanced technology devices that can be applied have been identified; where knowledge is available concerning which functions/devices in an existing equipment could be replaced by new devices; and where estimates can be made concerning device application factors.
2. A method suited to the case where design information is available with respect to previous equipments developed and estimates can be made relative to general advanced technology capabilities or trends.
3. A method suited to the case where only the most general information with respect to previous equipment developed is available and estimates can be made with respect to general advanced technology capabilities or trends.

Common to the application of all methods, in particular when dealing with microelectronic devices, are models which provide a relationship between device complexity and failure rate. This can be illustrated in the Mil-Hdbk-217E failure rate models for monolithic bipolar and MOS digital devices. The failure rate, λ_p, of an operating device is given by

$$\lambda_p = \Pi_q(C_1 \Pi_t \Pi_v + C_2 \Pi_e)\Pi_l \qquad (8\text{-}5)$$

where

λ_p = device failure rate in terms of failures per 10^6 hours
Π_q = device quality factor
Π_t = technology-based temperature acceleration factor
Π_v = voltage stress derating factor
Π_e = application environment factor
Π_e = device maturity factor
C_1 = circuit complexity factor based on gate count
C_2 = package complexity factor based on number of functional pins and package type

We can interpret:

$C_1 \Pi_t \Pi_v = \lambda_{pg}$ = the failure rate contributions due to gate complexity

$C_2 \Pi_e = \lambda_{pc}$ = the failure rate contribution due to packaging complexity

As a result (8-5) may be expressed as

$$\lambda_p = \Pi_q \Pi_l(\lambda_{pg} + \lambda_{pc})$$

Note that other failure rate models for microcircuits and other components/devices exist. These models are for the most part organization particular or "industry organization" particular. Virtually all take the same general form for most parts. Some, however, take forms which make failure rate independent of complexity and/or provide different weights with respect to other Π factors. Since Mil-Hdbk-217E is the most widely used failure-rate prediction standard, and since most of the other prediction models take forms similar to, or simpler than, Mil-Hdbk-217E models, we will use it as the basis of our discussion.

Provided the item to be developed was to be a "new" technology version of an existing product and if information was available concerning failure rate models and the design makeup of *both* the existing and the new product required, then

- λ_p could be calculated for each group (both existing and new) of microcircuits needed for the new design.
- Reduction in the numbers of devices required for the new design could be estimated (because the increased capability in terms of gates, etc., of each advanced device could be used singly and in combination to reduce the number of devices which was previously required).
- The differences between the failure rate of the parts eliminated and those introduced, and the failure rates of parts common to both designs but differing due to part reliability maturation processes could be estimated, and a prediction made.

Generally speaking, however, while it is possible to identify components with similarity in makeup or function to those to be developed, sufficient details relevant to the design of the new product are seldom available early in the development to allow the performance of the tasks listed above.

It is more likely that one of the remaining two methods would be most appropriate. If we had information relative to just the reliability prediction associated with the devices previously used, approximations could be employed to develop failure-rate estimates for application of generally enhanced devices. This would provide a reliability estimate not of the new item to be developed but of an updated version of an existing unit "similar" to the new product, and an extrapolation would be made to the new product.

Such an approach is developed in Appendix A9 based on information in Mil-Hdbk-217E. It provides for estimates of the failure rate, λ_{pz}, of a device of the same type of microcircuit having a factor Z (*i.e.*, $Z > 1$) increase in gates over a previously used version:

$$\lambda_{pz} \sim \Pi_q \Pi_l [K \Pi_t \Pi_v (ZG)^{1/2} + K_1 \Pi_e (W N_p)^{K_2} \Pi_e] \qquad (8\text{-}6)$$

where

λ_{pz} = the failure rate of the device with added gate complexity Z

G = the number of gates in the previously used device Π_q, Π_l,
Π_t, Π_e, Π_v are as defined before

K, K_1, K_2 = constants depending on device type (found in Mil-Hdbk-217E)

N_p = number of pins on the original device package that are connected to some substrate location

Z is the factor of increase in gates for the type of devices currently available, compared to the gate complexity of the same device used previously and equals the maximum device equivalence factor or the maximum number of devices potentially replaceable by a new device. For example, $Z = 2$ would indicate that new devices of the same type previously used have twice the number of usable gates of the older device.

As the number of gates per device increases, the potential increases for fewer devices being required to perform a given function. For example, in certain situations, a device containing 800 gates might be able to replace two 400-gate devices. While that strict proportionality might be unrealistic, certainly the idea that four 800-gate devices could replace six 400-gate devices is not unreasonable. As a consequence, we assume a savings in device number.

W is the factor of increase in packaging complexity associated with the new device expressed in terms of factor of increase in pins (for example, to take advantage of increases in gate complexity) compared to the same general type of device used previously.

Equation (8-6) is derived from (8-5), where, from Appendix 9, it is shown that C_1 and C_2 may be approximated by

$$C_1 \sim K(ZG)^{1/2}, \qquad C_2 \sim K_1(WN_p)^{K_2}$$

and

$\lambda_{pg} = C_1 \Pi_t \Pi_v = K(G)^{1/2} \Pi_t \Pi_v$ = failure rate contribution due to gate complexity for the particular type device in question

$\lambda_{pc} = C_2 \Pi_e = K_1(N_p)^{K_2} \Pi_e$ = failure rate contribution due to packaging complexity for the device type in question

Equation (8-6) can then be expressed as

$$\lambda_{pz} \sim \Pi_q \Pi_l [(Z)^{1/2} \lambda_{pg} + (W)^{K_2} \lambda_{pc}] \qquad (8\text{-}7)$$

If the equipment originally contained M devices of a given type, the total failure rate, λ_{pt}, contribution of those M devices is

$$\lambda_{pt} = M\lambda_p \tag{8-8}$$

With increased capability represented by Z, at best $(1/Z)M$ advanced technology units would be capable of performing the function previously performed by the M devices. The total failure rate contribution, λ_{ptz}, due to use of the advanced technology devices is

$$\lambda_{ptz} \sim \frac{1}{Z}M\lambda_{pz} \tag{8-9}$$

The proportional decrease in total failure rate, $P_{d\lambda t}$, attributed to the application of the advanced technology device can be expressed as

$$P_{d\lambda t} \sim \frac{\lambda_{pt} - \lambda_{ptz}}{\lambda_{pt}} = 1 - \frac{\lambda_{pz}}{Z\lambda_p} \tag{8-10}$$

If we assumed that the factor change in gate count Z is equal to the factor of change in pin count W, $W = Z$, (8-10) could be expressed as

$$P_{d\lambda t} \sim 1 - \frac{\lambda_{pg} + Z^{K_2 - 0.5}\lambda_{pc}}{\sqrt{Z}(\lambda_{pg} + \lambda_{pc})} \tag{8-11}$$

In addition if we were able to estimate the time, T, in years necessary to double gate count for given type of device, Z could be represented as (see Ref. 5)

$$Z = 2\frac{t}{T} \tag{8-12}$$

where t = time since the previous design was developed.

Estimates for the value of T can be generated empirically. Some sources have indicated T to lie in the range between 1 and 2. Other sources have estimated it as between 1 and 3.

If we had just general information available concerning the design detail of previous equipments developed, and hence were unable to provide information on λ_{pg} or λ_{pc}, the third alternative may be applied.

Analysis of the generic failure rates for this same type of microcircuit device in Mil-Hdbk-217E results in the finding that for virtually any environment the relationship between device failure rate, λ_p, and gate complexity (gate count), G, follows the approximate relationship

$$\lambda_p \sim K(G)^x \qquad (8\text{-}13)$$

where $x = $ a value empirically derived. Hence, λ_{pz} can be expressed as

$$\lambda_{pz} \sim K(ZG)^x \qquad (8\text{-}14)$$

As before, the total failure rate, λ_{pt}, of any M devices of that type in the equipment and the total failure rate, λ_{ptz}, attributed to the best case use of advanced technology units, can be represented as

$$\lambda_{pt} = M\lambda_p \qquad (8\text{-}15)$$

$$\lambda_{ptz} \sim \frac{1}{Z}M\lambda_{pz} \qquad (8\text{-}16)$$

Hence, the proportional decrease in total failure rate, $P_{d\lambda t}$, attributed to the application of the advanced technology device, can be expressed as

$$P_{d\lambda t} \sim \frac{\lambda_{pt} - \lambda_{ptz}}{\lambda_{pt}} = 1 - \frac{\lambda_{ptz}}{\lambda_{pt}} \qquad (8\text{-}17)$$

Substituting (8-15) and (8-16) into (8-17) results in

$$P_{d\lambda t} \sim 1 - Z^{(x-1)} \qquad (8\text{-}18)$$

Analysis of the generic data table in Mil-Hdbk-217E for this type of device indicates that in almost all cases within any given environment, x lies between 0.4 and 0.5. Defining a mid value for $P_{d\lambda t}$ when x lies between 0.4 and 0.5 as

$$P_{d\lambda t} = \tfrac{1}{2}[(1 - Z^{(0.4-1)}) + (1 - Z^{(0.5-1)})]$$
$$= 1 - \tfrac{1}{2}(Z^{-0.6} - Z^{-0.5}) \qquad (8\text{-}19)$$

and (8-19) can then be used to estimate $P_{d\lambda t}$ for a group of M extended-technology monolithic bipolar and MOS digital devices. The same type of procedure can be applied to many types of devices given the proper data base.

The above results represent an example of an approximate means to cope with extrapolation/evolution for a limited number of components. At the present, extrapolation/evolution is more of an engineering art than a science. The reasonableness and accuracy of any such extrapolation depends on the data and information directly available to the engineer from within his own organization and directly available or extractable from texts, standards, handbooks, technical reports, and data banks.

8.1.1.2 An Overview of the Initial Reliability Prediction Effort

Initial reliability prediction efforts are logically hampered not only by the fact that changes in part failure rate with respect to technology and maturation have to be accounted for (discussed in the previous section) even when there are design similarities, but by the fact that the new item can be significantly different in design from any so-called similar item. The reliability prediction process must take both into account.

The mechanics of the process are straightforward. Information will generally be available within the organization and from design engineers attached to the organization concerning similarity to, and differences between, the new item and previously developed similar items. At this time rough estimates of the order that "the new component will be about Y times more complex than a defined existing component" can be made.

Developer organization engineers will, even at this time, have some idea concerning the type of technology to be applied and how it will be used. As a consequence, from past records, information on reliability achieved in the past on similar components is available and information on the general design and complexity differences between the new and existing items has been articulated.

With the above given, the objective is to estimate the failure rate of the new item taking into account (1) the degree of application of advanced technology components and devices (i.e., VLSI, VHSIC, new ceramic applications, etc.), (2) the degree of application of more mature versions of previously used devices and components, (3) the degree of application of the same components and devices previously used, (4) the degree of additional complexity "estimated" to be required to meet new item needs.

Suppose, for simplicity, that a single component is taken as an example. The most similar component found has a failure rate equal to λ_{pt}. It was developed six years ago and meets some, but not all, of the new product requirements. From conversations with design engineers familiar with the requirements of both the existing and new component, it is estimated that the new component would be of the order of $Y\%$ more complex than the existing one. Furthermore, various portions of the design which provided an estimated $P_a\%$ of the original failure rate would apply a part technology not available six years ago. Other portions of the design which originally provided an estimated $P_m\%$ of the failure rate would utilize more mature versions of parts. The remainder of the component which provided an estimated $P_u\%$ of the original failure rate would use types of parts of the same quality (unchanged) that were used previously.

The failure rate of the new design, λ_{ptz}, can be estimated as

$$\lambda_{ptz} = (1 + Y_1)\lambda_{pt}[P_{a1}(1 - P_{d\lambda t}) + P_{m1}(1 - P_{d\lambda m}) + P_{u1}]$$

$$= (1 + Y_1)\lambda_{pt}(1 - P_{a1}P_{d\lambda t} - P_{m1}P_{d\lambda m}) \qquad (8\text{-}20)$$

where

$P_{a1} + P_{m1} + P_{u1} = 1$

$P_{a1}, P_{m1}, P_{u1} \quad = P_a/100, P_m/100, P_u/100$ respectively

$Y_1 \qquad\qquad = Y/100 =$ overall proportional increased complexity of the new item over the similar one

$\lambda_{pt} \qquad\qquad =$ failure rate of the similar component

$P_{d\lambda t} \qquad\qquad =$ proportional change (usually a decrease) in failure rate for the *functions* now to be performed by new/advanced technology parts

$\qquad\qquad\qquad = 1 - (\lambda_{ptz}/\lambda_{pt})$

$P_{d\lambda m} \qquad\qquad =$ proportional decrease in failure rate due to product maturity

$\qquad\qquad\qquad = 1 - (\lambda_{ptm}/\lambda_{pt})$

$\lambda_{ptz}, \lambda_{ptm} \qquad =$ estimate of new λ's due to technology and product maturity changes

Take, for example, a case where estimation of the reliability of a communications component is required. A company has an interest in being the developer of such an item. The general characteristics associated with the item to be developed are known or have been provided. Design information on similar types of communications components developed in the past by the developer's organization is available, including reliability information. It is clear that the exact performance characteristics of the component to be developed differ from those developed in the past. The form, fit, and electronic functions to be performed are similar, however, and the environment of operation is also the same. This product will have superior capabilities to those developed in the past by the development organization. However, it will also have design similarities to those products. Parts lists and reliability prediction results from prior programs are available and can be used as a basis for reference.

It is anticipated that new component part technology in signal and data processing will be applied. The previous contributions to overall failure rate from such functions were approximately 35% (i.e., $P_{a1} \sim 0.35$).

The application of more mature versions of semiconductors (diodes and general-purpose transistors) is also anticipated. The previous contribution to overall failure rate from such components was approximately 45% (i.e., $P_{m1} \sim 0.45$).

The use of certain parts and components identical to those used in previous designs and off-the-shelf components which have reached plateaus of quality is also anticipated. The previous contribution to overall failure rate from such parts/components was approximately 20% (i.e., $P_{u1} \sim 0.2$).

The proportional decrease in failure rate, $P_{d\lambda t}$, of the signal and data processing functions which would be associated with the application of new/advanced technology is estimated to be $P_{d\lambda t} \sim 0.4$.

The proportional decrease in failure rate, $P_{d\lambda m}$, in other items due to improvement in part component quality is estimated as $P_{d\lambda m} \sim 0.05$ per year. The equipment judged to be "most similar" contains parts of this nature, five years more mature than when last used. Hence the proportional decrease in failure rate of such equipment items over five years is estimated to be of the order of $P_{d\lambda m} \sim 0.25$. Predictions on the existing equipment indicate $\lambda_{pt} = 0.0044$.

If $\lambda_{pt} = 0.0044$, $P_{d\lambda t} = 0.4$, $P_{d\lambda m} = 0.25$, $P_{a1} = 0.35$, $P_{m1} = 0.45$, and $Y_1 = 0.15$, then (8-20) can be applied to evaluate λ_{ptz}:

$$\lambda_{ptz} = (1 + 0.15)(0.0044)[1 - (0.35)(0.40) - (0.45)(0.25)]$$
$$= (0.0051)(1 - 0.253) = 0.0038$$

Hence the first-cut estimate of the failure rate of the new item is 0.0038, or MTTF = 263 hours.

8.2 INTERMEDIATE DESIGN-LEVEL PREDICTIONS: MODULE COUNT METHODS

As described above, reliability extrapolation between existing and new design and between older and more advanced technology provides valuable information about general reliability boundaries for a new development. After development has been underway for a while, it will be possible to estimate the number of active devices (microcircuits, semiconductors, tubes, relays, pumps) necessary to perform: (1) those portions/functions of the new development hardware that constitute new design; (2) those portions/functions of the new development hardware that constitute modified design; (3) those portions/functions of the new development hardware that constitute existing design.

It may not be possible to provide schematics of each circuit/component/ function, but reasonable estimates of overall complexity in terms of the quantities of active devices which will make up the design can be furnished.

It is at this point that more accurate inferences about the actual reliability of the new product can first be made. The reliability prediction at this juncture is based on the quantities of active elements estimated to make up the design. This concept may be described as an active element group (AEG) means of prediction. The AEG prediction concept was first introduced in the early 1960s. Its application in early design is described in Mil-Std-756, "Reliability Modeling and Prediction." We will discuss the AEG concept covered in that document as well as other possible forms of part count reliability.

An AEG prediction is based on assigning a total failure rate to the *combination* of one active device and the number of passive devices necessary to support the active device. Specifically, we deal in terms of an average failure rate per AEG, where:

$(\lambda_{ei})/A_{ti} = \lambda_{aegi}$ = the average failure rate per AEG calculated from developer experiences or available from a standard or handbook

λ_{ei} = predicted total failure rate of equipment i

A_{ti} = number of active elements in the equipment

λ_{aeg} pro-rates the failure rates of all the passive elements associated with a given function, circuit or component (resistors, capacitors, inductors, connectors, plumbing, fittings, etc.) against the number of active devices associated with that function. The major assumptions associated with application of the concept are:

1. Within any design era, for any equipment i, $\lambda_{aegi} = \lambda_{ei}/A_{ti}$ is considered approximately constant over all equipment developed, normally distributed about a true value of, λ_{aeg}, or generally within a moderate size interval of a true value of λ_{aeg} (e.g., almost always within $\pm 30\%$ of the λ_{aeg} value). This is verifiable through comparison of the λ_{aegi} value chosen with values of average failure rate per active element calculated from other equipments developed during the design era. Seldom, however, will an organization have a sufficient quantity of data available on enough different equipments to perform such verification. In such instances, the fit to the assumption must be evaluated through more qualitative or indirect means. For example:

   a. If λ_{aegi} was calculated from a limited number of data points, the more similar in function and design past equipment developments have been, and the more similar the new equipment design is to the prior designs, the more reasonable the assumption. If your organization has developed airborne communications equipment in the past and calculates value of λ_{aegi} based on that experience, it would be more reasonable to accept that value if the new development was a communications equipment than if it was a radar. If the value of λ_{aegi} was developed on the basis of many different design programs where the equipments developed were heterogeneous (mixes of computers, communications, radars, etc.), the less likely it is to correspond to any one *specific* type of equipment.

   b. If the equipment being developed is being designed under the same design and part quality policies as were associated with prior development programs (which were used to calculate λ_{aegi}), the more reasonable the assumption.

2. Data is available on a sufficiently large group of active element groups over a wide enough sample of equipment to allow representative averages to be calculated.

3. All AEGs are connected in series such that a failure in any AEG will cause an equipment failure.

Information and tools to develop values of λ_{aegi} for this level of prediction are available from:

- Records within the developer's organization which provide information regarding detailed reliability predictions performed on previous development programs. Analyses of such records can provide relationships between active device quantities and total equipment failure rate. As technology and maturity advance, the analyses show trends with respect to average failure rate changes in AEGs.
- Open literature, technical reports, and papers which provide empirical information with respect to the relationship between active device quantity and total equipment failure rate.
- Standards issued by other organizations which define prediction techniques of this sort acceptable to the developer.

Of the three, the first, if available, is the most accurate for use *within* a given organization. This follows because each development organization may specialize in the design of a specific item (communications, radar, computers, pumps, different mechanical products). Each organization may have different general policies concerning such issues as derating, part quality, or general design procedures. Any tool available from the literature or used as a standard will either be attuned to the products and policies of a given organization or represent a composite average of many organizations. It may, in fact, represent a composite of many organizations and different items as well (e.g., communications, radar, computers).

In order to define a relationship geared to the products and policies of a particular organization, data should be collected on as many previously performed reliability predictions as possible. The greater the number, the more representative the result. For each product, note the number of active devices or components (i.e., for electronic products: integrated circuits, diodes, crystals, transistors, relays. For mechanical products, the active mechanical elements would be noted). It would also be desirable that digital and analog products be listed separately, as the failure rate of an analog active element can be significantly different from one that is digital. For each product, proceed as follows:

1. Note the year in which the design development was made.
2. Add all the active devices composing each product, A_{ti}.
3. Note the total predicted failure rate, λ_{ei}, for the product, or alternately,
4. Determine the actual failure rate of each product in field use and equate that to λ_{ei}. Note that there is a fundamental difference between basing a reliability estimation process on prediction data and basing it on field data. The former is intended to predict reliability as it is influenced by the design and parts in a given operating environment under conditions stated in, or interpreted from, the specifications and requirements. It assumes perfect maintenance and repair and that no operational influences, other than those covered in the prediction, affect reliability. Field data includes the effects of operation in unforeseen environments. It reflects the impacts of human factors or imprecise diagnostics, different and perhaps imperfect maintenance and repair policies, and operational influences not accounted for by the prediction. (See Ref. 6 which describes the differences in predicted vs. field reliability for a large number of equipments.)

Compute the average value of λ_{aeg}, $\bar{\lambda}_{aegj}$ for each year of experience:

$$\bar{\lambda}_{aegj} = \frac{\sum_{i=1}^{H_r} \lambda_{aegi}}{H_j} = \frac{\sum_{i=1}^{H_r} \lambda_{ei}/A_{ti}}{H_j} \tag{8-21}$$

where

$\bar{\lambda}_{aegj}$ = average value of failure rate per AEG in year j

H_j = number of equipment/components designed in year j

λ_{ei}, λ_{aegi}, A_{ti} are as defined previously

Plot $\bar{\lambda}_{aegj}$ on graph paper as in Fig. 8.1 to see whether a trend occurs there in value of average failure rate per active element group over time. Better yet, perform a simple two-variable regression analysis between $\bar{\lambda}_{aegj}$ and design year. The procedures of simple regression are available in most elementary statistics texts and built into even inexpensive hand calculators. The relationship that will result will provide a basis for failure rate extrapolation over time. While regression mechanics themselves are rather simple, transformation of variables is sometimes required and proper interpretation of results is not always straightforward. As a consequence, unless the relationship is relatively obvious (e.g., a plot of $\bar{\lambda}_{aegj}$ vs. design year is approximately linear), defer the analysis to your staff statistician.

For example, the reliability prediction data for 13 products an organization has developed over the past six years has been examined with results as shown in Chart 8-2. Plotting the values of $\bar{\lambda}_{aegj}$ for each year, the plot shown in Fig. 8-3 results. A line fitted to the data for the period of time in question shows a definite decreasing trend in average failure rate per active element group. The line has an approximate equation of

$$\bar{\lambda}_{aeg} = -(0.083 \times 10^{-6})(j - 1983) + 1.73 \times 10^{-6}$$

which could be used as an estimate of $\bar{\lambda}_{aeg}$ that includes the effects of application of more advanced/mature technology for the new product, where j = the design year to which $\bar{\lambda}_{aeg}$ is to be extrapolated. The trend line here is indicative of a projected increase in reliability with time.

However, what happens if data available is not extensive enough to provide a plot such as is shown in Fig. 8-3? In that case the data available could be averaged over all years to develop a grand average failure rate per active element, λ'_{aeg} such that

$$\lambda'_{aeg} = \frac{\sum\limits_{i=1}^{m} \lambda_{aegi}}{M} \qquad (8\text{-}22)$$

where M = number of equipments developed over a given period of time.

This quantity is representative of the average failure rate per active element produced over that time period. If the time period is long, covering a number of years, it will require adjustment relative to the current reliability state of the art. This is obviously less attractive than providing an estimate which includes considerations of the march of technology. However, it still is a reliability estimate representative of the types of equipment making up the

CHART 8-2 Average Failure Rate per Active Element over a 6-Year Period

Product	Year Developed	Number of Active Elements	Total Equipment Failure	Equipment MTBF	Average Failure Rate per Active Element
1	1984	120	200×10^{-6}	5000	1.66×10^{-6}
2	1984	900	1650×10^{-6}	606	1.83×10^{-6}
3	1984	600	900×10^{-6}	1110	1.5×10^{-6}
4	1985	250	380×10^{-6}	2630	1.51×10^{-6}
5	1985	750	1250×10^{-6}	800	1.67×10^{-6}
6	1986	500	720×10^{-6}	1390	1.44×10^{-6}
7	1986	800	1050×10^{-6}	950	1.32×10^{-6}
8	1987	1100	1500×10^{-6}	670	1.36×10^{-6}
9	1987	650	900×10^{-6}	1100	1.39×10^{-6}
10	1988	850	1100×10^{-6}	910	1.29×10^{-6}
11	1988	1200	1600×10^{-6}	625	1.33×10^{-6}
12	1989	800	980×10^{-6}	1020	1.23×10^{-6}
13	1989	1000	1280×10^{-6}	780	1.28×10^{-6}

Average overall per active element, λ_{AEG}: 1.45×10^{-6}

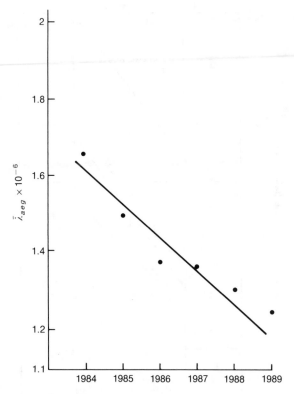

FIGURE 8-3.   Plot of $\bar{\lambda}_{aeg}$ against year.

sample as well as the particular organization development policies under which design will take place.

All the above assume that sufficient data is somehow available to the reliability engineer to allow for the generation of statistically representative values of λ_{aeg} needed. If such data is not available within the development organization, average failure rate information and prediction procedures are available in the literature.

One of the oldest general AEG prediction procedures is contained in Mil-Std-756B. Revised in 1981, it is subject to periodic revision, as all standards and handbooks are. As a consequence, the averages, which are the result of empirical and judgmental information and which form the basis of the prediction procedures, are subject to change. We will hence describe this AEG technique using the current version of the Mil-Std as a vehicle and refer the reader to the most current version available for application purposes.

The AEG method in Mil-Std-756B is based on field reliability data and

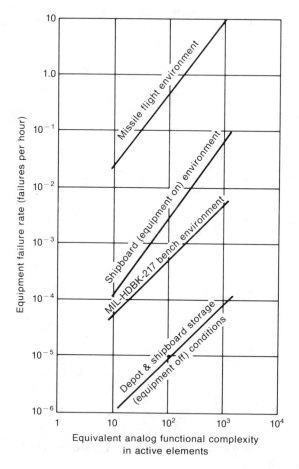

FIGURE 8-4.   Failure-rate estimation chart for electronic analog function.   [From Mil-Std-756B.]

relates the number of AEGs in an equipment in a given application environment to failure rates expected during use. The method is currently limited to reliability estimates for ship components and items. Figures 8-4, 8-5 and 8-6, respectively, show the AEG reliability prediction methodology for analog and digital electronics and for mechanical devices. Weighting factors (not shown) are included for AEGs based on analog, digital, microcircuit, power supply applications or the number of different mechanical elements. The bands around the predictive lines in the figures define the ranges of variation for the prediction. A fundamental difference exists between the Mil-Std-756 AEG process and the other AEG processes described above.

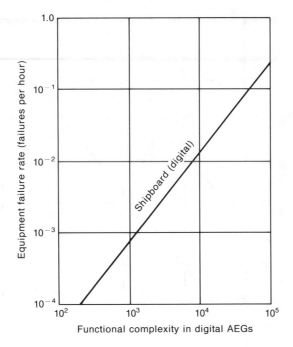

FIGURE 8-5.   Failure-rate estimation chart for digital electronic functions.   [From Mil-Std-756B.]

The former assumes that the value of λ_{aeg} (that is to say, the value of the average failure rate per active element group) changes as a function of the *number* of AEGs in an equipment. For example, from Fig. 8-5 an equipment made up of 200 AEGs would have associated with it a failure rate of $\lambda_e = 0.0001$ corresponding to an average of $\lambda_{aeg} = 0.0001/200 = 0.5 \times 10^{-6}$. An equipment composed of 3000 AEGs would have associated with it a failure rate of $\lambda_e = 0.003$ or an average of $\lambda_{aeg} = 0.003/3000 = 1 \times 10^{-6}$, double the value of the first. The operational influences acting on fielded equipment (maintenance, human, diagnostic, etc.) that make up the data base for the figures may indeed affect complex equipments more harshly than simpler items. There have been no recent studies, however, that either negate or verify that finding. The more detailed prediction procedures, for example, the stress analysis procedure of Mil-Hdbk-217E, do not add *extra* failure-rate penalties for overall equipment complexity.

Extrapolation factors must be applied to Mil-Std-756 prediction procedures to account for technology/maturity advances. Without considering extrapolation, the following example of application of the Mil-Std-756

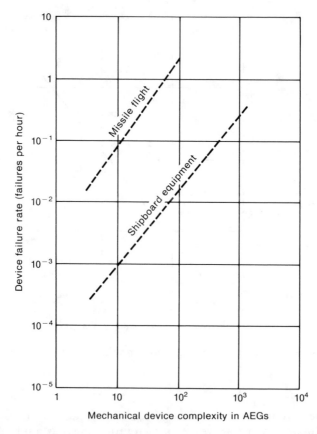

FIGURE 8-6.   Failure-rate estimation chart for mechanical devices.   [From Mil-Std-756B.]

prediction is provided. Assume that a developer has contracted to develop a navigation aid. A certain proportion of the design and components will constitute existing off-the-shelf items that will contribute a failure rate $\lambda_{oe} = 0.00040$. The remainder of the unit will constitute new design. Based on the best estimates of the company design engineers, the navigation aid, aside from off-the-shelf items, will require between 950 and 1100 active devices. Interpreting this to mean between 950 and 1100 AEGS (approximately 1000) will be required. Figure 8-5 indicates that after adjustments the value of $\lambda_{aeg} \sim 0.7 \times 10^{-6}$ would be a reasonable estimate for both cases. The failure rate of the portion of the equipment subject to advanced/more mature design, λ_{nd}, will be between

$$\lambda_{nd} = 950(0.7 \times 10^{-6}) = 0.00066$$

and

$$\lambda_{nd} = 1100(0.7 \times 10^{-6}) = 0.00077$$

Total equipment failure rate, λ_e, is estimated to be

$$\lambda_e = \lambda_{nd} + \lambda_{oe}$$

or between

$$\lambda_e = 0.00066 + 0.00040 = 0.00106 \qquad \text{or MTBF} = 943$$

and

$$\lambda_e = 0.00077 + 0.00040 = 0.00117 \qquad \text{or MTBF} = 855$$

8.3 INTERMEDIATE TO DETAILED APPROACHES TO PREDICTION. PART COUNT METHODS

As design progresses, estimation of the number of active and passive elements making up the design becomes more precise. First-cut schematic diagrams may be available. The design may not be firm enough to allow a detailed reliability prediction based on stress analysis to be performed, but more additional reliability visibility may be required. On the other hand, while such a detailed reliability prediction can be performed, there may be reasons for not performing one at this time. These reasons include the case that an immediate estimate is needed as quickly as possible, and the case that a detailed stress analysis prediction was not required. For these reasons, a prediction based just on total component part population, which would be one step more comprehensive than the AEG-based reliability estimate, may be either required or deemed desirable.

Like the AEG method discussed previously, part count methods can be developed based on the experience within a particular development organization. They are also available from the technical literature representing experiences of others, or in the form of standards developed by industrial or government organizations.

Part count procedures are the oldest form of reliability prediction and go back to the mid-1950s. A part count method basically assigns a failure rate to every class of device or component associated with a given type of item (e.g., relay, motor, pump, microcircuit, vacuum tube, transistor, resistor, inductor, connector). The failure rate is based on data acquired from field or test experience; detailed prediction results on other equipment; and more detailed failure-rate models after making assumptions with respect to typical stress conditions.

Average failure rates per part class may be calculated using information on equipments previously designed within a development organization, using reliability prediction results or field information on such products. Depending upon the quantity and detail of data available, failure rates may even be assigned to different types of devices within a given class (e.g., types of relays, pumps, microcircuits, transistors). The failure rates may even be particular to device/component operation in a given environment (ground, airborne, shipboard, etc.) or defined as a "base" failure rate which can be modified by correction factors to apply to a number of different environments. The total failure rate of the equipment, λ_e, is predicted by adding up all the part failure rates:

$$\lambda_e = \sum_{i=1}^{N} f_i \lambda_{pi} \tag{8-23}$$

where
λ_{pi} = the failure rate assigned to device/component part class i
f_i = the number of devices/components belonging to part class i in the equipment
N = the number of part classes making up the equipment

Prediction results and field and test failure data on which to base average part failure rates may be available within your own development organization. Figure 8-7, taken from Ref. 7, indicates the results of analyzing early C-5A electronic equipment data. Figure 8-8, taken from the same source, represents an extrapolation of that data to mid-1980s technology/maturity.

Note that even though the plot in Fig. 8-8 is performed on log–log paper, if the best fit line is examined closely it will be found that it is representative of a "constant" failure rate per part, $\bar{\lambda}_p = 2 \times 10^{-6}$.

Data is also available from technical reports. For example, "The RADC Nonelectronic Reliability Notebook"[8] provides estimates concerning failure rates and other characteristics of nonelectronic parts. Chart 8-3, extracted in part from that document, is an example of data documented for forced-balance, pendulum (linear), and pendulum (single axis) accelerometers.

Other reports from which generic failure rates on a part or component class/type can be extracted are provided by the Government Industry Data Exchange (GIDEP) (current location, Corona CA 91720-5000) and the Reliability Analysis Center (RAC) (current location, Rome, NY 13440-8200), a DOD information analysis center.

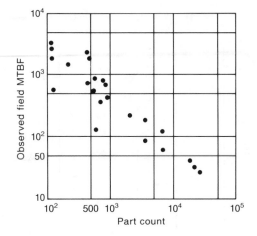

FIGURE 8-7.   Reliability of C-5A aircraft electronics.   [From Ref. 7, IEEE Copyright 1984.]

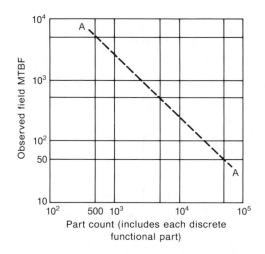

FIGURE 8-8.   Complexity–field reliability for transport aircraft electronics (mature systems).   [From Ref. 7, IEEE Copyright 1984.]

Part count prediction methods also either exist in, or can be generated from, more detailed reliability prediction standards, which may be either government- or industry-based. Examples of such standards include: Mil-Hdbk-217, "Reliability Prediction of Electronic Equipment (Gvt. Std.); BellCore "Reliability Prediction Procedure for Electronic Equipment"; British Telecom Reliability Handbook.

CHART 8-3 Failure Rates of Mechanical Items

Accelerometer	Environment Use	Lower Bound	Best Bound	Upper Bound
Forced balanced	Ground mobile	18.6×10^{-6}	26.6×10^{-6}	37.9×10^{-6}
Pendulum (linear)	Airborne uninhabited fighter	15.6×10^{-6}	30.4×10^{-6}	55.9×10^{-6}
Pendulum (single axis)	Airborne uninhabited fighter	3.8×10^{-6}	6.1×10^{-6}	9.6×10^{-6}

As indicated previously, the parts count method can take a variety of forms:

1. A *single* average value of failure rate, $\bar{\lambda}_p$, can be assigned to all parts:

$$\lambda_e = \bar{\lambda}_p M \qquad (8\text{-}24)$$

where
λ_e = failure rate of the equipment
M = number of parts/components in the equipment

Figure 8-8 shows the result of such an average that can be associated with every part in each equipment of the avionics suite.

2. A single typical value of failure rate may be assigned an entire *generic class*, e.g., microcircuits, diodes, resistors, transistors, relays, etc.
3. Different failure rates may be assigned to the different *types of parts within a class* when there are differences in part complexity, materials used, construction, etc., among the parts of that class (e.g., different types of resistors, relays, microcircuits).

Generally speaking, it would be logically expected that (2) would provide a more accurate estimate than (1) and that (3) would provide a more accurate estimate than (2). The reason for this is straightforward.

Each part class or type of part within a given class may have a significantly different failure rate from other classes or other parts within a class. The proportion of the different part classes or types necessary to effect a particular design determines the actual average failure rate per part of the design. As a consequence, the use of values of average failure rate per part is insensitive to the particular part mix of the design. The result could be a reliability prediction that is way off the mark.

A prediction method based on average failure rates per *part class* would, by its nature, be sensitive to *part mix* with respect to class and provide a generally more accurate assessment. It would, however, be insensitive to the failure rate differences occurring within different types of parts of the same class. As a consequence, a prediction of the form (3) would in general be even more sensitive to part mix and provide the most accurate estimate of the three.

The forms (2) and (3) can also be enhanced making them capable of distinguishing differences in reliability due to operational environment and the use of different-quality parts. This is usually accomplished through the use of multiplying factors applied to the average/typical values of failure rate assigned each part/class or type. (An alternative to this is to develop average/typical failure rate values particular to a given environment and quality level. This again is exemplified by Fig. 8-7, which provides an estimate of average failure rate per part for equipment containing primarily military type parts in an airborne flight environment.)

In order to take such factors into account, consider the failure rate assigned to a given part, λ_{pi}, described in (8-23), and define it as

$$\lambda_{pi} = \Pi_{ei} \Pi_{qi} \lambda_{bi} \tag{8-25}$$

where

λ_{bi} = the *base* failure rate associated with a given class of parts

λ_{pi} = failure rate of part i in a given environment when the part is of a given quality

Π_{ei} = failure rate multiplying factor associated with part i for operation in a given environment

Π_{qi} = failure rate multiplying factor associated with part i procured to a given quality level

As a consequence, (8-23) can be expressed as

$$\lambda_e = \sum_{i=1}^{N} f_i \Pi_{ei} \Pi_{qi} \lambda_{bi} \tag{8-26}$$

where all terms are defined as before.

The major assumptions associated with the general part count procedure are:

1. Within any design era the same average/typical failure rate per part (or per part class, or per part class type, depending on the form the method takes) can be associated with every equipment developed. The average part failure rate value assigned must satisfy one of the following conditions: (a) The average part failure rate(s) assigned/calculated is approximately constant over all equipment developed. (b) The average part failure rate(s) associated with each equipment developed must be approximately normally distributed around the values of the failure rates assigned by the prediction method. Or, more generally, they must almost always be within a moderate-sized interval centered on the values assigned.
2. Enough data is available to provide a sufficiently accurate estimate of each average failure rate.
3. All parts are considered connected in series such that the failure of any one will result in an equipment failure.

The first can be verified on the basis of past performance through examination and analyses of detailed prediction or field-use data on equipments developed over the design era. A reasonable approximation to the results achievable from a more detailed prediction is a form of verification. Seldom, however, will an organization have a sufficient quantity of developed equipments to provide the data necessary for a quantitative verification. In such instances, as in the case for AEG prediction process verification, the fit to the assumptions must be evaluated through more qualitative and indirect means. In this case, the conditions associated with such judgments vary with the form of the prediction method.

For form (1), the single average value of failure rate assigned to all parts, the same criteria described for the AEG prediction procedure apply. However, forms (2) and (3) can be made capable of taking into account different operational environments and quality levels. Moreover they are sensitive to part mixes which differ from those that may have been used in the past. As a result, similarity of the new design to other designs previously produced is a plus but is not as important as for form (1). Here, since quality and environmental factors can be treated separately, failure rate per part class (type) is largely dependent on how the part is applied. As a consequence, if each part of a given class or type is subject to consistent design application rules and policies with respect to stresses (for example, derating), then it is likely that the failure rate associated with each of its applications will be similar.

Data must be sufficient to allow an accurate estimate of each average failure rate. Average failure rates acquired from industry or government standards generally represent information from large data bases and certainly would be most likely to be accepted by those organizations. (If failure rate values differing from those are substituted, suitable justification and data would probably be requested, especially if such predictions require *approval* from such organizations.) Such standard failure rates, however, are usually calculated using data from many different sources, from many different applications. As a result, while they may represent a true average failure rate, the variance of failure rate can be high.

Assume that a part count reliability prediction is to be performed on the navigation system described in the last example. Design has now progressed to a point where the composition of each AEG, circuit, module, etc., can be estimated in terms of the quantities and types of parts it will contain. Design engineers have provided assessments of those quantities. It is now estimated that the equipment will contain approximately 3200 parts. Since information in this case is available not only on the total part quantity but also on the quantity of each part type contained in the system, the most detailed part count prediction procedure available will be exercised.

Note: All of the detailed reliability prediction standards and procedures mentioned previously either have a part count procedure of this type associated with them or are capable of being used to generate one.

The equipment in question contains:

- Two types of microcircuits which belong to the monolithic bipolar class: one type which contains 4000 gates, the other 2000 gates.
- Three types of transistors: silicon NPN, silicon PNP, and high-power.
- Two types of diodes: silicon general-purpose and germanium general-purpose.

CHART 8-4 Quantity and Reliability Characteristics of Each Part Type Slated for Use

	Quantity	Π_{ei}	Π_{qi}	λ_b^a	$f^i\lambda_{bi}^a$
Microcircuit, Monolithic Bipolar					
2000 Gates	555	3	1	0.08	133.2
4000 Gates	370	3	2	0.22	488.4
Transistors					
Type 1	45	14	1	0.03	18.9
Type 2	30	14	1.4	0.06	35.3
Type 3	20	15	1	1.40	420.0
Diodes					
Type 1	20	20	1.5	0.042	25.2
Type 2	35	20	1	0.60	420.0
Switches					
Type 1	45	8	1	0.002	0.72
Type 2	25	10	1.5	0.020	7.5
Resistors					
Type 1	360	3.5	1.5	0.002	3.8
Type 2	180	3.5	1.5	0.020	18.9
Type 3	180	3.5	1	0.040	25.2
Capacitors					
Type 1	675	3	1	0.006	12.2
Type 2	115	3	1.5	0.004	2.1
Type 3	425	3	1.5	0.008	15.3
Inductors					
Type 1	25	5	1	0.040	5.0
Type 2	90	5	1	0.080	36

[a] Expressed in units of 10^{-6}.
$\lambda = 1667.8 \times 10^{-6}$ hours.
MTBF $= 1/\lambda = 600$ hours.

- Two types of switches: toggle and thermal.
- Three types of resistors: composition, film, and power.
- Three types of ceramic capacitors.
- Two types of inductors: fixed and variable.

Chart 8-4 indicates the quantity of each type of part that is slated for use in the equipment, illustrative values of λ_{bi}, Π_{ei}, and Π_{qi} that could be associated with the part class types, and the total failure rate associated with each quantity of parts.

Applying (8-26) to the data on the chart, a prediction of equipment failure rate, λ_e, can be generated. For the equipment in question, λ_e is evaluated as

$$\lambda_e = \sum_{i=1}^{N} f_i \Pi_{ei} \Pi_{qi} \lambda_{bi} = 1667.8 \times 10^{-6}$$

which corresponds to MTBF $= 1/\lambda_e \sim 600$ hours.

8.4   DETAILED RELIABILITY PREDICTION APPROACHES

The classical reliability prediction approach is one based on the use of empirical data to define basic failure rates modified by factors which account for *application stresses*, environment, and quality. The factors themselves are based on a combination of empirical data, physical relationships, and engineering judgment.

Alternate approaches of a deterministic nature have also been suggested. These are based on the fundamental physical, chemical, and/or mechanical properties of devices, components, and equipments under stresses imposed by application and environmental operating conditions. Their objective is to develop a design product capable of surviving for a *given period of time* under a given set of stress conditions. We will discuss both such types of approaches.

8.4.1   Detailed Reliability Prediction: A Failure Rate Approach

Reliability prediction methods discussed thus far have shared a general trend with respect to assumptions inherent and required.

- The AEG method required the assumption that the actual failure rate associated with each AEG is within a small to moderate-size interval of a calculated/assigned value and: (1) that quality levels for the parts making up each AEG were more or less the same; (2) that the part mix making up each AEG was more or less the same from AEG to AEG and from equipment to equipment; (3) that the design and application factors associated with each AEG and its parts remained fixed over all items developed.

- The least sophisticated of the parts count methods required the assumption that the average actual failure rate per part in an equipment is within a small to moderate-size interval of the calculated/assigned value with conditions identical to (1) and (3) above, but not necessarily (2).
- The more sophisticated of the parts count methods required the assumption that the actual failure rate associated with each part class (type) is within a small to moderate-size interval of a calculated/assigned value. Since part quality and environment of operation can be treated individually, and since differences in part mix making up any given equipment can be coped with separately, only a single condition remains, which is (3).

The purpose of the detailed prediction procedure is to remove the need for that last condition. It is removed by taking into account the *specific* application and stress factors acting on a part and their effects on failure rate. This allows taking such factors into account on a part-by-part basis as opposed to assuming the same treatment for all parts.

The overriding assumptions for the detailed prediction procedure, then, are associated with the adequacy and quantity of the data base from which failure rates are generated; and the validity and accuracy of the relationships used to define the effects of the influences of quality, environment, and stresses on the failure rate of each part.

A number of detailed prediction procedures exist. These run the gamut of being company-, industry-, and large organization-particular. They in general all provide for computation of a base failure rate, provide means for taking into account the effects of application induced stresses, operational environments, quality levels, and more. They can differ with respect to the degree of weights associated with various prediction parameters (i.e., complexity, temperature, etc.). The present most commonly used prediction procedure for both U.S. Military and general commercial use is that detailed in Mil-Hdbk-217, "Reliability Prediction of Electronic Equipment" (see Section 3.5.5). Other industry prediction procedures in use include Bellcore "Reliability Prediction Procedure for Electronic Equipment" and British Telecom Reliability Handbook.

Base failure rates are based on either empirical data, physical and chemical reactions, engineering judgment, or a combination of these. The relationships governing the effects of quality, environment, stresses, etc., are similarly based. As time advances, the data base changes (e.g., new parts are included, more data becomes available, failure rates change due to maturity influences and technology advances). As a consequence, revisions to the basic prediction document are made over time; however, not all portions of a document might be changed during each revision. The discrete semiconductor section of Mil-Hdbk-217, for example, remained static between 1978 and approximately 1990.

The fact that detailed prediction techniques are defined in documents which are called "standards" has a special connotation that deserves comment. A standard provides a common basis for some action, in this case, reliability prediction. Use of a particular set of standard failure rates provides a common basis for comparing the results of two or more developers working on the same product. It provides a *standard* for the calculation of failure rate which can be used to assess the merits of alternate designs proposed by a particular developer. It provides a means common to both customer and developer which can be used to assess risk and progress.

However, a prediction is more likely to provide better information about the equipment's behavior during demonstration than an exact estimate of reliability in a field operational environment.[6,9-12] This occurs because usually an item's reliability in an actual operational environment is affected by such things as human factors, personnel training, testing faults, technical manual deficiencies, and the effects of operational and maintenance policies at different sites. None of these is adequately accounted for in prediction and they cannot be or are not duplicated during demonstration.

Demonstration/prediction usually reflects the ideal rather than an actual environment. Imperfections in the reliability design of the equipment are the focus of prediction/demonstration, not imperfections or permutations in external influences such as air conditioning, power, training, documentation, human error, or maintenance policies.

Detailed assessment procedures, as in Mil-Std-217E, for example, provide for reliability predictions which take into account part quality, operating environment, and stresses (electrical/electronic and thermal) on the vast majority of parts included in electronic equipment design. Chart 8-5 lists major part categories covered.

A base failure rate is associated with most parts along with multiplying factors associated with part quality, environment of operation, part applications, and stresses which can impact failure rates. Examples of such factors taken into account by the detailed prediction techniques include those listed in Chart 8-6. The particular factors associated with each part vary with part type.

Due to the facts that Mil-Hdbk-217E is considered both the most general and the most widely applied prediction procedure, illustrations of detailed reliability prediction models which follow will be based on that document. On that basis, part failure rate, λ_p, takes the form:

CHART 8-5 Major Part Categories Covered by Detailed Prediction Procedures

Microelectronics	Switches
Discrete semiconductors	Connectors
Tubes	Printed wireboards
Lasers	Connections
Resistors	Meters
Capacitors	Quartz crystals
Inductors	Lamps
Motors	Electronic filters
Relays	Fuses

CHART 8-6 Factors Taken into Account in the Prediction Process.

Part quality	Number of active pins
Learning/maturity	Complexity
Environment	Cooling
Applicable stresses:	Form
Thermal	Construction
Voltage	Device application
Power	
Cycling	
Frequency	

---

1. For monolithic bipolar devices (microcircuits),

$$\lambda_p = \Pi_q(C_1\Pi_t\Pi_v + C_2\Pi_e)\Pi_l \tag{8-27}$$

where

Π_q = a device quality factor dependent on the standards employed in part acquisition (see Chapter 6) qualification level of the device

Π_l = device technical maturity factor (see Section 8.1.1.1)

Π_t = a technology-based temperature acceleration factor

Π_v = a voltage stress factor

Π_e = a factor associated with operating environment (ground, airborne, shipboard, space, etc.)

C_1, C_2 = circuit/package complexity factors

2. For a silicon transistor,

$$\lambda_p = \lambda_b(\Pi_e \Pi_a \Pi_q \Pi_r + \Pi_{s2} \Pi_c) \tag{8-28}$$

where

λ_b = a base failure rate
Π_a = an application/use factor (e.g., linear or switching function)
Π_r = a power factor dependent on power rating of device
Π_{s2} = a factor associated with voltage stress
Π_e, Π_q = environmental and quality factors as defined before
Π_c = a factor associated with device construction

3. For a general-purpose vacuum tube,

$$\lambda_p = \lambda_b \Pi_e \Pi_l \tag{8-29}$$

where

λ_b, Π_e, Π_l = factors as defined before but having specific values associated with device type

4. For a fixed composition resistor,

$$\lambda_p = \lambda_b(\Pi_e \Pi_r \Pi_q) \tag{8-30}$$

where

Π_r = a factor associated with the resistance value of the device
λ_b, Π_e, Π_q = factors as defined before but having specific values associated with device type

5. For ceramic capacitors,

$$\lambda_p = \lambda_b(\Pi_e \Pi_q \Pi_{cv}) \tag{8-31}$$

where

Π_{cv} = a factor associated with the capacitance value of the device
λ_b, Π_e, Π_q = factors as defined before but having specific values associated with a device type

Tables and equations are provided in prediction standards particular to each device type which assign values to each factor. The factor values vary with part class/type and are dependent on two criteria: (1) The characteristics and parameters of the device, number of gates, power or voltage rating, construction, type of material, quality, etc. (2) The broad and specific application parameters associated with the part, operating environment, thermal and electrical stresses, cycling characteristics, etc. With knowledge of those factors, a value of λ_p can be calculated for each part and the sum of all the λ_p's of the parts making up the item will constitute λ_e, the total failure rate of the item. Note that λ_e represents the failure rate of an item under a given set of stress conditions.

In some cases a given part, component, or equipment may be subject to different sets of stresses at different times during use (discussed in Section 8.5, which follows). As a consequence, predictions of reliability may have to be performed for each such set of stresses experienced by the item.

Assuming for the moment that one prediction per part is all that will be required, equipment failure rate, λ_e, can be expressed as

$$\lambda_e = \sum_{i=1}^{n} \lambda_{pi} \qquad (8\text{-}32)$$

where
n = number of parts in the equipment
λ_{pi} = predicted failure rate of part i under a given set of conditions

If, for prediction purposes, the equipment is broken down into c major components connected in series (no fault tolerance), where components may be printed circuit boards, modules, assemblies, or mechanical units, then

$$\lambda_e = \sum_{j=1}^{c} \lambda_{ej} \qquad (8\text{-}33)$$

where

$$\lambda_{ej} = \sum_{i=1}^{n_j} \lambda_{pi} \qquad (8\text{-}34)$$

λ_{ej} = predicted failure rate of the jth equipment component
n_j = number of parts in the jth equipment component
λ_{pi} is as defined previously

As can be seen from the part failure rate models, λ_p is a function of many values, most of which are empirically derived. How representative of the actual failure rate is the value of λ_p calculated, and how does its accuracy effect λ_e, the estimate used to provide reliability design visibility?

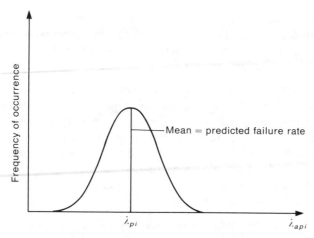

Frequency of occurrence

Mean = predicted failure rate

λ_{pi}

λ_{api}

FIGURE 8-9.   Distribution of achieved part failure rate for parts of the same type procured from different vendors (assuming a normal distribution).

Each value of λ_{pi} formed through use of a detailed failure-rate model represents an average estimate of the failure rate of the part. The estimate is based on information and data available from what is assumed to be a reasonable *cross section* of parts. Because it is a cross section parameter, it has an inherent variability. Populations of the same part from different vendors and even individual production lots of the same part from the same vendor, may exhibit different failure rates. As a consequence, if we were to (1) predict the failure rate λ_{pi} for any *given part using a detailed failure rate model*, (2) procure and test *(under the same conditions levied in the prediction) large individual populations* of that part procured from many different vendors, and from different production lots manufactured by the same vendor, and calculate the failure rate, λ_{api}, achieved for each population, then the plot of λ_{api} would take on some distribution, for example as illustrated in Fig. 8-9 with a mean equal to the predicted value λ_{pi}.

Obviously, then, since parts may be procured from any number of acceptable sources, from any part lot, a predicted value, λ_{pi}, will constitute an average failure rate which may be less than or greater than any actual value. Such variation will have an obvious impact on equipment level predictions of failure rate λ_e, since λ_e is the sum of all part failure rates. Fortunately, as will be seen, such variation is not additive, contrary to what intuition would at first conclude. It will be shown that if an equipment is made up of many different type parts, or parts supplied by many different vendors, the possible *relative* variation of equipment failure rate around the summed predicted value is less than that for any individual part type. The reason for this lies in the addition theorems for the mean and variance described in Section 5.9. Their application to this particular issue is shown in Appendix A10. A summary explanation is as follows.

The basic assumptions are:

- That the failure rate distributions associated with different part types are independent (Fig. 8-9 shows one such part type failure rate distribution).
- That distributions of failure rate for a given part type such as Fig. 8-9 represent variations in quality among different vendors.
- That the failure rate characteristics of all parts of a given type produced by a given vendor fall within a narrow range.
- That each predicted value λ_{pi} represents a true average of the cross section failure rate represented in Fig. 8-9.

The value of equipment failure rate actually achieved, λ_{ae}, can be expressed in the form

$$\lambda_{ae} = \sum_{i=1}^{m} Q_i \lambda_{ai} \qquad (8\text{-}35)$$

where

Q_i = number of parts of type i in the equipment—all received from the same vendor

M = number of different part types needed

λ_{ai} = actual failure rate for part type i

For simplicity assume that:

1. A failure rate distribution for each part type exists with mean equal to the predicted failure rate value, λ_{pi}, and a standard deviation $\sigma_{pi} = k_i \lambda_{pi}$, where $k_i > 0$. Further, each such part of that type is represented by a point on a distribution.
2. After a given vendor's product has been chosen, the failure rate associated with the product remains stable.

Each predicted part failure rate, λ_{pi}, is the mean value of failure rate associated with each part. Contracts are awarded to a group of vendors within a much larger cross section of vendors. Each member of the cross section is capable of providing the parts required. Choice of a particular vendor is tantamount to the random selection of a value of actual failure rate λ_{ai} for a part type.

The system level reliability prediction would yield a failure rate of λ_e. However, the actual total failure rate for the item would be λ_{ae} as expressed in (8-35).

If it is assumed as indicated that each value of λ_{ai} can be interpreted as being randomly selected from a distribution and then summed, λ_{ae} will take the form of a linear sum of random variables (see Section 5-9). It will take on a distribution with a *mean* $= \lambda_e$ and standard deviation, σ_{ae}, where

$$\sigma_{ae} = \left[\sum_{i=1}^{m} (Q_i \sigma_{pi})^2 \right]^{1/2} = \left[\sum_{i=1}^{m} (Q_i k_i \lambda_{pi})^2 \right]^{1/2} \tag{8-36}$$

where Q_i, m, σ_{pi}, k_i, λ_{pi} are as defined before.

The impact of the number of different part type populations, the quantity of each part type present, and the standard deviation of the failure rate of each part type population all play a role in prediction uncertainty. Prediction uncertainty can be visualized with the aid of the Tchebychev inequality (see Section 5.10) and under the special case where it is assumed that the distribution of part failure rates is normal in nature (see Section 5.9). The Tchebychev treatment will be discussed first.

Recall that the expression

$$P[(\mu + K_p \sigma) > x > (\mu - K_p \sigma)] \geq 1 - \left(\frac{1}{K_p} \right)^2 \tag{8-37}$$

denotes that the probability that a value of x in the range $(\mu \pm K_p \sigma)$ occurs is greater than $[1 - (1/K_p)^2]$, where

K_p = a number of standard deviations

μ, σ = the mean and standard deviation of *any* distribution

When applied to a single part type, (8-37) takes the form

$$P[(\lambda_{pi} + K_p \sigma_{pi}) > \lambda_{api} > (\lambda_{pi} - K_p \sigma_{pi})] \geq 1 - \left(\frac{1}{K_p} \right)^2 \tag{8-38}$$

When applied to an equipment, (8-37) takes the form

$$P[(\lambda_e + K_p \sigma_{ae}) > \lambda_{ae} > (\lambda_e - K_p \sigma_{ae})] \geq 1 - \left(\frac{1}{K_p} \right)^2 \tag{8-39}$$

For example, suppose that the standard deviations for each part type, σ_{pi}, were all large with respect to the mean values of λ_{pi}, which would signify a large degree of uncertainty with respect to the actual value of part failure

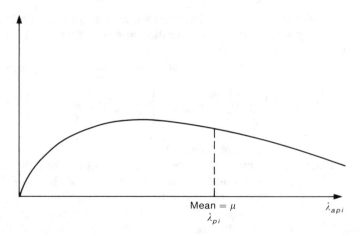

FIGURE 8-10.   Distribution of achieved part failure rate for parts of the same type procured from different vendors, exhibiting a value of $\sigma = 1.3\lambda_{pi}$ (when the distribution is not normal).

rate, λ_{api}, available for a given part type. (See Fig. 8-10 illustrating $\sigma_{pi} = 1.3\lambda_{pi}$.)

Consider first the case of a prediction of failure rate for a part of a given type, where $\mu = \lambda_{pi}$, $\sigma_{pi} = 1.3\lambda_{pi}$, and $K_p = 2$. Equation (8-43) yields

$$P[3.6\lambda_{pi} > \lambda_{api} > 0] \geqslant 0.75$$

(note that negative values of λ_{pi} are impossible) which yields the result that the probability is greater than 75% that λ_{api} for the part will be some value between zero and $3.6\lambda_{pi}$.

When prediction of *equipment* failure rate is the case and the equipment is comprised of quantities of parts representing a number of different part types, (8-39) can be applied to assess prediction uncertainty.

In order to illustrate the results simply, some assumptions will be made about the nature of the equipment:

1. Assume that each part type is present in the same quantity, which means that if m = the number of *different types* of parts in the equipment and n_t = the *total number* of parts in the equipment, the number of parts of type i in the equipment is

$$Q_i = \frac{n_t}{m}$$

2. Assume that the distribution for each part type has the same value of λ_{pi} and $\sigma_{pi} = 1.3\lambda_{pi}$; the same large value of variance as earlier.

In that case, retaining the value $K_p = 2$, as a basis for comparison (we will use the probability 0.75 as our base for comparing the range of coverage for the prediction),

$$\lambda_e = n_t \lambda_{pi}$$

$$\sigma_{ae} = \left[m\left(\frac{n_t}{m}\right)^2 (1.3\lambda_{pi})^2 \right]^{1/2}$$

$$\sigma_{ae} = 1.3\lambda_{pi}\left(\frac{n_t}{\sqrt{m}}\right)$$

The Tchebychev inequality can then be expressed in a more general form:

$$P[\lambda_e(1 + 2.6/\sqrt{m}) > \lambda_{ae} > \lambda_e(1 - 2.6/\sqrt{m})] \geqslant 0.75$$

If, for example, the equipment was composed of 10 ($m = 10$) different part groups, the probability is greater than 75% that the achieved value of λ_{ae} will lie between $1.82\lambda_e$ and $0.18\lambda_e$. This is significantly different from the part prediction variability figure for each part type of between 0 and $3.6\lambda_{pi}$.

If $m = 30$, the probability is greater than 75% that the achieved value of λ_{ae} will be between $1.47\lambda_e$ and $0.53\lambda_e$. Figure 8-11 shows a chart for this relationship under the assumptions outlined above for various values of m denoting total number of different part types used, where each part type is

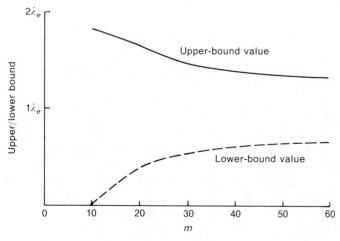

FIGURE 8-11.   0.75 Tchebychev bound on λ_{ae} as a function of m.

assumed to come from a *different* production line (each λ_{api} *per part type* is considered to be independent).

Hence, we see that the proportional degree of uncertainty associated with predictions of *equipment* failure rate is less than that associated with the failure rate prediction of a *single part* in the equipment. This occurs because a *single part* λ_{api} can take on a value very distant from its mean, λ_{pi}; see Fig. 8-9. If many different parts are chosen from the same distribution or from different distributions, an averaging process takes place, and it is unlikely that the summation of failure rates of parts from those sources will take on an extreme value.

Assuming the distribution of λ_{ae} is normal, prediction uncertainty can be treated using the characteristics of the normal distribution (as done in Section 6.6.3.1), i.e.,

$$[2p - 1] = P_z \text{ of the time } \lambda_{ae} \text{ would be found inside the range } \lambda_e \pm K_p \sigma_{ae}$$

where
    K_p = the coefficient of the normal distribution which defines a probability

$$p = \int_{-\infty}^{K_p} \phi(x)\, dx$$

$\phi(x)$ = a normal distribution

8.4.2 Detailed Reliability Prediction: Deterministic Approach

A deterministic reliability prediction is one which provides information on time to failure of an item based on the physical and/or chemical change processes associated with the item. Such processes act to change the character of the item and ultimately result in failure. Failure processes include fatigue, corrosion, diffusion, migration, erosion, elastic and inelastic actions, leakage, material breakdown, etc. The rate of progression of any given failure process is dependent on the character of the application and operational stresses imposed on an item as well as the defect state of the item itself. The number and types of defects associated with a device will determine when a failure threshold will be reached for any given failure process under a given set of operating conditions. The set of operating conditions themselves (i.e., the application and operational stresses), in the absence of outright defects, can cause failure resulting from oxidation or fatigue. In summary, we define a deterministic reliability prediction approach as one which is capable of translating application and environmental stresses on an item to their effects on the rate of progression of physical and chemical processes which result in item failure. The prediction process is dependent on the identification of the failure process; availability of tools, models, and data capable of linking the rate of progression of such processes to application and environmental stresses applied; and capability to identify and eliminate or limit the defects associated with a given item.

Reliability prediction by deterministic means through a reliability physics approach was suggested in the 1960s.[13-16] The objective was to relate changes at the part molecular level in a given environment, and under a given set of conditions, to changes in the characteristics of the part over time.

The information gained from such an approach served as a valuable means to improve part reliability. But the procedure was never practically refined for use as a prediction technique, per se, because:

1. Isolation and identification of all failure processes acting within a part proved to be difficult.
2. The screening procedures of that era for parts and assemblies resulted in the acceptance and use of items which contained undetected or undetectable defects (i.e., prone to failure processes or failure process rates which were not predictable). Parts found in this category contributed more to equipment total failure rate than the parts whose behavior was predictable.

More recently a deterministic prediction approach based on structural integrity considerations has been suggested to form the basis of a more generic reliability prediction process. Structural integrity procedures have been successfully applied in the past to engines and other structural components. But not until recently has their use been suggested for more general reliability prediction applications, specifically to electronics.[17-20] Research, development and validation of this procedure is currently under development.

The integrity approach is based on the assumption that most failures, even in electronic components (parts are made up of material bonds, layers, and connections), have a mechanical or metallurgical cause. The failure process is influenced by the mechanical stresses (thermal, shock, vibration, humidity, pressure, etc.) to which the item is exposed over a given service life, i.e., cycles or time. The structural integrity approach is considered to be a component of the overall design process that includes tests, analyses, inspections, feedback, and control of quality.

8.5   SPECIAL CONSIDERATIONS IN PREDICTING RELIABILITY

Sometimes reliability equipment prediction is not as straightforward as determining the failure rates of all parts in the equipment and summing them. In some cases the equipment/system will be:

- Subject to duty cycles or operation in different modes for different portions of mission time. These result in changes in application stresses and, hence, the failure rates associated with different part modes or states of operation.
- Charged with the performance of more than one function, where each function requires the support of different portions of the equipment/system. As a consequence, the failure rate of the equipment/system will be function-dependent.
- Subject to different stresses during different phases of a mission.
- In a dormant or nonoperating state during part of the mission. Such states exhibit different failure rates than when the equipment system is actively performing its functions.

All of these will affect the reliability process. Relatively simple means of coping with such situations will be discussed in the next subsections.

8.5.1   Duty-Cycle Mode Effects on Reliability

For many applications, all components (modules, assemblies, mechanical units) making up an item are not operating in the same mode during all portions of a given mission or use. As mode of operation changes, generally so do the application stresses on the component. The result is a change in failure rate. If such changes are relatively minor, they can usually be ignored. However, if they result in significant changes in failure rate, they must be considered. We handle such situations by calculating an *equivalent failure rate*, λ_{eq}. An equivalent failure rate may be defined as that single failure rate which, if acting throughout the entire span of the mission, would provide the same results as the different failure rates acting during different portions of the mission. In the event a calculation of failure rate were made over a large number of missions, the value calculated would actually represent an estimate of the *equivalent failure rate* (see also Section 5.3, Expected Value). For a component of an equipment subject to operation in more than one mode, component failure rate, λ_{eqj}, may be expressed as

$$\lambda_{eqj} = \sum_{i=1}^{h} P_{ji}\lambda_{eij} \qquad (8\text{-}40)$$

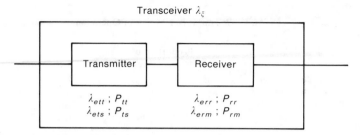

FIGURE 8-12.   Block diagram of an equipment broken down to two components each with two modes of operation.

$\lambda_{ett}, \lambda_{ets}$ = predicted failure rates of transmitter component in each of its modes of operation
$\lambda_{err}, \lambda_{erm}$ = predicted failure rates of receiver component in each of its modes of operation
P_{tt}, P_{ts} = proportion of time transmitter component is in each of its modes of operation
P_{rr}, P_{rm} = proportion of time receiver component is in each of its modes of operation

where
$\quad P_{ji}$ = proportion of mission time component j is in operating mode i
$\quad \lambda_{eji}$ = predicted failure rate of component j while in operating mode i
$\quad h$ = number of different operating modes associated with component j

Equipment equivalent failure rate for an equipment, λ_{eq}, made up of a number of components can then be calculated as

$$\lambda_{eq} = \sum_{j=1}^{c} \lambda_{eqj} \qquad (8\text{-}41)$$

where c = the number of components in the equipment.

Hence, λ_{eq} may be calculated through the use of (8-40) and (8-41) for an equipment/system comprised of a mix of components, which are subject to operation in more than one mode.

For example, take the simple transceiver depicted in Fig. 8-12 broken down into a two-component block diagram. For the transceiver to function, transmit and receive functions do not take place simultaneously. The transmitter will be in a "standby" state while the unit is in the receive mode. The receiver will also be in a receiver "standby" state while the unit is in a transmit mode. Each such state reflects a different failure rate for the component blocks.

Assume that the two modes of operation for the transmitter component are "transmit" and "standby," with respective predicted failure rates λ_{ett} and λ_{ets}. Assume that the two modes of operation for the receiver are "receive" and "mute," with respective predicted failure rates λ_{err} and λ_{erm}.

From the mission profile and contact with the customer, the proportion

CHART 8-7 Transceiver Modes of Operation

State	Proportion of Mission Time	Failure Rate
Transmitter–transmit	P_{tt}	λ_{ett}
Transmitter–standby	P_{ts}	λ_{ets}
Receiver–receive	P_{rr}	λ_{err}
Receiver–mute	P_{rm}	λ_{erm}

of the time P_{ji} that the component can be expected to be in each state of operation is determined as in Chart 8-7.

The equivalent failure rate for the transceiver may be represented as

$$\lambda_{eq} = \lambda_{eqt} + \lambda_{eqr} \tag{8-42}$$

where

λ_{eqt} = the equivalent failure rate for the transmitter
λ_{eqr} = the equivalent failure rate for the receiver
$\lambda_{eqt} = P_{tt}\lambda_{ett} + P_{ts}\lambda_{ets}$
$\lambda_{eqr} = P_{rr}\lambda_{err} + P_{rm}\lambda_{erm}$

For the transceiver in question, let $P_{tt} = 0.3$, $P_{ts} = 0.7$, $P_{rr} = 0.7$, $P_{rm} = 0.3$, $\lambda_{ett} = 0.002$, $\lambda_{ets} = 0.001$, $\lambda_{err} = 0.0005$, $\lambda_{erm} = 0.0006$. Taking into account the duty cycle factors,

$$\lambda_{eq} = 0.0013 + 0.0053 = 0.00183$$

or mean time to failure of 546 hours.

If one assumed that the system was in the transmitter-transmit and receiver-mute mode *all the time*, i.e., P_{tt}, $P_{rm} = 1$, the equipment failure rate would equate to a value of 0.0026, or a mean time to failure of 385 hours.

Bear in mind that we have performed no sleight of hand; if all the operating mode failure rates were predicted accurately and if the proportion of times spent in each mode are accurate, mean time to failure of 546 hours or a failure rate of 0.00183 is exactly what would be measured from experience results. As a consequence, a reliability prediction which ignores this factor can send the wrong message.

In general, if duty cycle/modes of operation are to be considered, we need first to provide ourselves with the visibility we need to fill

B_1			B_2			B_3			B_4			B_5		
Mode	λ_{ij}	P_{ij}	Mode	λ_{ij}	P_{ij}	Mode	λ_{ij}	P_{ij}	Mode	λ_{ij}	P_{ij}	Mode	λ_{ij}	P_{ij}
b_{11}	λ_{11}	0.30	b_{21}	λ_{21}	0.70	b_{31}	λ_{31}	1	b_{41}	λ_{41}	1	b_{51}	λ_{51}	0.50
b_{12}	λ_{12}	0.70	b_{22}	λ_{22}	0.30							b_{52}	λ_{52}	0.50

FIGURE 8-13.   Block diagram providing information on duty cycle/operating mode factors.
B_{ij} = the jth mode of operation associated with block i
λ_{ij} = the failure rate associated with the jth mode of operation of block j
P_{ij} = the proportion of mission time block i is in mode j

in all the blanks. Thus the first step is to develop an *equipment/ component mode of operation chart* similar to Fig. 8-13. This can be approached through use of a reliability block diagram as discussed in Section 7.1. The item is broken down to the level of detail most appropriate. At this point in the development, when a detailed prediction can be performed, a most detailed block diagram can be generated. Figure 8-13, a more expanded version of Fig. 7-1d, shows an equipment broken down to five blocks, $B_1, \ldots, B_5$. Determine the modes of operation b_{ij}, the failure rate associated with each mode of operation, λ_{ij}, and, based on a typical mission profile (discussed in Chapters 3 and 4), determine the proportion P_{ij} of mission time that a block is in each mode. In general, for each component block i,

$$\sum_{j=1}^{h} P_{ij} = 1$$

Using the data from the equipment/component mode of operation chart, (8-40) and (8-41) are applied to calculate λ_{eqj} and λ_{eq}.

The applicability of the above to component (block) failure distributions which are exponential is obvious and straightforward. However, in order to apply mode of operation considerations to distributions exhibiting non-constant hazard rates, the situation becomes significantly more complex. The complexity results from a number of factors, one of the most significant being that *different modes of operation or duty cycles can result in different times of operating use for parts and assemblies making up an equipment or component.* As a consequence, an operational history of t hours on an equipment will not necessarily relate to an operational history of t hours on all its components, assemblies, or parts. In fact, it may result in different operational histories for a number of different items within the same component. This will cause the original failure distribution of the component to change from that which would be expected even if all of its portions aged uniformly. See Appendix A11 for information on the application of this concept to components which exhibit Weibull distributions (a nonconstant hazard distribution) of failures when we assume that all components age homogeneously.

8.5.2 Multifunction Systems Reliability Prediction

Many modern-day systems are complex and contribute to more than one function during a given mission. Furthermore, a given mission may involve the performance of a number of different functions and a system might be called upon to perform a number of different missions at different times. A typical example is a military aircraft where functions performed during a typical mission can involve control, communication, navigation, fire control, data processing, etc. In addition the system may be called upon to perform different missions at different times. As a consequence, a reliability prediction is often required both on a function and on a mission basis for the same system.

Complicating the situation is the fact that many systems are subject to integrated design in order to meet their overall requirements in the most cost-effective manner. Integrated design involves the shared use of systems' capabilities and capacities such that all functions can be performed using the minimum of weight, volume, quantity, and cost of hardware. Examples may include use of a common power supply, or a computer or signal processor to perform several different functions.

The bottom line for all of the above is that reliability prediction is not always as simple as summing up all the failure rates of the parts in a given system. A better procedure to follow is best described by way of an illustration.

Assume that a system is comprised of six components, A to F, as shown in Fig. 8-14 with failure rates λ_a to λ_f, respectively. The system is capable of performing five different functions, f_1 to f_5. Each function *requires the utilization of the unique set of components shown*. The system is configured to undertake three different types of missions, types 1, 2, and 3, and the success of each mission type depends on the successful completion of a unique set of functions identified in the figure. Hence, mission 1 requires the successful performance of functions f_2 and f_4; mission 2 the successful performance of functions f_1, f_2, and f_5; mission 3 the successful performance of f_2, f_3, and f_4.

The reliability prediction is performed on three levels: the function level, the mission level, and the total system hardware level. In the event a shared component is required in order to perform two or more functions during a given mission, its failure rate is considered only once.

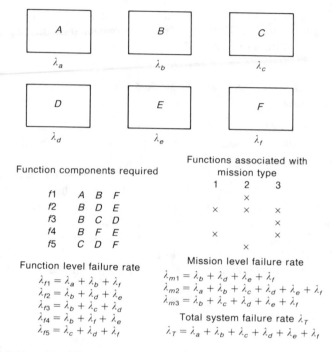

Function components required

f1	A	B	F
f2	B	D	E
f3	B	C	D
f4	B	F	E
f5	C	D	F

Functions associated with mission type

	1	2	3
f1		×	
f2	×	×	×
f3			×
f4	×		×
f5		×	

Function level failure rate

$$\lambda_{f1} = \lambda_a + \lambda_b + \lambda_f$$
$$\lambda_{f2} = \lambda_b + \lambda_d + \lambda_e$$
$$\lambda_{f3} = \lambda_b + \lambda_c + \lambda_d$$
$$\lambda_{f4} = \lambda_b + \lambda_f + \lambda_e$$
$$\lambda_{f5} = \lambda_c + \lambda_d + \lambda_f$$

Mission level failure rate

$$\lambda_{m1} = \lambda_b + \lambda_d + \lambda_e + \lambda_f$$
$$\lambda_{m2} = \lambda_a + \lambda_b + \lambda_c + \lambda_d + \lambda_e + \lambda_f$$
$$\lambda_{m3} = \lambda_b + \lambda_c + \lambda_d + \lambda_e + \lambda_f$$

Total system failure rate λ_T

$$\lambda_T = \lambda_a + \lambda_b + \lambda_c + \lambda_d + \lambda_e + \lambda_f$$

FIGURE 8-14.   A multifunction system.

The general procedure then can be summarized as follows:

1. Define the functions which must be performed by the system.
2. Develop a block diagram of the system to the level of detail needed to identify the components *necessary* to the performance of each function. One block might represent a computer, another might represent a signal-processing module.
3. Identify the set of component blocks which must perform satisfactorily in order to successfully perform each function.
4. Sum up the failure rates associated with each such set of components (redundant or fault-tolerant capabilities will be treated later). This will yield the measure of failure rate, λ_f, associated with a given *function* if a prediction of function reliability is required.
5. Identify the *functions* associated with the performance of each different mission.
6. Identify and list the component blocks which must perform satisfactorily in order to successfully meet *all* function objectives for each *mission*. In the event that one component block is required to perform more than one function for a given mission type, list it only once.

7. Sum up the failure rates associated with the items listed. This will yield the measure of mission-level failure rate, λ_m, for each mission if a prediction of mission reliability is required.

8.5.3   Unequal Operating Times Associated with System Components

Sometimes there are cases where *all* the components making up an item do not necessarily have to operate at the same time or through all the phases of a mission. Both situations militate against a simple summation of predicted failure rates for components to evaluate overall item failure rate. To put this in its proper perspective, imagine a system mission where the mission is broken down into a number of distinct phases as in Fig. 8-15. In order to perform the mission, certain components might have to function throughout all phases of the mission, others only at specific times during the mission. The solid lines indicate performance during the phases of the mission through which the component must operate. A dotted line indicates performance

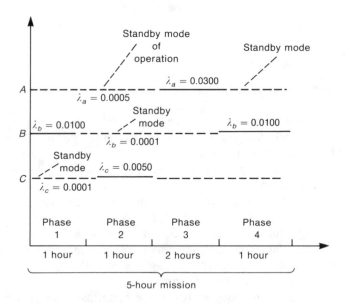

FIGURE 8-15.   Reliability taking into account mission phases.

$$\left.\begin{array}{l} \lambda_a = 0.0300 \\ \lambda_b = 0.0100 \\ \lambda_c = 0.0050 \end{array}\right\} \begin{array}{l} \text{Operating} \\ \text{modes} \end{array} \qquad \left.\begin{array}{l} \lambda_a = 0.0005 \\ \lambda_b = 0.0001 \\ \lambda_c = 0.0001 \end{array}\right\} \begin{array}{l} \text{Standby modes,} \\ \text{operation not} \\ \text{required} \end{array}$$

during phases where the component is not called on to work. This is designated as a standby state and a component can fail during a standby state. Failure rates (since failure can occur from both states) are associated with *each type of state*. A standby state can have either the same or a lesser failure rate than when the component is actively performing its function. There are instances where at specific times during the mission a component can fail after performing its function (when it is no longer needed) and not affect the mission at all. This occurs after phase 3 for component *A*, after phase 2 for component *C*. A failure at any time for component *B*, however, will render it incapable of performing throughout the entire mission and result in system failure.

As before, we compute an equivalent failure rate for each component and sum the failure rates for all components making up the system. We start by expressing a simple logic of operation and failure for each component of the system based on the following. If the component is required to operate *at least* over the last phase of the mission, then any failure in the component at *any* time during the mission duration, even before it is scheduled to function, will result in a component failure which will result in system failure. If, however, the component is required to operate during a mid-phase of the mission and is not required for later phases, then any failure which occurs in the component after its required use does not affect performance. In this case only failures between mission initiation and completion of its particular operation phase will result in relevant failure.

In the cases where a component is required to operate over the last phase of the mission, we must take into account the failure rates experienced over all parts of the mission. The equivalent failure rate, λ_{eqj}, of such a component may be represented as

$$\lambda_{eqj} = \frac{1}{T} \sum_{i=1}^{m} T_i \lambda_i \tag{8-43}$$

where the mission may be broken down into a series of consecutive phases $i = 1, 2, \ldots, m$ and

λ_i = the failure rate of the component during phase i
T_i = the time during the mission that the component is in phase i
m = the number of phases making up the mission
T = total mission time

The equivalent failure rate for a component subject to a mid-phase operation, for example, components such as *A* and *C*, can be represented as

$$\lambda_{eqj} = \frac{1}{T} \sum_{i=1}^{h} T_i \lambda_i \tag{8-44}$$

where h denotes the last phase of the mission that required the components' performance.

If a system was made up of a number of components, each required to operate during different portions of a mission, the equivalent failure rate of the system, λ_{eq}, could be represented as

$$\lambda_{eq} = \sum_{j=1}^{n} \lambda_{eqj} \qquad (8\text{-}45)$$

where
λ_{eqj} = the equivalent failure rate over the mission term for component j
n = number of components in the system

For example, take the system composed of three components A, B, and C with failure rates as indicated in Fig. 8-15. The mission in question is one which will take five hours and is made up of four distinct sequential phases. Components required, times of operation, and failure rates associated with each component are as shown. Component A must function during the third phase, and can cause a system failure if it fails prior to that time. If it fails after the third phase, it has no impact on system performance. Component B must function during the first and last phases and any failure in component B over mission length will cause system failure. Component C must function during the second phase, but can fail prior to that time. If it does, it will cause a system failure. If it fails after phase 2, it has no impact on system performance. What is the predicted failure rate for the system?

For component A:

$$\lambda_{eqa} = (1/5)[0.0005 + 0.0005 + 2(0.0300)] = 0.0122$$

For component B:

$$\lambda_{eqb} = (1/5)[0.0100 + 0.0001 + 2(0.0001) + 0.0100] = 0.0041$$

For component C:

$$\lambda_{eqc} = (1/5)(0.0001 + 0.0050) = 0.0010$$

System failure rate for the particular mission:

$$\lambda_{eq} = 0.0122 + 0.0041 + 0.0010 = 0.0173$$

8.5.4 Accounting for Stresses That Are Mission Phase-Dependent in Reliability Prediction

Prediction of mission reliability must sometimes take cognizance of the stresses introduced during different phases of a given mission which can affect failure rate. This is particularly true of space and missile systems, where stresses during blast-off/takeoff and reentry/recovery/landing are more severe than during climb and descent, and stresses during climb and descent more severe than during travel and system checkout. This situation may be treated in a manner similar to that which accounted for different modes of operation, except that the focus here is more directly centered on the stresses imposed on a given component during each mission phase. The failure rate profile for a given component may take a form shown in Fig. 8-16. That is to say, different phases of the mission may introduce stresses on the component which affect failure rate during those phases.

Let the mission be broken down into a series of consecutive phases $i = 1, 2, 3, \ldots, m$, with

λ_i = the failure rate of a component during phase i of the mission

T_i = the time the component is in phase i

m = the number of mission phases

$T = \sum_{i=1}^{m} T_i$ = total mission time

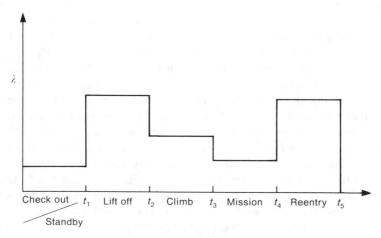

FIGURE 8-16.   Failure-rate profile per mission phase.

As before an equivalent failure rate can be computed as

$$\lambda_{eq} = \frac{1}{T} \sum_{i=1}^{m} T_i \lambda_i \tag{8-46}$$

For example, consider a space vehicle undertaking a mission consisting of the following five phases with associated failure rates and mission times.

Checkout/standby	$\lambda = 0.01$	$T = 1$ hour
Blastoff	$\lambda = 0.5$	$T = 0.17$ hour
Climb	$\lambda = 0.1$	$T = 0.25$ hour
Mission	$\lambda = 0.02$	$T = 3$ hours
Reentry	$\lambda = 0.1$	$T = 0.50$ hour

Mission duration (including checkout/standby) = 4.92 hours

$$\lambda_{eq} = \frac{1}{T} \sum T_i \lambda_i = (1/4.92)(0.010 + 0.085 + 0.025 + 0.060 + 0.050)$$

$$= 0.047$$

Hence, the system may be treated as an item having a failure rate of 0.047 over the mission.

8.5.5   Accounting for Failure Rates of Nonoperating Systems

Failure-rate requirements are usually levied on an equipment in a functioning, operating, or mission-use environment. As a rule, failure rates during nonoperating periods are considered negligible, if not nonexistent. There are times, however, when such failure rates must be accounted for in a prediction.

Most parts and components can be considered to be prone to the actions of chemical and physical failure mechanisms. These act at rates determined by the environments to which the item is exposed and the duration of exposure to such environments. Parts, components, and equipments have been found to fail during periods of storage or while in unenergized states. In situations when an unenergized item is in a benign environment (controlled temperature, humidity, no vibration or shock, etc.), especially when the ratio of the time spent in the unenergized, state to the time spent in an energized state is small, a failure rate of zero can usually be assigned to the nonoperating state. This does not necessarily mean that the failure rate is actually zero. It conveys the assumption that the value of nonoperating failure rate is so small compared to the operating failure rate that it can be ignored.

However, if an item is in an unenergized state but is either frequently or for long durations exposed to environmental stresses, or in unenergized states or in storage for long periods of time, then failure rate in the unenergized state cannot logically be ignored.

Nonoperating or domant failure rates have been a subject of concern and study since at least the late 1950s.[21] It has been of particular concern with respect to missiles subject to long-term storage[22-24] and of general concern for electronics in general.[25,26]

The literature contains references to particular situations where such failure rate is significant and to particular systems where nonoperating failure rate was practically nonexistent. A study analyzing the dormant failure-rate behavior of a tactical missile as discerned by nonoperational tests, "R & M 2000 Action Plan for Tactical Missiles" by J. Malcolm,[23] presented data which indicated that for storage periods exceeding five years, no significant nonoperating failure rate was observable. This finding was associated with both electronic and other portions of the missiles.

Studies of electronic subsystems installed in aircraft documented in two RADC reports, "Operational Influences on Reliability,"[26] and "Nonoperating Failure Rates for Avionics Study,"[27] indicated a potentially significant presence of a nonoperating failure rate.

Data on failure rates for parts in dormant states was collected and prediction models were developed. These were documented in the 1985 RADC report. "Impact of Non-operational Periods on Equipment Reliability."[28] The results of that effort formed the basis of the first quantitative approach to predicting nonoperating failure rate. The procedure developed was proposed for inclusion in future revisions of Mil-Hdbk-217. The prediction procedure developed was based on the quantification of four general model factors:

Π_{nq} = a nonoperating quality factor accounting for the effects of different quality levels

Π_{ne} = a nonoperating environmental factor accounting for the influence of all environmental stresses

Π_{nt} = a nonoperating temperature factor accounting for the effect of nonoperating ambient temperature

Π_{cyc} = an equipment on–off cycling factor accounting for the effects of transients due to equipment power cycling

λ_{nb} = a base failure rate associated with each part type.

Data identifying values of such factors for different part types, environments, stresses, and qualities are provided.[28]

The model generated to estimate nonoperating failure rate, λ_{np}, for each type part took the general form

$$\lambda_{np} = \lambda_{nb}\Pi_{nt}\Pi_{nq}\Pi_{ne}\Pi_{cyc} \tag{8-47}$$

Nonoperating failure rate, like operational failure rate, is a function of parts control, quality, part application, and stresses. Technical guidelines specific to such elements that can be used in design control of nonoperating failure rate are documented in the 1988 RADC report "Reliability/ Maintainability/Testability Design for Dormancy."[25]

Even if nonoperating failure rates are small, if they operate for long enough periods of time, they can result in proportionately higher incidences of failure for an item. Take, for example, a subsystem—say, a part of an aircraft—that is operated only $p\%$ of total time and is in a nonoperating state the remaining $(1 - p)\%$ of the time. Such a situation is not at all unusual for emergency backup systems or, for example, for a fighter aircraft which might spend only 100–200 hours per month in a flight status out of a total of 720 hours available (24 hours per day, 30 days per month). Let

λ_e = predicted failure rate of a item in its operating state

λ_{en} = predicted failure rate of an item in its nonoperational state

$d = \lambda_{en}/\lambda_e$

p = proportion of time a unit is in an operating state

Then it can be shown (see Appendix A12) that the proportional increase, ΔF, in failures over that predicted by λ_e is

$$\Delta F = \frac{d(1 - p)}{p} \tag{8-48}$$

For a value of $p = 0.2$, Fig. 8-17 shows the behavior of ΔF for various values of $d = \lambda_{en}/\lambda_e$. As can be seen, for values of d as low as 0.05, $\Delta F = 0.2$, indicating an increase in failures of 20% over that predicted. For values of $d = 0.2$, $\Delta F = 0.8$, indicating an increase in failure of 80%. See Ref. 29, which indicates similar effects.

Failures which occur during nonoperating periods are evident when the equipment is first energized. (Note: Differentiating between failures which occur due to shock effects when an equipment is first turned on and failures which occured prior to that time, is almost always impossible.) They *count as an item failure* and have *the same consequences as a failure which occurs*

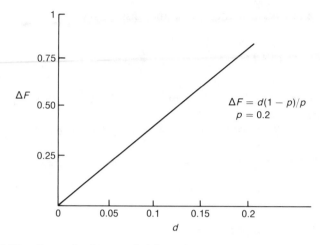

FIGURE 8-17.   Proportional increase in failures, ΔF, over that predicted due to nonoperating failure rate, $\lambda_{en} = d\lambda_e$.

during an operating state, especially if discovered when the equipment is energized just prior to mission start when no time is available to make repairs. As a consequence, where nonoperating failure rates exist, their effects must be blunted and/or taken into account in the prediction.

We will consider first blunting their effects such that they provide only an insignificant impact on *mission* reliability. One simple and popular approach to this is to perform checks on the item sufficiently in advance of mission initiation time to allow for repairs to the item, if necessary, prior to mission start. For those systems subject to mission initiation at a random point in time, a "check" performed, say, every T_c, hours could be implemented.

Under those circumstances, a maximum of T_c hours would remain prior to the initiation of the next mission. If the equipment was found to be failed, sufficient time would generally be available to make repairs. If the equipment was found to be operating satisfactorily, it has only to remain in that nonfailed, nonoperating state for a maximum of only T_c hours.

It can be shown (see Appendix A12) that the proportional increase in *mission failures*, ΔF_m (when nonoperating failures take on an exponential failure distribution), over that predicted by λ_e is equal to

$$\Delta F_m = \frac{d(1 - p)P(T_c)}{p} \tag{8-49}$$

where $P(T_c) = $ the proportion of total (nonoperating) failures that will occur between the last performance check and mission start:

$$P(T_c) \leqslant \frac{1 - \exp(-\lambda_{en} T_c)}{1 - \exp(-\lambda_{en} T_m)}, \qquad 0 < T_c \leqslant T_m \tag{8-50}$$

where nonoperating failures follow an exponential distribution; T_m = average time between missions.

When $\lambda_{en} T_c$, $\lambda_{en} T_m \leqslant 0.02$, which will usually be the case, $P(T_c)$ can be approximated as

$$P(T_c) = \frac{T_c}{T_m} \tag{8-51}$$

so that we can express ΔF_m as

$$\Delta F_m \sim \frac{d(1-p)T_c}{T_m p} \tag{8-52}$$

Figure 8-18 shows ΔF_m, the proportional increase in mission failures (over that predicted by λ_e) caused by nonoperating failure rates, as a function of $P(T_c)$.

Nonoperating failures will, of course, continue to occur and will impact logistics reliability as before. But, as can be seen, the effects of nonoperating

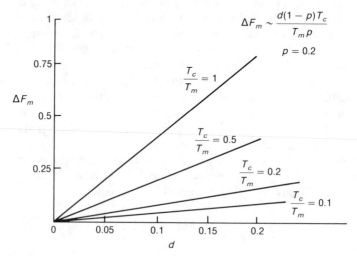

FIGURE 8-18.   Proportional increase in mission failures over that predicted as a function of $P(T_c) \sim T_c/T_m$.

failure rate with respect to mission reliability can be practically eliminated with proper monitoring policies. Note that with a nonoperating failure rate as high as 20% of an operating failure rate, ΔF can be reduced to less than 10%.

Adjustment of predicted failure rate, λ_e, to account for the additional failures which occur during nonoperating periods is straightforward. We must, however, make two types of adjustments to λ_e:

1. One adjustment to develop an equivalent failure rate, λ_{eq}, to be used for prediction of *mission* reliability. This is necessary to adjust predicted failure rate, λ_e, to account for those failures which occur during nonoperating periods and are not discovered until the item is energized at mission start. (As those failures, too, result in mission failure.) The adjustment modifies λ_e to account for the proportionate increase in mission failures ΔF_m due to nonoperating failure rate:

$$\lambda_{eq} = \lambda_e(1 + \Delta F_m) = \lambda_e + \frac{\lambda_e(d(1 - p)T_c)}{T_m p} \tag{8-53}$$

2. One adjustment to develop an equivalent failure rate, λ_{eq}, to be used for prediction of logistics reliability needs. This is necessary to adjust predicted failure rate, λ_e, to account for *all* failures which occur during nonoperating periods; those which are found during performance checks (and do not result in mission failure); and those which are first found when the item is energized at mission start. It is necessary to plan and adjust sparing levels required, modifying λ_e to account for the proportionate increase in failure incidents:

$$\lambda_{eq} = \lambda_e(1 + \Delta F) = \lambda_e + \frac{\lambda_e(d[1 - p])}{p} \tag{8-54}$$

For example, a subsystem associated with an aircraft has the following failure rate and operational characteristics:

$\lambda_e = 0.002$; $\lambda_{en} = 0.1$; $\lambda_e = 0.0002$; $\lambda_{en}/\lambda_e = d = 0.1$
Mission operating time per month = 122 hours
Mission nonoperating time per month = 598 hours
$p = 122/[122 + 598] = 122/720 \sim 0.17$
Time between missions $T_m \sim 20$ hours
Subsystem to be checked every 5 hours prior to mission start, i.e., equivalent to a subsystem check less than or equal to T_c hours prior to mission start

$$T_c = 5$$

What is the equivalent failure rate associated with mission reliability? What is the equivalent failure rate associated with logistics reliability?

λ_{eq} for mission reliability is

$$\lambda_{eq} \sim \lambda_e(1 + \Delta F_m) = \lambda_e + \frac{\lambda_e d(1-p)T_c}{T_m p}$$

$$= 0.002 + \{[(0.002)(0.1)(0.83)(5)]/[(20)(0.17)]\} = 0.0022$$

λ_{eq} for logistics reliability is

$$\lambda_{eq} = \lambda_e(1 + \Delta F) = \lambda_e + \frac{\lambda_e d(1-p)}{p}$$

$$= 0.002 + \{[0.002(0.1)(0.83)]/0.17\} = 0.0030$$

8.5.6   Accounting for Operational Influences on Reliability

As indicated earlier, predicted failure rate is a representation of reliability design capability in a defined environment where the item is maintained in accord with a given concept/policy and incurs no failures due to maintenance/operator or other outside source. The degrees to which such factors impact failure rates have yet to be defined except in subjective or qualitative terms. Studies and approaches in the past, dating at least from the 1968 paper "How Accurate Are Reliability Predictions,"[30] to the 1984 RADC report "Reliability/Maintainability Operational Parameter Translation,"[6] have attempted to develop relationships between predicted and field observed failure rates.

Such studies take into account the global effects of such idiosyncracies over a wide population of items in a wide variety of operational/maintenance scenarios. They can provide a statistical bound on failure rate behavior in general, but not one associated with a given set of operational/maintenance/ human factor interfaces. They include (or suffer from) the effects of characteristics unrelated to those introduced by operational and maintenance influences. Those extraneous characteristics cannot easily be isolated. Examples include the accuracy of the prediction itself and definition/ interpretation inconsistencies with respect to tagging and recording data.

A 1976 RADC effort, "Operational Influences on Reliability,"[26] had as its objective the development of a rudimentary prediction approach to allow better predictions of field failure rate to be made. The effort identified both definitional and operational factors which influenced the difference between predicted and field values of reliability and developed empirical relationships which could be used as first-order estimates of field failure rate for airborne electronic equipment.

The results of the effort were based on a statistical analysis of 16 different avionic equipments in 30 different application environments. Regression equations were developed for field MTBF (θ_0, where field failure rate $\lambda_0 = 1/\theta_0$), linking θ_0 with prediction results and "general" operational *maintenance* and *equipment* factors.

The actual values that the regression results yield cannot be relied on to represent equipment of later vintages, and as a consequence will not be included here. The fact that correlation was found between certain operational factors and θ_0 may well, however, be indicative of a more general relationship that can be of use today. It was found that a significant proportion of the difference between predicted and demonstrated values could be attributed to three factors:

1. Definitional elements such as definition of failure, establishment and interpretation of total field operating time.
2. Maintenance and handling elements. These include the total number of maintenance actions of all types an item is subjected to above and beyond its theoretical ideal. For example,
   a. The number of "no cause found" calls for maintenance.
   b. The number of scheduled maintenance and adjustment events per unit time.
   c. The number of times the unit must be removed to facilitate maintenance on another unit.
   d. The degree of modularization associated with the design.
3. Equipment use factors. These include
   a. Number of missions/flights per calender period.
   b. Duration of typical use interval or mission.
   c. Proportion of time spent in operation vs standby modes of operation, i.e., the effects of nonoperating failure rate.

In general it was found that as maintenance and handling increased, θ_0 decreased; as equipment use factors increased, θ_0 increased.

References
1. Mettler, G. and Walczak, M. (1982). "Assuring RMA Requirements for Standoff Weapon." Proceedings, 1982 Annual Reliability and Maintainability Symposium, IEEE, Piscataway, NJ.
2. Seidl, R. and Garry, W. (1990). "Pi Factors Revisited." Proceedings, 1990 Annual Reliability and Maintainability Symposium, IEEE, Piscataway, NJ.

3. IIT Research Institute (1988). "Reliability Prediction Models for Discrete Semiconductor Devices." RADC Report RADC-TR-88-97. Available from the Defense Technical Information Center. Report AD-A200529.

4. Reliability Analysis Center (1985). "Microcircuit Device Reliability Trend Analysis." RAC Report MDR-21.

5. Siewiorek, D. and Swarz, R. (1982). "The Theory and Practice of Reliable System Design." Digital Equipment Corporation (Digital Press), U.S.A.

6. Boeing Aerospace Company (1984). "Reliability/Maintainability Operational Parameter Translation." RADC Report RADC-TR-84-25. Available from Defense Technical Information Center. Report AD-B087426, AD-B087507.

7. Stovall, F. (1984). "A Management Guide to Reliability Predictions." Proceedings, 1984 Annual Reliability and Maintainability Symposium, IEEE, Piscataway, NJ.

8. Hughes Aircraft Company (1985). "RADC Nonelectronic Reliability Notebook." RADC Report RADC-TR-85-194. Available from the Defense Technical Information Center. Report AD-A163900.

9. IIT Research Institute (1981). "Correlation of Field Data with Reliability Prediction Models." RADC TR-81-329. Available from the Defense Technical Informationa Center. Report AD-A111258.

10. Hughes Aircraft Company (1983). "Study of Causes of Unnecessary Removals of Avionic Equipment." RADC-TR-83-2. Available from the Defense Technical Information Center. Report AD-A127546.

11. Westinghouse Electric Corporation (1988). "Operational and Logistics Impact on System Readiness." RADC-TR-118. Available from the Defense Technical Information Center. Report AD-A201346.

12. Hughes Aircraft Company (1976). "Operational Influences on Reliability." RADC-TR-76-366. Available from the Defense Technical Information Center. Report AD-A035016.

13. Bild, C. (1964). "Failure From a Materials Point of View." Proceedings, 10th National Symposium on Reliability and Quality Control, IEEE, Piscataway, NJ.

14. Vaccaro, J. and Smith, J. (1966). "Methods of Reliability Physics." Proceedings, 1966 Annual Symposium on Reliability, IEEE, Piscataway, NJ.

15. Moyer, E. (1967). "Device Failure Distributions From Failure Physics." Proceedings, 1967 Annual Symposium on Reliability, IEEE, Piscataway, NJ.

16. Earles, D. and Eddins, M. (1963). "Reliability Physics, The Physics of Failure." Proceedings, 9th National Symposium on Reliability and Quality Control, IEEE, Piscataway, NJ.

17. Halpin, J. (1988). "Avionics/Electronics Integrity Program (AVIP) Highlights." Ninth Annual IEEE/AESS Dayton Chapter Symposium on the Avionics Integrity Program.

18. Wong, K. et al. (1987). "Culprits Causing Avionic Equipment Failures." Proceedings, 1987 Annual Reliability and Maintainability Symposium, IEEE, Piscataway, NJ.

19. Bhagat, W. (1989). "R & M Through Avionics/Electronics Integrity Program." Proceedings, 1989 Annual Reliability and Maintainability Symposium, IEEE, Piscataway, NJ.

20. Benz, G. (1990). "Using Test Data to Predict Avionics Integrity." Proceedings,

1990 Annual Reliability and Maintainability Symposium, IEEE, Piscataway, NJ.

21. Advisory Group on Reliability of Electronic Equipment (AGREE) (1957). "Reliability of Military Electronic Equipment." U. S. Government Printing Office.

22. Cherkasky, S. (1970). "Long Term Storage and System Reliability." Proceedings, 1970 Annual Symposium on Reliability, IEEE, Piscataway, NJ.

23. Malcolm, J. (1988). "R & M 2000 Action-Plan for Tactical Missiles." Proceedings, 1988 Annual Reliability and Maintainability Symposium, IEEE, Piscataway, NJ.

24. IIT Research Institute (1985). "Impact of Nonoperating Periods on Equipment Reliability." RADC-TR-85-91. Available from the Defense Technical Information Center. Report AD-A158843.

25. Lockheed Electronics Company (1988). "Reliability/Maintainability/Testability Design for Dormancy." RADC-TR-88-110. Available from the Defense Documentation Center. Report AD-A202704.

26. Hughes Aircraft Company (1976). "Operational Influences on Reliability." RADC-TR-76-366. Available from the Defense Documentation Center. Report AD-A035016.

27. Hughes Aircraft Company (1980). "Nonoperating Failure Rates for Avionics Study." RADC-TR-80-136. Available from the Defense Documentation Center. Report AD-A087048.

28. IIT Research Institute (1985). "Impact of Nonoperating Periods on Equipment Reliability." RADC-TR-85-91. Available from the Defense Documentation Center. Report AD-A158843.

29. Kern, G. (1978). "Operational Influences on Avionics Reliability." Proceedings, 1978 Annual Reliability and Maintainability Symposium, IEEE, Piscataway, NJ.

30. Feduccia, A. and Klion, J. (1968). "How Accurate Are Reliability Predictions?" Proceedings, 1969 Annual Symposium on Reliability, IEEE, Piscataway, NJ.

9

Reliability Analysis and Evaluation of Series-Connected Systems

In the previous chapter, means available to predict component, equipment, and system reliability in terms of failure rate were described. This chapter addresses the transformation of failure-rate values at the component and equipment levels into the metrics most commonly used for reliability assessment of series-connected systems. These include the probability that an item will perform its intended function over an interval of time; the measures of mean life (i.e., mean time to failure, mean time between failure, steady-state mean time to failure, etc.); and availability measures.

When all components of an equipment or system must function satisfactorily for the item to perform its function, the components are considered for reliability purposes to be connected in series. Like an old-fashioned string of Christmas tree bulbs connected in series, any one failure in any bulb will cause failure of the entire string.

An example of such a series system is depicted in Fig. 9-1. In order for the system to function over an interval of time, T, each component must satisfactorily function over the same interval of time, T.

9.1 MISSION RELIABILITY MEASURES USED FOR EVALUATION

One of the most common expressions of reliability is the mission reliability measure. It is applicable to ordnance, electronic and mechanical components, and equipments and systems which must perform a function over a finite period of time. Mission reliability provides the customer and developer with information on the probability that the item will perform satisfactorily (not fail) during the time it is needed most.

286

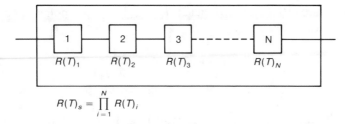

$$R(T)_s = \prod_{i=1}^{N} R(T)_i$$

FIGURE 9-1.   A system comprised of N component blocks connected in series.

In the simplest of terms, if a system is composed of a number of components and all must work in order for the system to function, and if (a) the probability of satisfactory performance over the particular time period, T, for each component of the system and (b) the number of components, n, making up the system, are known, then system reliability $R_e(T)$ can be expressed by

$$R_e(T) = R_e(T_2 - T_1) = \sum_{i=1}^{n} R_i(T_{2i} - T_{1i}) \qquad (9\text{-}1)$$

where

$R_e(T)$ $= R_e(T_2 - T_1) =$ probability that the system will satisfactorily perform over the *specific* operating time interval, T, between T_1 and T_2 which starts after the *system* has acquired T_1 hours of operating history and ends when the system has acquired T_2 hours of operating history (assuming it is operable at the start of the interval)

$R_i(T)$ $= R_i(T_{2i} - T_{1i}) =$ probability that component i will satisfactorily perform over a *specific* time interval, T, between T_{1i} and T_{2i} which starts after the *component* has acquired T_{1i} hours of operating history and ends when the component has acquired T_{2i} hours of operating history and was operable at T_{1i}

$(T_2 - T_1) = (T_{2i} - T_{1i}) = T$

T_{1i} $=$ the operating history on the ith component prior to the start of the interval of operation in question

T_{2i} $=$ the operating history on the ith component at interval end

T_{1i}, T_{2i} values may vary for different components

Equation (9-1) represents the most general form of the mission reliability measure. It is appropriate for use for any distribution of failure. It takes into account the fact that different components making up a system or equipment can have different operating ages (due to failures and replacement of components).

In Appendix A0 it is shown that $R(T_2 - T_1)$ for a component or a system can be expressed as

$$R(T) = R(T_2 - T_1) = \frac{R(T_2)}{R(T_1)} \qquad (9\text{-}2)$$

where
T_1 = the total operating time (history) on the item at the beginning of the mission (and assuming, of course, that the item is operable at the start of the mission)
T_2 = the total operating time (history) on the item at the end of the mission
T = $(T_2 - T_1)$ = mission time

While (9-1) provides the general process for generating an expression for $R_e(T)$, the exact form that the expression will take is largely dependent on the types of failure distributions associated with the system components. Generally speaking, if the system is made up of components with exponential distributions of times to failure (i.e., having constant hazard/failure rates), the expression for $R_e(T)$ will be relatively simple. On the other hand, if the distributions were ones in which hazard/failure rate varied as a function of operating time, the expression for $R_e(T)$ would be somewhat more complex. (We described hazard/failure rates associated with various distributions in Chapter 2.)

9.1.1 Components with Exponential Failure Distributions

When (9-2) is applied to assess mission reliability, $R(T)$, of a system or component which has an exponential failure distribution, $R(T)$ reduces to

$$R(T) = R(T_2 - T_1) = \frac{R(T_2)}{R(T_1)} = e^{-\lambda T} \qquad (9\text{-}3)$$

(see Appendix A13 for proof) where:
λ = hazard/failure rate associated with the system or component
T = $(T_2 - T_1)$ = the interval of time (mission) in question
$R(T)$, $R(T_2 - T_1)$ are defined as previously
T_2, T_1 are defined as previously
T_i = *any* interval of operating time (history)

Equation (9-3) signifies that mission reliability is not influenced by T_1. Only the value of T is an influence. As a consequence, any system or component with an exponential failure distribution can be considered as if it were brand new prior to each mission.

If all components of the system have exponential failure distributions, the reliability of each component i may be expressed as:

$$R_i(T) = \exp(-\lambda_i T) \tag{9-4}$$

where

$R_i(T)$ = component reliability, the probability that component i will perform satisfactorily over *any* interval (mission) of time of length T, given that the component was operable at the start of the interval. Failures, repairs and hours of component operation prior to mission start have no affect on $R_i(T)$. See Appendix A13 also for proof of these conditions

λ_i = hazard/failure rate associated with component i

T = interval of time (mission) in question (a period of operating hours measured from interval start)

Then through the multiplication law for probabilities (Section 5.7), we can develop an expression for systems reliability $R_e(T)$,

$$R_e(T) = \prod^n R_i(T) = \exp\left(-T \sum^n \lambda_i\right) = \exp(-\lambda_e T) \tag{9-5}$$

where

$R_e(T)$ = system reliability with definition analogous to that of component reliability above

n = number of components making up the system

λ_e = $\sum^n \lambda_i$

$R_i(T)$, λ_i, T are defined as before

Hence, for a system comprised of a number of components connected in series, where all components have exponential failure distributions, the failure rates of the components are summed to form the total failure rate λ_e and (9-5) is used to assess mission reliability.

Note that both $R_i(T)$ and $R_e(T)$ have the same equation form:

$$R_i(T) = R_e(T) = \exp(-a_j T) \tag{9-6}$$

where a_j = a constant based on failure rate of the item in question

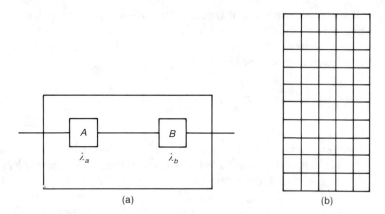

FIGURE 9-2.   (a) A system made up of two components. (b) A component made up of *m* replaceable modules.

An inelegant but effective way of determining whether or not the same failure distribution is common to a component and a system made up of such components is to examine the reliability expressions for the two. If they can be forced into the same general form just by a change or manipulation of constants, they can be considered to have the same distribution of failure.

Hence, it is obvious from (9-6) that if components have exponential distributions of times to failure, a system comprised of such components will also have an exponential distribution of time to failure.

The following examples will serve to illustrate the concepts discussed above. Assume that a system is made up of two components, A and B as in Fig. 9-2a, each with an exponential distribution of failure times.

If $\lambda_a = 0.01$ and $\lambda_b = 0.005$, what is the probability that the system will satisfactorily perform its function over a 20-hour interval/mission if we know both units are operable at the start of the interval and

- The system and all its components are brand new?
- The system and each component have acquired an operating history of 500 hours?
- The system has acquired 500 hours of operating history and has previously failed once? (Component A has been replaced and its replacement has acquired 100 hours of failure free operation, and component B has acquired a total operating history of 500 hours)

In order to illustrate that accrued operational history has no effect on reliability when components exhibit exponential distributions of failure, each problem will be solved separately, starting from the most general basis of (9-1).

If the system and all its components are brand new, then T_{1a}, T_{1b} and $T_1 = 0$; T_{2a}, T_{2b} and $T_2 = 20$. Since the most general expression for item reliability regardless of failure distribution takes the form

$$R_i(T_{2i} - T_{1i}) = \frac{R_i(T_{2i})}{R_i(T_{1i})}$$

then mission reliability for component A is

$$R_a(T_{2a} - T_{1a}) = R_a(T_{2a}) = R_a(20) = e^{-(0.01)(20)} = e^{-0.20} = 0.82$$

and mission reliability for component B is

$$R_b(T_{2b} - T_{1b}) = R_b(T_{2b}) = R_b(20) = e^{-(0.005)(20)} = e^{-0.10} = 0.90$$

Mission reliability for the system, from (9-1), is

$$R_e(T_2 - T_1) = R_e(T_2) = [R_a(20)][R_b(20)] = e^{-0.30} = 0.74$$

Note that these are the same results that would have resulted directly if the simpler relationships (9-4) and (9-5) keyed specifically to the characteristics of the exponential distribution had been applied.

If a system and each component have acquired 500 hours of operation and no failures have occurred, then T_{1a}, T_{1b}, and $T_1 = 500$; T_{2a}, T_{2b} and $T_2 = 520$. Again using the most general expression for item reliability,

$$R_i(T_{2i} - T_{1i}) = \frac{R_i(T_{2i})}{R_i(T_{1i})}$$

mission reliability for component A is

$$R_a(T_{2a} - T_{1a}) = R_a(520 - 500) = [R_a(520)]/[R_a(500)]$$
$$= [e^{-(0.01)(520)}]/[e^{-(0.01)(500)}] = e^{-0.20} = 0.82$$

and mission reliability for component B is

$$R_b(T_{2b} - T_{1b}) = R_b(520 - 500) = [R_b(520)]/[R_b(500)]$$
$$= [e^{-(0.005)(520)}]/[e^{-(0.005)(500)}] = e^{-0.10} = 0.90$$

Mission reliability for the system is

$$R_e(T_2 - T_1) = [R_a(520 - 500)][R_b(520 - 500)] = e^{-0.03} = 0.74$$

If the system has acquired 500 hours of operation and component A failed after 400 hours of operation and was replaced by a new one, the new unit has 100 hours of failure-free operation when the system has 500. The remaining component has been failure-free for the full 500 hours. Then $T_{1a} = 100$; T_{1b}, $T_1 = 500$; $T_{1a} = 120$; T_{2b}, $T_2 = 520$.
    Following the same procedures as above:
    Mission reliability for component A is

$$R_a(T_{2a} - T_{1a}) = R_a(120 - 100) = [R_a(120)]/[R_a(100)]$$

$$= [e^{-(0.01)(120)}]/[e^{-(0.01)(100)}] = e^{-0.20} = 0.82$$

Mission reliability for component B is

$$R_b(T_{2b} - T_{1b}) = R_b(520 - 500) = [R_b(520)]/[R_b(500)]$$

$$= [e^{-(0.0005)(520)}]/[e^{-(0.0005)(500)}] = e^{-0.10} = 0.90$$

Mission reliability for the system is

$$R_e(T_2 - T_2) = R_a(120 - 100)R_b(520 - 500) = e^{-0.03} = 0.74$$

Hence, it is shown that when components have an exponential distribution of times to failure, operating time acquired before a mission start does not affect reliability. The mission reliability for the system and for each of the components remains the same under all three scenarios.
    As long as we are aware of such equivalence, we could have solved each simply by applying (9-4) and (9-5). Since

$$\lambda_e = \lambda_a + \lambda_b = 0.01 + 0.005 = 0.015$$

$R_e(T)$ could simply be defined as

$$R_e(T) = e^{-(\lambda_a + \lambda_b)T} = e^{-(0.015)20} = 0.74$$

9.1.2 Components with Weibull Failure Distributions

If the distributions of failures for each component is *not* exponential (i.e., hazard/failure rate will vary with respect to operating time), the expressions for both $R_i(T)$ and $R_e(T)$ will take on more complex, detailed, and sometimes only approximate distribution forms. In the discussion which follows, the Weibull distribution will be used as an illustrative vehicle. The treatment of other distributions with nonconstant hazards rates will generally parallel that for the Weibull and the conclusions will be similar. Section 2.1 provides background information on various other distributions which can be used as a base.

Mission reliability for a component can be expressed as

$$R_i(T) = R_i(T_2 - T_1) = \exp[-(T_2^{\beta_i} - T_1^{\beta_i})/\alpha_i^{\beta_i}] \tag{9-7}$$

where

β_i, α_i = shape and scale parameters of the Weibull distribution associated with component i

$R_i(T), R_i(T_2 - T_1), T, T_1, T_2$ are defined as previously

Equation (9-7) signifies that, unlike the case of items which take on exponential distributions of failure, the degree of prior operating history, T_1, associated with the item has a very significant effect on mission reliability. This can be seen quite easily by calculating mission reliability for a mission of $T = 10$ hours for values of $T_1 = 200$, $T_2 = 210$ and $T_1 = 400$, $T_2 = 410$. The value of $R_i(T_2 - T_1)$ will be different for each pair. The values would be identical if the failure distribution were exponential.

The form of (9-7) also suggests that system reliability assessment will be some degrees more complicated than for a system having an exponential distribution of failure. For those reasons this section is organized to provide:

1. General information and facts that an engineer should be aware of concerning $R_e(T)$ and its characteristics.
2. An explanation of the logic associated with assessment procedures.
3. An appendix (A13) providing more detail on the derivation of some of the results.

The general results are that *for nonrepairable systems (i.e., upon failure the system will be replaced in its entirety with a new one),*

$$R_e(T) = R_e(T_2 - T_1) = \prod^n R(T_2 - T_1)_i = \exp\left[-\sum^n \left(\frac{1}{\alpha_i}\right)^{\beta_i}(T_2^{\beta_i} - T_1^{\beta_i})\right] \tag{9-8}$$

where

$R_e(T) = R_e(T_2 - T_1) =$ system reliability, the probability that the system will perform satisfactorily over a specific interval of operating time R (given that: the system has acquired an operating history of T_1 hours prior to mission start; has experienced *no failures* up to T_1; is operable at time T_1, and at the end of T will have acquired a total operating history equal to T_2 hours)

T_1 = operating time (history) on the system at the time of mission start. Think of it as a reading on a time meter installed on the system which records when the system is operating.

T_2 = the total operating time (history) on the system at end of the interval (mission)

T = $(T_2 - T_1)$ = time associated with the mission

n = number of components comprising the system

β_i, α_i = shape and scale parameters of the Weibull failure distribution associated with component i

If all components have Weibull distributions of time to failure with the same value of β, i.e., $\beta_i = \beta$, (9-8) will be reduced to

$$R_e(T) = R_e(T_2 - T_1) = \exp[-(T_2^\beta - T_1^\beta)/\alpha_n^\beta] \qquad (9-9)$$

which is a Weibull distribution where α_n is a new scale parameter,

$$\alpha_n = \left(\sum_i^n [1/\alpha_i]^\beta \right)^{-1/\beta} \qquad (9-10)$$

Hence, if $\beta_i = \beta$, the nonrepairable system will also have a Weibull distribution of failures.

If, however, the β values of the components differ, the resulting distribution of time to failure will not be Weibull. $R(T_2 - T_1)$ will not be able to be forced into the form of (9-7). In some situations where the values of β_i are not equal it may be possible to fit an "approximate" Weibull shape to the result, but a true Weibull distribution will not result.

Only when the system is brand new, i.e., $T_1 = 0$, with $T = T_2$ and β values the same, will system reliability take the simple form

$$R_e(T) = \exp\left(-\sum_i^n [T^\beta/\alpha_i^\beta] \right) = \exp(-T^\beta/\alpha_n^\beta) \qquad (9-11)$$

9.1.2.1 The Repairable System

Assume as before that all system components exhibit Weibull distributions of time to failure with the same value of the shape parameter β, but different values of the scale parameter α.

For a repairable system a unique situation develops. Strictly speaking, even if the system distribution of time to failure is Weibull initially, after the first failure and repair (replacement of one or more components), the failure distribution of the system will no longer be Weibull.

In that case, system reliability $R_{er}(T)$ may be represented as

$$R_{er}(T) = R_{er}(T_2 - T_1) = \exp\left[-(T_2^\beta - T_1^\beta) \sum_{i=1}^{n-r} \left(\frac{1}{\alpha_i}\right)^\beta - \sum_{j=1}^{r} \frac{(T_{cj}^\beta - T_{oj}^\beta)}{\alpha_j^\beta} \right]$$

$$(9\text{-}12)$$

where

$R_{er}(T) = R(T_2 - T_1) =$ system reliability, the probability that the system will perform satisfactorily over an interval (a mission) of time $T = (T_2 - T_1)$ (given that it has acquired an operating time (history) of T_1 hours and has experienced failures in r different components which were replaced by new components)

$n - r$ = number of components which have not failed during T_1 operating hours

r = number of different components which have failed during T_1. (Note: If the same component fails more than once it is only counted as "one.")

T_{oj} = the total operating time history associated with the current replacement j

T_{cj} = the total operating time on the current jth replacement

α_j = the scale parameter associated with the jth replacement

T = $(T_2 - T_1) = (T_{cj} - T_{oj}) =$ time associated with the mission, the interval of time in question

$T, T_1, n, \alpha_i, \beta$ are defined as before

which cannot be forced into a Weibull distribution form.

However, we can use a Weibull distribution (9-9) to approximate (9-12). If we assume

1. That $R_{er}(T_2 - T_1)$ is to be evaluated over early portions of service life, where only a very small proportion of components fail, where all components exhibit a similar order of reliability; or
2. That each component contains many, many parts, that a failure in a component requires the replacement of only a very small number of parts, and that each component j is put back into its system after repair

Then there will be a less than significant difference between the reliability values generated by (9-9) and (9-12) for any set of values T_1 and T_2, i.e., the value of

$$-\left\{\sum_{i=1}^{n-r} [(T_2^\beta - T_1^\beta)/\alpha_i^\beta] + \sum_{j=1}^{r} [(T_{cj}^\beta - T_{oj}^\beta)/\alpha_j^\beta]\right\} \sim -\sum_{i=1}^{n} [(T_2^\beta - T_1^\beta)/\alpha_i^\beta]$$

Hence time to failure can be considered approximately Weibull distributed. As a result $R_{er}(T)$ can be approximated as

$$R_{er}(T) = R_{er}(T_2 - T_1) \sim \exp\left(-\sum_{i=1}^{n} [(T_2^\beta - T_1^\beta)/\alpha_i^\beta]\right) \qquad (9\text{-}13)$$

where all terms are defined as previously.

If all components are failure-free to T_1 and have the same value of β, $\beta_i = \beta$, then $R_{er}(T)$ is equivalent to $R_e(T)$ and can be expressed as

$$R_{er}(T) = R_e(T) = R_e(T_2 - T_1) = \exp(-[(T_2^\beta - T_1^\beta)/\alpha_n^\beta]) \qquad (9\text{-}14)$$

In addition, since all values of β_i are assumed identical and all components have Weibull distributions of failure time, this system, like the nonrepairable system, will take on an initial Weibull distribution of time to failure (until the occurrence of the first system failure).

Unless time meters are associated with components, information on prior operating history will not be available. As a consequence, in most instances, (9-12) and (9-13) would be of limited practical utility. Equation (9-13) can, however, be used to approximate the different values of mission reliability that can be expected for different future periods of service life under the assumptions described previously.

Often such projections are required to assess the mission reliability characteristics of a given design over service life while an item is under development. Their use for this purpose and others will be illustrated in the examples that follow.

Take, for example, a system made up of two components, A and B, each with a Weibull distribution of failures and a common value of shape parameter, $\beta = 0.5$, and scale parameters, $\alpha_a = 100$ and $\alpha_b = 200$. Each component is composed of 50 replaceable modules which are of approximately the same complexity (see Fig. 9-2b). Once a component failure occurs, the *failed module(s)* is replaced within the component without removing the component from the system. Determine the exact or approximate probability that the system will perform satisfactorily over a 20-hour interval/mission under the same set of scenarios as in the last example. Pay particular attention to the way the third scenario is handled, i.e., when a component has been replaced some time prior to the mission.

If the system and all its components are brand new, then T_{1a}, T_{1b}, and $T_1 = 0$; T_{2a}, T_{2b}, and $T_2 = 20$;

$$R_{er}(T) = R_e(T) = R_{er}(T_2 - T_1) = \exp\left\{-(T_2^\beta - T_1^\beta)\sum_{}^{n}\left(\frac{1}{\alpha_i}\right)^\beta\right\}$$

$$R_e(T) = R_e(20) = \exp\{-(20)^{1/2}[(1/100)^{1/2} + (1/200)^{1/2}]\} = 0.47$$

If the system and all its components have acquired 500 hours of operation and no failures have occurred, then T_{1a}, T_{1b}, and $T_1 = 500$; T_{2a}, T_{2b}, and $T_2 = 520$:

$$R_{er}(T) = R_e(T) = R_e(T_2 - T_1) = \exp\left\{-(T_2^\beta - T_1^\beta)\sum_{}^{n}\left(\frac{1}{\alpha_i}\right)^\beta\right\}$$

$$R_{er}(T) = R_e(T) = R_e(520 - 500)$$
$$= \exp\{-(520^{1/2} - 500^{1/2})[(1/100)^{1/2} + (1/200)^{1/2}]\} = 0.93$$

(Note: As described before in Section 2.1.2, a $\beta < 1$ indicates a hazard/failure rate which decreases with operating time. As a consequence, the reliability after 500 hours of operating time will be greater than that experienced initially.)

Consider *component A as a system* comprised of 50 series-connected modules having a typical mission interval of 20 hours. It is in the development phase and there is a need to determine what the mission reliability will be after the system has been in operation for 500 hours. For convenience, assume that all modules have Weibull failure distributions with $\beta_i = 0.5$ and $\alpha_i = 25 \times 10^5$. (These will result in $\beta = 0.5$ and a value of $\alpha = 100$ for component A.) Application of (9-12) reduces to

$$R_{er}(T) = R_e(T) = R_e(T_2 - T_1) \sim \exp\left\{-(T_2^\beta - T_1^\beta)\sum_{}^{m}\left(\frac{1}{\alpha_i}\right)^\beta\right\}$$

$$R_e(T) = \exp\left\{-(520^{1/2} + 500^{1/2})\sum_{1}^{50}[1/[25 \times 10^5]]^{1/2}\right\} = 0.957$$

Assume that the system described above has *actually acquired* 500 hours of operation and has failed five times, resulting in the replacement of 5 of those 50 modules. The failures occurred at 50, 100, 200, 300, and 400 hours of operation, respectively, and no failures have been experienced since. The 45 modules in component A which have not failed have the characteristics $T_1 = T_{1a} = 500$, $T_2 = T_{2a} = 520$. The five replacement modules have the following values of T_{oj} and T_{cj} at $T_1 = 500$:

Replacement module number	T_{oj}	T_{cj}	β	α
1	450	470	$1/2$	25×10^5
2	400	420	$1/2$	25×10^5
3	300	320	$1/2$	25×10^5
4	200	220	$1/2$	25×10^5
5	100	120	$1/2$	25×10^5

Assume that all replacement modules were brand new when inserted and have the same parameters (α and β) as the modules being replaced.

Assess the reliability of the system over the next mission. Since $T_1 = 500$ hours and mission interval is 20 hours, $T_2 = 520$. Taking into account the five failures (replacements), application of (9-12) results in

$$R_{er}(T) = R_{er}(T_2 - T_1) = R_{er}(520 - 500)$$

$$= \exp\left\{-(520^{1/2} - 500^{1/2})\left[\sum_{i=1}^{45}(25 \times 10^5)^{-1/2}\right] - C\right\} = 0.955$$

where

$$C = [1/(25 \times 10^5)]^{1/2}\{[(470)^{1/2} - (450)^{1/2}] + [(420)^{1/2} - (400)^{1/2}]$$
$$+ [(320)^{1/2} - (300)^{1/2}] + [(220)^{1/2} - (200)^{1/2}]$$
$$+ [(120)^{1/2} - (100)^{1/2}]\}$$
$$= [1/(25 \times 10^5)]^{1/2}(3.173) = 0.006$$

which is a close fit to the result projected using the approximation.

If each such failure required the replacement of two modules, such that a total of 10 modules was replaced,

$$R_{er}(520 - 500) = 0.953$$

If each failure required the replacement of four modules, such that a total of 20, or 40%, of the modules were replaced,

$$R_{er}(520 - 500) = 0.949$$

While general conclusions should not be reached as a consequence of this rather contrived example, it does illustrate that the approximation can provide, rather simply, results that are accurate within reasonable bounds.

9.2   EVALUATING "MEAN TIME" MEASURES OF RELIABILITY

The "mean time" measures of reliability associated with series-connected systems can sometimes at first be confusing, at least semantically, to the average engineer (see Section 1.4). Of the "mean time" terms discussed in Section 1.4, those most appropriate for reliability design assessment purposes include mean time between failure (MTBF), mean time to first failure (MTFF), mean residual life (MRL) and mean time to failure (MTTF). Equivalence and differences among the terms and their applications depend largely on the failure distribution associated with the component or system.

9.2.1   "Mean Time" Measures Associated with an Exponential Distribution of Failures

In order to facilitate the discussions which follow, we change notation. We define

$$R_e(T) = R_e(T_2 - T_1) = R_e(t - t_0) \qquad (9\text{-}15)$$

where $t \equiv T_2$, $t_0 \equiv T_1$. As a result, when $t_0 = 0$,

$$R_e(T) = R_e(t) \qquad (9\text{-}16)$$

For items which exhibit an exponential distribution of time to failure there are four basic "mean time" measures: mean time between failure (MTBF); mean time to first failure (MTFF); mean residual life (MRL); mean time to failure (MTTF).

For both repairable and nonrepairable (items which are discarded in their entirety upon failure) components and systems, these terms relate to either or both of (a) the characteristics of *populations* of parts, components or equipments, (b) the characteristics associated with a given item.

If items or populations of items take on exponential distributions of times to failure, the following measures are all quantitatively equivalent:

$$MTBF \equiv MTFF \equiv MRL \equiv MTTF \tag{9-17}$$

Note: Some definition documents ascribe the use of MTTF as the preferred measure to be used for nonrepairable items (see Chapter 1). A significant portion of the literature, rightly or wrongly, uses it in a more general sense. This text, too, will treat the term in the general sense.

Under the distribution assumption above the "mean time" measures may be expressed in terms of

$$MTBF/MTFF/MRL/MTTF = \frac{1}{\lambda_e} \tag{9-18}$$

where λ_e = the total failure rate associated with a component, equipment or system, and (9-18) holds for both repairable and nonrepairable items.

To show the equivalence of these measures when the distribution of times to failure is exponential, recall from past discussions and the detail in Appendices A4 and A5 that, for *any* distribution, MTFF and MRL can be assessed through the relationships

$$MTFF = \int_0^\infty R(t)\, dt \tag{9-19}$$

$$MRL = \int_{t_0}^\infty R(t - t_0)\, dt = \frac{1}{R(t_0)} \int_{t_0}^\infty R(t)\, dt \tag{9-20}$$

Assume a system made up of n components each with an exponential distribution of failure, where component i exhibits a failure rate, λ_i. As in (9-2), system reliability $R_e(t)$ is

$$R_e(t) = \exp\left(-t \sum_i^n \lambda_i\right) \tag{9-21}$$

$$MTFF_e = \int_0^\infty R_e(t)\, dt = \int_0^\infty \exp\left(-t \sum_i^n \lambda_i\right) dt$$

$$= -\left(\frac{1}{\lambda_e}\right)[\exp(-\lambda_e t)]_0^\infty = \frac{1}{\lambda_e} \tag{9-22}$$

where

$$\lambda_e = \sum_i^n \lambda_i$$

$$\text{MRL} = \frac{1}{R_e(t_0)} \int_{t_0}^{\infty} R_e(t)\, dt = \exp\left(t_0 \sum_i^n \lambda_i \right) \int_{t_0}^{\infty} \exp\left(-t \sum_i^n \lambda_i \right) dt$$

$$= -\left(\frac{1}{\lambda_e} \right) \exp(-\lambda_e t_0) [0 - e^{(-\lambda_e t)}]_{t_0}^{\infty} = \frac{1}{\lambda_e} \tag{9-23}$$

where t_0 = the total operating time on the system at the beginning of the mission, which, as is shown in (9-23), has no bearing on the value of MRL when time to failure is exponentially distributed.

Therefore, MRL $= 1/\lambda_e$ for every value of t_0, and if after every failure (and repair/replacement) MRL is computed, MRL $= 1/\lambda_e$. Hence, MRL = MTFF. But if after each failure MRL has the same value, then MRL actually becomes the mean time *between* failures (MTBF). Further, since prior operating history has no bearing on reliability (see Section 9.1.1), the mean time to first failure is equal to the mean time to failure (MTTF) in general.

Generally speaking, MTBF is denoted by the symbol θ, where

$$\theta = \frac{1}{\displaystyle\sum_i^n \lambda_i} = \frac{1}{\lambda_e} \tag{9-24}$$

λ_e = total failure rate of an equipment or series connected system.
n = number of components making up the system

For example, a repairable system is made up of two components A and B as in Fig. 9-2a, each exhibiting an exponential time to failure distribution. If $\lambda_a = 0.01$ and $\lambda_b = 0.005$, compute its MTBF.

$$\theta = \frac{1}{\displaystyle\sum_i^n \lambda_i} = \frac{1}{\lambda_a - \lambda_b} = 1/(0.01 + 0.005) = 66.67 \text{ hours}$$

If as soon as one component fails the system is discarded, the system is termed nonrepairable. If a nonrepairable system is made up of two components and each has acquired an operational history of 500 hours, i.e., $t_0 = 500$, what is the mean time to failure of the system, measured from the

$t_0 = 500$ hours point in time? This is an example of an MRL measure and

$$\text{MRL} = \frac{1}{R(t_0)} \int_{t_0}^{\infty} R(t)\, dt = \exp[(\lambda_a + \lambda_b)(500)] \int_{500}^{\infty} \exp[-(\lambda_a + \lambda_b)]\, dt$$

$$= -\exp[(\lambda_a + \lambda_b)500]\left(\frac{1}{\lambda_a + \lambda_b}\right)[\exp[-(\lambda_a + \lambda_b)t]]_{500}^{\infty} = \frac{1}{(\lambda_a + \lambda_b)}$$

$$= \frac{1}{(0.01 + 0.005)} = 66.67$$

9.2.2 "Mean Time" Measures Associated with Items with Weibull Distributions of Failure

Again the Weibull distribution will be used as an example of a distribution of times to failures which exhibits a nonconstant hazard.

For repairable or nonrepairable components and systems exhibiting a Weibull distribution of failures, MTBF is largely inappropriate since the mean time between one failure and the next changes with the operating age of the population or component.

MTTF, depending on circumstances, can be interpreted in two ways. If the measurement period over which a mean is to be calculated starts when a component or a population has a zero operational history ($t_0 = 0$), then MTTF $\equiv$ MTFF. If it starts after an operational history ($t_0 \neq 0$) has been acquired, then MTTF $\equiv$ MRL. MRL, as will be illustrated, varies with the value of t_0.

The "mean time" measures for nonrepairable and repairable items are calculated through the use of similar relationships. They are similar, as opposed to identical, because if an item originally exhibited a Weibull distribution of time to failure it will at best exhibit only an approximation to that distribution after a series of failures and repairs (replacement of failed portions) (see Sections 2.1.1.2 and 9.1.2.1). For the purposes of this section, it will be assumed that repairable items take on an approximate Weibull distribution with their original values of β and α.

The following relationships can be applied to both repairable and nonrepairable systems. When dealing with repairable systems, assume that the equal sign is replaced by an approximation.

9.2.2.1 Repairable and Nonrepairable Systems

For repairable and nonrepairable components and systems exhibiting an original Weibull distribution of failures, MTTF and MRL are dependent on the prior operating history of the component/system and are equivalent, but

MTFF is equivalent in value to MTTF and MRL for only those components/systems with no prior operating history (i.e., $t_0 = 0$).

To show this, recall the relationships for MTTF and MRL in (9-19):

$$MTFF = \int_0^\infty R(t)\, dt$$

$$MRL = \frac{1}{R(t_0)} \int_{t_0}^\infty R(t)\, dt$$

Assume that the failure distribution associated with each component i is at least approximately Weibull and all component distributions have the same value of the shape parameter β. As before, assume that for complex repairable items the failure distribution remains approximately Weibull even after a number of failures and replacement of components/parts.

Section 2.1.2 and Appendix A5 indicate that for a Weibull distribution

$$MTFF = \alpha \Gamma\left(1 + \frac{1}{\beta}\right) \tag{9-25}$$

$$MRL = \frac{\alpha}{\beta}[Z^{r-1} + (r-1)Z^{r-2} + (r-1)(r-2)Z^{r-3} + \cdots + (r-1)!] \tag{9-26}$$

where, for (9-26),
$r = 1/\beta$, an integer
$Z = (t_0/\alpha)^\beta$

$$\alpha = \left[\sum_i^n (1/\alpha_i)^\beta\right]^{-1/\beta} \tag{9-27}$$

n = number of components in the system
α_i = scale parameter of the failure distribution associated with component i

When the component failure distributions have different values of β, the system failure distribution will not take on an exact Weibull distribution (see (9.8) and (9.9)). Its degree of similarity to the Weibull will depend on the characteristics of its components. The result will be a mixed distribution (see Ref. 1), the characteristics of which will not be covered in this text.

As can be observed, MTFF and MRL will be different for every value of t_0 except $t_0 = 0$. Only at $t_0 = 0$ is MRL equivalent to MTFF.

If MRL is evaluated after each failure, taking into account the value of

t_0 at that time, MRL will be seen to change with t_0. MRL is a measure of mean operating time between the current and the next failure and in that context is equivalent to MTTF if MTTF is defined in terms of mean operating time from one failure to the next. For example, an equipment is made up of two components A and B, both with Weibull distributions of time to failure with scale parameters $\alpha_a = 290$, $\alpha_b = 580$ and shape parameters $\beta_a = \beta_b = 0.5$. What is the MTTF for the system when it is brand new, i.e., at $t_0 = 0$? What is the MRL when the system has acquired an operating history of 200 hours ($t_0 = 200$); of 400 hours ($t_0 = 400$)?

From (9-27) the system α is

$$\alpha = \left[\sum_i^n (1/\alpha_i)^\beta \right]^{-1/\beta} = \left[(1/290)^{0.5} + (1/580)^{0.5} \right]^{-2} \sim 100$$

$$\text{MTTF} = \alpha \Gamma \left(1 + \frac{1}{\beta} \right) = 100\Gamma(3) = 200$$

$$\text{MRL} = \frac{\alpha}{\beta}[Z^{r-1} + (r-1)Z^{r-2} + (r-1)(r-2)Z^{r-3} + \cdots + (r-1)!]$$

where $Z = \left(\dfrac{t_0}{\alpha} \right)^\beta$, $r = 1/\beta = 2$.

When $t_0 = 0$, $Z = 0$ and $\text{MRL} = [100/0.5] = 200 = \text{MTFF}$. When $t_0 = 200$, $Z = 1.41$ and $\text{MRL} = 200[1.41 + 1] = 200[2.41] = 482$. When $t_0 = 400$, $Z = 2$ and $\text{MRL} = 200[2 + 1] = 600$.

9.3 EVALUATING AVAILABILITY MEASURES OF RELIABILITY

The previous sections dealt with equipment and systems reliability evaluation in terms of operation over a specific interval of time in an environment where (a) maintenance during a mission is not possible, (b) it was assumed that the item was operable at the start of a mission.

In many cases, such as traffic control systems, computer systems, and other systems subject to continuous use, necessary maintenance and repair normally is or can be performed during a mission period. For items such as aircraft and emergency systems which are subject to call to mission at random points in time, the primary requirement is that the item be available (operable) when needed. Reliability evaluation of such systems, because of objectives involved, takes forms different from those described earlier. In such cases the concern is with (1) the proportion of time over which the system can function, for example, the proportion of time a computer system will be capable of operating; (2) the probability that the system is operable at a specific point in time, for example, at an instant when an aircraft needs control guidance or when a system is called on to perform a mission at a random point in time.

The primary interests are whether or not the item will *be available* when needed and the proportion of time it will *remain available* over a mission or contemplated use duration.

The reliability evaluation metrics most appropriate to such systems are the availability measures. As implied previously, availability is applied to assess two related measures:

1. The probability that an item is in an operable state (capable of satisfactorily performing its function) at any random point in time, or
2. The proportion of time that the item will be capable of performing its function.

Evaluation of each of the two metrics can be made on a long-term (steady-state) basis and on a short-term (dynamic) basis.

The steady-state form is the one most familiar to reliability engineers. It is nonparametric in that it holds for any defined distribution of failure and for any distribution of repair time. It describes the probability that the item will be capable of satisfactorily performing its function at any random future point in time. Examples are the probability that a search radar will be properly performing when a target enters its area of coverage at a random point in time; and the probability that a critical communications system will function when called upon to perform at a random point in time.

It describes the proportion of time an item will be operable over a very long period of time, for example, the proportion of time the search radar will be operable over the next 1000 days. Due to the fact that steady-state evaluation is associated with performance over a very long period of time, both metrics (for the steady-state case), as will be shown later, take a common mathematical form.

The dynamic forms are failure time distribution and repair time distribution dependent. Two different expressions make up the dynamic forms of availability:

1. One that expresses the probability that the item will be in an operable state at a *specific* point in time.
2. One that expresses the proportion of time that an item will be in an operable state over a *specific period* of time.

Each availability metric is defined in terms of "states." This is a term and concept which has not been described previously and is very important to the comprehension and application of availability measures. These will be described next and the results will be used as the basis for the discussions which follow.

9.3.1 The System States: What They Are and How They Can Be Identified

Take a single component and assume it can only be in *two* conditions or states, working or failed (operable or not operable, up or down, it can be expressed in any number of ways). If a component B is in an operable state it will be denoted as B_o; if the component is in an inoperable state it will be denoted as B_f. At any time then, the state of the component can be described as either B_o or B_f, and it follows that if

A_b = probability that the component is in state B_o (operable)
at any point in time, or equals the proportion of time the
component is in state B_o

$\bar{A}_b$ = probability that the component is in state B_f
(inoperable) at any point in time, or equals the proportion of time
the component is in state B_f

then, since B_o and B_f are the only states possible for B, i.e., B is either operable or inoperable,

$$A_b + \bar{A}_b = 1$$

where we denote
  A_b is the availability of component B
  $\bar{A}_b$ is the unavailability of component B

If a system is composed of two components B and C each capable of being in only two states, operable and not operable, where C_o denotes that component C is operable and C_f denotes that C is not operable, and where

A_c = either the probability that component C is in state C_o,
or the proportion of time that component is in state C_o, i.e., A_c =
the availability of component C

$\bar{A}_c$ = either the probability that component C is in state C_f or
the proportion of time the component is in state C_f, i.e., $\bar{A}_c$ =
the unavailability of component C

and using the same reasoning as for component B,

$$A_c + \bar{A}_c = 1$$

then the *system* can take on the following possible states:

$B_o C_o$ = both components operable
$B_f C_o$ = one component operable, one inoperable
$B_o C_f$ = one component operable, one inoperable
$B_f C_f$ = both components inoperable

Since the above are the only states possible, i.e., the system *must* be in one of these four states, just as a coin must come up heads or tails. As a consequence,

$$A_b A_c + \bar{A}_b A_c + A_b \bar{A}_c + \bar{A}_b \bar{A}_c = 1$$

where
$A_b A_c$ = probability (proportion of time) system is in state $B_o C_o$
$\bar{A}_b A_c$ = probability (proportion of time) system is in state $B_f C_o$
$A_b \bar{A}_c$ = probability (proportion of time) system is in state $B_o C_f$
$\bar{A}_b \bar{A}_c$ = probability (proportion of time) system is in state $B_f C_f$

and if both components of the system must be operable in order for the system to operate, the availability, A_s, of the example system is

$$A_s = A_b A_c$$

The above represents a very simple form of an availability evaluation, but it illustrates the two initial steps required for *every* availability evaluation:

1. The identification of all possible different system states.
2. The definition of the probability associated with the system being in each system state.

When the system is made up of two or three components, the process of identifying all possible states and their composition is relatively easy, as illustrated in the example. However, when the system is composed of a large number of components, a procedure is required to ensure that no states are inadvertently omitted. One process which will easily identify all possible system states is illustrated in Chart 9-1. If n components make up the system and each component has two possible states, operable and not operable,

CHART 9-1. Process to Identify All Possible System States

State Identifier, A_{sj}	Component #1 $2^0 = 1$	Component #2 $2^1 = 2$	Component #3 $2^2 = 4$	Component #4 $2^3 = 8$	Component #5 $2^4 = 16$	$\ldots$	Component #N 2^{N-1}	Probability (Proportion of Time) that System is in State A_{sj}
0	0	0	0	0	0	$\ldots$	0	$\prod\limits_{i=1}^{n} A_i$
1	1	0	0	0	0		0	$\bar{A}_1 \prod\limits_{i=2}^{n} A_i$
2	0	1	0	0	0		0	$\bar{A}_2 A_1 \prod\limits_{i=3}^{n} A_i$
3	1	1	0	0	0		0	$\bar{A}_1 \bar{A}_2 \prod\limits_{i=3}^{n} A_i$
4	0	0	1	0	0		0	$\bar{A}_3 A_1 A_2 \prod\limits_{i=4}^{n} A_i$
$\vdots$								
m	1	1	1	1	1		1	$\prod\limits_{i=1}^{n} \bar{A}_i$

1 = Component i inoperable (unavailable, $\bar{A}_i$).
0 = component i operable (available, A_i).

make up a table with $n + 2$ columns. The first column is the state identifier, which will go from state 0 to state m, where $(m + 1)$ will be the total number of system states possible.

Note that it can be shown that the total number of system states possible,

$$(m + 1) = \sum_{i=0}^{n} \frac{n!}{(n - i)!i!}$$

but the above information is unnecessary because using the procedure illustrated the process of state identification will terminate at state m.

Denote the second column by 2^0 (representing the 1st component), the third by 2^1 (representing the 2nd component), the fourth by 2^2 (representing the 3rd component in the system), and so on, until the second to last column is identified by 2^{n-1} (representing the last component in the system). Represent each system state identifier number by its binary equivalent* by placing marks in the appropriate columns in each row: for state 0 place a zero in all the columns in the state 0 row; for state 1 place a one in the 2^0 column ($2^0 = 1$) in the state 1 row (and a zero in the remaining columns of the row); for state 2 place a one in the 2^1 column ($2^1 = 2$) in the state 2 row (and a zero in the remaining columns of the row); for state 3 place ones in the 2^0 column and the 2^1 column ($2^0 + 2^1 = 3$) in the state 3 row (and zeros in the remaining columns); for state 4 place a mark in the 2^2 column ($2^2 = 4$) in the state 4 row (and zeros in the remaining columns) and so on until a state (row) is reached where *all* columns denoted by 2^i will have ones in them. *This will be state m.*

For the series connected system in question associate a zero with an operable component; a one with a failed component. The ones and zeros in the row associated with each system state identifier, A_{si}, indicate the composition of that system state in terms of the states of each component making up the system. For example,

State 0   All components are operable
State 1   All components except component #1 are operable
State 2   All components except component #2 are operable
State 3   All components except components #1 and #2 are operable
State 4   All components except components #3 are operable

Hence Chart 9-1 generates all possible system states along with their composition (i.e., the operable and inoperable components associated with each system state).

The probability of the system being in a particular system state A_{sj} can

---

*$0000 = 0$; $1000 = 1$; $0100 = 2$; $1100 = 3$; $1111 = 15$.

be calculated in the final column of the chart, where

A_{sj} = the product of all the *component* availabilities, A_i, and unavailabilities, $\bar{A}_i$, associated with a given state

A_i = probability (proportion of time) *component i* is in an operable state. If for a given *system* state a column associated with a given component has a zero it is in an operable state and a value A_i is assigned

$\bar{A}_i$ = probability (proportion of time) *component i* is in an inoperable state. If for a given *system* state a column associated with a given component has a one it is in an inoperable state and a value $\bar{A}_i$ is assigned

and if we define all *possible* system states, A_{sj},

$$\sum_{j=0}^{m} A_{sj} = 1 \tag{9-28}$$

where m = the total number of system states − 1. (Note: total number of states = $m + 1$.)

For example, recall the illustrative system made up of two components B and C, where it was shown

$$A_b A_c + \bar{A}_b A_c + A_b \bar{A}_c + \bar{A}_b \bar{A}_c = 1$$

or

$$\sum_{j=0}^{3} A_{sj} = 1$$

Chart 9-2 shows an illustration of system state identification when $n = 4$. Also shown in the last column is the probability, A_{sj}, of being in each system state.

Develop such a table for $n = 3$.

9.3.2   Steady-State Availability in General

The steady-state measure of availability can be developed nonparametrically, that is to say, for any distributions of time to failure and time to repair which have means associated with them. Assuming that a component operates

CHART 9-2 Defining All System States for a Four Component System

State Identifier, A_{sj}	Component B 2^0	Component C 2^1	Component D 2^2	Component E 2^3	Probability (Proportion of Time) that System is in State A_{sj}
0	0	0	0	0	$A_b A_c A_d A_e$
1	1	0	0	0	$\overline{A_b} A_c A_d A_e$
2	0	1	0	0	$\overline{A_c} A_b A_d A_e$
3	1	1	0	0	$\overline{A_b}\,\overline{A_c} A_d A_e$
4	0	0	1	0	$\overline{A_d} A_b A_c A_e$
5	1	0	1	0	$\overline{A_b}\,\overline{A_d} A_c A_e$
6	0	1	1	0	$\overline{A_c}\,\overline{A_d} A_b A_e$
7	1	1	1	0	$\overline{A_b}\,\overline{A_c}\,\overline{A_d} A_e$
8	0	0	0	1	$\overline{A_e} A_b A_c A_d$
9	1	0	0	1	$\overline{A_b}\,\overline{A_e} A_c A_d$
10	0	1	0	1	$\overline{A_c}\,\overline{A_e} A_b A_d$
11	1	1	0	1	$\overline{A_b}\,\overline{A_c}\,\overline{A_e} A_d$
12	0	0	1	1	$\overline{A_d}\,\overline{A_e} A_b A_c$
13	1	0	1	1	$\overline{A_b}\,\overline{A_d}\,\overline{A_e} A_c$
14	0	1	1	1	$\overline{A_c}\,\overline{A_d}\,\overline{A_e} A_b$
15	1	1	1	1	$\overline{A_b}\,\overline{A_c}\,\overline{A_d}\,\overline{A_e}$

continuously over service life and that operation is only broken for repair or replacement to take place, let

A = availability, the proportion of time an item is in an operable state over service life

M = the aggregate mean time to failure of the item over its service life if it is repaired or the mean time to failure of an item which is replaced with a new item upon failure

MTTR = R = the mean time to repair or replace the item in the event it fails (for practical purposes R is often taken to represent the time from the detection of the failure to the time that repair is complete)

T = total operating hours to which the item will be exposed during service life

Assume that

1. The item is in continuous operation over service life. It is either operating or under repair over that period.
2. The item has a failure rate of zero when it is being repaired; hence, it cannot fail again while under repair.
3. Fault detection is instantaneous and a repair action starts immediately after detection.

T/M = expected number of failures over service life

$R(T/M)$ = total expected down time due to repair over service life

$T + R(T/M)$ = total service life in hours

$$A = \frac{T}{T + R(T/M)} = \frac{1}{1 + (R/M)} = \frac{M}{R + M} \qquad (9\text{-}29)$$

If A is the proportion of time an item is in an operable state and if the item cycles between operation and repair constantly, the probability that the item is in an operable state at any time chosen at random during service life is also equal to A. Imagine that you are walking down a very long corridor that has been carpeted using two colors, red and black. The carpet was laid in alternate color sections. The red sections on the average are 19 times the length of the black and, indeed, 95% of the corridor is carpeted in red. You have been walking a long time. Suddenly the fire alarm sounds. What is the probability you are standing on a section of red carpet?

The probability is 0.95 that you will be on a strip of red carpet at any random time in your journey and 0.05 that you will be on a strip of black carpet.

Think about availability as long periods of satisfactory operation punctuated with shorter periods of repair. The probability of being in an operable state is analogous to being on a red strip of carpet.

For convenience, in the rest of the discussion all values of A will be represented as

$$A = \frac{M}{R + M} \tag{9-30}$$

where A, R, M are defined as before.

Equation (9-30) holds exactly for any component (a) which is replaced in its entirety at failure where the average time for fault isolation and component replacement is R; (b) which has an exponential failure distribution and an exponential repair time distribution (the most common assumptions in availability evaluations); (9-30) can be used to assess the long-term availability of any repairable component with any distribution of failure which has an increasing or decreasing hazard over the availability assessment period (and a mean repair time). However, in that case, an *aggregate* value of mean time to failure over service life must be calculated.

More specifically, for components repairable or replaceable and having exponential distributions of time to failure, (9-30) can be expressed as

$$A_i = \frac{M_i}{R_i + M_i} = \frac{\theta_i}{R_i + \theta_i} \tag{9-31}$$

where

$M_i = \text{MTBF} = \theta_i = $ mean time between failures associated with component i

$R_i = $ mean time to repair (replacement) associated with component i

For components which are *replaceable* upon failure and have other than exponential distributions of time to failure, availability can be expressed as

$$A_i = \frac{M_i}{R_i + M_i} = \frac{\text{MTTF}_i}{R_i + \text{MTTF}_i} \tag{9-32}$$

Since in this case $\text{MTTF}_i = \text{MTFF}_i$,

$$\text{MTFF}_i = \text{MTTF}_i = \int_0^\infty R_i(t)\, dt = \text{mean time to failure for component } i \tag{9-33}$$

where

   R_i = mean time to repair or replace component i
   $R_i(t)$ = reliability of component i

For repairable components which have definable distributions of failure that are other than exponential, with hazard/failure rates which either increase or decrease with time, availability over a given long operating time span, t_s, can be approximated by

$$A(t_s) \sim \frac{M_i}{R_i + M_i} \qquad (9\text{-}34)$$

where

$$M_i = \tfrac{1}{2}[\mathrm{MTFF} + \mathrm{MRL}(t_s)] = \frac{1}{2}\left\{ \int_0^\infty R(t)\, dt + \frac{1}{R(t_s)} \int_{t_s}^\infty R(t)\, dt \right\}$$
$$(9\text{-}35)$$

which represents an approximate midpoint average on MTTF between mean time to first failure and the mean time to failure expressed by MRL measured from t_s.

Hence, for repairable components with nonexponential distributions of failure times, availability will vary with t_s.

9.3.2.1 Steady-State System Availability
The probabilities associated with particular system states were discussed earlier in this chapter. The availability of a system is dependent on the number of different system states in which the system can satisfactorily perform its function. If a system is made up of n components and all are required to perform a given system function, system availability A_s can be expressed through the multiplication law for probability as

$$A_s = \prod_i^n A_i \qquad (9\text{-}36)$$

where A_s = probability that the system is in an operable state (capable of satisfactorily performing its function) at any randomly chosen point in time, which is equal to the proportion of time the system will be capable of performing its function. (All components must be in operable states in order for the system to be in an operable state.)

Taking as an example the system in Chart 9-2, only one state, state 0, meets this criterion

$$A_s = A_{s0} = \prod_{i=0}^{4} \frac{M_i}{R_i + M_i} = \prod_{i=0}^{4} A_i = A_b A_c A_d A_e \qquad (9\text{-}37)$$

where

A_{s0} = probability (proportion of the time) the system is operable
$\equiv$ probability the system is in state 0

A_i = $M_i/(R_i + M_i)$
= probability (proportion of the time)
component i is operable

9.3.2.2   Things You Should Know About Availability Relationships

In the literature, (9-36) is the most commonly cited relationship for systems availability. What is not commonly stated is that it is based on two assumptions:

1. Each component has associated with it an available repair team which immediately initiates maintenance/repair action. For those instances where unlimited maintenance crews are not available, (9-36) will provide a slightly optimistic answer. As will be shown, however, there are means which can be used to assess the risk taken in using the assumption.
2. When a failure in one component occurs causing system failure, the remaining components continue to be operated while the repair is taking place. Consequently, additional failures are assumed to occur during this downtime, reducing system availability.

The latter assumption is invalid for any systems which are "turned off" or placed in a standby mode of operation during repair (such as some computer or radar systems). For those cases, relationship (9-36) will provide a slightly pessimistic answer. If the entire system is shut down (the components which were still operable as well as the failed component) while repairs are made on a failed component, it may be assumed that the system has virtually a zero failure rate in that state. As a result, additional failures will not occur. The system will be frozen in that failure state until repairs are completed.

Equation (9-36) evaluates system availability assuming that the system can take on all potentially possible states (16 states for the example in Chart 9-2). However, if the system is shut down after suffering only one failure, there are some states that it will be impossible for the system to assume. In the example in Chart 9-2, the system will be able to assume only states 0, 1,

2, 4, 8. It will be shut down immediately after reaching 1, 2, 4, or 8. At that time the system will exhibit a failure rate of 0, and hence be unable to enter the remainder of the failure states and will not be put in service again until it is in state 0.

When potential states are removed from consideration, the proportion of time the system spends in the remaining states changes. As a result, the value of A_s will be different from that generated through (9-36). If we define A_{sd} as probability (proportion of time) the system is in an operable state after assuming a shut down (failure rate = 0) for repair after entering any failed state (e.g., 1, 2, 4, or 8), it can be shown mathematically (see Appendix A14) that

$$A_{sd} = \frac{1}{1 + \sum\limits_{i=1}^{n} (R_i/M_i)} \qquad (9\text{-}38)$$

where n, R_i, and M_i are defined as previously.

9.3.2.3   The Impact of the Number of Repair Teams on Availability

The assumption of a repair team being immediately available to repair each component upon failure is not violated if (1) a given site has a single repair team dedicated to a particular single system, and (2) the system is shut down as discussed above, immediately after reaching a failure state. Hence, a single repair team can be considered to take immediate action at the first evidence of a failure. Since the team is dedicated to the system, it will be available at the first indication of failure. No additional component failure can occur while servicing the first failure as the system is shut down.

However, there is not always a repair team dedicated to each system. Furthermore, some systems will not or cannot be shut down completely while a repair is being made on an isolated component.

While A_s is computed under the assumption of a repair team being immediately available to repair each component, it does not necessarily mean that having less than n repair teams available to service n components (or n repair teams available to repair n systems) will always *significantly* affect the value of A_s or A_{sd}.

For a system made up of more than a few components, an exact assessment of the sensitivity of A_s to the number of repair teams is possible only from the development and solutions of a rather large number of differential equations. (The concept will be illustrated later in conjunction with systems containing a given number of components.) However, one can make both intuitive and approximate estimations of the impact of the number of repair crews on system availability.

Assume that a system was made up of n components and upon failure of one or more components, the still operable portions of the system were *not* shut down. If $M_i \gg R_i$ for each component, it would seem unlikely that it would be necessary to have one repair crew for each component. That is to say, n repair teams would be necessary only when n components were in failure states at the same time. Suppose that only rarely, perhaps once every 1000 years, would all components be expected to be in a failed state at the same time. As a consequence, if there were n components in a system (or n systems at a given site), then for all practical purposes $g < n$ repair teams would provide the same availability as n. The probability, P_{g+}, that a failed component will have to wait for a repair team (i.e., all available teams are busy), given that g teams have been assigned, will provide insight about the potential impact of g on availability. If the probability is very small, then the fact that $g < n$ will not impact availability significantly. Equation (9-36) then will provide an answer that, for all practical purposes, will be accurate.

A relatively simple means of assessing P_{g+} is provided below. It is based on the information contained in Chart 9-1. Sum all the state probabilities, A_{sj}, shown in Charts 9-1 and 9-2, *associated with $(g + 1)$ or more failed components*:

$$P_{g+} = \sum_{\Omega} A_{sj} \qquad (9\text{-}39)$$

where

P_{g+} = probability that $(g + 1)$ or more components will be under repair at the same time

      = the probability that more than g maintenance teams will be required at the same time.

A_{sj} = the probability the system is in state j

Ω = the set of all values of A_{sj} associated with $(g + 1)$ or more component failures

The value so calculated will correspond to the probability that more than g maintenance teams will be needed.

Alternatively, one could sum all the values of A_{sj} associated with g or fewer component failures and calculate P_{g+}:

$$P_{g+} = 1 - \sum_{\phi} A_{sj} \qquad (9\text{-}40)$$

where

P_{g+}, A_{sj} are defined as before

ϕ = the set of all values A_{sj} associated with (g) or fewer component failures

For the system depicted in Chart 9-2, for $g = 2$,

$$P_{2+} = A_{s7} + A_{s11} + A_{s13} + A_{s14} + A_{s15}$$

For $g = 1$,

$$P_{1+} = A_{s3} + A_{s5} + A_{s6} + A_{s9} + A_{s10} + A_{s12} + A_{s7} + A_{s11} + A_{s13}$$
$$+ A_{s14} + A_{s15}$$

Example problems dealing with steady state availability evaluation follow.
For the four-component example in Chart 9-2 assume $M_i = 200$, $R_i = 10$; what is the probability that two repair teams will be sufficient? From the illustration above,

$$P_g = 0.000103 + 0.000103 + 0.000103 + 0.000103 + 0.000005 = 0.000314$$

Hence only 0.000314 of the time would three or more repair teams be busy at the same time. As a result there would be no significant effect on availability if $g = 2$.
Take a two-unit system, B and C,
For B,

$$M_b = 200, \qquad R_b = 10$$

For C,

$$M_c = 100, \qquad R_c = 15$$

Compute the availability of the system, A_s, when upon failure of one component the remaining component continues to function and where each unit has its own repair crew.
Compute the availability of the system, A_{sd}, when upon failure of one component the remaining component is shut down (and has $\lambda = 0$ while shut down) and only one repair team is associated with the system.

$$A_s = \frac{M_b}{R_b + M_b} \frac{M_c}{R_c + M_c} = \frac{(200)(100)}{(210)(115)} = 0.828$$

$$A_{sd} = \frac{1}{1 + \sum_1^2 (R_i/M_i)} = \frac{1}{1 + (10/200) + (15/100)} = 0.833$$

Assume that the system was composed of 10 units each with $M_i = 500$ and $R_i = 10$ and 10 units each with $M_i = 1000$, $R_i = 50$. Compute its availability A_s, A_{sd}.

$$A_s = \prod_{i=1}^{20} A_i = (500/510)^{10}(1000/1050)^{10} = 0.504$$

$$A_{sd} = 1/(1 + 0.2 + 0.5) = 1/1.7 = 0.588$$

Consider a system made up of components having exponential distributions of time to failure that is in a standby mode of operation until it is called upon to perform at a random point in time. While in a standby mode of operation, it has an MTBF = 50 hours. Failures are detected as soon as they occur and a repair crew is always available to provide repair. The mean time to repair is 2 hours. What is the probability that the system will be ready to perform when called on to operate at a random point in time?

$$A = \frac{M}{R + M} = 50/(50 + 2) = 0.962$$

9.3.3 Dynamic Measures Associated with Availability

Dynamic measures of availability evaluate:

1. $A_i(t)$, the probability that an item, i, is in an operable state *at* a specific point in time (e.g., 100 hours from now). Note the difference between $A_i(t)$ and mission reliability $R_i(t)$. $R_i(t)$ expresses the probability that an item will perform satisfactorily over an entire interval of time, t. $A_i(t)$ assesses the probability of an item being operable *at* t, the interval end point. Hence, in general,

$$A_i(t) > R_i(t)$$

2. $A_i(T)$, the proportion of time that an item, i, is in an operable state over a specific period of operating time (e.g., for the next 100 hours of operation).

Though similar, the measures have different forms and derivations. Both will be discussed.

In reliability engineering the dynamic measures for component and system availability are most commonly developed assuming that time to failure and time to repair for components are exponentially distributed. For reasons of commonality and simplicity those same assumptions will be applied here.

9.3.3.1   A(t), The Probability That a System is Operable at a Given Time, t

Since exponential distributions for time to failure and repair have been assumed, a Markov process can be employed to establish a set of differential equations which can be solved to determine $A_i(t)$. The Markov process is the procedure most commonly used to perform availability analyses. Its characteristics and applications are explained in most operations research texts. It is employed to develop sets of differential equations which when solved will provide evaluations of the availability measures required. This text will explain its mechanism in the detail necessary to allow its use in developing expressions for $A_i(t)$ without going deeply into its underlying theory. The important thing is to realize that the tool exists; that it can be applied for availability analysis; and that the engineer should be familiar with its characteristics for that purpose. It will be used in Chapter 10 when the reliability and MTFF of redundant systems are discussed.

To provide the reader with a rudimentary background, a brief discussion of classical Markov analysis mechanisms as applied to system availability analysis follows. The technique will be applied to a two-component system for illustrative purposes. A simpler shortcut technique to system availability evaluation based on application of the process to single components* and using those single-component availability results to derive system level availability evaluations will also be provided.

In order to apply the classical Markov process directly to a system, all possible system states must be identified. For each component in the system, generally only two states are considered—operable and inoperable (failed). For a system comprised of many components, there will be many possible system states. (See Chart 9-2.) Such states and how they may be identified were discussed in Section 9.3.1. After such identification, the next step is the definition of all the ways a component/system may transition from one state to another. The last step is the generation and solution of a set of differential equations.

The definition process proceeds in accord with certain logical rules: (1) Only one component can fail at a given instant. A fraction of a second may separate two failures, but they do not fail simultaneously. (2) If two or more separate components are under repair at the same time, only one will have its repair completed at a given instant. (3) One component cannot fail at the same instant that another has a repair completed. (4) It is possible for nothing to happen at a given instant, i.e., neither a failure nor a repair.

In order to clarify what is meant by a given instant, it will be defined as the time between t and $(t + \Delta t)$, where Δt can be considered as infinitesimally small.

---

*System 9.3.3.2 will provide expressions for single component availability.

One means which can be used to define all the ways a component/system can transition from one state to another is to set up a chart similar to Chart 9-3 and fill in the blanks for the system in question. For a system made up of two components, for example, the technique of Chart 9-3 would provide the information for the first two columns as: state $0 = B_oC_o$, state $1 = B_fC_o$, state $2 = B_oC_f$, state $3 = B_fC_f$. The remaining columns provide information on the ways such states can be reached

where

   B_o, C_o indicates the component is in an operable state
   B_f, C_f indicates the component is in an inoperable (failed) state

According to the rules above:

1. The only way the system can be in state 0 at time $(t + \Delta t)$ is for it:
   a. To have been in state 1 at time t and complete the repair of component B in Δt; or
   b. To have been in state 2 at time t and complete the repair of component C in Δt; or
   c. To have been in state 0 at time t and have no failures occur during Δt.
2. The only way the system can be in state 1 at time $(t + \Delta t)$ is for it:
   a. To have been in state 0 at time t and have component B fail in Δt; or
   b. To have been in state 3 at time t and complete the repair of component C in Δt; or
   c. To have been in state 1 at time t and have nothing happen (i.e., neither a failure or repair) in Δt.
3. The only way the system can be in state 2 at time $(t + \Delta t)$ is for it:
   a. To have been in state 0 at time t and have component C fail in Δt; or
   b. To have been in state 3 at time t and complete the repair of component B in Δt; or
   c. To have been in state 2 at time t and have nothing happen (i.e., neither failure or a repair) in Δt.
4. The only way the system can be in state 3 at time $(t + \Delta t)$ is for it:
   a. To have been in state 1 at time t and have component C fail in Δt; or
   b. To have been in state 2 at time t and have component B fail in Δt; or
   c. To have been in state 3 at time t and have no repairs occur during Δt.

Such expressions simply outline all the possible mutually exclusive ways (see Section 5.6) through which transitions to a given state can take place.

If a probability could be associated with each possible transition, the probability of being in each state could be calculated. That would take the

CHART 9-3 An Aid in Defining Ways a System Can Transform from One State to Another

State No.	State Makeup	Entered By Way Of		Experiencing No Repairs or Failures in Current State	
		A Repair While in State i, of Component j	A Failure While in State k, of Component j	Components Remaining in Repair States	Components Not Failing
0	$B_o C_o$	i { 1: $j[B$ 2: $C]$	k { none: $j[none$]	Not applicable	B, C
1	$B_f C_o$	i { 3: $j[C$]	k { 0: $j[B$]	B	C
2	$B_o C_f$	i { 3: $j[B$]	k { 0: $j[C$]	C	B
3	$B_f C_f$	i { $j[none$]	k { 1: $j[C$ 2: $B]$	B, C	Not applicable

following general form. The probability of being in state j at time $(t + \Delta t)$ is:

$$P_j(t + \Delta t) = a + b + c$$

where

    $a =$ the probability, $P_{k/rj}(t)$, of being in a state k, at time t, a state where *one component repair (or one of a group of possible repairs)* would place the system in state j, times the probability, $\mu_{k/rj}\Delta t$, that that repair will be completed in Δt

    $b =$ the probability, $P_{h/fj}(t)$, of being in a state h, at time t, a state where *one component failure (or one of a group of possible failures)* would place the system in state j, times the probability, $\lambda_{h/fj}\Delta t$, that the particular component will fail in Δt

    $c =$ the probability of already *being in state j* at time t, times the probability, $[1 - (\lambda_f + \mu_r)\Delta t]$, that no transitions (repairs on inoperable components or failures in-operable components) occur in Δt.

If we assume exponential distributions of time to failure and repair for each component, then

$$R_i = 1/\mu_i = \text{MTTR of component } i$$
$$\theta_i = 1/\lambda_i = \text{MTBF of component } i$$
$$\lambda_i = \text{failure rate of component } i$$
$$\mu_i = \text{repair rate of component } i$$

the following probability statements can be made for each component:

    $\mu_i \Delta t =$ the probability of (failed) component i being repaired in Δt

    $1 - \mu_i\Delta t =$ the probability of component i still being inoperable (in repair) at the end of Δt

    $\lambda_i \Delta t =$ the probability of component i failing in Δt

    $1 - \lambda_i\Delta t =$ the probability of component i not failing in Δt

Consistent with the above, the following system-level terms can be generated:

$$1 - \left(\sum_\phi \lambda_i + \sum_\upsilon \mu_i \right) \Delta t =$$ probability that no transitions occur from a given system state

$\phi =$ the number of components satisfactorily operating (operable) in a given state $0 \leqslant \phi \leqslant n$

$\upsilon =$ the number of components under repair (inoperable) in a given state $0 \leqslant \upsilon \leqslant n$

$\lambda_{h/fj} =$ the failure rate associated with a specific component operating in state h, which on failure would transition the system to state j, or

$=$ the sum total failure rate of the group of components operating in state h, where the failure of any one would transition the system to state j

$\lambda_{h/fj} \, \Delta t =$ the probability, given that the system is in state h at time t, that it will transition to state j in Δt due to a failure in one or more of its components

$\mu_{k/rj} =$ the repair rate of the specific component under repair in state k which, if repair is completed, will transition the system to state j

$=$ sum total repair rate of a group of components under active repair while the system is in state k, where the repair of any one would transition the system to state j

$\mu_{k/rj} \, \Delta t =$ the probability, given that the system is in state k at time t, that it will transition to state j in Δt due to repair of one of its inoperable components

The probability of being in state j at time $(t + \Delta t)$ can now be quantitatively expressed as

$$P_j(t + \Delta t) = P_{h/fj}(t)\lambda_{h/fj} \, \Delta t + P_{k/rj}(t)\mu_{k/rj} \, \Delta t$$

$$+ P_j(t)\left[1 - \left(\sum_\phi \lambda_i + \sum_\upsilon \mu_i \right) \Delta t \right] \tag{9-41}$$

(See how the terms of (9-34) correlate with Chart 9-3.) But (9-41) can be expressed in the form of a differential equation:

$$\frac{P_j(t + \Delta t) - P_j(t)}{\Delta t} = \dot{P}_j(t) = P_{h/fj}(t)\lambda_{h/fj} + P_{k/rj}(t)\mu_{k/rj}$$

$$- \left(\sum_\phi \lambda_i + \sum_\upsilon u_i \right) P_j(t) \qquad (9\text{-}42)$$

$(m + 1)$ different states will yield $(m + 1)$ differential equations, and since

$$P_j(t) \equiv A_{sj}(t) \qquad (9\text{-}43)$$

$$\sum_{j=0}^{m} P_j(t) = 1$$

which allows solution by way of LaPlace transforms.

As an example, consider the two-component system described earlier and characterized in Chart 9-3.

$$P_0(t + \Delta t) = P_1(t)\mu_b\, \Delta t + P_2(t)\mu_c\, \Delta t + P_0(t)[1 - (\lambda_b + \lambda_c)\, \Delta t]$$
$$P_1(t + \Delta t) = P_0(t)\lambda_b\, \Delta t + P_3(t)\mu_c\, \Delta t + P_1(t)[1 - (\lambda_c + \mu_b)\, \Delta t]$$
$$P_2(t + \Delta t) = P_0(t)\lambda_c\, \Delta t + P_3(t)\mu_b\, \Delta t + P_2(t)[1 - (\lambda_b + \mu_c)\, \Delta t]$$
$$P_3(t + \Delta t) = P_1(t)\lambda_c\, \Delta t + P_2(t)\lambda_b\, \Delta t + P_3(t)[1 - (\mu_b + \mu_c)\, \Delta t]$$

This series of expressions, as indicated previously, can be expressed in terms of differential equations:

$$\dot{P}_0(t) = \mu_b P_1(t) + \mu_c P_2(t) - (\lambda_b + \lambda_c)P_0(t)$$
$$\dot{P}_1(t) = \lambda_b P_0(t) + \mu_c P_3(t) - (\lambda_c + \mu_b)P_1(t)$$
$$\dot{P}_2(t) = \lambda_c P_0(t) + \mu_b P_3(t) - (\lambda_b + \mu_c)P_2(t)$$
$$\dot{P}_3(t) = \lambda_c P_1(t) + \lambda_b P_2(t) - (\mu_b + \mu_c)P_3(t)$$

We apply LaPlace transforms (whose explanation and application are available in almost any engineering mathematics text) and prepare for solution of the series of equations by assigning initial conditions at $t = 0$. At time 0, the probability is 1 that the system is in state 0, the probability is 0 that it is in state 1, 2, or 3. As a consequence, the differential equations above are transformed into the following set of simultaneous equations:

$$-1 = \mu_b P_1(s) + \mu_c P_2(s) - (s + \lambda_b + \lambda_c)P_0(s)$$

$$0 = \lambda_b P_0(s) + \mu_c P_3(s) - (s + \lambda_c + \mu_b)P_1(s)$$

$$0 = \lambda_c P_0(s) + \mu_b P_3(s) - (s + \lambda_b + \mu_c)P_2(s)$$

$$0 = \lambda_c P_1(s) + \lambda_b P_2(s) - (s + \mu_b + \mu_c)P_3(s)$$

To simplify the calculations assume $\mu_b = \mu_c = \mu$, $\lambda_b = \lambda_c = \lambda$. Then $P_1(s) = P_2(s)$ and the relationships may be expressed as

$$-1 = -(s + 2\lambda)P_0(s) + 2\mu P_1(s)$$

$$0 = \lambda P_0(s) - (s + \lambda + \mu)P_1(s) + \mu P_3(s)$$

$$0 = 2\lambda P_1(s) - (s + 2\mu)P_3(s)$$

Solving for $P_0(s)$ using determinants,

$$A_s(t) = P_0(t) =$$

$$= \frac{\begin{vmatrix} 2\mu & 0 & -1 \\ -(s + \lambda + \mu) & \mu & 0 \\ 2\lambda & -(s + 2\mu) & 0 \end{vmatrix}}{\begin{vmatrix} 2\mu & 0 & -(s + 2\lambda) \\ -(s + \lambda + \mu) & \mu & \lambda \\ 2\lambda & -(s + 2\mu) & 0 \end{vmatrix}}$$

$$= \frac{s^2 + 3\mu s + \lambda s + 2\mu^2}{s(s + \mu + \lambda)[s + 2(\mu + \lambda)]} \tag{9-44}$$

Expressing the result in terms of partial fractions to facilitate reversing the LaPlace transform results in

$$\frac{s^2 + 3\mu s + \lambda s + 2\mu^2}{s(s + \mu + \lambda)[s + 2(\mu + \lambda)]} = \frac{D}{s} + \frac{E}{(s + \mu + \lambda)} + \frac{F}{s + 2(\mu + \lambda)}$$

or

$$s^2 + 3\mu s + \lambda s + 2\mu^2 = D(s + \mu + \lambda)[s + 2(\mu + \lambda)]$$

$$+ Es[s + 2(\mu + \lambda)] + Fs(s + \mu + \lambda)$$

Setting

$$s = -(\mu + \lambda); \qquad E = \frac{2\lambda\mu}{(\mu + \lambda)^2}$$

$$s = -2(\mu + \lambda); \qquad F = \frac{\lambda^2}{(\mu + \lambda)^2}$$

$$s = 0; \qquad D = \left(\frac{\mu}{(\mu + \lambda)}\right)^2$$

Equation (9-44) can then be expressed as

$$\frac{s^2 + 3\mu s + \lambda s + 2\mu^2}{s(s + \mu + \lambda)[s + 2(\mu + \lambda)]} = \frac{\mu^2}{s(\mu + \lambda)^2} + \frac{2\lambda\mu}{(s + \mu + \lambda)(\mu + \lambda)^2}$$

$$+ \frac{\lambda^2}{[s + 2(\mu + \lambda)](\mu + \lambda)^2}$$

Reversing the Laplace transform,

$$A_s(t) = P_0(t) = \frac{\mu^2}{(\mu + \lambda)^2} + \frac{2\lambda\mu e^{-(\mu + \lambda)t}}{(\mu + \lambda)^2} + \frac{\lambda^2 e^{-2(\mu + \lambda)t}}{(\mu + \lambda)^2} \qquad (9\text{-}45)$$

9.3.3.2 An Alternate Approach to Determining $A(t)$

As can be seen, the work needed to develop an expression for $A_s(t)$, even for a simple two-component system where each component has identical μ and λ characteristics, can require some effort. However, there is a significantly simpler alternative approach which can be taken. In order to apply the procedure, the expression for $A_i(t)$ is developed for each component and as before system availability, $A_s(t)$ is defined as

$$A_s(t) = \prod_{i=1}^{n} A_i(t) \qquad (9\text{-}46)$$

where $A_i(t)$ = dynamic (time dependent) availability of component i.

A general expression for $A_i(t)$ can be developed using a very simple Markov process. As before, it is assumed that a component can be in only two states: an operable state, which is defined as state 0; and an inoperable state signifying that the component has failed and is under repair, which is defined as state 1. Assume as before that as soon as a failure occurs a maintenance team is available for repair, that μ = the repair rate, and that λ = the failure rate of the component when it is in an operable state.

Applying the procedures discussed in the last section, component state equations can be developed expressing the probability $P_0(t + \Delta t)$ that a component is in state 0 at time $(t + \Delta t)$ and the probability $P_1(t + \Delta t)$ that a component is in state 1 at time $(t + \Delta t)$:

$$P_0(t + \Delta t) = P_1(t)\mu \, \Delta t + P_0(t)(1 - \lambda \, \Delta t)$$
$$P_1(t + \Delta t) = P_0(t)\lambda \, \Delta t + P_1(t)(1 - \mu \, \Delta t)$$

where
$P_0(t + \Delta t)$ = the probability that the component is in state 0 at $(t + \Delta t)$
= the probability that component i is in state 1 (under repair) at time t and is repaired in Δt plus the probability that component i is in state 0 at time t and no failures occur in Δt.
$P_1(t + \Delta t)$ = the probability that the component is in state 1 at $(t + \Delta t)$
= the probability that component i is in state 0 at time t but fails in time Δt plus the probability that component i is in state 1 at time t and no repairs are completed in Δt.

Forming differential equations as before:

$$\dot{P}_0(t) = \mu P_1(t) - \lambda P_0(t)$$
$$\dot{P}_1(t) = -\mu P_1(t) + \lambda P_0(t)$$

Applying Laplace transforms and assigning initial conditions at $t = 0$ such that $P_0(t) = 1$, $P_1(t) = 0$,

$$-1 = \mu P_1(s) - (s + \lambda)P_0(s)$$
$$0 = -(s + \mu)P_1(s) + \lambda P_0(s)$$

Simultaneous solution of

$$(s + \mu) = -\mu(s + \mu)P_1(s) + (s + \lambda)(s + \mu)P_0(s)$$
$$0 = -\mu(s + \mu)P_1(s) + \mu\lambda P_0(s)$$
$$(s + \mu) = P_0(s)[(s + \lambda)(s + \mu) - \mu\lambda]$$

yields

$$A_i(t) = P_0(s) = \frac{s + \mu}{s(s + \mu + \lambda)} = \frac{D}{s} + \frac{E}{s + \mu + \lambda}$$

Solving for D and E as before,

$$E = \frac{\lambda}{\mu + \lambda}, \qquad D = \frac{\mu}{\mu + \lambda}$$

and

$$A_i(t) = P_0(t) = \frac{\mu}{\mu + \lambda} + \frac{\lambda e^{-(\mu + \lambda)t}}{\mu + \lambda} \qquad (9\text{-}47)$$

Using (9-47) *as a general result for any* component, we can then express system availability $A_s(t)$, for any system without having to undertake a Markov analysis as

$$A_s(t) = \prod_{i=1}^{n} A_i(t) = \prod_{i=1}^{n} \left[\frac{\mu_i}{(\mu_i + \lambda_i)} + \frac{\lambda_i}{(\mu_i + \lambda_i)} \exp[-(\mu_i + \lambda_i)t] \right]$$

$$(9\text{-}48a)$$

Evaluating the availability of a two-unit system made up of two components B and C with associated repair and failue rates λ_b, λ_c, μ_b, μ_c results in

$$A_s(t) = \frac{\mu_b \mu_c + \lambda_c \mu_b \exp[-(\mu_b + \lambda_b)t] + \lambda_b \mu_c \exp[-(\mu_c + \lambda_c)t]}{(\mu_b + \lambda_b)(\mu_c + \lambda_c)}$$

$$+ \frac{\lambda_b \lambda_c \exp[-(\mu_b + \mu_c + \lambda_b + \lambda_c)]t}{(\mu_b + \lambda_b)(\mu_c + \lambda_c)} \qquad (9\text{-}48b)$$

When, as before, $\lambda_b = \lambda_c = \lambda$, $\mu_b = \mu_c = \mu$, (9-49) takes the form

$$A_s(t) = \frac{\mu^2}{(\mu + \lambda)^2} + \frac{2\lambda\mu \exp[-(\mu + \lambda)t]}{(\mu + \lambda)^2} + \frac{\lambda^2 \exp[-2(\mu + \lambda)t]}{(\mu + \lambda)^2}$$

which is identical to the result shown in (9-45).

The same procedure (i.e., using the result of (9-47) to define $A_i(t)$, then applying (9-48)), is appropriate for defining $A_s(t)$ for any series-connected system.

9.3.3.3 Using the Steady-State Value of Availability as an Approximation for the Dynamic

When we consider t very large, the ratio of λ/μ very small, or some combination of both, the dynamic measure of availability, $A_i(t)$, approaches the steady-state measure of reliability, A_i:

$$A_i(t) \rightarrow \frac{\mu}{\mu + \lambda} = \frac{M}{R + M} = A_i \qquad (9\text{-}49)$$

where

$$M = \text{MTBF} = \frac{1}{\lambda}; \qquad R = \text{MTTR} = \frac{1}{\mu}$$

In actuality, in most cases considering most typical ratios of λ/μ or R/M, t does not really have to be excessive for $A_i(t) \rightarrow A_i$.

The exact degree of fit of A_i to $A_i(t)$ can be determined for any combination of the ratio (R/M) and t in terms of a factor z, where

$$z = \frac{A_i(t) - A_i}{A_i}$$

= the proportion of deviation of A_i from $\qquad (9\text{-}50)$

and where $A_i(t)$ can be expressed as

$$A_i(t) = A_i(1 + z) \qquad (9\text{-}51)$$

As $z \rightarrow 0$, $A_i(t) \rightarrow A_i$, and the smaller z, the better the fit between the two quantities.

If we let

$$k = \frac{\lambda}{\mu} = \frac{R}{M}, \qquad LM = \frac{L}{\lambda} = t$$

where

LM = operating time expressed as a multiple of M

M = MTBF = $\dfrac{1}{\lambda}$

L = a multiple of MTBF

t = operating time

R = $\dfrac{1}{\mu}$ = mean time to repair

μ = repair rate

we can then express (9-47) as

$$A_i(t) = \frac{1}{1+k}\left[1 + k \exp\left(-\frac{(1+k)L}{k}\right)\right] \qquad (9\text{-}52)$$

Substituting k for (R/M) in (9-49), $A_i = 1/(1 + k)$. Substituting that result and the result of (9-52) in (9-51)

$$\frac{1}{1+k}\left[1 + k \exp\left(-\frac{(1+k)L}{k}\right)\right] = \frac{1}{1+k}(1 + z)$$

Simplifying,

$$z = k \exp\left(-\frac{(1+k)L}{k}\right) \qquad (9\text{-}53)$$

Figure 9-3 depicts values of z associated with different mission times, expressed in multiples (L) of MTBF, for various selected values of k. As can be seen, for a value of $k = 0.01$ which, considering the current state of the

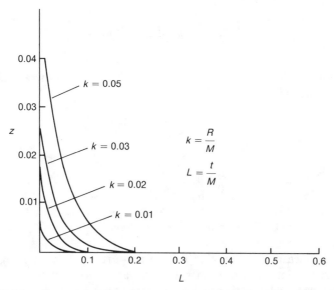

FIGURE 9-3.   Error, z, introduced by approximating $A(t)_i$ by A_i for various values of k and L. L = length of t expressed in multiples of component MTTF, M.

art, is pessimistic (high), the deviation is approximately on the order of one-tenth of one percent for a mission duration equal to 2% of the MTBF of the component.

In principle, as

$$A_i(t) \to A_i, \qquad \text{for large } t$$

$$A_s(t) \to \prod A_i = A_s \tag{9-54}$$

However, for any value of t, the accuracy of (9-54) depends not only on the values of z but on the number of components making up the system as well. If, for example, z were as small as $z = 0.001$ for each component; in other words, if

$$\frac{A_i(t)}{A_i} = (1 + z) = 1.001$$

then

$$A_s(t) = \prod_{}^{n} A_i(t) = \prod_{}^{n} A_i(1 + z) = (1 + z)^n A_s \tag{9-55}$$

and

$$\frac{A_s(t)}{A_s} = (1 + z)^n = (1.001)^n$$

If n took on the following values, A_s would understate the actual value of $A_s(t)$ by the values indicated:

n	Deviation
10	1%
20	2%
30	3%
40	4%

9.3.3.4   A(T), The Proportion of Time an Item is Operable

The second measure of availability—the proportion of time, $A(T)$, that a component/system is in a given state over an interval of time T—is *not* equivalent to $A(t)$, but rather is derived from that quantity.

Let $g(t)$ be a function linked to the availability of an item, $A(t)$, in such a way that when the item is operable at t, $g(t) = 1$; when the item is under repair (failed) at t, $g(t) = 0$. It follows then, that over an interval of time T, T_1 to T_2, the proportion of time that the item will be operable, $A(T)$, can be expressed as

$$A(T) = \frac{1}{T} \int_{T_1}^{T_2} g(t)\, dt \tag{9-56}$$

But since $A(t) =$ probability that the item is operable at time t, the proportion of the time that $g(t)$ equals 1 at time t equals $A(t)$ and

$$\int_{T_1}^{T_2} g(t)\, dt = \int_{T_1}^{T_2} A(t)\, dt$$

As a result (9.56) can be expressed as

$$A(T) = \frac{1}{T} \int_{T_1}^{T_2} A(t)\, dt \tag{9-57}$$

or

$$A(T) = \frac{1}{T(\mu + \lambda)} \int_{T_1}^{T_2} \{\mu + \lambda \exp[-(\mu + \lambda)t]\}\, dt$$

which when integrated, since $T = T_2 - T_1$, results in

$$A(T) = \frac{\mu}{\mu + \lambda} + \frac{\lambda}{(\mu + \lambda)^2 T} \{1 - \exp[-(\mu + \lambda)T]\} \tag{9-58}$$

It follows, then, that if $A_i(T)$ represents the proportion of the interval T that a component i can be expected to be operable, then the proportion of time that a system made up of n independent components can be expected to operate over the interval T is $A_s(T)$ where

$$A_s(T) = \prod_{i}^{n} A_i(T) \tag{9-59}$$

When T is very large,

$$A_i(T) \rightarrow \frac{\mu}{\mu + \lambda} = \frac{M}{R + M} = A_i \tag{9-60}$$

which shows the relationship between steady-state availability and the last form of dynamic availability.

The steady-state value can be utilized as an approximation to $A_i(T)$ as it was used to approximate $A_i(t)$. In this case the accuracy of the approximation is a function of T and the (R/M) ratio characteristics of the component. As a result a term, z, can be defined which represents the proportion of error involved in approximating $A_i(T)$ by A_i. Similarly to before, z takes the form

$$z = \frac{A_i(T) - A_i}{A_i}$$

and

$$A_i(T) = A_i(1 + z) \tag{9-61}$$

As before define $LM = T =$ the interval of time such that correspondingly $L = T/M$ and $k = R/M$.

Substituting L and k into the expressions for A_i and $A_i(T)$ and applying (9-61), the following results:

$$\frac{1}{1+k}\left(1 + \frac{k^2}{(1+k)L}(1 - e^{-(1+k)L/k})\right) = \frac{1}{1+k}(1 + z)$$

Simplifying,

$$\frac{k^2}{(1+k)^2 L}(1 - e^{-(1+k)L/k}) = z \tag{9-62}$$

Figure 9-4 depicts the values of z associated with different mission times, expressed in multiples (L) of MTBF and various values of k. As can be seen, $A_i(T)$ approaches A_i in a much slower fashion than does $A_i(t)$. For values of $k = 0.01$, the deviation is on the order of one-tenth of one percent over a mission duration equal to 10% of the MTBF of the component.

In principle, as

$$A_i(T) \to A_i, \qquad \text{for large } T$$

$$A_s(T) \to \prod^{n} A_i = A_s$$

However, for the same reasons as described earlier in discussing the steady-state characteristics of $A_s(t)$, the values of z for each component and the number of components making up the system have a significant effect on the accuracy of applying the steady-state results to system evaluation. As before, when A_i differs from $A_i(T)$ by only a small factor, z,

$$A_s(T) = \prod^{n} A_i(1 + z) = (1 + z)^n A_s \tag{9-63}$$

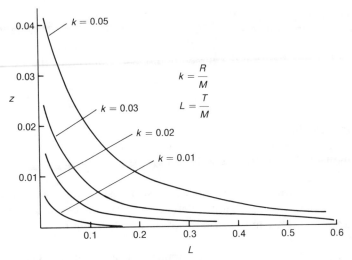

FIGURE 9-4.   Error, z, introduced by approximating $A(T)_i$ by A_i for various values of k and L. $L =$ length of T expressed in multiples of component MTTF, M.

and if z is 0.001 for example,

$$\frac{A_s(T)}{A_s} = (1 + z)^n = (1.001)^n$$

and if n were to take on the same values as illustrated earlier, the deviations of A_s from the actual value of $A_s(T)$ would accumulate to the same results as shown earlier.

Three components B, C, and D, each with respective MTTF of $M_b = 100$, $M_c = 200$, $M_d = 500$, and mean times to repair of $R_b = 1$, $R_c = 1$, $R_d = 2$ are combined to form a series system to be used in aiding the landing of a long-endurance aerospace vehicle. Each component is to run continuously, interrupted only by failures and the time required to correct the failures. Because an emergency might require a landing at any time, each component has associated with it a dedicated repair crew and failure is detected immediately.

The results of the analysis are to be provided to higher management. It would be desirable to present those results in the simplest, most logical terms. The steady-state availability metrics fit those criteria more appropriately than do the dynamic measures. As a consequence, the system must be examined to see how acceptable and accurate steady-state measures are.

The availability measure in question is the probability that the system will be operable at the time the vehicle will land. Evaluate that measure under the repair team assumptions above. Would availability be enhanced if the system were shut down after any component failure (assume $\lambda = 0$ when shut down; a probability of 100% associated with successful restart)?

First, z is computed for the three components, for use of the A_i vs the $A_i(t)$ measure:

$$z = ke^{-(1+k)L/k}$$

For component B: $k = 0.01$, $L = 5$, and $z_b \sim 0$.
For component C: $k = 0.005$, $L = 2.5$, and $z_c \sim 0$.
For component D: $k = 0.004$, $L = 1$, and $z_d \sim 0$.

Each component obviously approaches its steady-state value very quickly. As a consequence, no significant difference will exist between steady-state and dynamic measures and, hence, steady-state values will be utilized in the analysis.

The availability of the system, assuming that a maintenance team was available for each component and that upon failure of the system only the faulty component were shut down, is

$$A_s = \prod^n A_i = [100/(100 + 1)][200/(200 + 1)][500/(500 + 2)]$$

$$= (100/101)(200/201)(500/502) = 0.9812$$

The availability of the system when upon failure of any component the entire system is shut down (requires only one maintenance crew) is

$$A_{sd} = \frac{1}{1 + \sum^n R_i/M_i} = 1/[1 + (1/100) + (1/200) + (2/500)] = 1/1.019$$

$$= 0.9814$$

Hence, availability would be enhanced marginally by shutting down the system after each component failure, and that action would require less overall crew.

What is the probability that one repair crew will be sufficient, assuming that after a component failure the remaining components will remain in an energized state? In order to determine whether or not one repair crew will be adequate, the different states requiring at most one crew are defined. Let

State 0 = the state where the system is operable, that is to say, the state where all components are operating. We know that the probability that the system is in such a state is equal to the availability A_s. As a result:

$$A_0 = A_s = 0.9812$$

State 1 = the state where component B is undergoing repair and the remaining components are operable.

$$A_1 = \bar{A}_b A_c A_d = (1/101)(200/201)(500/502) = 0.00981$$

State 2 = the state where component C is undergoing repair and the remaining components are operable

$$A_2 = \bar{A}_c A_b A_d = (1/201)(100/101)(500/502) = 0.00005$$

State 3 = the state where component D is undergoing repair and the remaining components are operable

$$A_3 = \bar{A}_d A_b A_c = (2/502)(100/101)(200/201) = 0.00004$$

Since the probability of needing more than one repair crew is equal to the probability that the system is in a state where more than one component is failed and undergoing repair, the probability that one repair crew will be adequate for any one mission is

$$(1 - P_{g+}) = A_0 + A_1 + A_2 + A_3 = 0.991$$

Your company has accepted an incentive-type contract to run and maintain a large computer system. The system must be operational 22 out of each 24 hours. It must be shut down 2 hours each day for maintenance. It has a mean time to failure of $\theta = 100$ hours. In the event of a failure, the mean time to repair is 5 hours. For every hour of operation up to 22 hours per day, the company will receive $1000. For each hour the system is down, outside of the 2 hours set aside for maintenance the company must pay the customer $3000. Over a long period of time, what is the average amount of money your company can expect to net per day, assuming that failure detection is immediate and that a repair crew is always available?

In this case, $k = R/M = 5/100 = 0.05$ and $L = 22/100 = 0.22$. Looking at Fig. 9-4 we see that a steady-state solution would result in an answer that would be conservative by a factor of about one per cent. As a consequence, we elect to provide a dynamic-type evaluation of the expected proportion of the time per operating day, $A_i(T)$, that the computer is operable.

$$A_i(T) = \frac{1}{1+k}\left(1 + \frac{k^2}{(1+k)L}(1 - e^{-(1+k)L/k})\right)$$

$$= 0.952 + 0.0103(0.9901) = 0.962$$

The expected number of hours per day the system will be operable is $22(0.962) = 21.164$, and, hence, the expected payment from the customer for operation will amount to $1000(21.164) = \$21\,164$ per day.

The expected number of hours per day the system will be inoperable is $22(0.038) = 0.836$ and, hence, the expected penalty payment *to the customer* will amount to $3000(0.836) = \$2508$ per day.

The expected net per day is $\$21\,164 - \$2508 = \$18\,656$.

9.3.4 Availability When Maintenance Is Not Performed at Failure

Virtually all of the discussions in the previous sections assumed that failures were immediately detected and maintenance teams were immediately available. In the event that an item has no automatic fault-detection capability or the item cannot be monitored constantly, failures cannot be detected immediately nor can maintenance be initiated at failure. Therefore, other strategies for availability enhancement and evaluation must be considered.

Standby systems and components, in particular, fall into this category. Items in a standby mode of operation usually have limited fault-detection capabilities but experience a failure rate while in that mode. As a result, failures can occur which are not detected until the item is subject to a mission call at a random point in time. At that time the item is found to be unavailable due to failure and the mission may be aborted.

One strategy to improve availability in such a situation is to check (test) the item periodically to see whether it is operable. If found inoperable, it is then repaired and not tested again until the next period of test or mission call, whichever comes first.

If we represent

λ = failure rate associated with the item while in a standby mode (an exponential distribution of failure times is assumed for the item)

T_C = the standby operating time between item checks

then the probability that the item will not fail in the interval between checks and, hence will be available for use through the period, equals its reliability, $R(T_C)$:

$$R(T_C) = \exp(-\lambda T_C) \tag{9-64}$$

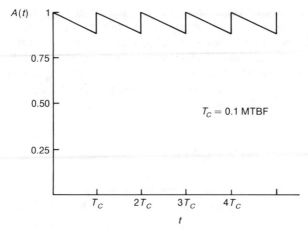

FIGURE 9-5.   Effect of check every T_C hours on probability that the item will be operable at t (instant repair assumed).

Hence, $R(T_C)$ can be considered the minimum availability of the item, the minimum probability that an item will be capable of operating when called on at a random point in time $t \leqslant T_C$. That is to say, availability can be expressed as

$$\text{Availability over } (t \leqslant T_C) \geqslant R(T_C) \tag{9-65}$$

and when steady-state availability, A, is considered over many time intervals T_C

$$A > R(T_C) \tag{9-66}$$

See Fig. 9-5, which shows the effect of checks every T_C hours on the probability that the item will be operable at any time t.

Since (9-57) defines $A_i(T)$ as

$A_i(T)$ = the average proportion of time that the item will be
       operable (or capable of operation) over an interval of time T

if T is equated to T_C, then

$A_i(T_C)$ = the average proportion of time that the item will be
       capable of operation between checks made every T_C hours

and

$T_C \cdot A_i(T_C)$ = the *expected time* that the item will be
       operable over an interval of time T_C

This value represents the average value of operating time that would result if the item were put through many cycles of operation of specific length T_C where at the beginning of each cycle the item was in an operable condition.

Since in this case failure detection and repair is not possible *during* the interval T_C (only at the end of each interval T_C), the expression for $A_i(T)$ shown in (9-58) can be modified to represent that fact by letting $R = \infty$, i.e., $\mu = 0$. The relationship shown in (9-58) can be expressed as

$$A_i(T_C) = \frac{1}{\lambda T_C} [1 - \exp(-\lambda T_C)] \qquad (9\text{-}67)$$

which equals the proportion of time the item will be in an operable state over the time interval T_C when no repair is possible over T_C, where

$$\frac{1 - \exp(-\lambda T_C)}{\lambda} = \text{the expected time the item will be operable over}$$
$$\text{an interval of time } T_C \qquad (9\text{-}68)$$

Part of an emergency flight system for an aircraft is located at an inaccessible place which can be practically checked only as part of a more general inspection every 500 hours. If its failure rate is $\lambda = 0.0001$, what is the minimum probability it will be available if needed?

$$A \geqslant R(T_C) = \exp(-\lambda T_C)$$
$$A = e^{-(0.0001)500} = 0.951$$

A standby backup power system for a large plant is in a remote portion of an industrial site and can be checked only every 7 days. Its standby failure rate is $\lambda = 0.0002$. What is the probability it will be available if called on to operate at a random point in time?

$$A \geqslant R(T_C) = \exp(-\lambda T_C)$$
$$A \geqslant e^{-(0.0002)(24)(7)} = 0.967$$

9.4 COMBINED MEASURES OF RELIABILITY AND AVAILABILITY EFFECTIVENESS

When mission reliability was discussed, the definition statement was always prefaced with "given that the item is operable" at the start of the mission or call to perform. As was seen in the discussion of availability in the last sections, an item may well be unavailable for operation when asked to perform at a random point in time. An item which is not available when called for can have the same effect as a failure during the course of task performance. Task or mission reliability in such instances must take on a joint availability–reliability nature. This is particularly true of items which (a) have standby modes of operation when not in active use (see Section 8.5.3), and (b) have nonoperationg failure rates associated with them (see Section 8.5.5).

The "reliability" measure of concern for such instances is one of effectiveness, where effectiveness is defined as the probability that an item can successfully meet an operational demand. This includes being available when the demand is received and performing successfully through the demand period. Applicability of effectiveness measures spans both military and commercial systems, from safety controls on nuclear power plants, to missile systems, to military and emergency vehicles, to any item or system in general which is in a nonoperating mode of operation until called upon to perform a function.

Effectiveness, E, may be defined as the product of availability and reliability:

$$E = A(t_s)R(T) \tag{9-69}$$

where

$A(t_s)$ = the probability the item will be operable at the end of a standby period of length t_s when a demand is received

$R(T)$ = the probability the item performs satisfactorily over the demand/task/mission interval

We will assume that time to failure and repair both take exponential distributions and that immediate failure detection and access to maintenance is possible during standby operation:

$$E = \left[\left(\frac{\mu}{\mu + \lambda_s} \right) + \left(\frac{\lambda_s}{\mu + \lambda_s} \right) \exp[-(\mu + \lambda_s)t_s] \right] \exp(-\lambda T) \tag{9-70}$$

or, where t_s is large,

$$E = \frac{\mu}{\mu + \lambda_s} e^{-\lambda T} \tag{9-71}$$

where

λ_s = failure rate of the item in its standby state
λ = failure rate of the item in its operating state
t_s = average period of time the item is in a standby state
T = mission time
$\mu = 1/R$ = repair rate for the item
R = mean time to repair (MTTR) of the item

Similarly, when fault detection and maintenance are not performed immediately after failure occurs but in accord with an established schedule, i.e., every T_C hours,

$$E \sim \exp(-\lambda_s T_C) \exp(-\lambda T)$$

or

$$E \sim \exp[-(\lambda_s T_C + \lambda T)] \tag{9-72}$$

where all terms are defined as previously.

A critical military system is in a standby manned state until its use is called for at a random point in time. Its standby failure rate is $\lambda_s = 0.001$, its operational failure rate is $\lambda = 0.04$; its mean time to repair is $R = 5$ hours; and its mission is 3 hours in length. Mission success requires that the item be available at mission start, and that it perform satisfactorily over the mission duration. What is the probability that the mission will be successful provided fault detection and maintenance during standby is immediate?

$$E = AR(T) = \frac{\mu}{\lambda_s + \mu} \exp(-\lambda T)$$

$$E = [0.2/(0.01 + 0.2)]e^{-0.04(3)} = 0.845$$

An emergency backup system located in an inaccessable portion of a power-generation facility can be practically checked only every 7 days. Its standby failure rate is $\lambda_s = 0.0001$; its operational failure rate is $\lambda = 0.0007$; and, if energized, it must operate for 100 hours. What is its overall effectiveness per potential cycle of use?

$$E \sim \exp[-(\lambda_s T_C + \lambda T)]$$

$$E \sim e^{-[(0.0001)(7)(24)+(0.0007)100]} = e^{-0.0868} = 0.917$$

Reference

1. Mann, N., Schafer, R. and Singpurwalla, N. (1974). *Methods for Statistical Analysis of Reliability and Life Data*. Wiley, New York.

10

Reliability Analysis and Evaluation of Redundant or Fault-Tolerant Systems

In order to cope with technological developments of the past thirty years, systems have been compelled to expand in both size and complexity at a rapid rate. Of equivalent importance to the need for this growth has been the coincident need for greater system reliability and availability. As industrial and space requirements necessitated the construction of ever more complex systems, the philosophy of simply increasing reliability by making parts more reliable became unrealistic. The only answer to such situations has been recourse to redundancy, or as it is sometimes termed, fault tolerance. In Section 4.2 we described briefly the role of redundancy as a reliability multiplier.

Redundancy exists when one or more components of a system can fail and the system can still perform its function satisfactorily with those that remain. The system may adapt to component failures by performing the required function with the components remaining, or it may "switch in" available spare components to take the place of the failures. The components used to take the place of the failed ones might be functionally redundant (i.e., not identical in form to the failed unit but capable of compensating for the failed function; for example, one processor taking over the role of a different type of processor, a VHF transceiver taking over the function of a UHF transceiver) or it could be an exact duplicate of the failed component. All or only certain of the components making up the system could be redundant.

There are many different redundancy strategies. Some work in conjunction with maintenance and/or corrective action performed during the mission or periodically. Some are applied when no maintenance or corrective action can be performed. Each has its advantages (e.g., reliability gain) and its

343

disadvantages (e.g., the number of duplicative elements, which impacts on total system weight, cost, and volume). A system may employ a single means of redundancy throughout, a mixture of redundant and nonredundant components, or a mixture of different types of redundancy. If we were to break down a system to a set of required subsystem functions, and determine the probability $R_i(t)$ that each subsystem would perform satisfactorily, taking into account the particular form of redundancy applied to the subsystem, or the absence of redundancy, then analogously to the relationships discussed in Chapter 9, system reliability, $R_e(t)$, can be expressed simply as

$$R_e(t) = \prod_{i=1}^{n} R_i(t) \tag{10-1}$$

and as before

$$\text{MTFF}_e = \int_0^{\infty} R_e(t)\, dt \tag{10-2}$$

where
$R_e(t)$ = system reliability
$R_i(t)$ = subsystem reliability

Depending on the type of redundancy applied, $R_i(t)$ can take many different forms. $R_e(t)$ and MTFF can likewise take different forms. The objective of this chapter is to provide rationale and guidance for development of the forms that $R_i(t)$, $R_e(t)$, MTFF_i and MTFF_e take for the most common types of redundancy.

For purposes of organization, discussion will be broken down into consideration of three basic classes of redundant systems/subsystems:

1. Systems which are redundant and which are nonmaintained until a system failure occurs (failed components of a redundant system are not repaired or replaced until the system fails). This situation is common to unattended applications—an unmanned field site, a satellite, etc.
2. Systems which are redundant and cannot be maintained during mission or use intervals. However, the systems can be subjected to scheduled/ periodic maintenance, or maintenance can be performed after mission or use periods. Such maintenance repairs or replaces all failed components. This situation is common to most aircraft, remote sites or locations which are periodically maintained.
3. Systems which are redundant and can be *continuously maintained* (failed components or a redundant system are repaired and replaced as they

occur). This situation is common to attended applications, i.e., manned sites, any system where maintenance can be performed during usage.

For each class several of the most common types of redundancy are considered. *In all cases it will be assumed that the components all have times to failure which follow an exponential distribution described by*

$$f(t, \lambda) = \lambda e^{-\lambda t} \qquad t > 0, \qquad \lambda > 0 \tag{10-3}$$

where
λ = failure rate of the component
t = operating time in question
$R(T) = 1 - \int_0^T f(t, \lambda) \, dt = e^{-\lambda T}$, the probability that the component will operate satisfactorily over the interval, T (given it is operable at the start of the interval).

For notational convenience, $R(t)$ will be substituted for $R(T)$ in discussions which follow.

10.1   REDUNDANT SYSTEMS WHICH ARE NOT MAINTAINED

First consider the class of redundancy for which no immediate or scheduled maintenance is performed to repair/replace failed components. In this case the system is put into operation with all components performing satisfactorily. The system continues to operate until it can no longer perform its function satisfactorily. This is a form of redundancy associated with systems in satellites or other space applications or systems operated in remote or inaccessible areas.

The principal measures of reliability for nonmaintained systems are:

1. The probability that the system will perform satisfactorily over an interval of time t given that the system is in a particular operating state at the start of the interval. Operating state in this instance refers to which components and how many of them are in an operable state at the start of the interval. For the purposes of our discussion, unless otherwise specified, it will be assumed that *all* components are in an operable state at the start of a mission.
2. The mean time to first failure, MTFF, or the mean time to failure, MTTF. For redundant systems made up of components which have *exponential distributions* of time to failure, whether or not the components are repairable or discardable upon failure, the values of MTFF and MTTF

are equivalent. The "mean" values represent the *average* time to system failure measured from the point in time that the system is put into operation with *all* components operable to the point in time the system next fails (the point at which the system is unable to perform satisfactorily). For example, a system is put into operation with all components operable. Since it is redundant, it continues to operate even though a number of *components* fail. Since the system is nonmaintained, we make no attempt to repair such components until the system reaches a failed state. When the system reaches a failed state, all failed components are repaired/replaced and the system is put into operation again. MTFF and MTTF represent the average time between operational start and system failure.

Applications of several different types of redundancy will be considered at both subsystem and system levels:

- Full-on redundancy with perfect and imperfect failure sensing and switching
- Standby redundancy with perfect and imperfect failure sensing and switching

No matter what types of redundancy are applied at the subsystem level, system reliability can always be evaluated through the use of (10-1) and (10-2). This applies to situations where different forms of redundancy are employed for each subsystem as well as when the same form of redundancy is applied throughout the system.

10.1.1   Full-On Redundancy: The Single-Survivor Subsystem; Perfect Sensing and Switching

One of the most widely discussed forms of subsystem redundancy in the literature is a parallel arrangement of n redundant components as shown in Fig. 10.1. In this type of redundancy, all n components are identical and *continuously energized*. (Hence the term full-on redundancy). It is assumed that as long as at least one component is functioning properly in a subsystem, that subsystem will perform satisfactorily. The reliability of such a redundant subsystem is evaluated as follows.

Assume perfect failure sensing and that upon failure of a component the failed component is switched off so that it can cause no disturbances. If required, adjustments are made automatically to the remaining components. Let

n = number of continuously energized components comprising each subsystem

λ_j = failure rate of each redundant component

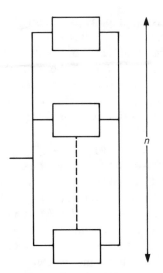

FIGURE 10-1.   A subsystem made up of n redundant components.

Then, given the reliability of each component, $R_j(t)$, and assuming that all components share the same failure rate λ_j:

- The probability of any single component (j) failing in a time interval t is $1 - R_j(t) = 1 - \exp(-\lambda_j t)$.
- The probability of all n components in a given subsystem failing in a time interval t is $[1 - \exp(-\lambda_j t)]^n$.
- The probability that at least one component out of n will *not* fail over the time interval t is $1 - [1 - \exp(-\lambda_j t)]^n$.

But if at least one component of a subsystem operates satisfactorily over the interval t, the subsystem will operate satisfactorily over t. Subsystem reliability, $R_i(t)$, can then be expressed as

$$R_i(t) = 1 - [1 - \exp(-\lambda_j t)]^n \qquad (10\text{-}4)$$

where $R_i(t) =$ the probability that the subsystem i will satisfactorily perform over the time interval t.

Suppose a *system is composed of L subsystems* of that same type:

1. All of which are connected in series as in Fig. 9-1.
2. All of which are full-on redundant, single-survivor types.
3. All of which have the same number of components, n.
4. All of which have components which have the same value of failure rate $\lambda_j = \lambda$.

System reliability $R_e(t)$ is equal to the probability that the system will operate satisfactorily over a time interval t. But this is equal to the probability that each subsystem will have at least one component which will perform satisfactorily over time t. If $R_i(t)$ = the probability that at least one component in subsystem i will perform satisfactorily over the time interval t, then the probability that each subsystem will have at least one component which will perform satisfactorily over t is $R_e(t)$ and

$$R_e(t) = [1 - (1 - e^{-\lambda t})^n]^L \tag{10-5}$$

Equation (10-5) was developed under the assumption that each subsystem contains the same number of components, n, and that each component in *every* subsystem had the same value of λ. More general expressions can be developed which are not sensitive to all such assumptions.

For a system composed of L subsystems each containing redundancies of this type in which each subsystem contains a different number of components n_i and where the failure rates of the components within a given subsystem are identical but component failure rates among different subsystems differ,

$$R_e(t) = \prod_{i=1}^{L} \{1 - [1 - \exp(-\lambda_i t)]^{n_i}\}_i \tag{10-6}$$

where
  L = number of subsystems
  n_i = number of components in the ith subsystem
  λ_i = the common failure rate shared by all components belonging to subsystem i

The most general forms of $R_e(t)$ allow each component to take on a separate failure rate, λ_{ij}. In those cases

$$R_e(t) = \prod_{i=1}^{L} \left\{1 - \prod_{j=1}^{n_i} 1 - \exp(-\lambda_{ij} t)] \right\}_i \tag{10-7}$$

where λ_{ij} = the failure rate of the jth component in subsystem i and all other terms as defined previously.

Mean time to first failure (MTFF) for subsystems and systems employing this type of redundancy is considered next (some portions of the literature refer to it as mean time to failure, MTTF).

For subsystems containing identical components which have exponential distributions of time to failure and which are all either replaced or repaired (assuming they have failed) at *subsystem* failure, MTFF or MTTF (see Appendix A4) can be expressed as

$$\text{MTTF}_i = \text{MTFF}_i = \int_0^\infty R_i(t)\, dt = \int_0^\infty [1 - (1 - e^{-\lambda t})^n]\, dt \qquad (10\text{-}8)$$

Equation (10-8) after integration (see Appendix A15) reduces to

$$\text{MTTF}_i = \text{MTFF}_i = \frac{1}{\lambda} \sum_{s=1}^{n} \left(\frac{1}{s} \right) = \frac{1}{\lambda} + \frac{1}{2\lambda} + \frac{1}{3\lambda} + \cdots + \frac{1}{n\lambda} \qquad (10\text{-}9)$$

For systems (i.e., subsystems of the type above connected in series) that are either discarded at failure or repairable, system mean time to failure, MTTF_e, or mean time to first failure, MTFF_e, can be expressed as

$$\text{MTTF}_e = \text{MTFF}_e = \int_0^\infty R_e(t)\, dt \qquad (10\text{-}10)$$

Except for special cases (10-10) does not lend itself to a closed form solution or to as neat a closed form solution as does (10-9).

For the special case represented by (10-5), after some manipulations (see Appendix A15) (10-10) reduces to the following simple form:

$$\text{MTTF}_e = \text{MTFF}_e = \int_0^\infty [1 - (1 - e^{-\lambda t})^n]^L\, dt$$

$$= \frac{1}{\lambda} \sum_{K=1}^{L} (-1)^{K+1} \binom{L}{K} \sum_{s=1}^{K_n} \frac{1}{s}; \qquad \binom{L}{K} = \frac{L!}{(L-K)!\,K!} \qquad (10\text{-}11)$$

where

λ = the common failure rate of each component
L = the number of subsystems making up the system
n = the number of components in each subsystem

For the more general case, when $R_e(t)$ is expressed under the conditions of (10-6), (10-10) takes the form

$$\text{MTTF}_e = \text{MTFF}_e = \int_0^\infty \left\{ \prod_{i=1}^{L} [1 - (1 - \exp(-\lambda_i t))^{n_i}]_i \right\} dt \qquad (10\text{-}12)$$

a simple closed form solution as (10-11) is not possible. The evaluation of the integral is, however, less complex than it is tedious. The quantities which must be integrated after $R_e(t)$ is expanded are elementary in nature and follow a generally common form. In many cases they may be integrated by observation. The following examples will provide an illustration of that characteristic.

Examples   A system is made up of two subsystems, A and B which are connected in series. The functions that both perform are necessary to the system function. Both A and B are designed to be redundant. A is made up of two components; only one need operate for its function to be performed. The failure rate of each component is $\lambda_a = 0.01$. B is made up of three redundant components of which only one need operate for its function to be performed. The failure rate of each component is $\lambda_b = 0.02$.

- What is the reliability of the system over a 20-hour mission?
- What would the reliability be if B contained only one component; that is to say, only one part of the system was redundant?
- What would the reliability of the system be under the last condition if each component of subsystem A had a different failure rate, e.g., $\lambda_{a1} = 0.01$, $\lambda_{a2} = 0.03$?
- What is the MTFF of the original system?

In order to evaluate the reliability of the system with

$$\lambda_a = 0.01, \qquad \lambda_b = 0.02, \qquad n_a = 2, \qquad n_b = 3, \qquad t = 20$$

since in this case each subsystem contains components which are identical and unique to each subsystem,

$$R_i(t) = 1 - (1 - e^{-\lambda t})^n$$

The reliabilities of subsystems A and B are,

$$R_a(t) = (1 - 0.0329) = 0.9671$$

$$R_b(t) = (1 - 0.0358) = 0.9642$$

$$R_e(t) = \prod_i^L \{1 - [1 - \exp(-\lambda_i t)]^{n_i}\}_i$$

$$R_e(t) = (1 - 0.0329)(1 - 0.0358) = 0.9325 = \text{reliability of the original system}$$

If B contained only one unit,

$$R_b(t) = (1 - 0.3297) = 0.6703; \qquad R_a(t) = 0.9671$$

as before.

$$R_e(t) = (1 - 0.0329)(1 - 0.3297) = 0.6483$$

$$= \text{reliability if } B \text{ contained only one unit}$$

Hence, the subsystem and system-level relationships can be used to evaluate the reliability of subsystems and systems that are only partially redundant or not redundant at all. In such cases they reduce directly to the basic relationship. For example, they hold when $n_i = 1$ and $L = 1$ as well as for any values of n_i, $L > 1$.

Assuming that each component of subsystem A had a different failure rate as indicated earlier and that subsystem B was made up of a single component, using equation (10-7)

$$R_e(t) = [1 - (0.1813)(0.4512)][1 - 0.3297] = 0.6155$$

$$= \text{the reliability if } B \text{ contained only one component and } A \text{ contained}$$
$$\text{components having different failure rates.}$$

The MTFF of the original system is evaluated as

$$\text{MTFF}_e = \int_0^\infty \prod \{1 - [1 - \exp(-\lambda_i t)]^{n_i}\}_i \, dt$$

Letting $\exp(-\lambda_i t) = R_i(t) =$ the common reliability of each *component* in subsystem i,

$$\text{MTFF}_e = \int_0^\infty \left\{ \prod_i [1 - (1 - e^{-\lambda_i t})^{n_i}] \right\} dt$$

$$= \int_0^\infty [1 - (1 - 2R_a(t) + R_a(t)^2)]$$

$$\times [1 - (1 - 3R_b(t) + 3R_b(t)^2 - R_b(t)^3)] \, dt$$

$$= \int_0^\infty [6R_a(t)R_b(t) - 6R_a(t)R_b(t)^2 + 2R_a(t)R_b(t)^3$$

$$- 3R_a(t)^3 R_b(t) + 3R_a(t)^2 R_b(t)^2 - R_a(t)^2 R_b(t)^3] \, dt$$

While the relationship is long, each term is simple enough to be integrated by inspection.

$$\text{MTFF}_e = (6/0.03) - (6/0.05) + (2/0.07) - (3/0.04) + (3/0.06) - (1/0.08) = 71$$

10.1.2 Full-On Redundancy: Multiple-Survivor
Subsystem/System; Perfect Sensing and Switching

This type of redundancy is the more general case of the one described above at the subsystem level. The subsystem may be considered, as before, as composed of n fully energized parallel components but in this case requires not one but a minimum of d $(d \leqslant n)$ operating units (nonfailed units) in order to perform its function. It is a formulation appropriate for evaluating the performance of a group of signal processors, or a phased array radar.

Bear in mind that the distinction between what is construed to be a subsystem and a system can sometimes be very fine. As a consequence, what is considered a subsystem to some is a system to others.

This type of subsystem can be described by the binomial distribution discussed previously (see Chapter 2). If $R(t) =$ the probability of a component performing satisfactorily over a time period, t, where

$n =$ the number of components in a system or subsystem
$d =$ the minimum number of components which must be operable in order
    for a function to be performed satisfactorily

then based on the binomial distribution, we can describe the probability of at least d out of n components performing satisfactorily over the time period t when all components have the same failure rate λ and are operable at the start of the time interval.

$$R_i(t) = \sum_{k=d}^{n} \frac{n!}{(n-k)!\,k!} R(t)^k [1 - R(t)]^{n-k} \qquad (10\text{-}13)$$

or alternately,

$$R_i(t) = \sum_{k=0}^{n-d} \frac{n!}{(n-k)!\,k!} [1 - R(t)]^k R(t)^{n-k} \qquad (10\text{-}14)$$

where $R_i(t) =$ probability the subsystem will incur $(n - d)$ or fewer failures over a period of time t given that all components were operable at the start of t.

Two expressions for $R_i(t)$ are provided because, under some circumstances, computation is significantly reduced if one is used rather than the other. For either expression, $R(t)$ is defined as

$$R(t) = \exp(-\lambda t)$$

where $\lambda =$ the common failure rate of a component.

Note that when $d = 1$, it can be shown that

$$\sum_{k=1}^{n} \frac{n!}{(n-k)!\,k!} R(t)^k [1 - R(t)]^{n-k} = 1 - (1 - e^{-\lambda t})^n$$

Hence, (10-14) and (10-4) are equivalent for $d = 1$.

The reliability of a system, $R_e(t)$, made up of L subsystems of this type will take a form similar to that described in the last section. When all components in the system have the same failure rate λ and require the same number of operating components d,

$$R_e(t) = \left\{ \sum_{k=d}^{n} \frac{n!}{(n-k)!\,k!} R(t)^k [1 - R(t)]^{n-k} \right\}^L \qquad (10\text{-}15)$$

where

$n = $ number of components in every subsystem

$d = $ minimum number of components necessary for every subsystem to function

When all components within a subsystem have a common failure rate λ_i, but (a) components associated with different subsystems have different failure rates, and (b) different subsystems require different minimum numbers of operating components d_i in order to perform satisfactorily,

$$R_e(t) = \prod_{i=1}^{L} \sum_{k=d_i}^{n_i} \frac{n_i!}{(n_i - k)!\,k!} R_{ci}(t)^k (1 - R_{ci}(t))^{n_i - k} \qquad (10\text{-}16)$$

where

$R_e(t) = $ reliability of a system

$R_{ci}(t) = \exp(-\lambda_i t)$,

$\quad = $ reliability of a component in subsystem i

$\lambda_i = $ common failure rate of each component in subsystem i

$d_i = $ number of components necessary for subsystem i to perform its function

$n_i = $ total number of components in subsystem i

$L = $ number of series-connected subsystems making up the system

Assuming that all components have exponential distribution of times to failure and all components are operable at mission start, the MTTF and MTFF for subsystem and system, respectively, can be defined as

$$\text{MTTF}_i = \text{MTFF}_i = \int_0^{\infty} R_i(t)\,dt = \frac{1}{\lambda} \sum_{k=d}^{n} [1/k] \qquad (10\text{-}17)$$

and

$$\text{MTTF}_e = \text{MTFF}_e = \int_0^\infty \left[\prod_{i=1}^L R_i(t) \right] dt \qquad (10\text{-}18)$$

As before, closed-form solution for MTTF_e or MTTF_e is difficult except for special cases, but, as before, its solution is more involved and tedious than it is difficult.

For the special case represented by (10-15),

$$\text{MTTF}_e = \text{MTFF}_e = \int_0^\infty R_e(t)\, dt = \frac{1}{d\lambda} \sum_{k=1}^L \frac{L!}{(L-k)!\,k!}(-1)^{k+1} \sum_{s=1}^{k(n+d+1)} \frac{1}{s}$$

$$(10\text{-}19)$$

where n, d, λ are defined as in (10-15).

For the more general case represented by (10-16),

$$\text{MTTF}_e = \text{MTFF}_e = \int_0^\infty R_e(t)\, dt$$

$$= \int_0^\infty \prod_{i=1}^L \sum_{k=d_i}^{n_i} \frac{n_i!}{(n_i-k)!\,k!} R_i(t)^k (1 - R_i(t))^{n_i-k}\, dt \qquad (10\text{-}20)$$

Each term must be determined and integrated.

Example   An array of signal processors performs a specific function. A total of 100 processors make up the array. The failure rate of each processor is $\lambda = 0.0005$. The system (subsystem) can perform satisfactorily as long as at least 90 processors are operating.

- What is the probability that the subsystem will operate satisfactorily over a 100-hour period, given that all components were operable at the start?
- What is the mean time to failure for the subsystem?
- What would the above reliability characteristics be if a minimum of 95 processors must operate?

$$R_i(t) = \sum_{k=0}^{n-d} \frac{n!}{(n-k)!\,k!}[1 - R(t)]^k R(t)^{n-k}$$

$$R_i(t) = e^{-\lambda t} = e^{-0.0005(100)} = e^{-0.05} = 0.95$$

Even though it is a better choice than (10-13), solution of (10-14) with the use of the average pocket calculater is tedious and long. Recall from

Chapter 2 that we can approximate the binomial distribution by the Poisson distribution when n is moderately sized and p, i.e., $(1 - R(t))$, is small:

$$R_i(t) = \sum_{k=0}^{n-d} \frac{a^k e^{-a}}{k!}$$

where $a = p \cdot n = 0.05(100) = 5$. Hence, the probability that the subsystem will operate satisfactorily over a 100-hour period i.e., will incurr 10 or fewer failures, is

$$R_i(t) = \sum_{k=0}^{10} \frac{(5)^k e^{-5}}{k!} = 0.0067 + 0.0337 + 0.0842 + 0.1404 + 0.1755$$

$$+ 0.1755 + 0.1044 + 0.0653 + 0.0363 + 0.0181 = 0.98$$

The MTFF of the subsystem (or MTTF) is

$$\text{MTFF}_i = \frac{1}{\lambda} \sum_{k=d}^{N} \frac{1}{k}$$

$$\text{MTFF}_i = \frac{1}{\lambda} \sum_{90}^{100} \frac{1}{k} = 2000(0.0111 + 0.0111 + 0.0109 + 0.0108$$

$$+ 0.0106 + 0.0105 + 0.0104 + 0.0103 + 0.0102 + 0.0101 + 0.01$$

$$= (2000)(0.1160) = 232$$

10.1.3 Full-On Redundancy: An Alternate Way of Understanding and Evaluating Mean Time to First Failure

An alternate means of understanding and of quickly calculating mean time to first failure, one which is more logical than it is mathematical, is to consider the concept of system states and the concept of transition from one state to another.

The subsystem starts out initially with n components operating. Since under this concept *no repair is made until a subsystem failure occurs*, the subsystem will experience a component failure and be reduced to $(n - 1)$ operating components; it will eventually experience a second unit failure and be reduced to $(n - 2)$ operating components, and so on until only d components are operating. The next failure which occurs results in only $(d - 1)$ units operating and will cause the subsystem to fail.

Let $E_k [(k = 0, 1, \ldots, (n - d + 1)]$ represent the state where the subsystem has k units failed. For example, if $n = 4$ and $d = 2$, the subsystem would start out in state E_0, signifying that no components were failed and four were operating. Given a failure of a component, it would automatically enter E_1, a state where one component was in a failed state and three were operating. Given the next failure, it would enter E_2, the minimum operating state. Upon the next failure, it would enter E_3, a failed state. Since no component repairs will be made until subsystem failure, the subsystem will always go from state E_k to state $E_{[k+1]}$ and so on in order, until fewer than d components are operable. At that time a subsystem failure will occur.

Each state has associated with it an average time, $\theta_{k/[k+1]}$, to transition into the next possible state (E_k to $E_{[k+1]}$) given by

$$\theta_{k/[k+1]} = \frac{1}{\lambda(n - k)} \tag{10-21}$$

which is the *mean time to next component failure* for the operating components associated with a given state. Hence, average time to the next transition is equivalent to the mean time to next failure for the group of operating components, where λ = failure rate of each component.

Since the transition must occur in sequence, the mean time for transition from state E_0 to a subsystem failed state, E_{n-d+1}, is simply the average time in state E_0 + average time in state $E_1 + \cdots +$ average time in state E_{n-d}, or

$$\text{MTFF}_i = \sum_{k=0}^{n-d} \theta_{k/[k+1]} = \sum_{k=0}^{n-d} \frac{1}{\lambda(n - k)}$$

$$= \frac{1}{\lambda} \sum_{k=d}^{n} \frac{1}{k} \tag{10-22}$$

which corresponds to equation (10-19).

Example   A subsystem is initially made up of five components. A minimum of three components must operate in order to perform the functions required of the subsystem. All components are operating all the time and each has a failure rate $\lambda = 0.01$. What is the mean time to failure for the subsystem?

The subsystem starts operation in a state with five components operating (i.e., $k = 0$). The total failure rate for the five components = 5λ, where λ = the failure rate of a single component,

$$5\lambda = \sum_{i=1}^{5} \lambda = 0.05$$

and the average time to a component failure (mean time to transition to the next operating state) is

$$\theta_{k/[k+1]} = \theta_{0/1} = 1/5\lambda = 20 \text{ hours}$$

That is to say, on the average a subsystem made up of these five components will operate for 20 hours before it fails, and when it fails it will then be in state 1 ($k = 1$) with only four components operating.

Total failure rate during that state is

$$4\lambda = \sum_{i=1}^{4} \lambda = 0.04$$

and the average time to the next component failure (mean time to next transition) is

$$\theta_{1/2} = 1/4\lambda = 25 \text{ hours}$$

That is to say, on the average a subsystem made up of these four components will operate for 25 hours before it transitions to state 2. When that occurs, it will then be in a state with only three components operating, state 3 ($k = 3$). The next transition will be to a subsystem failure state, since fewer than the minimum number of three components will be operable.

The average time the subsystem will spend in state 3 before transition to the subsystem failure state, state 4 ($k = 4$), will be

$$\theta_{3/4} = 1/3\lambda = 33.33 \text{ hours}$$

If on the average it takes 20 hours to go from the first operating state to the second; 25 hours to go from the second operating state to the third; 33.33 hours to go from the third operating state to a failed state; then the average time to go from the first operating state to a failed state is

$$20 + 25 + 33.33 = 78.33 \text{ hours}$$

10.1.4 Full-On Redundancy: Taking Into Account Imperfect Sensing and Switching

The previous sections provided an upper-bound relationship for reliability potential due to the application of full-on redundancy by assuming perfect detection of failures; perfect isolation or switch-off of the output of the faulty component; and perfect rebalance of the system (if necessary) to compensate for loss of a system load.

If not detected and isolated (switched off), the output from a malfunctioning component can cause disruption of system function. In addition, for a great many redundant systems, automatic or manual adjustments are required when a previously operating component is "removed" from the system.

If it is assumed that

1. When those detection, switching and balance functions are not performed properly, the subsystem and hence the system will experience failure;
2. Due to reliability as well as hardware and/or software design characteristics associated with the detection/switch/balance mechanisms, there exists a probability, $(1 - P)$, that upon a subsystem component failure such mechanisms will not function as required;

then the reliability $R_i(t)$ of a subsystem which requires d out of n components, can be expressed as

$$R_i(t) = \sum_{k=0}^{n-d} \frac{n!}{(n-k)!\,n!} R_j(t)^{n-k}(1 - R_j(t))^k P^k \qquad (10\text{-}23)$$

where

d, n, are defined as previously
$R_j(t)$ = the reliability of component j in subsystem i
P = the probability of a successful detection, switch and balance
        = a combined value of fraction of failures detectable and probability that the switching and balancing circuitry will work given that detection has been accomplished
P^k = the probability that k component failures have been met by k successful detections/switches/balances

$$\frac{n!}{(n-k)!\,k!}[R_j(t)]^{n-k}[1 - R_j(t)]^k$$

        = the probability that the subsystem will experience k failures over a time interval t

Hence (10-23) defines the compound probability that

1. $k \leqslant (n - d)$, that is to say, the number of units that fail over t must be less than or equal to $(n - d)$. More than $(n - d)$ failures would reduce the number of operable components to less than d, the minimum number of operable components necessary to maintain the subsystem function. For example, if $n = 5$ and $d = 3$, the maximum number of failures tolerable over t would equal 2.

2. k successful detections/switches/balances must have occurred, one for each failure.

If at any time a detection/switch/balance fails to properly occur, the subsystem fails.

The value associated with P provides practical boundaries for reliability improvement due to application of redundancy and sheds light on some aspects of redundancy not usually considered. In order to examine this more closely, a reasonable state-of-the-art value of $P = 0.95$ will be assumed, and some comparisons provided.

Figure 10-2 shows plots in dashed lines of $R_i(t)$ when $d = 1$, $n = 2, 3, 4$, and $P = 1$; that is to say, for perfect detection, switching, balance, etc. for a full-on redundant subsystem. These represent plots of (10-14) or (10-4) which, as indicated previously, assume perfect failure detection, switching, etc. In solid lines on the same figure we keep $d = 1$, $n = 2, 3, 4$ for a full redundant subsystem as before, but a value of $P = 0.95$ is assigned and (10-23) is applied. As can be seen, even for a value as large as $P = 0.95$ a significant difference results between the reliability gained with perfect failure detection and switching and that gained when failure detection/switching is imperfect. Also

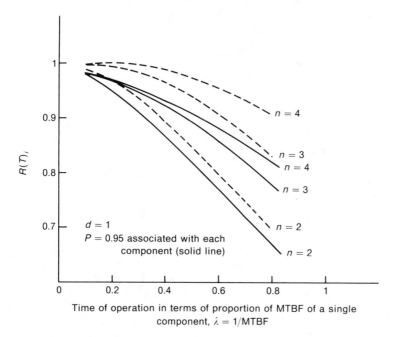

FIGURE 10-2.   Comparison of reliability improvement considering perfect and imperfect failure detection and switching.

note that as the number of redundant components increases, the difference is more pronounced. As a matter of fact, while it is hard to detect on the figure, for some times of operation, given imperfect detection and switching, $R_i(t)$ is actually *less* for subsystems which have a *greater* degree of redundancy. This is shown in Fig. 10-3, where $R_i(t)$ is plotted against time of operation for subsystems comprised of 2, 3, 4, and 5 components where $P = 0.95$ is associated with each component. Note, for example, that for a mission/use time between approximately 0.2 and 0.4 multiples of the MTBF of a single component, the value of $R_i(t)$ is greater for a subsystem containing three redundant components than for one containing five components.

The basic reason this occurs is that, for a full-on redundant subsystem, the more redundant components operating at a given time, the greater the *total component failure rate* and hence the greater rate of occurrence of *component* failures. While such failures themselves will not cause the system to fail (they are redundant, after all), each time a component fails it gives

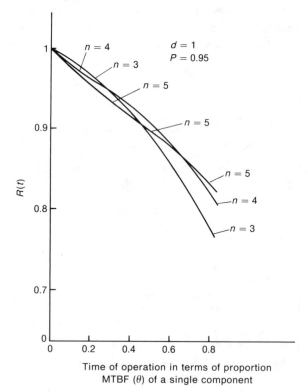

FIGURE 10-3.   Reliabilities of redundant systems with less than perfect failure detection and switching.

the imperfect detection/switching apparatus an opportunity to cause a system failure. Depending on the value of P and the mission time, there are instances where a lesser degree of redundancy will provide a greater value of reliability than a greater degree of redundancy. Taking an extreme example as an illustration, a comparison will be made between the reliability of two subsystems: A, comprised of a single component with a failure rate $\lambda_a = 0.001$; B, made up of two redundant components each with a failure rate $\lambda_b = 0.001$ (only one need work to perform the subsystem function). If detection, etc., were perfect, then obviously system B would have a greater reliability than system A. Assume, however, that the second subsystem, B, has an imperfect failure detection/switching apparatus associated with it such that the value of $P = 0.2$ (we said this was an extreme example). The mission time t for both subsystems is $t = 2$ hours.

The reliability, $R_i(t)$, for subsystem A is

$$R_a(t) = \exp(-\lambda_a t) = \exp[-(0.001)2] = 0.9980$$

The reliability, $R_i(t)$, for subsystem B is

$$R_b(t) = \sum_{k=0}^{1} \frac{2!}{(2-k)!\,k!}(e^{-\lambda t})^{n-k}(1 - e^{-\lambda t})^k P^k$$

$$= e^{-2(0.001)(2)} + 2e^{-0.001(2)}(1 - e^{-0.001(2)})(0.2) = 0.9968$$

Hence, the single system is more reliable than the redundant system for a 2-hour mission. While not apparent in this extreme example, but apparent in Fig. 10-3, reliability $R_i(t)$ is also clearly a function of mission time. For example, in Fig. 10-3, for a mission use interval $t = 0.5$ of the MTBF of a component, a redundant subsystem containing five components is less reliable than a redundant subsystem containing four components. But if $t = 0.8$ of the MTBF of a component, a subsystem containing five components is more reliable than one containing four.

In summary, in the practical world, more redundancy is not always synonymous with greater reliability. The value of redundancy as a multiplier of reliability is a function of failure detection/switching capability and mission duration. For more information on the subject see Ref. 1.

Mean time to first failure for a subsystem of this type is evaluated in the usual way:

$$\text{MTFF}_i = \int_0^\infty R_i(t)\,dt$$

$$\text{MTFF}_i = \frac{1}{\lambda}\sum_{k=d}^{n} \frac{P^{(n-k)}}{k} = \frac{1}{\lambda}\sum_{k=0}^{n-d} \frac{P^k}{(n-k)} \tag{10-24}$$

where
    d = the minimum number of components necessary to maintain satisfactory subsystem performance
    λ = component failure rate
    n = number of components making up the subsystem

10.1.5   Standby Redundancy: The Single-Survivor Subsystem; Perfect Sensing and Switching

Another common type of redundancy is one which possesses a similar configuration to the system previously described but employs switched-in redundancy as illustrated Fig. 10-4.

In this instance only one component in each subsystem is activated at a time. Upon failure of that component the next component of the subsystem will automatically be switched into operation. Until such a switch occurs, the standby components are not energized and hence a failure rate of $\lambda = 0$ is assumed for each standby unit. When the operating component fails and no operable components are available to be switched in, the subsystem has failed. In order to hypothesize an upper bound for the reliability of such a system, the detection/switching apparatus is considered to be perfect and its failure rate is assumed to be 0.

An exponential distribution of failure times for each component is assumed. As a result: (1) The failure rate of a component after t hours of operation is

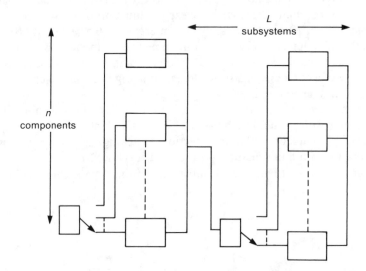

FIGURE 10-4.   Switched-in (standby) redundancy example.

the same as the failure rate of the component when it is brand new. (2) The failure rate of a repaired component is the same as the failure rate of the component when it is brand new. As a consequence, operating a component until it fails and replacing it immediately with an identical new component is *equivalent* to operating a component until it fails and instantly resurrecting (repairing) it. In this case the number of resurrections possible is equal to $(n - 1)$ when n = the number of components in the subsystem. Analogously to the cat that has nine lives and determining the probability that fewer than nine will be expended by a given time, we determine the probability that a given subsystem made up of one operating component will suffer *less than n failures* over a period of time t. That probability is *the reliability*, $R_i(t)$, of the subsystem.

In order to express $R_i(t)$, recall the discussion of the Poisson distribution in Chapter 2. The probability of x events, $P(x)$, was defined as

$$P(x) = \frac{a^x e^{-a}}{x!}$$

(10-25)

where
x = number of events which occur
a = expected number of events

and the probability that x is less than a number n, $P(x < n)$, can be expressed as

$$P(x < n) = \sum_{x=0}^{n-1} \frac{a^x e^{-a}}{x!}$$

(10-26)

Since $a = \lambda t$ = the expected number of failures in a time period t, and if $(n - 1)$ failures or fewer occur, the subsystem, like the cat who has experienced fewer than nine lives, will still perform its function. Subsystem reliability, $R_i(t)$, can then be expressed as

$$R_i(t) = \sum_{x=0}^{n-1} \frac{(\lambda t)^x e^{-\lambda t}}{x!}$$

(10-27)

where
λ = failure rate of each component in the subsystem
t = time interval of operation for the subsystem
n = number of redundant components making up each subsystem

If the entire system were comprised of a group of L subsystems connected in series, all with this form of redundancy, and if each subsystem contained the same number of components n, where each component had the same failure rate λ, the following special case expression for $R_e(t)$ would result:

$$R_e(t) = \left(e^{-\lambda t} \sum_{x=0}^{n-1} \frac{(\lambda t)^x}{x!} \right)^L \tag{10-28}$$

A more general form for $R_e(t)$ results when each subsystem contains different numbers of components n_i; when each component within the same subsystem has the same failure rate λ_i,

$$R_e(t) = \prod_{i=1}^{L} \sum_{x=0}^{n_i-1} \frac{(\lambda_i t)^x \exp(-\lambda_i t)}{x!} \tag{10-29}$$

where
   λ_i = failure rate of each component in subsystem i
   n_i = number of redundant components in subsystem i
   L = number of subsystems
   t = time interval of operation for the system

For subsystems made up of components having exponential distributions of time to failure, which are discarded upon failure (nonrepairable) or repairable,

$$\text{MTFF}_i = \text{MTTF}_i = \int_0^{\infty} R_i(t)\, dt$$

(Note that for subsystems made up of components having any distribution of time to failure, mean time to first failure is defined in the same way, as $\text{MTFF}_i = \int_0^{\infty} R_i(t)\, dt$.)

Therefore, subsystem mean time to failure (or mean time to first failure), when all components have the same failure rate λ, takes the form

$$\text{MTTF}_i = \text{MTFF}_i = \int_0^{\infty} \sum_{x=0}^{n-1} \frac{(\lambda t)^x e^{-\lambda t}}{x!}\, dt \tag{10-30}$$

Since from almost any book of integrals it can be found that

$$\int_0^{\infty} \frac{e^{-\lambda t}(\lambda t)^x}{x!}\, dt = \frac{1}{\lambda} \tag{10-31}$$

(10-30) reduces to

$$\mathrm{MTTF}_i = \mathrm{MTFF}_i = \sum_{x=1}^{n} \frac{1}{\lambda} = \frac{n}{\lambda} \qquad (10\text{-}32)$$

where $\lambda_i = \lambda$, the failure rate of the ith component.

For the special case where all components throughout the system have the same failure rate λ and where each subsystem contains the same number of components n, an expression for the system-level measures of MTTF_e or MTFF_e can be generated:

$$\mathrm{MTTF}_e = \mathrm{MTFF}_e = \int_0^\infty \left(e^{-\lambda t} \sum_{x=0}^{n-1} \frac{(\lambda t)^x}{x!} \right)^L dt = \int_0^\infty e^{-L\lambda t} \left(\sum_{x=0}^{n-1} \frac{(\lambda t)^x}{x!} \right)^L dt \qquad (10\text{-}33)$$

Expressing

$$\left(\sum_{x=0}^{n-1} \frac{(\lambda t)^x}{x!} \right)^L = \left[1 + \left(\sum_{x=1}^{n-1} \frac{(\lambda t)^x}{x!} \right) \right]^L \qquad (10\text{-}34)$$

where $n > 1$ and recalling the discussion of the binomial expansion in Appendix A15, the same general procedure used, transforms (10-34) to

$$\left(1 + \sum_{x=1}^{n-1} \frac{\lambda t}{x!} \right)^L = \sum_{k=0}^{L} \left(\frac{L!}{(L-k)!\,k!} \right) \left(\sum_{x=1}^{n-1} \frac{\lambda t}{x!} \right)^k \qquad (10\text{-}35)$$

(10-33) can be expressed as

$$\mathrm{MTTF}_e = \mathrm{MTFF}_e = \int_0^\infty e^{-L\lambda t} \sum_{k=0}^{L} \left[\left(\frac{L!}{(L-k)!\,k!} \right) \left(\sum_{x=1}^{n-1} \frac{\lambda t}{x!} \right)^k \right] dt \qquad (10\text{-}36)$$

which can be evaluated easily through the use of elementary but tedious computations.

For example, if $n = 2$, after expanding and generalizing (10-36) reduces to

$$\mathrm{MTTF}_e = \mathrm{MTFF}_e = \sum_{k=0}^{L} \frac{L!}{[\lambda L^{k+1}(L-k)!]} \qquad (10\text{-}37)$$

For the more general case where it is assumed that each subsystem is made up of components with different failure rates λ_i and different numbers of components, n_i:

$$\text{MTTF}_e = \text{MTFF}_e = \int_0^\infty \left\{ \prod_i^L \sum_{x=0}^{n_i-1} \frac{(\lambda_i t)^x \exp(-\lambda_i t)}{x!} \right\} dt \qquad (10\text{-}38)$$

where
   L = number of subsystems
   λ_i = failure rate of a component belonging to subsystem i
   n_i = number of components in subsystem i

Equation (10-37) can be evaluated through use of a computer and can even be evaluated by tedious (but nevertheless elementary) calculation.

For example, a subsystem is made up of two subsystems, A and B, connected in series. Both subsystem functions are required to perform the system function. Both A and B are designed to employ single survivor standby redundancy. A is made up of two redundant components, one operating and one in standby. Only one operates to perform the required subsystem function. The failure rate associated with each component of A is $\lambda_a = 0.01$ when the component operates.

B is made up of three redundant components, one operating and two in standby. Only one operates to perform the required subsystem function. The failure rate associated with each component of B is $\lambda_b = 0.02$ when the component operates.

- What is the reliability of each subsystem over a 20-hour mission?
- What is the reliability of the system over a 20-hour mission?
- What is the MTFF of each subsystem?
- What is the MTFF of the system?

$$R_i(t) = \sum_{x=0}^{n-1} \frac{(\lambda t)^x e^{-\lambda t}}{x!}$$

$$R_a(t) = \sum_{x=0}^{1} [(0.01)(20)]^x \frac{e^{-0.01(20)}}{x!} = (0.819 + 0.164) = 0.983$$

$$R_b(t) = \sum_{x=0}^{2} [(0.02)(20)]^x \frac{e^{-0.02(20)}}{x!} = (0.670 + 0.268 + 0.054) = 0.992$$

$$R_e(t) = \prod_i^L \sum_{x=0}^{n-1} \frac{(\lambda_i t)^x \exp(-\lambda_i t)}{x!}$$

$$R_e(t) = R_a(t)R_b(t) = (0.983)(0.992) = 0.975$$

$$\text{MTFF}_i = \sum_{x=1}^{n} \frac{1}{\lambda} = \frac{n}{\lambda}$$

$$\text{MTFF}_a = 2/0.01 = 200 \text{ hours}$$

$$\text{MTFF}_b = 3/0.02 = 150 \text{ hours}$$

$$\text{MTFF}_e = \int_0^\infty \prod_i^L \sum_{x=0}^{n_i-1} \frac{(\lambda_i t)^x \exp(-\lambda_i t)}{x!} dt$$

$$\text{MTFF}_e = \int (e^{-\lambda_a t} + \lambda_a t e^{-\lambda_a t}) \left(e^{-\lambda_b t} + \lambda_b t e^{-\lambda_b t} + \frac{(\lambda_b t)^2}{2} e^{-\lambda_b t} \right) dt$$

$$= \int \left[e^{-(\lambda_a + \lambda_b)t} + \lambda_b t e^{-(\lambda_a + \lambda_b)t} + \frac{(\lambda_b t)^2}{2} e^{-(\lambda_a + \lambda_b)t} + \lambda_a t e^{-(\lambda_a + \lambda_b)t} \right. $$

$$\left. + \lambda_a \lambda_b t^2 e^{-(\lambda_a + \lambda_b)t} + \frac{\lambda_a \lambda_b^2 t^3}{2} e^{-(\lambda_a + \lambda_b)t} \right] dt$$

Since in general $\int_0^\infty (e^{-\lambda t}(\lambda t)^x/x!) \, dt = 1/\lambda$, forcing each term above into that form and integrating each term one at a time, we have

$$\text{MTFF}_e = \frac{1}{\lambda_a + \lambda_b} + \frac{\lambda_b}{(\lambda_a + \lambda_b)^2} + \frac{\lambda_b^2}{(\lambda_a + \lambda_b)^3} + \frac{\lambda_a}{(\lambda_a + \lambda_b)^2} + \frac{2\lambda_a \lambda_b}{(\lambda_a + \lambda_b)^3}$$

$$+ \frac{3\lambda_a \lambda_b^2}{(\lambda_a + \lambda_b)^4}$$

$$= (33.333 + 22.222 + 14.815 + 11.111 + 14.815 + 14.815) = 111.111$$

10.1.6 Standby Redundancy: The Multiple-Survivor Subsystem; Perfect Sensing and Switching

A variation of the single-survivor standby redundancy configuration is the multiple-survivor subsystem. In this case the subsystem contains n identical components. A set of d of the n components are required to operate in order to maintain the function of the subsystem. The remaining $(n - d)$ components are in a standby mode of operation with a failure rate of $\lambda = 0$ while in that standby mode. When one of the d operating components fails, the failure is detected and the failed component is replaced by one of the standby units. The subsystem continues to perform its function until the store of standby units is exhausted and fewer than d nonfailed operating components remain. At that time, it fails.

This redundancy configuration represents a more general form of the standby redundancy single-survivor configuration discussed in the last section. This follows because, if we define $d = 1$, the multiple-survivor case will reduce to the single-survivor case.

Operating d components, all with exponential distributions of time to failure, until one fails (it is assumed that no more than one failure in the same subsystem can occur at the same instant) and replacing it immediately with a new component, is equivalent to operating *a unit made up of d components* until it fails and immediately resurrecting it. The reliability of such a *unit* is dependent on the number of resurrections possible. In this case the number of resurrections possible is dependent upon the number of standby units available as replacements. As a result, the reliability of the subsystem over an interval of time t is equal to the probability that it fails fewer than $(n - d + 1)$ times over that interval.

Since d components are operating all the time while the subsystem performs its function, and since each component has an operating failure rate λ, the failure rate λ_i of the operating subsystem made up of d components is

$$\lambda_i = d\lambda$$

and the expected number of failures a, over an interval time t, is

$$a = \lambda_i t = d\lambda t$$

The Poisson distribution can then be applied to evaluate the reliability $R_i(t)$ of the subsystem. Reliability can be expressed as the probability that the subsystem will satisfactorily perform its function over a period of time t. In this case the probability that the subsystem will satisfactorily perform its function over an interval of time t equals the probability that some number, x, less than or equal to $(n - d)$ component failures will occur over $tP[x \leqslant (n - d)]$. $P[x \leqslant (n - d)]$ can be expressed as before in terms of the Poisson distribution.

Assuming all components per subsystem have the same operating failure rate λ,

$$R_i(t) = P[x \leqslant (n - d)] = \sum_{x=0}^{n-d} \frac{(d\lambda t)^x e^{-d\lambda t}}{x!} \qquad (10\text{-}39)$$

where
λ = failure rate of operating component
d = number of components necessary to maintain subsystem function
t = interval of operating mission/use time

Note that when $d = 1$, (10-39) is equivalent to (10-27). Hence the multiple-survivor model can be thought of as a more general case of the single-survivor model.

System reliability, $R_e(t)$, in the special case where the system is made up of L subsystems of the same type connected in series, all with the same values of n and d and all with components with the same λ, can be expressed as

$$R_e(t) = \left(\sum_{x=0}^{n-d} \frac{(d\lambda t)^x e^{-d\lambda t}}{x!} \right)^L \tag{10-40}$$

For the more general case where the values of d and λ vary with subsystem,

$$R_e(t) = \prod_{i=1}^{L} \sum_{x=0}^{n_i-d_i} \frac{(d_i \lambda_i t)^x}{x!} \exp(-d_i \lambda_i t) \tag{10-41}$$

where

d_i = the minimum number of units which must operate for subsystem i to perform its function

n_i = the total number of components making up subsystem i

λ_i = the failure rate of each component in subsystem i

L = the number of subsystems connected in series

For subsystems made up of components having exponential distributions of time to failure which are discarded upon failure (nonrepairable) or repairable, where it is assumed all components are operable when use begins,

$$\text{MTTF}_i = \text{MTFF}_i = \int_0^\infty R_i(t)\, dt$$

$$\text{MTTF}_i = \text{MTFF}_i = \int_0^\infty \left[\sum_0^{n-d} (d\lambda t)^x e^{-d\lambda t}/x! \right] dt \tag{10-42}$$

After integrations similar to the operations performed on (10-30), (10-42) reduces to

$$\text{MTTF}_i = \text{MTFF}_i = \sum_0^{n-d} 1/d\lambda = \sum_1^{n-d+1} 1/d\lambda = (n - d + 1)/d\lambda \tag{10-43}$$

For systems made up of L subsystems, each of which exhibit this form of redundancy, under the same assumptions as in the past,

$$\text{MTTF}_e = \text{MTFF}_e = \int_0^\infty R_e(t)\, dt$$

When all subsystems are made up of components with the same failure rates, and values of d, and n,

$$\text{MTTF}_e = \text{MTFF}_e = \int_0^\infty \left[\prod_1^L \sum_{x=0}^{n-d} \frac{(d\lambda t)^x e^{-d\lambda t}}{x!} \right] dt \qquad (10\text{-}44)$$

which can be evaluated using the same procedures as for (10-33).

When each subsystem is made up of different numbers of components n_i, have different component failure rates λ_i, and require different values of d_i,

$$\text{MTTF}_e = \text{MTFF}_e = \int_0^\infty \left[\prod_{i=1}^L \sum_{x=0}^{n_i-d_i} \frac{(d_i\lambda_i t)^x}{x!} \exp(-d_i\lambda_i t) \right] dt \qquad (10\text{-}45)$$

where L, n_i, d_i, λ_i are defined as for (10-41).

Both (10-44) and (10-45) require either tedious, relatively elementary term-by-term expansion and integration or the use of a computer.

Examples   Two subsystems A and B make up a system. Subsystem A is a three-component standby redundant configuration. Two components are required to operate in order to perform the subsystem function. When in operation each component has a failure rate of $\lambda_a = 0.01$. The remaining component is in inactive standby with a failure rate of zero when in that state. Upon failure of any operating component, the standby component will be switched in to take the place of an operating unit. When a failure occurs and there is no standby unit available, the subsystem is in a failed state. Subsystem B is a standby redundant configuration made up of four components. Three of these are required to operate in order to perform the subsystem function. When in operation each component has a failure rate of $\lambda_b = 0.02$. The remaining component is in an inactive standby state with a failure rate of zero. The operation of the subsystem and failure definition parallels that of subsystem A.

- What is the reliability of each subsystem over a period of operating time of $t = 20$ hours?
- What is the reliability of the system over that period?
- What is the MTFF of each subsystem? Of the system?

$$R_i(t) = \sum_{x=0}^{n-d} \frac{(d\lambda t)^x e^{-d\lambda t}}{x!}$$

$$R_a(t) = \sum_{x=0}^1 [2(0.01)20]^x \frac{e^{-[2(0.01)20]}}{x!} = 0.6703 + 0.2681 + 0.9384$$

$$R_b(t) = \sum_{x=0}^{1} [3(0.02)20]^x \frac{e^{-[3(0.02)20]}}{x!} = 0.3012 + 0.3614 = 0.6626$$

$$R_e(t) = \prod^{L} R_i(t) = (0.9384)(0.6626) = 0.6218$$

$$\text{MTFF}_i = \sum_{1}^{n-d+1} \frac{1}{d\lambda} = \frac{n-d+1}{d\lambda}$$

$$\text{MTFF}_a = \sum_{0}^{1} 1/[2(0.01)] = 2/[2(0.01)] = 100 \text{ hours}$$

$$\text{MTFF}_b = \sum_{0}^{1} 1/[3(0.02)] = 2/0.06 = 33.33$$

$$\text{MTFF}_e = \int_{0}^{\infty} R_e(t)\, dt = \int_{0}^{\infty} \left[\prod_{i=1}^{L} \sum_{x=0}^{n_i-d_i} \frac{(d_i\lambda_i t)^x \exp(-d_i\lambda_i t)}{x!} \right] dt$$

$$\text{MTFF}_e = \int_{0}^{\infty} [e^{-(d_a\lambda_a + d_b\lambda_b)t} + d_b\lambda_b t e^{-(d_a\lambda_a + d_b\lambda_b)t} + d_a\lambda_a t e^{-(d_a\lambda_a + d_b\lambda_b)t}$$
$$+ d_a d_b \lambda_a \lambda_b t^2 e^{-(d_a\lambda_a + d_b\lambda_b)t}]\, dt$$

Each term above can be found in tables of definite integrals in almost any book of tables. The evaluation of the integrals is left as an exercise.

10.1.7 Standby Redundancy: An Alternate Way of Understanding and Evaluating Mean Time to Failure

An alternate means of developing, understanding, and quickly calculating mean time to first failure for standby redundant subsystems is to consider the concept of system states and the concept of average time required to transition from one state to another. The same concepts as were used in Section 10.1.3 dealing with full-on redundancy configurations will be applied here.

The subsystem starts out initially with d components operating. Immediately upon failure of any working component, the failed component will be switched out and a standby unit switched in, maintaining the operation of d components. This will continue until a component failure occurs and no standby unit is available. In that case the subsystem would be in a failed state.

Let $E_{d/k}$ represent the state that the subsystem has d components in operation and k available standby components [$k = 0, 1, \ldots, (n-d)$]. The subsystem starts off in state $E_{d/(n-d)}$, that is to say, d components operating and $(n-d)$ components in standby. Upon the first failure, assuming perfect detecting/switching, the subsystem enters state $E_{d/(n-d-1)}$, where d components are operating and $(n - d - 1)$ standby units are available. It continues to operate through states $E_{d/(n-d-2)}$, $E_{d/(n-d-3)}$, and so on, until it reaches state $E_{d/0}$ where no standby components remain. The next component failure from state $E_{d/0}$ will cause the subsystem to enter a failed state. The state progression follows a logical process from one state to the next to subsystem failure.

Since each operable state of the subsystem contains d operating components, the total failure rate associated with each state is $d\lambda$. As a result, provided $k \geqslant 0$, the average time, $\theta_{k/(k-1)}$, to transition from any state $E_{d/k}$ to the next possible state $E_{d/(k-1)}$ is always

$$\theta_{k/(k-1)} = \frac{1}{d\lambda} \tag{10-46}$$

The MTFF or MTTF for the subsystem, assuming subsystem start when all components are operable, will equal the sum of the average times required to transition from state $E_{d/k}$ to state $E_{d/(k-1)}$ to state $E_{d/(k-2)}$, etc., until a failed state is reached. Since the average time to transition from one state to the next is the same, and $(n - d + 1)$ transitions are required to reach a failed state,

$$\text{MTFF}_i = \frac{n - d + 1}{d\lambda} \tag{10-47}$$

which is the same as (10-43) derived through a different approach.

For example, consider a standby redundant subsystem with $n = 4$ and $d = 2$, where each operating component exhibits a failure rate of $\lambda = 0.01$ and where each standby component has a failure rate of $\lambda = 0$. The subsystem would start out in state $E_{2/2}$, indicating that two components were operating as required and two spares were available. Given a failure of a component and assuming that detection and switching is automatic and perfect; the system would next enter state $E_{2/1}$; after the next failure it would enter $E_{2/0}$, and the next failure would produce a subsystem failed state—fewer than two units operating and no replacement available.

Since in each functioning state two components must operate, all functioning states have the same value of total failure rate, 2λ. As a result, the average time to transition from state $E_{2/k}$ to state $E_{2/(k-1)}$ is the same, no matter what the state, as long as the subsystem is operable and

$$\theta_{k/(k-1)} = \frac{1}{d\lambda} = 1/[2(0.01)] = 50 \text{ hours}$$

$$\text{MTTF}_i = \frac{n - d + 1}{d\lambda} = 150$$

10.1.8 Standby Redundancy: Taking into Account Imperfect Sensing and Switching

The previous sections provided upper-bound relationships for reliability due to the application of standby redundancy by assuming perfect detection of failures and perfect switching functions to switch out the failed unit and to switch in the standby unit. More obviously than for the full-on redundancy configuration, these two functions must be carried out after each component failure or the subsystem will fail.

The reliability of a standby redundant subsystem which requires d out of n components to function, and which is subject to less than perfect detection and switching, can be expressed as

$$R_i(t) = \sum_{x=0}^{n-d} \frac{(d\lambda t)^x e^{-d\lambda t}}{x!} P^x \qquad (10\text{-}48)$$

where

λ $\quad$ = failure rate of an operating component

d $\quad$ = number of components necessary to maintain subsystem function

P $\quad$ = the probability of a successful detection and switch

$\quad$ = combined value of fraction of failures detectable and probability that the switching functions will be implemented satisfactorily, given detection has been achieved. A successful detection and switch must follow each component failure in order for the subsystem to remain in an operable state

t $\quad$ = mission or operating time interval

$\dfrac{(d\lambda t)^x e^{-d\lambda t}}{x!}$ = probability that the subsystem experiences x failures

P^x $\quad$ = probability that the subsystem experiences x successful detections/switches, each triggered by a component failure

Taking the value of P into account provides a degree of insight into the practical boundaries of improvement due to the application of redundancy and sheds light on some aspects of this type of redundancy not usually considered. As we did for full-on redundancy, a value of $P = 0.95$ will be chosen and comparisons will be shown. Figure 10-5 shows the plots, in dashed lines, of $R_i(t)$ for a standby redundancy subsystem using equation (10-48) when $d = 1$ and $n = 2, 3, 4$ and $P = 1$ is assumed (note that when $P = 1$, (10-48) reduces to (10-39)), that is to say, perfect detection and switching. In solid lines on the same figure the results for $d = 1$, $n = 2, 3, 4$, but $P = 0.95$ are plotted from the application (10-48). Note that as n increases,

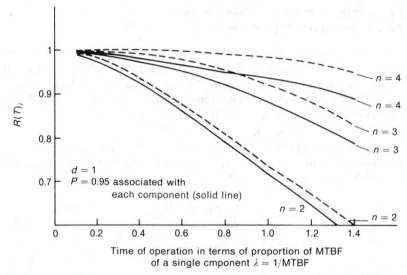

FIGURE 10-5. Comparison of reliability improvement for standby subsystems considering perfect and imperfect failure detection and switching.

the greater the difference between the results assuming perfect detection/switching and those assuming imperfect detection/switching.

As before, $MTFF_i$, or $MTFF_i$ assuming all components are operable at the start of each use, is evaluated through

$$MTFF_i = \int_0^\infty R_i(t)\, dt$$

When (10-48) is expanded, $R_i(t)$ and $MTFF_i$ can be expressed as

$$MTFF_i = \int_0^\infty e^{-d\lambda t}\left(1 + Pd\lambda t + \frac{(Pd\lambda t)^2}{2!} + \cdots + \frac{(Pd\lambda t)^{(n-d)}}{(n-d)!}\right) dt \quad (10\text{-}49)$$

Recalling that

$$\int_0^\infty \frac{(\lambda t)^x e^{-\lambda t}}{x!}\, dt = \frac{1}{\lambda}$$

(10-49) when integrated takes on the form

$$MTFF_i = \frac{1}{d\lambda}\sum_{x=0}^{n-d} P^x \quad (10\text{-}50)$$

10.1.9   Comparing Full-On and Standby Redundancy Effects

The form of redundancy that can be applied at a given subsystem level is usually governed by the design characteristics of the subsystem as opposed to being an open choice. In some cases, however, there is opportunity for choice, and in those instances the engineer should be aware of the general consequences of applying one alternative or the other.

In the event that either full-on or standby redundancy can be employed, generally speaking the standby configuration will provide the greater reliability payoff. This follows from the fact that a standby unit usually exhibits a lower failure rate than a unit which is fully energized while waiting to be used.

In order to quantitatively assess the magnitude of this difference, we compare equation (10-19), which defines the MTFF_i of a full-on redundant configuration as

$$\text{MTFF}_i = \frac{1}{\lambda} \sum_{k=d}^{n} \frac{1}{k}$$

with equation (10-43), which defines the MTFF_i of a standby redundant configuration as

$$\text{MTFF}_i = \frac{n - d + 1}{d\lambda}$$

In either case the term $\lambda \cdot \text{MTFF}_i$ equals the ratio of the mean time to first failure of the subsystem to the MTBF of a single component in the subsystem, which equals a measure of the reliability benefit due to application of redundancy.

By expressing (10-19) and (10-43) in terms of λMTFF_i, the following result:
For the full-on subsystem:

$$\lambda \text{MTFF}_i = \sum_{k=d}^{n} \frac{1}{k} \tag{10-51}$$

For the standby subsystem:

$$\lambda \text{MTFF}_i = \frac{n - d + 1}{d} \tag{10-52}$$

These are plotted on Fig. 10-6 for the same values of n and for $d = 1$. As can be seen, the standby redundant configuration provides a higher general payoff than a full-on redundant configuration.

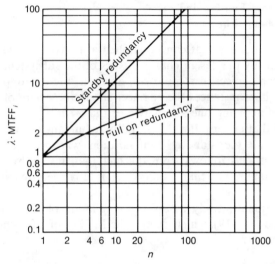

FIGURE 10-6.   Comparison of standby and full-on redundancy.
n = number of redundant units in a subsystem
λ = failure rate of any active subsystem
MTFF_i = mean life of subsystem

10.2   REDUNDANT SUBSYSTEMS WHICH ARE PERIODICALLY MAINTAINED

In the previous sections, only redundant subsystems which were not subject to any corrective maintenance until a subsystem failure occurred were considered. In this section redundant subsystems which are periodically maintained will be discussed. While nonmaintained redundancy provides reliability improvement over *nonredundant* configurations, periodic maintenance on a redundant configuration can provide significant reliability improvements over *nonmaintained* redundant configurations. Many types of subsystems fall into this category, for example:

- Subsystems aboard aircraft which suffer *component* (but not subsystem) failures during use which are repaired between missions
- Systems and subsystems placed in operation, left unattended, but visited periodically to repair all component failures.

It is assumed in both instances that, just prior to the start of a new period of operation, T, all components have been repaired/replaced and are operable.

The "mean" measures of reliability for periodically maintained subsystems take two unique forms: one based on mean life over a given mission duration or cycle of use, and the second based on mean operating life prior to failure, over many cycles of use.

10.2.1   Mean or Average Uninterrupted Life Over An Operating Period (AUL)

Knowledge that a nonredundant or redundant subsystem has a nonmaintained mean time to failure of 1000 hours when the subsystem's mission period is only 10 hours (and is then completely restored to its original state between use periods) does not provide information needed to assess subsystem reliability. Subsystems subject to operation under such conditions are contained in such systems as aircraft, space vehicles, and ships. A "mean" measure more suited in such instances is *average uninterrupted life over a given use period or mission* (AUL). For example, if a subsystem was tasked to perform a repetitive mission of T_m hours and between missions all component failures suffered on the last mission were repaired, AUL for a subsystem is calculated as

$$\text{AUL} = \frac{xT_m + \sum_{i=1}^{f} T_i}{N} \qquad (10\text{-}53)$$

where

T_m = typical mission duration

T_i = time that a *subsystem failure* occurred (measured from the start of the mission that incurred the failure), $T_i < T_m$

N = total number of missions performed

f = total number of subsystem mission failures occurring in N missions

x = total number of missions performed that incurred no subsystem failures
 = $N - f$

(Note that while we used an example of a single subsystem performing N missions, an example of N subsystems performing one mission each will yield the same result. In that case, AUL represents the average uninterrupted operating time per member of the *population* of subsystems.)

As a result, the best possible value of AUL can approach T_m but logically can never exceed it.

While (10-53) represents the way AUL is calculated from data, it does not provide a means to predict the measure based on information on T_m and knowledge of subsystem component failure rates. However, using such intelligence, AUL can be evaluated quite simply through the results derived in Appendix A5:

$$\text{AUL} = \int_0^{T_m} R_i(t)\, dt \qquad\qquad (10\text{-}54)$$

where
$R_i(t)$ = subsystem reliability assuming that the subsystem is nonmaintained, i.e., defined as in previous sections
T_m is defined as before

The illustrations which follow will make the logic of the measure clearer. Assume that a subsystem is made up of two components in a standby redundancy configuration where one is sufficient to maintain acceptable subsystem performance. During a given mission, only one component may fail, but at times between use the subsystem will be restored to a state where a full two components will be operable for the start of the next mission.

In these circumstances subsystem measures of MTTF or MTFF, based on the *assumption that no maintenance will be performed until an actual subsystem failure occurs*, would not be appropriate.

Assume that 1000 missions are to be performed. In 990 cases the subsystem survives the T_m hour mission. In 10 cases the *subsystem* fails during T_m: one at T_1 hours into a mission, one at T_2 hours into another mission, ..., one at T_{10} hours into a tenth mission, where $T_i < T_m$. If such data were available from actual experience, AUL would be calculated as

$$\text{AUL} = \frac{990 T_m + \sum_{i=1}^{10} T_i}{1000}$$

If $T_m = 10$ and $\sum_{i=1}^{10} T_i = 80$, then AUL = 9.98, which represents the average uninterrupted operating time for the subsystem over the mission period of T_m hours.

Using the principles described in Appendix A5, we can assess AUL during the design phase through

$$\text{AUL} = \int_0^{T_m} R_i(t)\, dt$$

where $R_i(t)$ is the reliability of the subsystem, which equals the probability that the *subsystem* will perform satisfactorily over an interval t, given that all components are operable at the start of the interval, which is equal to *the proportion of missions over which the subsystem will perform satisfactorily given the components are operable at the start of the interval.*

For a single nonredundant component (the AUL concept can be applied to both redundant and nonredundant items) with reliability

$$R(t) = e^{-\lambda t}$$

$$\text{AUL} = \int_0^{T_m} e^{-\lambda t}\, dt = \frac{1}{\lambda}[1 - \exp(-\lambda T_m)]$$

where AUL represents the average component operating time over a mission or use period T_m.

A subsystem made up of two components in a single-survivor full-on redundancy configuration performs a function in an aircraft, each component has a failure rate of $\lambda = 0.01$. The average aircraft usage time is $t = 10$ hours. After each use, the aircraft is maintained and all faulty components are repaired/replaced. Assuming each component has an exponential distribution of failure time:

- What is the reliability of the average subsystem 10-hour mission?
- What is the mean uninterrupted operating time of the subsystem AUL over the 10-hour mission?

$$R_i(t) = 1 - (1 - e^{-\lambda t})^n$$

For $T = T_m$:

$$R_i(T) = [2e^{-\lambda T_m} - e^{-2\lambda T_m}] = 1.8097 - 0.8187 = 0.9910$$

$$\text{AUL} = \int_0^{T_m} R_i(t)\, dt$$

$$\text{AUL} = 2\int_0^{T_m} e^{-\lambda t}\, dt - \int_0^{T_m} e^{-2\lambda t}\, dt = -\frac{2}{\lambda}[\exp(-\lambda T_m) - 1]$$

$$+ \frac{1}{2\lambda}[\exp(-2\lambda T_m) - 1]$$

$$= 9.9691 \text{ hours}$$

10.2.2 Mean Uninterrupted Total Operating Time to First Failure (MUOT)

The mean time to failure or mean time to first failure measures discussed in Section 10.1 assumed that no maintenance was performed until after the subsystem failed. Suppose a redundant subsystem is put into operation at $t = 0$ and all components are known to be operable at that time. *Every T_s* hours of operation the subsystem is inspected, necessary repairs/replacements are made, and the subsystem is put into operation again. Sometimes at the end of each cycle when the subsystem is inspected:

1. It is still capable of performing its function and no components need to be repaired or replaced.
2. It is still capable of performing its function, but one or more components must be repaired or replaced.
3. The *subsystem* will be in a failed state.

What must be determined is the mean operating time of a subsystem between subsystem operation start when all components are operable and the time when an inspection or other intelligence indicates that a *subsystem failure* has occurred (MUOT). If an exponential distribution of time to failure is assumed for all components, MUOT is also synonymous with mean time to first failure, MTFF, for the subsystem.

It is possible for a subsystem to fail before the first cycle $(0, T_s)$ is complete, or it is possible that the subsystem not fail until the nth cycle is complete.

If a large number (X) of such subsystems were in the field and it was intended to maintain these over a long period of time over many cycles, where $R_i(t) = R_i(T_s)$, i.e., $T_s = t$,

$R_i(T_s)$ = proportion of subsystems surviving the first cycle with no failure

$R_i(T_s)^2$ = proportion of subsystems surviving the first two cycles with no failures

$R_i(T_s)^3$ = proportion of subsystems surviving the three cycles with no failures

$$\vdots$$

$R_i(T_s)^N$ = proportion of subsystems surviving the first N cycles with no failures.

Therefore, of the original X subsystems, in

Cycle 1 X subsystems would start the cycle and operate an average of AUL hours each before failure during the cycle (AUL, the average uninterrupted operating time over T_s is computed as in the last section where $T_s = T_m$)

Cycle 2 $R_i(T_s) \cdot X$ subsystems would start this cycle and operate an additional AUL hours each

Cycle 3 $R_i(T_s)^2 \cdot X$ subsystems would start this cycle and operate an additional AUL hours each

$$\vdots$$

Cycle N $R_i(T_s)^{N-1} \cdot X$ subsystems would start this cycle and operate an additional AUL hours each

and the average or mean total uninterrupted operating time to failure, or to first failure, per subsystem is

$$\text{MUOT} = \frac{X \cdot \text{AUL} + \sum_{i=1}^{N-1} R_i(T_s)^i \cdot X \cdot \text{AUL}}{X} = \text{AUL}\left[1 + \sum_{i=1}^{N-1} R_i(T_s)^i\right] \quad (10\text{-}55)$$

But $1 + \sum_{i=1}^{N-1} R_i(T_s)^i$ is equal to a progression of the form $a, ar, ar^2, \ldots, ar^n$, where $a = 1$, with sum

$$S_N = \frac{R_i(T_s)^N - 1}{R_i(T_s) - 1}$$

Since $R_i(T_s) < 1$, then $S_N \to 1/[1 - R_i(T_s)]$ as N gets arbitrarily large. Hence, average uninterrupted total time to failure, MUOT, provided that all components of each subsystem surviving the previous cycle are made operable for the start of the next cycle, is

$$\text{MUOT} = \frac{\text{AUL}}{1 - R_i(T_s)} \quad (10\text{-}56)$$

In order to get an idea of the factor of improvement gained by introduction of periodic maintenance, the ratio of MUOT_i to MTFF_i is compared, where MTFF_i is the mean time to first failure for nonmaintained subsystems. Recalling from Sections 10.1.1 and 10.1.5 that

For single-survivor full-on redundancy:

$$\text{MTTF}_i = \frac{1}{\lambda} \sum_{s=1}^{n} \frac{1}{s}$$

$$R_i(t) = 1 - (1 - e^{-\lambda t})^n$$

For single-survivor standby redundancy:

$$\text{MTTF}_i = \frac{n}{\lambda}$$

$$R_i(t) = \sum_{x=0}^{n-1} \frac{(\lambda t)^x e^{-\lambda t}}{x!}$$

for full-on redundancy the factor of improvement due to performance of scheduled maintenance is equal to

$$\frac{\text{MUOT}_i}{\text{MTFF}_i} = \frac{\int_0^{T_s} [1 - (1 - e^{-\lambda t})^n] \, dt}{[1 - \exp(-\lambda T_s)]^n (1/\lambda) \sum\limits_{s=1}^{n} (1/s)} \tag{10-57}$$

For standby redundancy the factor of improvement due to performance of scheduled maintenance is equal to

$$\frac{\text{MUOT}_i}{\text{MTFF}_i} = \frac{\int_0^{T_s} \left[\sum\limits_{x=0}^{n-1} (\lambda t)^x e^{-\lambda t}/x! \right] dt}{\left[1 - \sum\limits_{x=0}^{n-1} (\lambda T_s)^x \exp(-\lambda T_s)/x! \right] (n/\lambda)} \tag{10-58}$$

For example, if $n = 2$, (10-57) reduces to

$$\frac{\text{MUOT}_i}{\text{MTFF}_i} = \frac{\int_0^{T_s} [2e^{-\lambda t} - e^{-2\lambda t}] \, dt}{(3/2\lambda)(1 - 2e^{-\lambda T_s} + e^{-2\lambda T_s})} = \frac{3 + e^{-2\lambda T_s} - 4e^{-\lambda T_s}}{3(1 - 2e^{-\lambda T_s} + e^{-\lambda T_s})} \tag{10-59}$$

and (10-58) reduces to

$$\frac{\text{MUOT}_i}{\text{MTFF}_i} = \frac{\int_0^{T_s} [e^{-\lambda t} + \lambda t e^{-\lambda t}] \, dt}{(2/\lambda)[1 - e^{-\lambda T_s} - \lambda T_s e^{-\lambda T_s}]}$$

A plot of (10-59) for a range of periodic maintenance times between 0.1θ and θ is shown in Fig. 10-7, where $\theta = 1/\lambda$ = the MTBF/mean time to failure of a single component. As can be seen, large factors of improvement in reliability accrue as time between scheduled maintenance decreases and levels of redundancy increase. When, for example, maintenance is performed every 0.4θ hours, even when only one spare component ($n = 2$) is available, the factor of improvement is over 2. For the case where two spare components are available ($n = 3$), the factor of improvement is over 6. And, of course, when T_s is chosen smaller and smaller, the factor of improvement increases. When for example, maintenance is performed every 0.2θ hours and $n = 2$, the factor of improvement is over 4; when $n = 3$, the factor of improvement is over 18.

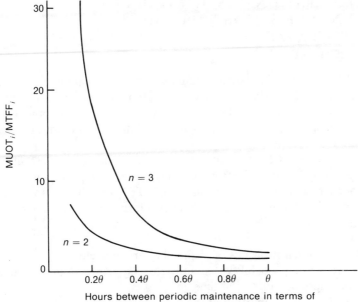

FIGURE 10-7.   Factor of reliability improvement as a consequence of periodic maintenance where $\theta = 1/\lambda$ of each unit.

Example   For the example in the last section, calculate MUOT and the ratio $\mathrm{MUOT}_i/\mathrm{MTFF}_i$.

$$\mathrm{MUOT} = \frac{\mathrm{AUL}_i}{1 - R_i(t)}$$

From the last example,

$$\mathrm{AUL}_i \sim 9.97, \qquad R_i(t) \sim 0.99, \qquad \mathrm{MUOT}_i \sim 1108$$

MTFF_i calculated for the last example is

$$\mathrm{MTFF}_i = \frac{1}{\lambda} \sum_{s=1}^{n} \frac{1}{s}$$

$$\mathrm{MTFF}_i = [1/0.01][1 + (1/2)] = 150 \text{ hours}$$

$$\frac{\mathrm{MUOT}_i}{\mathrm{MTFF}_i} \sim 7.38$$

10.3 OPERATIONALLY MAINTAINED REDUNDANT SUBSYSTEMS AND SYSTEMS: AVAILABILITY AND RELIABILITY

In the last section redundant subsystems which were maintained periodically were considered. In this section consideration will be given to redundant subsystems which are constantly maintained (when any component fails). This recourse can provide even greater reliability gains than the strategy of periodic maintenance.

As discussed in Chapter 9 for nonredundant items with constant hazards (failure rates), the capability to perform maintenance impacts item availability but not such reliability measures as $R(t)$, MTFF, or MTTF. However, when such operationally maintained components form redundant subsystems and systems:

1. The traditional measures of availability are impacted
   a. $A(t)$ = the probability that subsystem/system is in an operable state at a specific point, t, in time.
   b. $A(T)$ = the proportion of the time that the subsystem/system will be in an operable state over a specific interval of time, T.
   c. A = steady-state availability = the probability that the subsystem or system will be in an operable state at a random point in time = the proportion of the time that the subsystem or system will be in an operable state over a long period of time.
2. The traditional reliability measures $R(t)$, MTFF, and MTTF are impacted, and an additional measure, *steady-state MTBF* (M_{ss}), becomes relevant as well.

Of the reliability and availability measures mentioned above, the ones most commonly applied to or used as specification parameters for operationally maintained subsystems and systems are steady-state availability, steady-state MTBF, and MTTF/MTFF. We will contain our discussion to these.

10.3.1 Steady-State Availability for Full-On Redundant Subsystems and Systems

When steady-state availability A is defined for a subsystem, A is the probability that a subsystem will be in an operable state (capable of satisfactorily performing its function) at any random point in time (or similarly in terms of proportion of time). The phrase "operational state" connotes that at least the minimum number of components necessary to maintain subsystem function is operable.

As in Chapter 9, steady-state availability A_j of the jth component is defined as

$$A_j = \frac{M_j}{R_j + M_j} = \frac{\mu_j}{\mu_j + \lambda_j}$$

where component j has exponential distributions of time to failure and time to repair, and

A_j = steady state availability of component j as defined above
M_j = MTBF/MTTF/MTFF associated with component j
R_j = mean time to repair associated with component j
$\mu_i = 1/R_j$ = repair rate for component j
$\lambda_j = 1/M_j$ = failure rate for component j

Assume a full-on redundant subsystem made up of n identical components such as that in Fig. 10-1, which requires a minimum of d operable components in order to perform the subsystem function. The binomial distribution can be applied as in Section 10.1.2 to assess subsystem availability A_s:

$$A_s = \sum_{k=d}^{n} \frac{n!}{(n-k)!\,k!} A^k (1 - A)^{n-k} \qquad (10\text{-}60)$$

where
A_s = subsystem steady-state availability
= the probability that the subsystem will be *operable* (will have d or more components in operation) at any random point in time
= the proportion of time that the subsystem will be *operable* over a long period of time
d = minimum number of components required for the subsystem to perform its function
A = steady-state availability of each component, $M/(M + R)$ and where the term

$$\frac{n!}{(n-k)!\,k!} A^k (1 - A)^{n-k} \qquad (10\text{-}61)$$

= probability that at a random point in time the subsystem is in a state where k components are operating and $(n - k)$ are failed (under repair), which equals the proportion of the time that the subsystem is in a state where k components are operating and $(n - k)$ are failed (under repair).

Equation (10-60) assumes that a repair crew is available for each component as it fails, or more practically that sufficient maintenance teams are available such that waiting time for maintenance may be assumed to be zero for each failure. The impact of this assumption in a particular case can be evaluated by determining the probability that $g < n$ repair teams can provide this result for all practical purposes. This can be assessed through use of the procedures in Section 9.3.2.1 when all components in the full-on redundant subsystem are assumed to be connected in series and a fixed number of repair teams, g, are assigned. For example, if the probability is less than 0.01 that more than g maintenance teams will be required at any one time, then more than 99% of the time g repair crews assigned will be able to address each component failure as it occurs. Hence, no significant change in the value of A_s will occur.

Equation (10-60) also assumes that the subsystem *is not shut down* when $(d - 1)$ or fewer components are operable. If, as before, the subsystem *is shut down* whenever less than d components are operable (the remaining operable components have $\lambda = 0$ when shut down), the subsystem availability A_{sd} is defined as

$$A_{sd} = A_s \bigg/ \left[1 - \sum_{k=0}^{d-1} \frac{n!}{(n-k)!\, k!} A^k (1 - A)^{n-k} \right], \quad \text{(see Appendix A14)} \quad (10\text{-}62)$$

System availability, A_{ss}, for any system made up of L subsystems connected in series (a) regardless of whether or not the subsystems are redundant and (b) regardless of the type of redundancy employed, can be expressed as

$$A_{ss} = \prod_{i=1}^{L} A_{si} \tag{10-63}$$

where
$\quad A_{si} =$ the availability of the ith subsystem (can denote A_s or A_{sd})
$\quad L =$ number of subsystems

Example   A subsystem is made up of three identical components in a full-on, *single-survivor* ($d = 1$) redundant configuration where only one is required to perform the subsystem function. The failure and repair rates of each component are, respectively, $\lambda = 0.01$, $\mu = 0.015$.

- What is the probability that the subsystem will be capable of performing its function at any random point in time in the long term future?
- What proportion of time will the subsystem be capable of performing its functions?

In both cases the answer will be identical. The availability A of each component is

$$A = \frac{M_j}{M_j + R_j} = \frac{\mu_j}{\lambda_j + \mu_j} = 0.015/0.025 = 0.6$$

$$A_s = \sum_{k=1}^{3} \frac{n!}{(n-k)! \, k!} A^k [1-A]^{n-k} = 3A(1-A)^2 + 3A^2(1-A) + (A)^3$$

$$= 0.288 + 0.432 + 0.216 = 0.936$$

10.3.2 Steady-State Availability for Standby Redundant Subsystems and Systems

Unlike the previous case, there is no simple expression for the availability of a single component in a standby configuration which can serve as a foundation for subsystem evaluation. In standby redundant systems, the availability of each component must take into account not two states (operating and failed/under repair), but three—operating, failed/under repair, and standby—and is dependent on the number of components making up the subsystem.

Instead, a more general approach must be taken which will nonetheless yield a relatively simple relationship that can be used to evaluate subsystem availability for subsystems of this type.

The most common and powerful procedure that can be employed to evaluate the availability or reliability of a maintained redundant subsystem is a Markov process. This was discussed in a simpler context in Chapter 9, when it was applied to the development of $A(t)$ and $A(T)$ for a single component which was subjected to repair actions as soon as a failure occurred. Complete descriptions of the application of Markov processes to redundant, maintained subsystems can be found in most of the more theoretical texts on reliability.

While a Markov process was used to develop the following general expression necessary to evaluate A_s, knowledge of Markov processes is not required to apply the expression. As a consequence, an explanation of the development of the expression and assumptions necessary to apply the Markov process are included in Appendix A16 but will not be included in the text.

Subsystem availability A_s may be expressed as

$$A_s = P_n \sum_{i=0}^{n-d} \frac{n!}{i!} \left(\frac{\mu}{d\lambda} \right)^{n-i} \left(\prod_{j=Z_i}^{Z_{(n-1)}} j \right)^{-1} \tag{10-64}$$

where

$$P_n = 1 \bigg/ \left[1 + \sum_{i=0}^{n-1} \frac{n!}{i!} \left(\frac{\mu}{d\lambda} \right)^{n-i} \left(\prod_{j=Z_i}^{Z_{(n-1)}} j \right)^{-1} \right]$$

= probability that all components are in a failed state

A_s = subsystem steady-state availability

= the probability that the subsystem will be *operable* (will have d or more components operable) at any random point in time

= the proportion of the time that the subsystem will be *operable* over a long period of time

$Z_i = (n - i)/d,$ if $(n - i)/d < 1$

$Z_i = 1,$ if $(n - i)/d \geqslant 1$

d = number of components required for the subsystem to perform its function

n = number of components making up the redundant subsystem

When $d = 1$ (a single-survivor configuration), (10-64) takes the simpler form

$$A_s = P_n \sum_{i=0}^{n-d} \frac{n!}{i!} \left(\frac{\mu}{\lambda} \right)^{n-i} \tag{10-65}$$

where

$$P_n = 1 \bigg/ \left[1 + \sum_{i=0}^{n-1} \frac{n!}{i!} \left(\frac{\mu}{\lambda} \right)^{n-i} \right]$$

Equation (10-64) assumes that a repair crew is available for each component, or more practically, that sufficient maintenance teams are available such that waiting time for maintenance is essentially zero. The impact of this assumption in a particular case can be evaluated by assuming that g maintenance teams have been assigned and determining the probability of *more than g components* being in repair at the same time. For example, if that probability is less than 0.01, then more than 99% of the time g repair crews will be sufficient to maintain the subsystem. Hence, no significant change in the value of A_s will occur when g, as opposed to n, repair teams are assigned to the subsystem. The probability, $P(g+)$, that more than g repair crews will be required at any point in time to immediately address all failures can be expressed as

$$P(g+) = 1 - P_n \sum_{i=0}^{g} \frac{n!}{i!} \left(\frac{\mu}{d\lambda} \right)^{n-i} \left(\prod_{j=Z_i}^{Z_{(n-1)}} j \right)^{-1} \tag{10-66}$$

Equation (10-64) also assumes that the subsystem *is not shut down* when $(d - 1)$ or fewer components are operable.

If the subsystem *is not shut down* upon failure (in that case the remaining operable components will have $\lambda > 0$), fewer than d operable components continue to be energized during subsystem repair. Availability, A_{sd}, under the assumption that the subsystem is shut down when $(d - 1)$ components are operable takes the form

$$A_{sd} = A_s \bigg/ \left[1 - P_n - P_n \sum_{i=(n-d+2)}^{n-1} \frac{n!}{i!} \left(\frac{\mu}{d\lambda}\right)^{n-i} \left(\prod_{j=Z_i}^{Z_{(n-1)}} j\right)^{-1} \right] \qquad (10\text{-}67)$$

Example   The two-component subsystem illustrated in the last section has been subjected to a redesign. It has been decided to reconfigure the subsystem into one which employs standby, rather than full-on redundancy. What is the probability that the subsystem will be capable of performing its function at a random point in time in the future?

Compare the availability for the single survivor full-on redundant design illustrated previously to a single survivor standby redundant design:

$$A_s = P_n \sum_{i=0}^{n-1} \frac{n!}{i!} \left(\frac{\mu}{\lambda}\right)^{n-i}$$

$$P_n = 1 \bigg/ \left[1 + \sum_{i=0}^{n-1} \frac{n!}{i!} \left(\frac{\mu}{\lambda}\right)^{n-i} \right]$$

$$P_n = 1/[1 + 20.25 + 13.5 + 4.5] = 0.0255$$

$$A_s = 0.0255(20.25 + 13.5 + 4.5) = 0.9745$$

10.3.3   Steady State MTBF:
A Subsystem/System Reliability Measure

The MTTF and MTFF measures discussed previously all assume that after a redundant subsystem fails it is not brought back into operation again until *all* components which malfunctioned are repaired or replaced. In many practical instances of redundant subsystem operation, however, after a failure occurs at subsystem level, the subsystem does not remain in maintenance until all components are operable. It is rather put back into service as soon as it can perform its function. Over an arbitrarily long period of operating time under such an operating policy, the average time between failures will not be equal to the MTTF or MTFF measures discussed previously but to another measure defined as steady-state MTBF (M_{ss}), where $M_{ss} \neq$ MTFF, MTTF.

An example will make the distinction between MTTF, MTFF, and steady-state MTBF clearer. Consider a subsystem made up of two components A and B as in Fig. 10-8a. The subsystem operates in a full-on, single-survivor

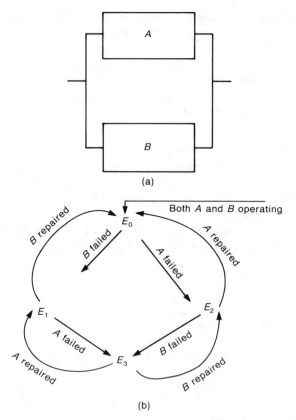

FIGURE 10-8. (a) Two redundant components. (b) Possible state transitions for the subsystem in (a).

redundant configuration. The subsystem can perform its function when just one component is operating. The subsystem will fail when both components are in a failed state at the same time.

Any subsystem in operation over time is a dynamic entity. As failures occur, its state changes; as repairs are completed, the state changes again. The possible operating states for this particular subsystem are:

E_0 = both components A and B operating, neither under repair

E_1 = component A operating, B failed and under repair or awaiting repair

E_2 = component B operating, A failed and under repair or awaiting repair

E_3 = both components A and B failed and under repair or awaiting repair

Thus, upon entering state E_3, the subsystem is in a failed state.

If it is assumed as in Chapter 9 that it is impossible for more than one action to occur simultaneously—that they may be separated by $1/100 \times 10^{-6}$ second but cannot occur at exactly the same time—then it is impossible for two or more failures, repairs or combinations of failures and repairs to occur at exactly the same instant. This means the subsystem cannot transition from state E_0 directly to state E_3, and cannot transition from state E_3 to state E_0 directly. In both instances the subsystem must first transition to state E_1 or E_2 before transitioning to E_3 or E_0.

As a result, if the subsystem is in state E_3 and the policy is to get the subsystem into operation as soon as possible, the subsystem would be put into use whenever it reached state E_1 or E_2, whichever occurred first. Bear in mind that after entering state E_1 or E_2, there are probabilities that:

1. The subsystem will next directly transition to state E_3 again.
2. The subsystem will transition next to E_0, and given that eventuality, probabilities that it may ping-pong between E_0 and both E_1 and E_2 a number of times before ending up in E_3 again.

Steady-state MTBF defines over a long period of operation the average time between when the system enters state E_1 or E_2 after a repair has been made to restore the minimum degree function required and *next* enters E_3, taking all such possible transitions into account.

We can depict this graphically if we employ the convention that a straight-line arrow indicates a transition from one state to another due to a failure, and a curved-line arrow indicates a transition from one state to another due to repairs. Logic tells us, then, that a state transition map associated with Fig. 10-8a takes the form shown in Fig. 10-8b.

Hence in summary, M_{ss} defines the average time between when the subsystem is put into operation in state E_1 or E_2 and the time it next enters state E_3. MTTF and MTFF, on the other hand, defines the average time between when the subsystem enters operation in state E_0 until it next enters state E_3.

10.3.3.1 Evaluating Steady-State MTBF
For Full-On Redundant Subsystems

For a mathematical derivation of steady-state MTBF, see Ref. 2 and Appendix A17. More mathematical discussions related to the concept can be found in Refs. 3 and 4. For the purposes of this text, a simpler but equivalent procedure will be described which can be used to evaluate steady-state MTBF. For those not particularly interested in the detail of the derivation, see equations (10-77) and (10-78), which define M_{ss} for a subsystem of this type.

Chart 10-1 depicts all possible states associated with a full-on redundant subsystem which requires a minimum of d components to perform a given

CHART 10-1 All Possible States Associated with a Full-On Redundant Subsystem Requiring a Minimum of d Units to Perform its Function

State	Description	
0	n units operating	0 failed
1	$(n-1)$	1
2	$(n-2)$	2
3	$(n-3)$	3
$\vdots$		
$(n-d)$	d units operating	$(n-d)$ failed
$(n-d+1)$	$(d-1)$	$(n-d+1)$
$\vdots$		
n	0 units operating	(n) failed

function. [For ease of discussion of redundant subsystems, we have adopted a simplified means of state depiction. State i denotes the number of components in repair (inoperable).] As can be seen, assuming a start at state 0, a failed state can result only after the subsystem has undergone a number of transitions. It can ping-pong from state to state, but because only one repair or failure can occur at exactly the same time, the subsystem can ping-pong (transition) only between adjacent states as shown in Fig. 10-9. As is evident from the figure, the subsystem can *only fail from state* $(n-d)$. Once failed it will be considered in operating condition again only at such time that sufficient repairs occur and it enters state $(n-d)$ again from $(n-d+1)$.

A transition rate may be associated with each subsystem state, (the expected rate at which a subsystem will change states given that it is in a given state at a given time). However, as indicated before, only a *specific* transition from the $(n-d)$ [to the $(n-d+1)$] state can result in subsystem failure. All other *nonfailed* states have more than d components operating.

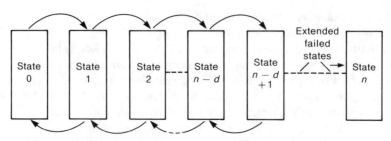

FIGURE 10-9. Representing the only paths and their order that transitions can take for full-on redundant/maintainable subsystems.

A single component failure in any one of such "other" states will still leave the subsystem with at least d operating components, which represents a transition to another *operating state but not a subsystem failed state.*

As a consequence, since it is *impossible* for the *subsystem* to directly fail from a state $i < (n - d)$, a *subsystem* failure rate $\lambda = 0$ will be associated with such states. Since subsystem failure is possible from state $(n - d)$, a *subsystem* failure rate > 0 can be ascribed to that state.

As was discussed in Chapter 8, when a subsystem or item is exposed to different modes (states) of operation over its life span, each with a different associated failure rate, its average or expected value of failure rate, $E(\lambda)$, can be calculated based on the proportion of time P_i, spent in each mode (state), and the subsystem failure rate, λ_i, associated with that state:

$$E(\lambda) = \sum_{i=0}^{n} \lambda_i P_i \qquad (10\text{-}68)$$

where P_i = proportion of the time the subsystem is in state i when it is operated for an arbitrarily long period of time under a maintenance policy as depicted in Fig. 10-9.

If one were to define the expected proportion of time, $E(P)$, that the subsystem will be in an *operable* state,

$$E(P) = \sum_{i=0}^{n-d} P_i \qquad (10\text{-}69a)$$

considering that

$E(\lambda)T$ = expected number of subsystem failures over an arbitrarily long period of time T

$E(P)T$ = expected total subsystem operating hours in an operable state over an arbitrarily long period of time T

then

$$\frac{E(\lambda)T}{E(\lambda P)T} = \frac{E(\lambda)}{E(P)}$$

= number of failures divided by subsystem operating time = $\bar{\lambda}$

= average failure rate over a long period of time which results in

$$M_{ss} = 1/\bar{\lambda}$$

= a measure of the average time to failure for the subsystem over an arbitrarily long period of time

Since all λ_i except $\lambda_{(n-d)}$ have the value 0,

$$E(\lambda) = \lambda_{(n-d)} P_{(n-d)} \tag{10.69b}$$

and the failure rate $\lambda_{(n-d)}$ associated with state $E_{(n-d)}$ equals the sum of the failure rates of all operating components in that state:

$$\lambda_{(n-d)} = \sum_{j=1}^{d} \lambda_j$$

where

λ_j = the failure rate associated with the jth operating (nonfailed) component in state $(n-d)$

n = number of components in the redundant subsystem

d = minimum number of components required for subsystem operation

For full-on redundant subsystems made up of identical components, $P_{(n-d)}$—the proportion of time that the subsystem is in state $(n-d)$ (i.e., $(n-d)$ components under repair and d operating)—can be ascertained using (10-61). Equation (10-61) also provides the means to calculate P_i for any state i.

The proportion of time the subsystem is in state $(n-d)$ is

$$P_{(n-d)} = \frac{n!}{(n-d)!\,d!} A^d (1-A)^{n-d} \tag{10-70}$$

More generally, the proportion of time spent in any state i is

$$P_i = \frac{n!}{(n-i)!\,i!} A^{n-i} (1-A)^i \tag{10-71}$$

or where $P_i = P_{n-k}$, $i = (n-k)$,

$$P_{n-k} = \frac{n!}{(n-k)!\,k!} A^k (1-A)^{n-k} \tag{10-72}$$

$E(\lambda)$ can be represented as

$$E(\lambda) = \frac{n!}{(n-d)!\,d!} A^d (1-A)^{n-d} \lambda_{(n-d)} \tag{10-73}$$

Applying (10-69a) yields

$$E(P) = \sum_{k=d}^{n} \frac{n!}{(n-k)!\,k!} A^k (1-A)^{n-k} \tag{10-74}$$

or its equivalent

$$E(P) = \sum_{i=0}^{n-d} \frac{n!}{(n-i)!\,i!} A^{n-i} (1-A)^i \tag{10-75}$$

For any subsystem requiring d or more components for successful operation, $\bar{\lambda} = E(\lambda)/E(P)$ takes the form

$$\bar{\lambda} = \frac{n!\,d\lambda^{n-d+1}\mu^d}{d!\,(n-d)!} \Bigg/ \sum_{i=0}^{n-d} \frac{n!\,\mu^{n-i}\lambda^i}{(n-i)!\,i!} \tag{10-76}$$

and steady state MTBF, M_{ss}, is defined as

$$M_{ss} = \frac{1}{\bar{\lambda}} = \sum_{i=0}^{n-d} \frac{n!\,\mu^{n-i}\lambda^i}{(n-i)!\,i!} \Bigg/ \frac{n!\,d\lambda^{n-d+1}\mu^d}{d!\,(n-d)!} \tag{10-77}$$

Simplifying and cancelling like terms,

$$M_{ss} = \frac{d!\,(n-d)!}{d\lambda^{n-d+1}\mu^d} \sum_{i=0}^{n-d} \frac{\mu^{n-i}\lambda^i}{(n-i)!\,i!} \tag{10-78}$$

which is a different form but equivalent to the evaluation equation for M_{ss} calculated in Appendix A17, repeated below, using a different derivation process. (Their equivalence is shown in the appendix.)

$$M_{ss} = \frac{1}{d\lambda} + \sum_{k=0}^{n-d-1} \prod_{j=k+1}^{n-d} j\mu \Bigg/ \prod_{j=k}^{n-d} (n-j)\lambda \tag{10-79}$$

Examples   A subsystem contains four components, a minimum of two of which are required to perform the subsystem function. The subsystem is intended for long term usage in a maintained full-on configuration environment. When the subsystem fails (has fewer than two operating components), it will be put back into operation as soon as two components are operable. Compute its steady-state MTBF when the failure and repair rates for each component are, respectively, $\lambda = 0.01$ and $\mu = 0.015$.

Using (10-78),

$$M_{ss} = \frac{d!(n-d)!}{d\lambda^{n-d+1}\mu^d} \sum_{i=0}^{n-d} \frac{\mu^{n-i}\lambda^i}{(n-i)!\,i!}$$

$$M_{ss} = \{2!(2)!/[2(0.01)^3(0.015)^2]\}[(0.015)^4/4! + (0.015)^3(0.01)/3!$$

$$+ (0.015)^2(0.01)^2/2!\,2!)] = 118.75 \text{ hours}$$

10.3.3.2 Evaluating Steady-State MT BF
For Standby Redundant Subsystems

For a mathematical derivation of steady-state MTBF for this class of system, see Ref. 2 and Appendix A18. For the purposes of this text, an equivalent expression will be derived to evaluate steady-state MTBF taking a more elementary approach. For those not particularly interested in the detail of the derivation, see equations (10-86) and (10-87) which define M_{ss} for a subsystem of this type.

The previous descriptions of a standby redundant subsystem hold. That is to say, the subsystem is comprised of n redundant identical components (each with identical values of λ and μ); failure and repair times of each component take exponential distributions; d components must operate in order to maintain the subsystem function and when operating each has a failure rate of λ; $(n - d)$ components are originally in a standby unenergized state and when in that state have a failure rate of 0. When a failure in one of the d working components occurs, one of the standby units is engaged to take its place and the failed component is put under repair. When repair is complete, the repaired unit is placed in the standby pool of units. A subsystem failure occurs when one of the d operating components fails and no component is available in the standby pool to take its place (that is to say, $(n - d)$ units are under repair when one of the d operating units fails). Thus, the subsystem is in a failed state once at the state $(n - d + 1)$. Since we are dealing with the steady-state MTBF measure, once in a failed state the subsystem is considered operational again when it *first* transitions back into a minimum operational state, i.e., state $(n - d)$.

The system states take the form shown in Chart 10-2 and transitions occur only between adjacent states as depicted in Fig. 10-9.

As before, since it is impossible for the *subsystem* to directly fail from a state $i < (n - d)$, a *subsystem* failure rate $\lambda = 0$ is associated with such states. Since failure is possible from state $(n - d)$, a subsystem failure rate can be ascribed to state $(n - d)$.

CHART 10-2 All Possible States for a Standby Redundant Subsystem

State	Units Operating	Units in Standby	Units in Repair
0	d	$(n-d)$	0
1	d	$(n-d-1)$	1
2	d	$(n-d-2)$	2
$\vdots$			
$(n-d)$	d	0	$(n-d)$
$(n-d+1)$	$(d-1)$	0	$(n-d+1)$
$\vdots$			
n	0	0	n

We define, as in the last section, the ratio

$$E(\lambda)T/E(P)T = E(\lambda)/E(P)$$

$$= \text{average number of failures/subsystem operating hour}$$

which equals $\bar{\lambda}$, average failure rate, where, as in the last section,

$$E(\lambda) = \sum_{i=0}^{n} \lambda P_i = \text{average or expected value of subsystem failure rate over an arbitrarily long period of time}$$

$$E(P) = \sum_{i=0}^{n-d} P_i = \text{proportion of time that the subsystem will be in an operable state over an arbitrarily long period of time}$$

P_i = proportion of the time the subsystem is in state i when it is operated for an arbitrarily long period of time under the conditions depicted in Fig. 10-9

λ_i = failure rate associated with state i

T = an arbitrarily long period of time

and as before,

$1/\bar{\lambda}$ = the measure of steady state MTBF for the subsystem over an arbitrarily long period of time

Since all λ_i except $\lambda_{(n-d)}$ have a value of 0,

$$E(\lambda) = \lambda_{(n-d)}P_{(n-d)}$$

and, as in the previous section,

$$\lambda_{(n-d)} = \sum_{j=1}^{d} \lambda_j$$

where
λ_j = the failure rate associated with the jth operating (nonfailed component) in state $(n-d)$
n and d are defined as before

If all components are identical,

$$\lambda_{(n-d)} = d\lambda$$

The values for $P_{(n-d)}$ and P_i in general, for standby redundant subsystems made up of identical components, can be calculated using the expression derived in Appendix A16:
In the general case from Appendix A16

$$P_i = P_n \left(\frac{n!}{i!}\right)\left(\frac{\mu}{d\lambda}\right)^{n-i}\left(\prod_{j=Z_i}^{Z_{(n-1)}} j\right)^{-1} \tag{10-80}$$

where
P_i = the proportion of time the subsystem is in state i
P_n = the proportion of time the subsystem is in state n

$$P_n = 1 \bigg/ \left[1 + \sum_{i=0}^{n-1}\left(\frac{n!}{i!}\right)\left(\frac{\mu}{d\lambda}\right)^{n-i}\left(\prod_{j=Z_i}^{Z_{(n-1)}} j\right)^{-1}\right] \tag{10-81}$$

$Z_i = (n-i)/d$ if $(n-i)/d < 1$; $Z_i = (n-i)/d$
$Z_i = (n-i)/d$ if $(n-i)/d \geqslant 1$; $Z_i = 1$

For the particular case of $P_{(n-d)}$

$$P_{(n-d)} = P_n \left(\frac{n!}{(n-d)!}\right)\left(\frac{\mu}{d\lambda}\right)^{d}\left[\prod_{Z_{(n-d)}}^{Z_{(n-1)}} j\right]^{-1}$$

since all values of Z_i will equal 1 for $i = (n-d)$, the above reduces to

$$P_{(n-d)} = P_n \left(\frac{n!}{(n-d)!}\right)\left(\frac{\mu}{d\lambda}\right)^{d} \tag{10-82}$$

$E(\lambda)$ can then be represented as

$$E(\lambda) = d\lambda P_n\left(\frac{n!}{(n-d)!}\right)\left(\frac{\mu}{d\lambda}\right)^d \tag{10-83}$$

And from (10-80), $E(P)$ is expressed as

$$E(P) = \sum_{i=0}^{n-d} P_i = P_n \sum_{i=0}^{n-d} \left(\frac{n!}{i!}\right)\left(\frac{\mu}{d\lambda}\right)^{n-i} \tag{10-84}$$

$$\frac{E(\lambda)}{E(P)} = \bar{\lambda} = \left(\frac{d\lambda n!}{(n-d)!}\right)\left(\frac{\mu}{d\lambda}\right)^d \Bigg/ \sum_{i=0}^{n-d} \left(\frac{n!}{i!}\right)\left(\frac{\mu}{d\lambda}\right)^{n-i} \tag{10-85}$$

After some manipulation,

$$M_{ss} = \frac{1}{\bar{\lambda}} = (n-d)! \sum_{i=0}^{n-d} \frac{\mu^{n-d-i}}{i!(d\lambda)^{n-d+1-i}} \tag{10-86}$$

which is equivalent to the result reached using the technique in Appendix A18:

$$M_{ss} = \sum_{k=0}^{n-d} \frac{(n-d)!\,\mu^k}{(n-d-k)!(d\lambda)^{k+1}} \tag{10-87}$$

Use whichever of the two is easiest for you.

Example   Assume that the four-component subsystem used in the last example is now configured as a standby redundant subsystem which requires two components in order to function. Compute its steady-state MTBF:

$$M_{ss} = (n-d)! \sum_{i=0}^{n-d} \frac{\mu^{n-d-i}}{i!(d\lambda)^{n-d+1-i}}$$

$$= 2\left(\frac{\mu^2}{(d\lambda)^3} + \frac{\mu}{(d\lambda)^2} + \frac{1}{2(d\lambda)}\right) = 2(90.63) = 181.3$$

10.3.3.3 Evaluating MTFF for Full-On and Standby Redundant Systems Using M_{ss} Principles

In many instances the MTFF of complex subsystems can be more simply evaluated as a function of M_{ss} than directly calculated using the procedures discussed earlier in this chapter. This holds particularly for subsystems which can be maintained during usage. Using M_{ss} quantities, MTFF can be assessed without recourse to integration, use of Laplace transforms, or partial fractions.

Assume as before that all components making up the subsystem have exponential distributions of times to failure and repair. As indicated previously, MTTF or MTFF for such subsystems is calculated under the assumption that, after each subsystem failure, the subsystem is only put back into operation after *all* components are in operable states. This contrasts with steady-state MTBF, which assumes that the subsystem is put back into operation when the *minimum* number of components necessary to maintain subsystem functions are operable.

MTTF/MTFF for such subsystems can be evaluated by simple extensions of the procedures used to define steady state MTBF.

1. Appendix A17 and Appendix A18 show the derivation of such relationships from M_{ss} metrics. MTFF (MTTF) derived from (10-79) for a full-on redundant subsystem takes the form

$$\text{MTFF (MTTF)} = \sum_{i=d}^{n} \left\{ [1/i\lambda] + \sum_{k=0}^{n-i-1} \left[\prod_{j=k+1}^{n-i} ju \middle/ \prod_{j=k}^{n-i} (n-j)\lambda \right] \right\} \qquad (10\text{-}88)$$

where d, n, μ, and λ are defined as before.

MTFF (MTTF) derived from (10-87) and described in Appendix A18 for a standby redundant subsystem takes the form

$$\text{MTFF (MTTF)} = \sum_{i=0}^{n-d} \sum_{k=0}^{i} \frac{i!\,\mu^k}{(i-k)!\,(d\lambda)^{k+1}} \qquad (10\text{-}89)$$

2. The alternative M_{ss} relationships derived in the text can also be simply and logically extended to evaluate MTFF (MTTF). If M_{ss} defines the average time to transition from state $(n-d)$ to the failed state $(n-d+1)$, "pretend" that $(n-d)$ is a failed state that can only be reached from state $(n-d-1)$ and compute a value of $M_{ss} = M_{ss1}$. For that situation, M_{ss1} will then define the average time to transition from state $(n-d-1)$ to state $(n-d)$. Similarly, "pretend" next that $(n-d-1)$ is a failed state that can only be reached from state $(n-d-2)$ and compute a value of $M_{ss} = M_{ss2}$. M_{ss2} will define the average time to transition from $(n-d-2)$ to $(n-d-1)$. If the above procedure is continued until a value of M_{ssj} is reached, which defines the average time to transition from state 0 to state 1, the sum of all values of M_{ssj} will be equal to MTFF (MTTF):

$$\text{MTFF (MTTF)} = \sum_{i=0}^{n-d} (\text{time to transition from state } i \text{ to state } i+1) \qquad (10\text{-}90)$$

Using this rationale, MTFF (MTTF) relationships can be derived by simply letting the definition of failed state range from its actual value of $(n - d + 1)$ to 1 and summing the values of M_{ss} generated. Under those conditions, MTFF/MTTF derived from (10-78) for a full-on redundant subsystem takes the form

$$\text{MTFF (MTTF)} = \sum_{i=d}^{n} \left(\frac{i!(n-i)!}{i\lambda^{n-i+1}\mu^i} \right) \sum_{j=0}^{n-i} \frac{\mu^{n-j}\lambda^j}{(n-j)!j!} \tag{10-91}$$

MTFF (MTTF) derived from (10-86) for a standby redundant subsystem takes the form

$$\text{MTFF (MTTF)} = \sum_{j=0}^{n-d} j! \sum_{i=0}^{j} \frac{\mu^{j-i}}{i!(d\lambda)^{j+1-i}} \tag{10-92}$$

Examples Determine the MTTF for the example in the section on steady-state MTBF for full-on subsystems.

Proceed by first defining all possible states for that example:

State	Units operating	Failed (in repair)
0	4	0
1	3	1
2	2	2
3	1	3
4	0	4

The M_{ss} value calculated in the example provided us with the average time required to go from state 2 to state 3 as 118.75 hours.

We now assume that the failure state is state 2 and we compute the average time to transition from state 1 to state 2. Under those conditions, M_{ss} is computed pretending that $n = 4$ and $d = 3$. (This corresponds to pretending that state $(n - d) =$ state 2 is a failed state.)

Computing M_{ss} for that situation, the result below is generated:

$$M_{ss} = [3!/3(0.01)^2(0.015)^3][(0.015)^4/4! + (0.015)^3(0.01)/3!]$$

$$= 45.83$$

signifying that on the average it will take 45.83 hours to transition from state 1 to state 2.

Assume now that the failure state is state 1 and compute the average time to transition from state 0 to state 1. Under those conditions, pretend that

$n = 4$ and $d = 4$:

$$M_{ss} = \frac{d!}{d\lambda\mu^4} \frac{\mu^4}{n!} = \frac{d!}{d\lambda n!}$$

$M_{ss} = 1/d\lambda$, which is equal to the MTTF for a subsystem made up of four components connected in series. Since in this case $d = 4$, $\lambda = 0.01$,

$$M_{ss} = 1/0.04 = 25$$

signifying that on the average it will take 25 hours to transition from state 0 to state 1.

We can now evaluate MTFF (MTTF) as

$$\text{MTFF (MTTF)} = 25 + 45.83 + 118.75 = 214.58$$

Example Compute the MTTF for the example in the section on steady-state MTBF for standby subsystems. Defining all possible states for that example:

State	Units operating	In standby	Failed (in repair)
0	2	2	0
1	2	1	1
2	2	0	2
3	1	0	3
4	0	0	4

The M_{ss} value calculated previously for this configuration provided us with the average time required to go from state 2 (state $(n - d)$) to state 3 (state $(n - d + 1)$) as 181.3 hours.

Assume now that the failure state is state 2 and compute the average time to transition from state 1 (state $(n - d - 1)$) to state 2 (state $(n - d)$). Applying (10-86) as before, and substituting $(n - d - 1)$ for $(n - d)$,

$$M_{ss} = (n - d - 1)! \sum_{i=0}^{n-d-1} \frac{\mu^{n-d-1-i}}{i!(d\lambda)^{n-d-i}} = \frac{\mu}{(d\lambda)^2} + \frac{\mu^2}{(d\lambda)^3}$$

$$= [0.015/(0.02)^2] + [(0.015)^2/(0.02)^3]$$

$$= 37.50 + 28.13 = 65.63 \text{ hours}$$

Assuming now that the failure state is state 1, compute the average time to transition from state 0 (state $(n - d - 2)$) to state 1 (state $(n - d - 1)$).

Applying (10-86) as before, and substituting $(n - d - 2)$ for $(n - d)$,

$$M_{ss} = (n - d - 2)! \sum_{i=0}^{n-d-2} \frac{\mu^{n-d-2-i}}{i!(d\lambda)^{n-d-1-i}} = \frac{1}{d\lambda} = 1/0.02 = 50 \text{ hours}$$

We can now evaluate MTFF (MTTF) as

$$\text{MTFF (MTTF)} = 50 + 65.63 + 181.3 = 296.93 \text{ hours}$$

References
1. Born, F. (1988). "Reliability Impact of Diagnostics on Fault Tolerance." Proceedings, 1988 ATE and Instrumentation East Symposium.
2. Klion, J. (1977). *A Redundancy Notebook.* RADC-TR-77-287. Report available from the Defense Documentation Center, Report AD-A050837.
3. Einhorn, S. and Plotkin, M. (1963). *Reliability Prediction for Degradable and Non-degradable Systems.* ESD Technical Documentary Report ESD-TDR-63-642. Also see *IEEE Transactions on Reliability*, March 1965.
4. Applebaum, S. (1965). "Steady State reliability of Systems of Mutually Independent Subsystems." *IEEE Transactions on Reliability*, March 1965.

11

Reliability Design Technologies

Reliability design technology is composed of those reliability engineering tasks necessary to translate reliability objectives, decisions made with respect to the parts program, and results of reliability prediction and evaluation into the final product to be tested and demonstrated. It includes performance of failure modes and effects analysis, sneak circuit analysis, thermal analysis, part derating processes, environmental stress screening, and reliability growth efforts. Each such task has, through the years, evolved into almost a unique speciality. While the general reliability engineer defines the needs of the program with respect to each task and may indeed be capable of performing each task in a rudimentary fashion, the actual performance of each task is usually performed by specialist designers in each particular task area.

Over the years a significant body of literature has developed for each such speciality task, from government standards to technical reports and texts. A detailed technical discussion of each task would be beyond both the scope of this text and the needs of its average reader. Instead, the approach taken in what follows will be:

- To provide a description of the task in terms of its payoff to the reliability programs.
- To provide the basic mechanics associated with task performance such that the reader will be able to perform at least rudimentary exercises of each task and understand the efforts of the "specialists."
- To provide key references associated with each task area.

11.1 FAILURE MODES AND EFFECTS ANALYSIS (FMEA)

The failure modes and effects analysis (FMEA) is basically a design review process which provides for a systematic examination of all conceivable failures. The need for the review process is a direct result of the growing complexities and makeup of modern-day systems and their development processes. Separate engineering teams are assigned the responsibilities for the design of different portions of a given system or subsystem. The products developed by each team must provide outputs and receive inputs from portions of the system developed by other teams. Even with the best of integration specifications, the probability is significant that a failure mode in one portion of a system can have unforseen effects on the performance of other parts of the system. Even for those item designs which are the responsibility of a single engineering team, the consequences of a given failure mode within the item can go unrecognized.

The need for a FMEA is not equally critical for all items. Need is dependent on the nature of the item, its intended use, and its characteristics. These include:

- Safety considerations: the degree to which the item is capable of causing damage or injury to personnel or other items or facilities. Obviously, *a low-voltage item* with *no outputs* to other items would have a lesser need for a FMEA than a high-voltage power supply.
- Interface and integration characteristics: the degree to which the item interfaces and provides outputs to other items (i.e., its criticality).
- Repair/discard at failure policy. If an item is to be discarded at failure, there is less need for a FMEA *on the components making up the item* than when the item is to be repairable (e.g., elimination of secondary failures or fail-safe provisions to protect expensive components within the item are no longer necessary).
- Design complexity and number of parts making up the item. The simpler the design, the fewer the parts, the less the probability that the consequences of a failure mode have been ignored. By the same token, the more complex the design, the more parts, the more the integration of resources and functions within the item, the greater the probability that failure mode consequences have been missed.
- Logistics, maintenance and test system plans/programs needs. While such needs are outside the direct purvue of reliability, the organizations responsible for those functions may rely on the reliability program to provide FMEA data necessary for certain of their functions.

The necessary content, scope and detail of the FMEA effort for each application depend on:

- The phase of development of the item. To be sure, a FMEA can be performed during any phase, but its character, detail and cost will vary in accord with the phase. A FMEA for a system can be performed as early as *concept formulation* and account for only functional failure modes among different subsystems. The results can be used to determine which subsystems require redundancy or can support integrated design safely, fail-safe design, or greater attention to reliability than others. As the development process continues, more explicit failure mode and effect analyses can be applied at the subsystem, equipment, component, and part levels and determinations and findings become more design-characteristic and hardware-specific.
- The level to which the FMEA is to be performed. The lower the indenture level, the greater the quantity of items which must be subjected to the FMEA process and the more effort required to acquire and operate on failure mode data and generate failure effects information.
- The degree of new and existing design associated with the system. Most existing design will have documented failure modes by virtue of prior use or prior FMEAs performed.

The FMEA is one of the most controversial tasks in the area of reliability. It can be costly and time-consuming. It has been charged with having little or no impact on the design/development program and with being subject to duplication of effort. It requires the use of engineering judgments and some failure modes can be unintentionally omitted. Adhering to a few simple criteria, however, can help in avoiding such pitfalls:

1. Do not perform a FMEA to a lower level than necessary for the needs of a program.
2. Do not perform a FMEA if its results will be available too late to make design changes, unless engineering change proposals are encouraged.
3. Think twice about performing a FMEA when all the data needed is not readily available or cannot be practically generated, and when too many unsubstantiated assumptions must be made with respect to failure modes and effects (e.g., parts, VLSI, microcomputers) or failure rates.
4. FMEAs are commonly called out not only for reliability, but for safety, logistics testability, and maintainability programs as well. Often, several independent FMEAs are performed on the same item instead of a single coordinated FMEA.

One factor which often causes confusion in the interpretation and performance of a FMEA is the definition of a *failure mode* and that of a *failure effect*. We commonly assume that each failure of a part or connection changes the *output* of the lowest-level block containing the part/connection (e.g., a circuit). The output is the *failure effect*. The failure effect is propagated in the form of inputs to other blocks, which in turn propagate other unique failure effects at their outputs due to the acquisition of an incorrect input, and so on. This process continues until the results of the part/connection failure mode are represented as an ultimate observed failure effect at the system or equipment output. When a FMEA is performed *starting at a level higher than the part/connector level*, we are, in fact often hypothesizing not failure modes but rather the characteristics of the failure effect *output* of another system block.

Hence, in actuality, we are performing a *failure characteristic and effects analysis* at those levels. This is nevertheless termed a FMEA.

When a FMEA is applied properly, it identifies failures which cause secondary malfunctions and critically impact mission success, personnel, and system safety. (See Refs. 1 to 4 for more extensive and detailed discussions of FMEA details than will be covered here.)

As alluded to previously, FMEA is a required task under many program plans associated not only with reliability but with maintainability, test system design, and logistics. As a matter of fact, it is called out in current military standards in the area of safety, reliability, maintainability, test system design, and logistics. It is the subject of a standard itself, Mil-Std-1629 described previously in Section 3.5.

Over the years, the techniques associated with performing FMEA have taken on many different forms and, indeed, software programs for the implementation of specific FMEA techniques are generally available. All techniques, however, require that engineering judgments be independently made and used as inputs. Most are variations of the two most commonly applied procedures, the *tabular FMEA* and the *fault tree analysis*. The two procedures have the following in common:

1. Each requires the development of sets of functional block diagrams to provide the visibility needed to perform the analysis. Each such set represents a particular level of system indenture, i.e., the subsystems making up the system, the equipments making up each subsystem, the components making up each equipment, the parts making up each component. Each block diagram must depict the operational dependencies and interdependencies among the blocks making up the diagram. This can be represented in the form of inputs and outputs, where an output (see Fig. 11-1) signifies either performance of a function or a transfer of data, etc., from that block; an input signifies a transfer of a signal, energy or data, etc., to the block which is necessary in order to perform one or more of the block's needed functions.

2. Each assumes that a failure may affect the performance of a block.
3. Each can be initially applied at any level of indenture of the system. As a result, each can be applied early in the design to provide general system-level guidance or during circuit-level design in full-scale engineering development.
4. Each can take either a hardware approach or a functional approach or an approach which is a combination of the two, where:
   a. A hardware approach is keyed to specific hardware items (blocks) and the specific failure modes of the hardware (i.e., a resistor fails open or short).
   b. A functional approach is one which is keyed to the specific functions which are provided by a block where a "failure mode" results in a loss or diminished capability to provide one or more such functions (i.e., signal transfer cannot be effected).

However, the two take diametrically opposed approaches to achieving their objectives. The tabular FMEA takes a *bottom-up* approach. That is to say, given a particular failure mode in one of the components in Fig. 11-1, what is the effect of the failure mode on the output of the failed component, on the performance of the components which receive that output, or on the output of any block diagram? The fault tree analysis takes a *top-down* approach. That is to say, given an unsatisfactory effect at the output of the

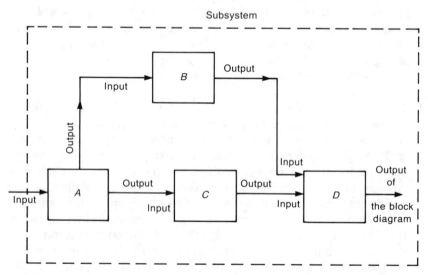

FIGURE 11-1.   Block diagram of four equipments making up a subsystem indenture level.

block diagram, what combination of failure modes in the components could have produced it?

Both procedures provide the same answers, and a combination of the two procedures can be used in performing an FMEA on the same system.

11.1.1   The Tabular FMEA: Makeup and Process

Given that the block diagram with its functional dependencies has been developed, a systematic process of linking the modes of failure of each component with their effects on system performance is begun. This is generally accomplished through a FMEA worksheet. While there is no one standard composition for the FMEA worksheet, most include the factors and general format of that in Chart 11-1.

Column 1   Provides the serial number/nomenclature provided a given block in the subject block diagram set

Column 2   Identifies the function that the block provides

Column 3   Lists all the failure modes that the block can take, all those conceivable, including those caused by operational and environmental stresses. Information on such stresses may be found in the mission profile and environmental profile which were discussed earlier.

Column 4   Identifies the potential causes within the block that can result in each failure mode identified in the preceeding column. Initially a great deal of specificity is not required. If the block, for example, were composed of a number of equipments, A_1 to A_n, the column could have the notation "failures in equipments A_1 or A_2 or A_3"; or if insufficient information on block makeup were available, we could simply say that single failures in one or more of the equipments making up the block could result in that mode. Later FMEAs to lower levels will link that mode of failure of the block to specific lower-level items and their failure modes. At that time the worksheet will be updated. The entry in this column will ultimately provide us with a track for top-down visibility.

Column 5   Identification of different mission phases and different operating environments, for example, takeoff, flight, and landing for an aircraft, or surface and undersea cruise for a submarine. The criticality of a failure will change as operating environment and mission phase change. A given failure mode when a submarine is surfaced may have little impact on mission performance or

CHART 11-1 Failure Mode and Effects Analysis Work Sheet

Block ID Number	Block Function	Failure Modes	Potential Causes	Mission Phase	Failure Effects		Compensating Provisions	Severity	
					Local Effects	End Effect		With	Without

safety. The same failure mode when it is submerged can be critical
to both.

Column 6 Identifies the effects of the failure modes:

     a. *Locally*: the effect on the function of the immediate recipient
       (s) of the output

     b. *Distantly*: the effect on the output of the block diagram. In
       the event that no effect will be apparent due to a compensating
       provision (i.e., a feature such as redundancy, acceptability of
       reduced performance for short periods of time) or if the effect
       can be minimized in one way or another, it should be so noted
       here and described in the next column.

Column 7 Identifies the compensating provisions in the design or operational
scenario, if any, which are associated with a particular failure
mode. It may include special provisions for maintenance, inspection, or maximum-quality parts.

Column 8 Identifies the severity of the effect both with and without
consideration of the compensating provision. For example, Mil-
Std-1629, "Failure Modes Effect Analysis," suggests four classes
of failure effects: actual loss, probable loss, possible loss, and no
effect. If desired, a narrative describing criticality consequences
may be provided.

Identification of all potential failure modes or failure effect characteristics
associated with each block and the effect of such modes on the performance
of each recipient of outputs from the failed block is accomplished with the
aid of the system/equipment/component designers and/or FMEA specialists.

Examples of the types of "failure modes" that may be identified include:
premature operation; failure to operate on time; intermittent operation;
failure to cease operation; loss of power; loss of speed; noise or distortion;
degraded or out of tolerance output; open; short.

The effort necessary to provide the information needed by the worksheet
forms the makeup of the FMEA process. The worksheet documents the
results of the review process and its findings. Conclusions as to criticality of
each failure mode for each block and what compensating provisions, if any,
are included in the design or elsewhere to mitigate its effect, are found in
the last column. These may be used to develop a FMEA summary of all
block failure modes and their potential impacts.

In some cases, as will be seen in the example depicted in Chart 11-2, a
ninth and tenth column can be added indicating the failure rate associated
with each block and the proportion of that block's (say block i's) failures
attributable to a given failure mode j, denoted as P_{rij}. If that information is

CHART 11-2 Example of a Completed FMEA Worksheet With Compensating Provisions Omitted

Block ID #	Block Function	Failure Modes	Potential Cause	Failure Effects Local	Failure Effects End	Operating Environment	Severity of Effect	Block λ	Probability of Failure Mode
A	Data acquisition	No signal	A_1 failed	No input to B	System will not function	IFR	1	0.001	0.2
						VFR	0.2		
		Continuous noise	A_3 or A_4 failed	No input to B	System will not function	IFR	1		0.1
						VFR	0.2		
		Signal/noise ratio low	A_2 failed	Function B must ask for repeats saturating the system, and providing complete sets of data to C slower	System slows down	IFR	0.5		0.3
						VFR	0		
		Works intermittently	A_4 or A_1 failed	Function B not affected	System slows down	IFR	0.5		0.4
						VFR	0		
B	Data processing	Capacity reduced	B_1 or B_2 failed	Function C receives information slower	System slows down	IFR	0.5	0.0005	0.3
						VFR	0		

Subsystem	Failure mode	Cause	Effect		IFR/VFR value		
	Speed reduced	B_1 or B_2 failed	Function C receives information slower	System slows down	IFR 0.5 / VFR 0		0.2
	No data processed	B_1 and B_2 or B_3 and B_4 failed	No input to C	System will not function	IFR 1 / VFR 0.2		0.2
	Data out of synch	B_3 or B_4 failed	Function C receives garbled or erroneous information	System functions erroneously	IFR 1 / VFR 0.5		0.3
C Display/ communication	Communication degraded	C_3 or C_4 failed		System slows or parts will not function	IFR 0.4 / VFR 0	0.001	0.4
	Communication loss	C_3 and C_4 failed		System will not function	IFR 1 / VFR 0.2		0.3
	Loss of major display portion of S&S	C_1 failed		System works slower, less effectively with local displays	IFR 0.6 / VFR 0.1		0.3

available, then the following can be calculated:

1. The probability P_{ij} of the occurrence of that failure mode (i.e., the proportion of item failures attributable to failure mode j of block i) within the item represented by the block diagram set.
2. The probability P_E of the occurrence of any particular failure effect that can result from that failure mode or any combination of failure modes from different blocks.

For a series-connected system with exponential distribution of time to failure, where

λ_i/λ_e = given a failure in the block diagram set, the proportion of failures that will be in block i

P_{rij} = the proportion of failures in block i that take the form of failure mode j

the proportion of block diagram set failures associated with failure mode j *in block i, P_{ij},* is equal to

$$P_{ij} = \frac{P_{rij}\lambda_i}{\lambda_e} \tag{11-1}$$

where

λ_i = failure rate of block i
$\lambda_e = \sum^n \lambda_i$ = total failure rate of block diagram set
n = number of blocks in block diagram set

Assume that it is possible that more than one "failure mode" associated with *a given block output* can result in the *same failure effect* on the output of the block diagram set. Assume that it is possible that "failure modes" associated with the outputs of *more than one block* can result in the *same failure effect* at the output of the block diagram set. Then the proportion of failures which result in a given effect or which result in a given severity classification, P_E, can be evaluated as

$$P_E = \sum_i^y \sum_j^{h_i} \frac{P_{rij}\lambda_i}{\lambda_e} \tag{11-2}$$

or

$$P_E = \frac{1}{\lambda_e} \sum_i^y \sum_j^{h_i} P_{rij}\lambda_i$$

where

P_E = the probability that the item represented by the block diagram set, will experience a given failure effect or severity level

y = the number of blocks that have failure modes that can result in the given failure effect or severity level

h_i = the number of different failure modes within block i that will result in the given failure effect or severity level

Equation (11-2) can be used (1) to determine the probability of occurrence of failure *severity classes* equal to and above a given level, and (2) as a basis for determining the most cost-effective course of action to take in reducing the overall criticality associated with a given item or subsystem; in evaluating the payoff associated with various compensating features which are capable of reducing the value of P_E (see Section 7.2.3).

For example, take the system depicted in Fig. 11-2, a rudimentary traffic control system comprised of three subsystems: A is a data acquisition subsystem such as a radar, which in turn is made up of four components, A_1, A_2, A_3, and A_4. B is a data processing subsystem which collects and processes the data for A and transfers it in proper format to subsystem C. B is made up of four components, B_1, B_2, B_3, and B_4. C is a display and communications subsystem that receives information from B, acts on it and controls traffic through a communications link. C is made up of four components, C_1, C_2, C_3, and C_4. A FMEA and criticality analysis of the system will be performed for two conditions of use, one an environment requiring use of instrument flight rules (IFR), the other an environment allowing the use of visual flight rules (VFR). Severity or criticality of a given effect will be ranked between 0 and 1, where the most severe effect = 1.

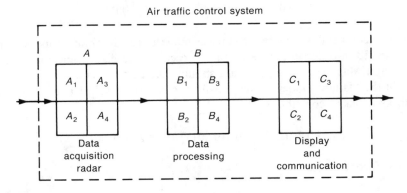

FIGURE 11-2.   Rudimentary air traffic control system.

Chart 11-2 shows an example of the content and organization of a subsystem level FMEA worksheet for such a system where the critical event is defined as *system incapable of handling traffic load.*

Note that P_E will vary as a function of condition of use. That is to say, the proportion of system failures that will result in a failure severity level of 1 during IFR operation is greater than that which would be experienced during VFR operation. This illustrates an important characteristic; that criticality is as much a function of use as it is of design. An item may provide safe, effective, and cost-efficient operation in one environment and the exact opposite operation in another.

When a value, P_E, associated with a too severe effect is too large, the block or blocks which contribute the greatest to that value must be identified and design compensations considered. The contribution of block i to P_E equals

$$P_{Ei} = \frac{1}{\lambda_e} \sum_j^{h_i} P_{rij}\lambda_i \tag{11-3}$$

where

$$P_E = \sum_i^y P_{Ei}$$

In this case, the proportion of system failures, P_E, that will result in a failure severity level of 1 during IFR operation is equal to

$$P_E = \frac{1}{\lambda_e} \sum_i^y \sum_j^{h_i} P_{rij}\lambda_i = [1/0.0025][0.2(0.001) + 0.1(0.001)$$

$$+ 0.2(0.0005) + 0.3(0.0005) + 0.3(0.001)] = 0.34$$

In the event that this is considered unacceptable and engineering compensation is required in order to reduce P_E to an acceptable level, a determination of how and where such compensation should take place is required.

Of the "failure modes" which result in a severity level of 1 for all three blocks, block A contributes

$$[0.2(0.001) + 0.1(0.001)]/0.00025 = 0.12$$

block B contributes

$$[0.2(0.0005) + 0.3(0.0005)]/0.0025 = 0.10$$

and block C contributes

$$[0.3(0.001)]/0.0025 = 0.12$$

Hence, all things equal, effort to reduce the severity of failure effects in blocks A or C would provide a better overall payoff than if B were chosen. Compensating provisions which could be considered include use of redundancy, application of better quality parts, elimination of that failure mode by design efforts.

Extensions to basic FMEA procedures have been developed over the years, for example, *matrix* FMEA.[5] Matrix FMEA is a systematic trace process. It assigns modes of failure to each part and connection; determines their effect on the characteristics of the output of the lowest-level block; defines a finite set of output characteristics at that block level consistent with all modes of part/connection failure; considers those as inputs to the next higher-order blocks which in turn generate outputs based on such inputs, etc., until the impacts of the failure modes on the parts/connections are traced ultimately to their effect on system output. Computer programs to facilitate the application of matrix FMEA procedures as well as the basic forms of FMEA are readily available through a number of commercial reliability software organizations.

11.1.2   The Fault Tree Analysis: Makeup and Process

Fault tree analysis (FTA) by its nature takes a more general, functional form than a tabular FMEA. However, for systems which have fault-tolerant features or multiple causes resulting in the same failure effect, it is more sensitive to combinational effects of malfunction than is a tabular FMEA approach. In all cases it allows for a better direct visibility of a failure cause given a failure effect than does tabular FMEA.

Using the block diagram and its functional dependencies, a systematic process of linking all important *failure effects* in a given block (or block diagram set) with the types and locations of failures and failure modes which can result in such failure effects is instituted. When completed, the results are represented in a form that resembles a tree and its branches. The "fault tree" constitutes a graphical representation of cause and effect. In simple terms it depicts the inputs necessary to define a given output. If we define a *failure effect* as an *output* of a block diagram set (BDS), it describes the particular failures and combinations of failures in the members of a BDS necessary (as inputs) to produce that failure effect. If all discernible output failure effects are identified, the process will identify all combinations of potential failures and failure modes within the BDS required to produce each.

A block's status, whether it is failed or nonfailed, special conditions present to allow for proper use of a block's output, and other conditions which affect the output characteristics of the BDS and its blocks are all defined as "events." They are so termed because *in the event* that they occur or do not occur, the output is affected. Events are connected by a Boolean logic consisting of gate functions (AND, OR, Priority AND, Exclusive OR, etc.) which define the scenario of events associated with a given failure effect.

Chart 11-3A provides some common examples of "events" routinely used in a fault tree analysis (FTA). Chart 11-3B provides common examples of logical gate functions frequently used in FTA.

The FTA process starts at the output of each BDS. A failure effect, $F(P_0)$, is chosen which represents a failure to function or an undesirable state for the BDS. Examination of the specification governing the function to be performed by the BDS can provide the information needed to define such failure effect(s). Next, the immediate events which can cause the effect are identified. Both of these first steps are sometimes best done by a group effort, where the group contains representatives from the design team responsible for the BDS development and from system engineering. The result of these steps provides the information for initial FTA formulation.

For example, one general failure effect $F(P_0)$, has been determined for the air traffic control center example discussed earlier:

$$F(P_0) = \text{"System is Incapable of Handling Traffic Load"}$$

The group effort has defined three immediate causes: block A wil not function; block B will not function, or block C will not function. $F(P_0)$ can be expressed in Boolean form as

$$F(P_0) = \bar{A} + \bar{B} + \bar{C}$$

that is to say, the system will not function if A has failed or B has failed or C has failed, where $\bar{A}, \bar{B}, \bar{C}$ denote failed and can be depicted as the FTA in Fig. 11-3.

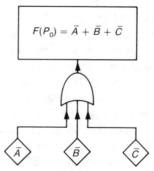

FIGURE 11-3.   Fault tree analysis to the first level.

CHART 11-3A Terms/Symbols Commonly Used in Fault Tree Analysis

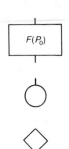

An event that results from a combination of failed and/or operating blocks.
$F(P_0)$ represents the postulated output of the block diagram set.

A basic failure event, the failure of a part of a connection.

An immediate failure event reflected in the output of block diagram set caused by failure of a lower-level block, or a basic failure event.

CHART 11-3B Examples of Logic Gate Functions Commonly used in Fault Tree Analysis

AND function gate—the failure effect would have resulted only if all outputs had failure events somewhere in their chain.

OR function gate—the failure effect would have resulted only if input 1 or 2 or 3 *out of a possible n* had a failure event somewhere in their chain.

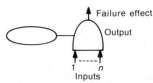

Priority AND functional gate—the failure effect would have resulted only if all inputs had a failure event in their chain and if, for example, input #1 failed first.

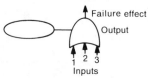

Exclusive OR functional gate—the failure effect would have resulted only if 1 or 2 or 3 had failed, where for example, only the first input to contain a failure in its chain provides the desired output.

Extending the FTA further, treating each block as an individual BDS, the "group" comes to the conclusion that in order for block A, B, or C to provide a "nonfunctioning" failure effect at their respective outputs:

- For block A, A_1 or A_3 or A_4 must be failed.
- For blocks B, B_1 and B_2 or B_3 and B_4 must be failed.
- For block C, C_3 and C_4 must be failed.

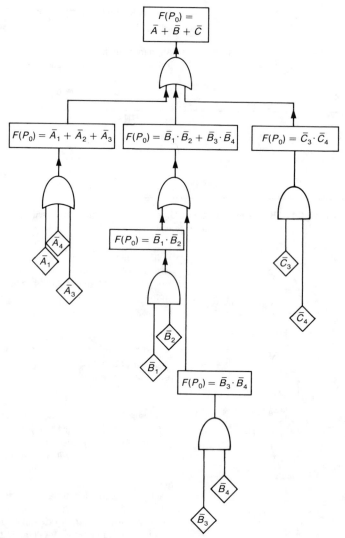

FIGURE 11-4.   Fault tree analysis to a second level.

As before, denoting failed sub blocks as $\bar{A}_i$, $\bar{B}_i$, $\bar{C}_i$ the fault tree of Fig. 11-3 is extended to result in Fig. 11-4. Note that the FTA makes it very obvious that a failure in unit A_1 or unit A_2 or unit A_3 will singly cause the system to cease its function. How obvious was that fact from the FMEA performed on the same system?

Fault tree analysis can be performed using various degrees of rigor, from sophisticated procedures which apply operations research techniques and algorithms that appear frequently in reliability journals (see Refs. 6, 7, and 8) to more qualitative procedures such as that illustrated above.

Software is available both from commercial sources and in the public domain to assist in FTA. Some examples of software which have been mentioned in other publications as being in the public domain which are applicable to performing qualitative FTA include PREP, ELRAFT, and SETS (see Ref. 4 for additional information).

11.2 SNEAK CIRCUIT ANALYSIS: ITS LINK TO RELIABILITY; WHAT IT IS AND ITS PAYOFF

Sneak circuit analysis (SCA) is an engineering tool used to identify inadvertently designed-in characteristics which inhibit desired item actions and functions and/or produce unwanted or unexpected actions and functions. While part of the reliability program (see Mil-Std-785, "*Reliability Program Plan*," discussed earlier), it is not employed directly for traditional reliability purposes. While relatable by many to a type of FMEA process[4] and integratable with the FMEA program,[9] it is not based on determining the effects of failure or on the identification of failures which result in undesirable effects.

SCA is related to the objectives of the reliability program by analysis, identification, and connection of unintended modes of operation in an item, in the absence of actual component or part failures, where such modes of operation can *have the same effect on item function as an actual failure*.

A sneak circuit can be catastrophic or helpful. Both types result when the item is subjected to combinations of inputs and conditions which may be directly operator induced; induced through the operation of other subsystems/ components; induced through a particular combination of operating environments; or induced through any combination of the above.

A catastrophic sneak circuit can cause the inadvertent shutdown of an important function at a critical time, or cause an unexpected loss of a critical portion of a control system (e.g., nuclear power control, weapon system control, aircraft control system, traffic control system), or cause aberrations in subsystem behavior at a given point in time.

A helpful sneak circuit can allow for alternate means for performing a function which could be used either as a redundant means to perform the same function or as rationale for the elimination of unnecessary parts. It can also be used to enhance design if it provides for the performance of additional functions that were not planned.

SCA provides payoffs in terms of identifying missed or overlooked problems which would otherwise result in increased downstream engineering change costs, incur delays to the program schedule, or reduce reliability or safety. Its pitfalls, however, are similar to those of FMEA:

- It, too, can be expensive. The RADC report, "Sneak Analysis Application Guidelines,"[10] tabulated the costs of over 100 hardware and software SCAs. The cost ranged from 0.0001% to 0.4% of total program cost.
- It can be performed to an inappropriate depth. SCA functions best at the most detailed level possible. Often system functional block diagrams, which are available relatively early in the design, represent just a general concept of information transfer and flow from block to block as opposed to the actual transfer and flow represented by circuit schematics and wire lists. Information needed for useful SCA application can be available only when it is too late to take practical action.
- Last, but not least, the fact that a SCA has been performed provides no guarantee that all sneak paths have been identified.

Nonetheless, performed in accord with some simple engineering rules and logic, it is possible to assess general SCA applicability, avoid the worst pitfalls, and accrue the payoffs intended.

To begin, let us describe by way of example what a sneak circuit is. Figure 11-5 provides a simple illustration of a sneak circuit. Assume a laboratory equipped with an air conditioning/heating system with a dedicated power source and a switch, S_1 for control, and an electrostatic air cleaner with a dedicated power source and switch, S_2; a separate humidity control unit is also available so that it can be used with the air conditioning/heating subsystem, the air cleaner, or both. Switches S_3 and S_4 control the application of that unit.

If it is desired to have just the humidity control system and the air cleaner subsystem working, switches S_2 and S_4 are closed. However, if S_3 is inadvertently left closed, the air conditioning/heating system will function without regard to the position of S_1. Similarly, if it is desired to have just the air conditioning/heating subsystem and the humidifier function, switches S_1 and S_3 are closed. However, if S_4 is inadvertently left closed, the air cleaner will function without regard to the position of S_2. If either sneak circuit occurs, its associated power line will become overloaded and a circuit breaker will be actuated, shutting down the power to a portion of the system.

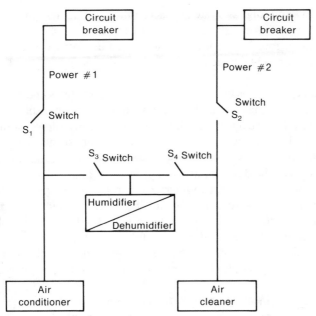

FIGURE 11-5.   A sneak circuit example.

11.2.1 Applications of Sneak Circuit Analysis—On What and When?

Although SCA may be applied to both the hardware and software functions of a system, here discussions will be confined only to hardware applications. For more detail on both the subject and the information which follows, see Refs. 10 and 11, which contain the results of lessons learned on over a hundred SCA projects.

Historically, SCA was introduced in the mid-1960s to support crew safety in manned spacecraft operations, in particular to reduce problems which occur due to latent "sneak" paths in electrical designs. Since that time, SCA application has been extended to such uses as electrical power generation/ distribution/control; aircraft and space systems and subsystems—flight control, landing, propulsion, avionics, weapons control, etc.; missiles; computer systems and networks; environmental control systems; general vehicle operation; subsystem functions—telemetry, instrumentation, etc. Within these categories, SCA has been performed on a total system basis or performed selectively on items or functions within a given system or subsystem. In general, the *typical candidate* for SCA is a system, subsystem, or component which performs an *active function*. This includes systems containing relay logic, analog and digital systems, electromechanical systems, integrated circuits, and microprocessor-based functions.

Some criteria to be taken into account when assessing need, practicality, and priorities for SCA application include:

1. The number of interfaces between and among subsystems and components. Each such item may be, and usually is, developed by a different design team guided by an overall specification which usually allows for maximum flexibility for all design teams. This produces a situation which is conducive to sneak paths.
2. Availability of the data required for a SCA. A high-indenture-level (less difficult) SCA with respect to *interfaces* can usually be performed late in the validation phase or early in full-scale engineering development. A detailed SCA to the component level will usually not be able to be performed until at least midway into full-scale engineering development when detailed design is established. The specific data needs for a detailed component SCA include circuit schematics, prototype drawings, and wire lists.
3. The rate of engineering changes to the baseline design. The greater the number of engineering changes per development phase, the greater the likelihood of sneak path occurrences.
4. The complexity and "newness" of the system. The more complex, the more unique, the greater the opportunity for sneak paths.
5. The criticality of the item to safety and mission performance.

11.2.2 Sneak Circuit Analysis: The Process

Sneak circuit analysis performance is a combination of art and science. It is an art inasmuch as experience and data gained as a result of past SCA projects facilitates and plays a key role in performance. It is a science inasmuch as it lays a network logic over the experience factors. Computer programs have been developed to provide traceability of paths, switches, and connections to functions performed, taking into account various contingencies. In general, it is not for the neophyte. For this reason, many SCAs are subcontracted to organizations which have both experience and the software available to perform SCA effectively. This does not mean that the developing organization can do nothing to minimize sneak circuits; just the opposite is true and it will be discussed in the following section.

In the simplest of terms, SCA involves the logical analysis of energy flow through a network to determine the existence of unintentional energy paths, undesired or unanticipated activation of given components or functions. The analysis involves the definition of topological network trees to represent the flow of energy through a circuit/component/module. The information from which such trees evolve results from the analysis of detailed circuit schematics, prototype production drawings (if available), and wire lists. Energy and signal flow to, from, and within each component are represented on the network tree. Experience has shown that no matter how complex the network tree, it can almost always be broken down into combinations of the five particular patterns shown in Fig. 11-6a: the single line; the ground dome; the power dome; the combination dome; and the H pattern.

Of the five patterns, the one that has been found to contain the most significant number of sneak paths is the H pattern (see Refs. 9 and 10). Figure 11-6b shows a larger network which is partitionable into two single line patterns and one each of the remaining four.

Associated with each pattern or combination of patterns, derived from experience and usually proprietary to the source, are clues, guides, and questions on which the analyst focuses. These pertain to designs, part, and operational characteristics of each pattern and are structured to provide visibility concerning the capability of experiencing a sneak circuit in any component in the pattern.

When a formal SCA is performed, either within the development organization or under a separate subcontract, the following minimum ground rules pertain:

1. Make sure before it starts that a criterion is in place to determine which paths will be analyzed.
2. Provide traceability of each component to its path and traceability of each path to its network tree(s).
3. Document the results of each path analysis and provide identification of "sneak" manifestations found.

11.2.3   Sneak Circuit Analysis: Minimizing Occurrences

Performing an analysis to find sneak circuits and providing engineering changes for their correction is after-the-fact engineering. It would be far more desirable to design a system with no sneak circuits the first time around than to perform a SCA in order to find oversights made earlier. While that objective may be unrealistic at this time, it is not unrealistic to develop guidelines which can be applied during basic design phases and which are capable of reducing the incidences of sneak circuits.

A step in this direction was fostered in 1989 with the publication of the RADC report "Sneak Circuit Analysis for the Common Man."[11] The guidelines in the report were structured to be used by a *design engineer* to *avoid* the most commonly encountered sneak circuits and as an aid to

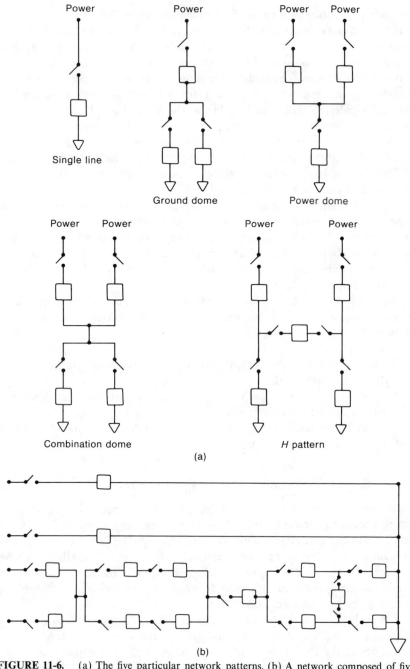

FIGURE 11-6.   (a) The five particular network patterns. (b) A network composed of five particular patterns.

identifying potential sneak circuits. It was not intended to be used as a substitute for a formal SCA, but rather as a simplified method for minimizing the occurrence of sneak circuits. It provides rules/guidelines for avoiding sneak circuits during design; identifying sneak circuits at the functional level; and identifying sneak circuits at the device level. Specifically it provides seven rules for sneak circuit design avoidance, five basic guidelines to help in identifying functional networks commonly associated with sneak conditions, and six basic guides to help in identifying designs commonly associated with sneak conditions.

Condensing the rules for sneak design avoidance, the most common design problems which result in sneak circuits were identified as:

1. Power distribution from two or more sources of power and/or ground
2. Power distribution with ground-side current interruption other than connectors
3. Power distribution with ground-side current interruption including connectors
4. Power connectors
5. Power distribution to more than one load
6. Power distribution from two or more sources of power to essential loads
7. Manually controlled switching devices

11.3 THERMAL MANAGEMENT AND ANALYSIS: THEIR ROLES IN RELIABILITY PROGRAMS

As was seen in Chapter 8, the temperature to which a part is exposed has an influence on its failure rate. With respect to reliability needs, the purpose of the thermal management/analysis program is to provide visibility and control concerning thermal stresses which affect part, component, and system failure rate. The achievement of such objectives requires that the efforts of thermal analysts be integrated into reliability efforts from planning to prediction, design, and test.

A report published in 1982, "RADC Thermal Guide for Reliability Engineers," [12] provides an introduction to the basic background, information, and tools necessary for evaluating and managing those aspects of a thermal design program appropriate to reliability needs. Much of the technical detail covered in this subsection is extracted from that report. The purpose of this section is:

• To provide procedures which can be utilized to assess the impact of temperature on the failure rate of any given part or component.

- To provide guidance on how thermal analysis may be integrated into the reliability program.
- To provide some rudimentary insights, guides, and rules about applicability and use of the most common thermal design procedures and strategies which may be helpful when interacting with a thermal analyst.

11.3.1 The Relationship of Temperature to the Reliability of Electronic Parts*

In Chapter 5 we discussed detailed failure rate prediction methods which defined part failure rate, λ_p, as a function of different factors which influenced reliability. For each such prediction relationship at least one such factor was representative of the impact of part temperature on failure rate.

In order to determine a first-cut estimate of the sensitivity of the failure rate of a particular part to temperature and in order to provide information on thermal boundaries necessary to achieve a given range of failure rate, a logical place to start would be the detailed part failure-rate equations.

Illustrations will be provided through the use of failure-rate equations for microcircuits, transistors, and resistors extracted from Mil-Hdbk-217E.

11.3.1.1 Failure Rate Sensitivity to Temperature for Monolithic MOS Digital Devices
Representing part failure rate by λ_p and using the microcircuit failure-rate equation from Mil-Hdbk-217E,

$$\lambda_p = \Pi_q \Pi_l (c_1 \Pi_t \Pi_v + c_2 \Pi_e) \qquad (11\text{-}4)$$

where
Π_l = a learning factor
Π_q = a quality factor
Π_t = a temperature-based failure rate acceleration factor based on, t_j, an estimated/measured value of *worst-case junction temperature*
Π_v = a voltage stress derating factor (sometimes influenced by t_j)
Π_e = an application environment factor
C_1 = a circuit complexity factor
C_2 = a package complexity factor

---

*Different detailed prediction procedures (see Section 8.4.1) put different emphasis on temperature effects on device failure rates. As a result, failure-rate sensitivity to temperature effects is no less debatable than the differences among prediction procedures. As indicated previously, the present most commonly used prediction procedure for both the U.S. Military and general commercial use is that detailed in Mil-Hdbk-217, "Reliability Prediction of Electronic Equipment." As a result, the prediction models associated with that document will be used for illustrative purposes.

If we define

λ_{pj} = the failure rate at any given worst-case value of t_j
λ_p = the failure rate at $t_j = 25°C$
Π_{tj} = the value of Π_t at any given value of worst-case t_j
Π_{to} = the value of Π_t at a worst-case $t_j = 25°C$

then the ratio

$$\frac{\lambda_{pj}}{\lambda_p} = \frac{C_1 \Pi_{tj} \Pi_v + C_2 \Pi_e}{C_1 \Pi_{to} \Pi_v + C_2 \Pi_e} \tag{11-5}$$

denotes the proportional increase in failure rate due to junction temperature over that experienced at $t_j = 25°C$.

If Mil-Hdbk-217E is applied and it is assumed, for example, that the device is hermetic ASTTL, CML, TTL, HTTL, FTTL, DTL, ECL, or ALSTTL, and that

$\Pi_v = 1$, indicating a maximum supply voltage requirement of less than
    12 volts
$\Pi_e = 3$, indicating an airborne transport inhabited environment
$C_1 = 0.02$, indicating a device complexity between 100 and 1000 gates
$C_2 = 0.009$, indicating a package complexity failure rate factor associated
    with 24 functional pins
$\Pi_{to} = 0.1$ when $t_j = 25°C$

then

$$\lambda_{pj}/\lambda_p = \frac{0.02\Pi_{tj} + (0.009)(3)}{0.02(0.1) + (0.009)(3)}$$

$$= \frac{0.02\Pi_{tj} + 0.027}{0.002 + 0.027}$$

and failure rate variations with junction temperature can be assessed.

Mil-Hdbk-217E provides tables equating values of t_j to Π_{tj}. Table 11-1 provides a condensed table of values for this particular device type.

Figure 11-7a represents λ_{pj}/λ_p values as a function of t_j. As can be seen, failure rate increases by 16% as t_j goes from 25°C to 50°C but increases by almost 50% as t_j approaches 70°C. It can also be seen that the failure rate at a t_j of 95°C is approximately twice that of the failure rate at $t_j = 55°C$.

Similar procedures can be applied to any microcircuit model by defining approximate C_1, C_2, Π_e, Π_v factors and applying (11-5). While we have used the failure-rate model from Mil-Hdbk-217E, a similar procedure can be developed for any model which has temperature sensitive factors.

TABLE 11-1 Π_{t_j} **Values Associated with** t_j **Worst Case Junction Temperature for Monolithic Hermetic ASTTL, CML, TTL, HTTL, FTTL, DTL, ECL, and ALSTTL devices**

$t_j(°C)$	Π_{t_j}
25	0.10
30	0.13
35	0.17
40	0.21
45	0.27
50	0.33
55	0.41
65	0.63
70	0.77
85	1.40
95	1.90
105	2.70
120	4.30
135	6.60

11.3.1.2 Failure Rate Sensitivity to Temperature for Transistors

Again representing part failure rate by λ_p and using the conventional transistor failure rate model from Mil-Hdbk-217E,

$$\frac{\lambda_{pj}}{\lambda_p} = \frac{\lambda_{bj}(\Pi_e \Pi_a \Pi_q \Pi_r \Pi_{s2} \Pi_c)}{[\lambda_{bo}(\Pi_e \Pi_a \Pi_q \Pi_r \Pi_{s2} \Pi_c)]}$$

$$= \frac{\lambda_{bj}}{\lambda_b} \qquad (11\text{-}6)$$

where

λ_{bj} = the base failure rate of the transistor, which is a function of device *ambient* temperature T_j

λ_{bo} = the base failure rate of the transistor when ambient operating temperature = 20°C

For example, for a silicon NPN transistor as represented in Mil-Hdbk-217E, λ_{bj} will vary with ambient temperature as in Table 11-2.

Figure 11-7b represents λ_{pj}/λ_p values as a function of ambient temperature T_j, when the power stress ratio on the transistor is 0.3. As can be seen, as ambient temperature increases from 20°C to 40°C, failure rate increases by

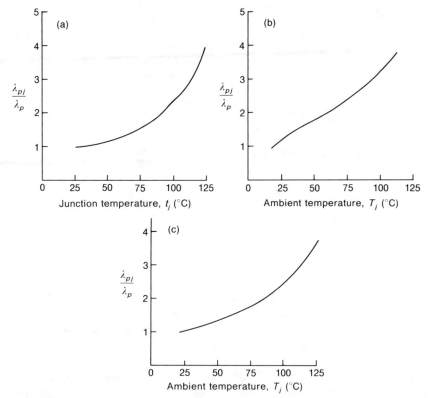

FIGURE 11-7. (a) λ_{pj}/λ_p as a function of junction temperature t_j for a microcircuit. λ_{pj} = failure rate at any given value of t_j (junction temperature); λ_p = failure rate at $t_j = 25°C$. (b) λ_{pj}/λ_p as a function of ambient temperature for a silicon NPN transistor. λ_{pj} = failure rate at any ambient temperature; λ_p = failure rate at ambient temperature of 20°C. (c) λ_{pj}/λ_p as a function of ambient temperature for a composition resistor, λ_{pj} and λ_p as for (b).

approximately 60%; as temperature increases from 30°C to 50°C, failure rate increases by approximately 60%; as temperature increases from 30°C to 60°C, failure rate increases by approximately 77%.

11.3.1.3 Failure Rate Sensitivity to Temperature for Resistors

Representing part failure rate as λ_p, the resistor failure rate model in Mil-Hdbk-217E in terms of the ratio λ_{pj}/λ_p can be represented as

$$\frac{\lambda_{pj}}{\lambda_p} = \frac{\lambda_{bj}(\Pi_e \Pi_r \Pi_q)}{\lambda_{bo}(\Pi_e \Pi_r \Pi_q)}$$

$$= \frac{\lambda_{bj}}{\lambda_b} \qquad (11\text{-}7)$$

TABLE 11-2   Condensed Table of
Base Failure Rates for
Silicon NPN
Transistors as a
Function of Ambient
Temperature T and a
Power Stress on the
Transistor of 0.3

$T(°C)$	λ_{bj}
20	0.00063
30	0.00071
40	0.00079
50	0.00089
60	0.00099
80	0.0012
100	0.0016
130	0.0025

where

λ_{bj} = the base failure rate of the resistor under a stated value of power stress at a given ambient temperature T_j

λ_{bo} = the base failure rate of the resistor under the same stated power stress at an ambient temperature of 20°C

Mil-Hdbk-217E provides tables equating λ_{bj} to various ambient temperatures under given values of power stress. Table 11-3 provides a condensed table

TABLE 11-3   Values of Base
Failure Rate λ_{bj} for
Fixed Composition
Resistors, Associated
with Various Ambient
Temperatures Under
a Stress Level of 0.3 of
its Rated Wattage

$T(°C)$	λ_{bj}
20	0.00022
30	0.00031
40	0.00045
50	0.00066
70	0.00140
90	0.00290
110	0.00600

of values for a fixed-composition resistor subjected to a stress value of 30% of its rated wattage.

Figure 11-7c represents λ_{pj}/λ_p values as a function of ambient temperature. As can be seen, failure rate increases by 25% as the ambient temperature increases from 20°C to 40°C and by approximately 40% as the temperature increases from 30°C to 60°C.

11.3.2   Incorporating Thermal Design/Analysis into the Reliability Program: Theory and Practice

As the circuit design team and the part specialists influence the reliability of a product, the thermal analysis and design engineer does likewise. The thermal design effort encompasses not only the local temperature effects at a given point on a printed circuit board but the cooling strategies to be employed at the system packaging levels as well. As a consequence, defined linkages among reliability evaluation/prediction, part selection/application, the establishment of thermal requirements for each level of hardware indenture, and the use of thermal analysis results must be defined early in the development program. Inputs to the thermal design program must include information from the reliability program with respect to the temperature criticality of specific components and parts (i.e., the results of analyses such as described in Section 11.3.1) as well as general reliability characteristics and requirements which may have particular relevance to thermal design.

While we will not dwell in depth on the makeup and details of the thermal design program, it is important to recognize that it should ordinarily incorporate the following tasks which are particularly relevant to the reliability program:

1. Identification of possible alternative thermal designs and strategies. Definition of the impacts of each with respect to effects at system, subsystem, component, printed circuit board, and part levels.
2. Performance of tradeoffs among such alternatives, taking into account reliability as well as other tradeoff factors.
3. Performance of thermal surveys and analyses to provide thermal profiles to the levels needed for reliability assessment. Such analyses should include the effects of part placement, layout, and mounting. Schedule for completion should track with reliability prediction schedule.
4. Performance of tests to verify survey and analysis results.
5. Submittal of reports documenting the results of the above and the thermal design goals to be achieved.

These tasks insure that a reasonably detailed thermal design/analysis program has been established for the effort, one that includes consideration of design alternatives and tradeoffs, performance of thermal surveys and analysis, and verification tests to validate analysis results.

In many instances, a given development may undergo a thermal design/ analysis which, for one reason or another, is less than rigorous in the performance of all the tasks and analyses outlined above. Reference 12 cites four common alternative approaches to meeting the thermal needs of a system. They are listed next in amended form along with their advantages and disadvantages:

1. *Overdesign of the cooling system.* In some instances the need for a detailed thermal analysis can be avoided by obvious overdesign of the basic cooling system. Examples are providing more or larger blowers, providing more area for heat dissipation, or putting the operating system in a "special environment." Disadvantages include a loss of the capability to fine-tune temperature control and loss of thermal visibility throughout the system. Additionally, since a detailed thermal analysis was not performed, there will be no assurance that the cooling system will be adequate for every *portion* of the system. Some of the disadvantages could be overcome by a test verification program, but unless there is a commitment to make engineering changes upon finding a problem, the test program simply provides information on mistakes made.

2. *Concentration of effort on portions of the system which are either known to be thermally sensitive or dissipate high values of temperature.* In this case, thermal analysis is selectively performed. For example, part-level thermal analysis could be confined only to semiconductors and/or moderate to high power level parts. Component areas containing lower power use or non-temperature-sensitive parts could be assigned average or worst-case values of temperature. The basic disadvantage of the approach is that one or more potential thermal problem areas may be overlooked. The risk associated with that, however, is less than when alternative (1) is chosen.

3. *Concentration of effort on thermal verification as opposed to analyses.* In cases where it is obvious that the system and its components will be operating in a benign environment, where the parts are known to be operating under very low stresses, where thermal activity is known to be low throughout the system, or where there is virtual certainty that an analysis will provide nothing worse than validation of worst-case thermal design goal expectations, then a verification program involving limited sample measurements (measurement locations to be selected at least in part by the customer) could be substituted for analyses. Since only small numbers of measurements would be taken, the verification test cost would be low. Generally this course of action would be considered a high-risk test to the developer unless confidence was high enough that a test would only serve to verify worst-case thermal design expectations.

Employing this alternative, worst-case thermal design expectations would be used in reliability evaluations/predictions.

4. *Analyses by comparison with other hardware.* When the system being developed is a modification of an existing system or uses off-the-shelf items and if (a) the thermal design of such items is considered satisfactory, and (b) reliability predictions or field data are available for the existing system components, then thermal analysis data on only those modifications or new interface units would be required for reliability purposes. Disadvantages are that one or more potential thermal problem areas may be overlooked. That risk however, is no more and probably less than when alternative (2) is chosen.

11.3.3 General Insights, Guides, and Rules About Thermal Design That a Reliability Engineer Should Know

The reliability engineer must usually rely on the judgment of a thermal analyst for guidance as to feasibility, practicality, and general causes and effects of various thermal design features. There are however, certain factors about which the reliability engineer should be aware when evaluating thermal designs (see Ref. 12). For example:

1. Limitations on cooling techniques for components and modules. Any given cooling technique is limited by the ratio of the energy dissipated, measured in watts, to the volume of the box in which the hardware is placed, measured in cubic meters. Charts 11-4A, 11-4B and 11-4C, extracted from Ref. 12, provide rough rules of thumb on limitations for
   a. Various cooling techniques applied to equipment used at sea level
   b. Forced air cooling techniques applied to module microelectronic parts
   c. General limitations of other cooling techniques

CHART 11-4A Cooling Techniques for Equipment Used at Sea Level

Dissipation Technique that Will Suffice for Most Applications	Dissipation per Unit Volume (W/m³)
Free convection to ambient air	0–11 000[a]
Forced air	11 000[a]–35 000
Custom design; thermal considerations should have top priority in physical design	>35 000

[a] 3500 W/m³ instead of 11 000 W/m³ if box is poorly ventilated and thermally sensitive parts are mounted horizontally.

CHART 11-4B Limitations on Forced-air Cooling Techniques for Module Microelectric Parts

Cooling Technique	Maximum Cooling Capacity (W/m^2)
Impingement	800
Coldwall	1500
Flow-through	3400

CHART 11-4C Limitations on Various Cooling Techniques

Thermoelectric Coolers
Heat-sink temperature $\leqslant 100°C$
Cooling load $\leqslant 300$ W

Vapor-Cycle Refrigeration
Power requirement = 250–1000 W
per 1000 W of refrigeration
Ambient temperature $\leqslant 71°C$
($\leqslant 200°C$ for specially designed
vapor-cycle equipment)

Expendable Evaporant Cooling
Heat sink temperature $\geqslant 93°C$
Duration of operation $\leqslant 3$ hr

Source: Ref. 12.

---

2. The placement/layout of parts to reduce thermal stress:
   a. Provide as much separation as possible between dissipating parts
   b. Use large areas
   c. Use materials having high thermal conductivity
   d. When contact interfaces are used, minimize contact thermal resistance
   e. Circuit cards with dissipation > 2 watts need a copper ground plane
   f. Use materials having similar values of coefficient of thermal expansion
3. Blower selection/installation:
   a. Use fixed-speed blowers only at altitudes < 10 000 ft. Use variable speed blowers at altitudes > 10 000 ft
   b. Cooling blower dissipations should not exceed 10% of thermal load
   c. Install blower such that it pulls rather than pushes the air through the equipment

CHART 11-5 Cooling Methods for Various Types of Parts

				Type of Parts				
Application	Microelectronics	Power Transistors	Capacitors	Resistors	Transformers	Oscillators	Gyros	Traveling Wave Tubes (TWT)
Airborne-fighter	B, C, D, E	E, F, G, H	A, B, C, D, E, F, G	A, B, C, D, E, F, G, H	B, C, D, E, H	A	A, B	G, I
Airborne-cargo	B, C, D, E	C, F, G, H	A, B, C, D, E, F, G	A, B, C, D, E, F, G, H	B, C, D, E, H	A	A, B	G, I, J
Missile	A, G	A, G, H	A, G	A, G, H	A, G, H	A, G	A, G	G, I
Space	G, L	G, L	G, K, L	G, K, L	G, K, L	K	G, K, L	G, K, L
Ground	A, B, D, E, F, G	A, B, D, E, F, G, H	A, B, D, E, F, G	A, B, D, E, F, G, H	A, C, J, H	A	A, B	G, I, J
Shipboard	A, B, D, E, F, G	A, B, D, E, F, G, H	A, B, D, E, F, G	A, B, D, E, F, G, H	A, B, D, E, F, G, H	A	A, B	G, I, J

Type of cooling:

A: Free convection to ambient air and radiation
B: Air impingement, non-card-mounted
C: Air impingement, card-mounted
D: Card-mounted, air-cooled cold walls
E: Card-mounted, flow-through modules
F: Cold-plate-mounted, air-cooled
G: Cold-plate-mounted, liquid cooled
H: Free convection, liquid heat sink
I: Flush-cooled
J: Liquid evaporation
K: Radiation to space
L: Space radiator with liquid transport loop

Source: Ref. 12.

    d. Do not place the inlet of a blower directly downstream of the exhaust of another blower

    e. Leave a clearance of at least 1×10^{-2} m downstream of the coolant air blower

    f. Do not put two different blowers in series or in parallel unless the flow rates and head pressures in the system are balanced

    g. Account for short duration on thermal overstress

4. Coolant flow passage design:

    a. Do not obstruct coolant flow over parts

    b. Use perforated covers as opposed to solid ones on cabinets of electronic equipment cooled by ambient air

    c. Use splitters and vanes to direct flow to parts

    d. Do not use sudden enlargement or contractions in flow passages.

    e. For free convection in cooled equipment, install circuit boards and cooling fins vertically.

In order to provide a first approximation of the types of cooling needs which should be considered for various types of components in various environments, Chart 11-5, extracted from Ref. 12, is provided.

11.4   RELIABILITY PARTS DERATING: ITS LOGIC AND PRACTICE

Parts derating is a reliability development task that results from the efforts of the part specialist, the thermal specialist, and the reliability engineer. It is a function of the parts chosen, the application factors associated with the parts, their performance objectives, stresses imposed, and failure-rate objectives, and can be directly linked with the thermal design program discussed in the last section and the parts program discussed in Chapter 6, although it is discussed here as a separate task.

Derating is the imposition of limits on the types of design stresses (mechanical, thermal, electronic/electrical) that can be imposed by designers in order to enhance reliability. Derating policies are instituted to meet two basic objectives:

1. To act as a de-facto detriment to operating items beyond their specified maximum stresses in order to achieve a particular performance or design goal (a) with a particular part basically unsuited for a given level of performance, (b) with a minimum number of parts.

2. To improve the reliability of a design by "forcing" the application of design-imposed stresses which are less than the specified maximum levels for each part, hence reducing the failure rate of the part.

Derating procedures generally concentrate on the latter objective. Such procedures require establishment of organization design policies which are reasonable in terms of minimizing the potential for increase in complexity/cost and are enforceable.

Derating policies have been established by many development organizations for internal use. Some DOD organizations have published design guides for that purpose.[13,14,15]

The "optimum" derating for a part occurs:

- At a point above that required for proper part function.
- At a point below that level of stress at which failure rate per unit of applied stress increases rapidly.
- At a point which provides an effective cost/benefit ratio for the development program in question in terms of performance, complexity, and resources required.

11.4.1 Defining Derating Levels: Establishing a Derating Window

One means to provide visibility relative to derating levels is through establishment of a *derating window*. A derating window provides visibility relative to a range of different practical stress levels and the impact of such levels on failure rate. Derating windows for a given part can be developed through use of detailed part failure-rate prediction models and the same general procedure can be applied to *any* part failure-rate prediction model. For illustrative purposes, Mil-Hdbk-217E models will be utilized.

Section 8.4, dealing with detailed failure-rate prediction procedures, discussed the definition of base failure-rate factors, λ_b, and failure-rate multiplying factors, Π_v, which were functions of electrical/electronic and/or thermal stresses imposed on a part. In Section 11.3 the specific impact of thermal stresses on reliability was discussed. Here that discussion will be extended to the establishment of more general derating criteria.

Virtually every part failure-rate model includes consideration of factors which deal with the influences of design controllable stresses. For example:

1. For monolithic bipolar MOS digital devices the failure-rate model, λ_p,

$$\lambda_p = \Pi_q(c_1\Pi_t\Pi_v + c_2\Pi_e)\Pi_l \qquad (11\text{-}8)$$

includes the factors

Π_v = the stress related factor taking into account the effect of voltage applied and worst-case junction temperature

Π_t = the temperature acceleration factor taking into account technology and worst-case junction temperature

(In Mil-Hdbk-217E, worst-case junction temperature is defined as a linear function of device power dissipated. Hence, junction temperature acts as a parameter for power dissipation effects.)

2. For conventional transistors the failure-rate model, λ_p, is

$$\lambda_p = \lambda_b(\Pi_e\Pi_a\Pi_q\Pi_r\Pi_s\Pi_c) \qquad (11\text{-}9)$$

where

λ_b = the base failure-rate factor, which takes into account ambient temperature and the power stress ratio (ratio of applied to maximum rated power)

Π_s = the factor associated with the ratio of applied to rated voltage

(See Table 11.4a for a condensed table of λ_b versus ambient temperature and power stress ratio, and Table 11-4b for a condensed table of Π_s versus voltage stress ratio.)

3. For composition resistors the failure-rate model, λ_p, is

$$\lambda_p = \lambda_b(\Pi_e\Pi_r\Pi_q) \qquad (11\text{-}10)$$

where

λ_b = the base failure-rate factor, which takes into account the effects of ambient temperature and power stresses

Generally, each λ_p relationship can be expressed as a function of stress variables, s_i, and of values that can be construed as constants, b_i. For example, for the microcircuit model,

$$\lambda_p = f(\Pi_v, \Pi_t, b_1, b_2, b_3, b_4, b_5) \qquad (11\text{-}11)$$

where

Π_v and Π_t are stress related variables or factors

b_1, b_2, b_3, b_4, b_5, are factors unrelated to stress and hence are treated as constants or given.

And in general λ_p for any part failure-rate model can be expressed as

$$\lambda_p = f(\Pi_i, \Pi_{i+1}, \ldots, b_1, \ldots, b_n) \qquad (11\text{-}12)$$

TABLE 11-4

(a) Condensed Table of Base Failure Rate vs. Temperature and Power Stress for Silicon NPN Transistors

Temperature (°C)	Power Stress				
	0.2	0.4	0.6	0.8	1
0	0.00067	0.00084	0.0012	0.0017	0.0027
20	0.00075	0.0010	0.0015	0.0023	0.0043
40	0.00093	0.0013	0.0019	0.0034	
60	0.0012	0.0017	0.0027		
80	0.0015	0.0023	0.0043		
100	0.0016	0.0034			

(b) Table Showing Effect of Voltage Stress[a] on the Failure Rate Acceleration Factor π_s for Conventional Transistors

S(%)	π_S
100	3.0
90	2.2
80	1.62
70	1.2
60	0.88
50	0.65
40	0.48
30	0.35
20	0.30
10	0.30
0	0.30

[a] Voltage stress, $S = \dfrac{\text{applied}(V_{CE})}{\text{rated}(V_{CEO})} \times 100$

where

Π_i = a stress-related variable or factor

b_i = a parameter or factor unrelated to or independent of applied stress

For each situation where it is desired to construct a window, first determine the maximum stress or worst-case value for each stress-related parameter *within part specification limits* and apply (11-12) to evaluate the value of λ_p *associated with such worst-case/maximum values*. Denote that value of λ_p as λ_{pw}.

$$\lambda_{pw} = f(\Pi_{\max(i)}, \Pi_{\max(i+1)}, \ldots, b_1, \ldots, b_n) \qquad (11\text{-}13)$$

where

λ_{pw} = the maximum failure rate associated with the part when operated at its limits

$\Pi_{max(i)}$ = maximum or worst-case value of Π_i associated with worst-case/maximum values of stresses

Then

$$\frac{\lambda_p}{\lambda_{pw}} = \frac{f(\Pi_i, \Pi_{(i+1)}, \ldots, b_1, \ldots, b_n)}{f(\Pi_{max(i+1)}, \ldots, b_1, \ldots, b_n)} \tag{11-14}$$

where

λ_p / λ_{pw} = the proportion of maximum failure rate which will result if any values of Π_i less than $\Pi_{max(i)}$ are employed. λ_p / λ_{pw} can now be expressed in an alternate form

$$1 - \frac{\lambda_p}{\lambda_{pw}} = k \tag{11-15}$$

where k = *the factor of failure rate reduction achieved by choosing stress parameters less rigorous than the maximum or worst-case value*, i.e., $k = 0.9$ denotes that failure rate was reduced by 90% from the maximum.

Rearrange (11-15) such that

$$\lambda_p = \lambda_{pw}(1 - k) \tag{11-16}$$

For a particular part:

1. Define its maximum within-specification failure rate, λ_{pw}.
2. Choose a range of values for k (i.e., $k_1 = 0.99$; $k_2 = 0.95$; $k_3 = 0.90, \ldots, k_n = 0.15$, etc.).
3. Apply (11-16) to calculate a value λ_p for each value of k.
4. Determine the various combinations of stresses which will yield that value of λ_p for the part.

After that has been done for all values of k, a tableau of design stress combinations associated with k results. If many *different* stresses impact the values of Π_i, a computer will be required to search the tableau and determine the optimum range of stresses (derating ranges) appropriate for goals. When only two or three different stresses are to be contended with, a graphical solution can be applied which makes the potential derating ranges visible in a window.

Take, for example, the conventional transistor model (11-9) where an ambient temperature of 30°C has been verified by thermal analysis. From Tables 11-4a and 11-4b, λ_{pw} can be calculated as

$$\lambda_{pw} = \lambda_b \Pi_s b, \ldots, b_n$$

$$\lambda_{pw} = \lambda_b \Pi_s C$$

$$\lambda_{pw} = 0.0049(3)C*$$

where $C =$ some known constant based on parameters which are not stress-related. Assume for convenience that $C = 1$. Applying (11-16),

$$\lambda_p = 3(0.0043)(1 - k) = 0.0129(1 - k)$$

Choosing first a value of $k = 0.95$, $\lambda_p = 0.000735$.

From Tables 11-4a and 11-4b, the following approximate voltage and power stress combinations can be found to yield a value of $\lambda_p = 0.000735$:

Power stress (ratio of applied to rated power)	Voltage stress (ratio of applied to rated voltage)
0.82	0.20
0.70	0.30
0.52	0.40
0.35	0.50
0.20	0.60
0.05	0.70

Graphing the results, the $k = 0.95$ plot in Fig. 11-8 results. Each point on the plot represents a power/voltage stress combination capable of achieving a value of $\lambda_p = 0.000735$.

In a similar fashion, k is taken equal to 0.90, 0.85, 0.75, and 0.50 and plotted as shown in Fig. 11-8. The derating window results when

1. A vertical line is superimposed on the figure representative of the minimum voltage required for the device to function; for example, if the minimum operating voltage $= 0.4$ of its rated voltage.

---

* Assume that a $\lambda_b = 0.0049$ is associated with a power stress of 100% and a temperature of 30°C.

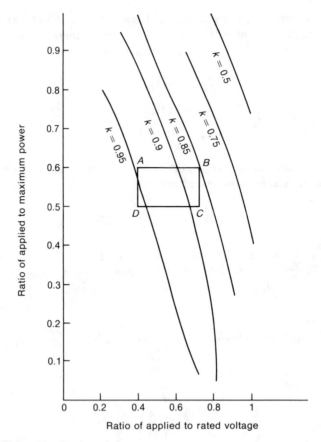

FIGURE 11-8.   The derating window.

2. A horizontal line is superimposed, representing the estimated maximum power needs of the device for its design application; for example if a power need = 0.6 of its maximum rated power.
3. A vertical line is added parallel to the first, starting at the intersection of the horizontal line and the plot representing the minimum acceptable value of k; for example $k = 0.85$, going in a downward direction.
4. A horizontal line is added to form a rectangle or window.

The result is window $ABCD$ in Fig. 11-8 which provides visibility of the ranges of power and voltage stress combinations (derating values) which will yield a failure-rate result such that

$$0.95 > k > 0.85$$

The various values of k contained in the window represent the failure-rate results consistent with device application needs and provide visibility as to the derating alternatives and their ramifications on failure rate.

11.4.2   Choosing Derating Levels: Rules of Thumb

Even with the window to provide visibility, judgment must be applied to choose an approximate derating level for a particular application. Information and general guidance to that end for particular components has been developed in at least two RADC studies, "Reliability Derating Procedures," [14] "Reliability Parts Derating Guidelines," [13] and included in "The RADC Reliability Engineers Toolkit." [15].

Those studies suggest the use of three general levels of derating (stresses) for parts depending on application:

1. A *maximum* level, denoted as level 1, appropriate for application in components, equipments, or systems where failure would significantly jeopardize life or accomplishment of a critical function, or where repair is unfeasible or impractical. Stress levels *below* those chosen will provide negligible reliability improvement and incur inordinate design cost penalties.
2. A *moderate* level, denoted as level 2, appropriate for application in components, equipments, or systems where failure would degrade overall function performance, result in inconvenience (such as partial, short term power outage), or incur significant repair/correction costs. Stress levels below these are still possible and would provide significant levels of reliability improvement, but would also incur significant increases in design or cost penalties.
3. A *minimum* level, denoted as level 3, appropriate for application in components, equipments, or systems where failure is less critical than in the above. Failure in these items would not significantly degrade overall function performance and can be quickly and inexpensively repaired. Extreme stress ranges are purposely avoided.

Table 11-5 extracted from the "RADC Reliability Engineers Toolkit" [15] provides rule-of-thumb derating guidelines for specific types of parts. The data resulted from a series of studies, reports, and analyses performed by RADC. They represent "suggested" derating levels for a diverse sampling of parts keyed to the derating categories described above.

TABLE 11-5   Derating Guidelines for Specific Types of Parts

Part Type	Derating Parameter	Derating Level I	II	III
		Microcircuits		
Linear	Supply voltage	70%	80%	80%
	Input voltage	60%	70%	70%
	Output current	70%	75%	80%
	Max. $T_j(°C)$	80	95	105
Digital	Supply voltage	±3%	±5%	±5%
	Frequency	80%	90%	95%
	Output current	80%	85%	90%
	Max. $T_j(°C)$	85	100	110
	(F, % absolute max.)			
Hybrid	Thick film*	<50	<50	<50
	Thin film*	<40	<40	<40
	Max. $T_j(°C)$	85	100	110
	*(Power density, W/in²)			
Complex (LSI, VLSI, VHSIC)	Supply voltage	±3%	±5%	±5%
	Frequency	75%	80%	90%
	Output current	70%	75%	80%
	Fan out	70%	75%	80%
	Max. $T_j(°C)$	85	100	125
	(F, % absolute max.)			
Memory	Supply voltage	±3%	±5%	±5%
	Output current	70%	75%	80%
	Max. $T_j(°C)$	85	100	125
		Transistors		
Bipolar	Power dissipation	50%	60%	70%
	Breakdown voltage	60%	70%	70%
	Max. $T_j(°C)$	95	105	125
Field Effect	Power dissipation	50%	60%	70%
	Breakdown voltage	60%	70%	70%
	Max. $T_j(°C)$	95	105	125
		Diodes		
Power Rectifier	Forward current	50%	65%	75%
	Reverse voltage	70%	70%	70%
	Max. $T_j(°C)$	95	105	125
Voltage Regulator	Power dissipation	50%	60%	70%
	Max. $T_j(°C)$	95	105	125

TABLE 11-5 *Continued*

Part Type	Derating Parameter	Derating Level		
		I	II	III
	Diodes			
Voltage	Current I_{zt}	[a]	[a]	[a]
Reference	Max. $T_j(°C)$	95	105	125
	(P, % rated)			
Signal/switch	Forward current	50%	65%	75%
(axial lead)	Reverse voltage	70%	70%	70%
	Max. $T_j(°C)$	95	105	125
	Tubes			
	Power output	80%	80%	80%
	Power reflected	50%	50%	50%
	Duty cycle	75%	75%	75%
	Inductors			
Pulse transformers	Operating current	60%	60%	60%
	Dielectric voltage	50%	50%	50%
	Temp. (°C) hot spot	40	25	15
Coils	Operating current	60%	60%	60%
	Dielectric voltage	50%	50%	50%
	Temp. (°C) hot spot	40	25	15
	Relays			
	Current (res/cap)	50%	75%	75%
	Current (ind)	35%	40%	40%
	Contact power	40%	50%	50%
	Temp. (°C)	20	20	20
	(T from max. limit)			
	Resistors			
Composition	Power/temp. (°C)	50%/30	50%/30	50%/30
Film	Power/temp. (°C)	50%/40	50%/40	50%/40
Variable	Power/temp. (°C)	30%/45	50%/35	50%/35
Thermistor	Power/temp. (°C)	50%/20	50%/20	50%/20
	(T from max. limit)			
Wirewound accurate	Power/temp. (°C)	50%/10	50%/10	50%/10
Wirewound power	Power/temp. (°C)	50%/125	50%125	50%/125
	(T from max. limit)			

TABLE 11-5   *Continued*

Part Type	Derating Parameter	Derating Level		
		I	II	III
	Capacitors			
Film Mica Glass Ceramic	Voltage/temp. (°C)	50%	60%	60%
Electrolytic aluminum	Voltage/temp. (°C)	NR[b]	NR[b]	80%
Electrolytic tantalum	Voltage/temp. (°C)	50%	60%	60%
Variable piston	Voltage/temp. (°C)	40%	50%	50%
Variable ceramic	Voltage/temp. (°C)	30%	50%	50%
	Rotating			
	Bearing load	75%	90%	90%
	Operating temp. (°C) (L, % rated) (T from max. limit)	40	25	15
	Switches			
	Current (res./cap.)	50%	75%	75%
	Current (ind.)	35%	40%	40%
	Contact power	40%	50%	50%

Source: RADC Reliability Engineers' Toolkit.[15]
[a] Fixed test current, do not derate.
[b] NR, not recommended for use.

11.5 FAILURE REPORTING AND CORRECTIVE ACTION SYSTEM (FRACAS)

The failure reporting and corrective action system (FRACAS) is one of the integral parts of a reliability program. It provides the means to find and correct problems in design, material, and specifications during development, production, and field use phases of the system life cycle. The FRACAS will generally take the form outlined in Fig. 11-9.

All FRACASs are not created equal. They can vary with respect to:

- The depth of analysis which will be provided, i.e., failure analysis of the part, the circuit or assembly.
- Limitations and constraints on corrective actions allowed, i.e., change circuit design, modify stresses, choose different parts, change item specification.

- Anticipated payoff, for example, a FRACAS program to analyze field failure data on a warranted item. In that case, since the repair of the item is the contractor's responsibility, FRACAS would provide a larger payoff to the contractor than if the item's repair were not the contractor's responsibility.
- Scope of FRACAS activities, whether FRACAS efforts will be open-ended or limited to certain items or to certain types of failures.

No matter what the depth or scope, when performed as defined, it serves to provide a means for auditing and tracking failure trends, and impacts of overall design changes on reliability, and for assessing the effectiveness of specific reliability design changes.

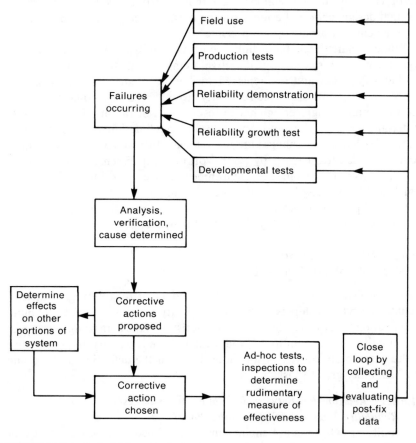

FIGURE 11-9.   Basic failure reporting and corrective action system functioning.

11.6 RELIABILITY GROWTH:
A MEANS FOR IMPROVING
RELIABILITY DESIGN

Reliability growth or, as it is sometimes called, Test Analyze and Fix (TAAF), is defined as that degree of reliability improvement which can occur as the result of tests performed to identify design, material/part and specification deficiencies, and the *permanent correction of such deficiencies through engineering changes*. It is applied near the end of full-scale engineering development to improve the inherent reliability of an item prior to production by finding and correcting unanticipated reliability design problems. Examples of actions triggered by reliability growth tasks include providing design changes to eliminate circuit transients, providing a higher degree of derating in selected portions of the design, eliminating circuit hotspots by locally improved thermal design, choice of a more rugged part, or change in the specification requirements for a part or component.

While apparent reliability improvements *over time* may occur in some equipments and systems as defective items (or items containing poor workmanship) fail and are replaced by "good" items (see Ref. 16), such improvement is not considered reliability growth, because no inherent design, specification, or engineering change results. In general, any failures due to material defects which cannot be reduced through design, basic material, or material specification changes are not a focus of the reliability growth task; neither are failures caused by poor workmanship. Such failure types are reduced through burn-in and environmental stress screening (ESS) tests, which will be discussed in a following section.

General procedures for performing a reliability growth test are contained in Mil-Hdbk-189, "Reliability Growth Management."

11.6.1 Reliability Growth:
Scoping a Reliability Growth Test Task

Factors which influence the scope and conduct of reliability growth tests include the relationship between the reliability growth test and reliability qualification test; the schedule; the characteristics of the item being developed; the quantities of the item to be produced; and the capability and experience of the developer. See Ref. 17 for more detail on the definition and scoping of a reliability growth task.

11.6.1.1   Relationship Between Reliability Growth Tests
and Reliability Demonstration/Qualification Tests
Reliability growth tests are engineering tasks aimed at improvement of the reliability design, not at assessing its quality. Its objective is to generate

failures caused by design or specification deficiencies and make corrections. Accordingly, the more such failures are generated and corrected the better. The purpose of reliability demonstration/qualification testing is to provide proof that the contractor has met development requirements. Accordingly, the fewer failures which occur due to any cause the better.

Under the best of conditions a reliability growth test can serve as a demonstration/qualification test. Such conditions include:

- Absence of economic or schedule pressure on the developer's part to reduce or cut short the growth test process.
- Clearly defined growth objectives and procedures to show they have been achieved within a given confidence band.
- Clearly defined and exercised control over the reliability growth test environment to insure that it is representative of the item's operational environment.
- Clearly defined sets of rules, definitions, and procedures for the analysis and treatment of failures.
- Procedures for recommending, approving, implementing, and verifying corrective actions.

A growth test and a demonstration test can also be used in a complementary fashion, i.e., a growth test and a demonstration test both of shorter durations than if performed without the other. Under such a strategy, lower, more stringent, statistical demonstration test risks may be associated with items having undergone shorter growth tests than longer ones. (Test risks will be discussed in more detail in the next chapter.)

11.6.1.2 Schedule Considerations
A reliability growth test can have an impact on schedule. A study of the literature performed in conjunction with the "Reliability Growth Testing Effectiveness" study[17] found that recommended growth test duration ran between 5 to 100 multiples of MTBF. The schedule should provide time not only for the performance of the test, but for the implementation of corrective actions recommended and their verification.

11.6.1.3 Characteristics of the
Item Development Program
The need for and the effectiveness and utility of a reliability growth test are affected by program decisions made previously and by the makeup and characteristics of the item under development. For example:

- *Off-the-shelf vs. new development items.* Off-the-shelf buys provide little opportunity to make design changes suggested by test. Items under development provide for maximum opportunities to make design changes.

- *State-of-the-art technology involved.* New state-of-the-art technology can translate into a greater number of unanticipated reliability design problems than current state-of-the-art technology. Hence, a growth program is more important for the former than for the latter.
- *Item complexity.* The more complex an item, the more likely the presence of unanticipated reliability design problems. Hence, a growth program will be more important to a complex item than to a simpler one.

11.6.1.4   Production Quantity in Question
A reliability growth test costs money and the decisions made affect the life-cycle cost of support for the items produced. A reliability growth test which reduces the costs of $2x$ items by $p\%$ provides a bigger payoff than one which reduces the costs of x items by $p\%$. As a result, the greater the production quantity, the more benefit accrued by a reliability growth test.

11.6.1.5   The Capability of the Developer
The success of a growth program is significantly linked to:

- The test facilities available to the developer; the number of items that can be tested at the same time (in order to reduce test time and accommodate schedule constraints); the time allotted to use of the test facilities for reliability growth test purposes.
- The scope of the FRACAS activity planned. The degree of depth to which failure analyses can be performed.
- Experience of the developer in performing reliability growth programs; data available showing performance of growth trends for other items developed; estimates of growth rates, starting points; knowledge of stresses to be associated with growth tests. All of these are necessary to the planning and implementation of the test and interpretation of test results.

11.6.2   Reliability Growth: Its General Mechanics and Considerations

The concept and mechanics of reliability growth are relatively simple.

1. There exist two types of failures which contribute to failure rate:

1. Those that are caused by randomly situated latent defects and by failure mechanisms in parts and materials. These manifest themselves as a function of accumulated stress over time *where stresses are not inordinate and redesign or changes in specification would provide little relief.* These are the types of failures which will appear to occur randomly in time during the growth test and during operational life after other redesign and design corrections have been accomplished.

2. Those that can be reduced by redesign or reduction of stresses, or through improved specifications on components, parts, and material. These, too, occur during growth test and operational use.

The failure rate of an item is a function of the two. The purpose of the growth test is to identify as many of the latter as possible and to make and implement fixes such that failures of that type are minimized and overall failure rate is reduced.

2. Items undergoing growth test are operated at a stress level at or slightly above that which would occur in an actual use environment. Sometimes items are cycled on and off to simulate a logical number of missions, leaving only sufficient times between cycles for the item to stabilize thermally. This to simulate the stresses created by on–off cycling encountered during use. In some instances much more stringent "accelerated" stresses are applied to the item under test to provide more failures. The last can, however, introduce failure mechanisms and modes of failure which would otherwise not be relevant. As a consequence, care must be taken in the choice of any so-called "accelerated" stress.

3. Each failure which occurs is analyzed to determine whether:

- Overstressed for its application.
- Exposed to too a high a thermal stress, vibration, humidity, shock, etc.
- A given part parameter must be within a narrower range than that to which it is presently specified.
- Protective circuitry required.
- Wrong type part chosen.
- Interface requires modification.
- Caused by a defect in a part which is identifiable through part screening.
- Caused by a defect in a part which is not identifiable through part screening.
- Caused by workmanship defect.

4. Corrective action takes place which will either reduce or eliminate future failures from that cause.

It is important to distinguish among failures of both the types described above in order to avoid masking true growth characteristics and general quality problems.

For example, failures resulting from defects in parts and materials can make the item appear to take on an increasing or decreasing failure rate with time. If growth test data shows an increasing failure rate with time with respect to failures resulting *from such defects*, a serious quality problem is evident which must be recognized and corrected. On the other hand, if the data shows a decreasing failure rate with time and it is not realized that it was caused by the "weeding out" of defects, a false or exaggerated indication of growth may be portrayed.

For this reason it is desirable to break down reliability growth analysis into several analyses to provide the visibility required, one taking into account all failures, one omitting non-design-deficiency failures.

The physical and engineering analyses performed on each failure are accomplished through the FRACAS task discussed in the last section.

11.6.3   Reliability Growth: The Analysis Process

The fact that reliability growth occurs as the failure rate of an item is reduced by redesign is logical and easy to grasp. If an organization's objectives are just to track the reliability growth of a particular item under development, and provide approximate worst-case estimates of the failure rate attained at the end of the test and projections of future failure rate, then the analyses associated with reliability growth process can take on a relatively simple, straightforward process.

The most simple procedure applicable to such analyses involves a simple plot of the cumulative failure rate, $\lambda(T)$, over test time, T, as shown in Fig. 11-10, where $\lambda(T) =$ average rate of failure per hour over time interval T (see Chapters 1 and 2).

$$\lambda(T) = f/T$$

$$f = \text{number of failures in test time } T$$

$$T = \text{interval of test time}$$

It is assumed that failure rate will decrease over the course of the growth test as changes are implemented and will remain constant at the value attained after test termination.

Bear in mind that $\lambda(T)$ represents the average value of failure rate observed over T, not the value attained at test termination, which will be represented henceforth as $h(T)$. In general, sample size will be inadequate to make a satisfactory estimate of $h(T)$. However, if reliability growth has occurred,

$$\lambda(T) > h(T)$$

and $\lambda(T)$ will be used as a conservative estimate of $h(T)$.

With a process as crude as that, variations in $\lambda(T)$ or MTTF = MBTF = $1/\lambda(T)$ versus test time can be observed. Worst-case approximations of $h(T)$ can be made at the end of various test periods; projected values of $h(T)$ can

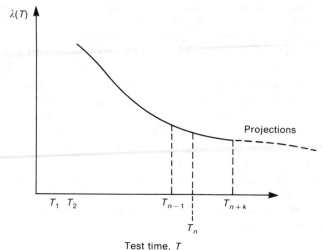

FIGURE 11-10.   Reliability growth as failure causes are identified and fixes are implemented.

be made for near-future test time values, $> T_{n+k}$, assuming the current trend of the growth process continues.

Procedures such as that above are inappropriate for: (a) Reliability growth test planning purposes and providing insights about how future reliability growth programs will behave. (b) Providing more than the most crude estimates of current and projected values of $\lambda(T)$ and $h(T)$ for a specific product under test.

In order to cope with such problems, inferences about the generic nature, form, or behavior of the growth process itself must be made. If the behavior of the growth process can be defined, then guides and more accurate estimates can be developed for (a) and (b).

Over the years, various growth-modeling processes have been suggested. The reports "Reliability Growth Testing Effectiveness,"[17] "Reliability Growth Studies,"[16] and Mil-Std-189, "Reliability Growth Management," describe, at least in summary, approximately a dozen such models. Two models, however, have experienced more interest and application in general than the rest: *the Duane Model* and the *AMSAA Model*. These will be described next.

11.6.4 Reliability Growth: Performance and Analyses—The Duane Model (Method)

The first and still among the most widely used procedures for simple reliability growth depiction, visibility, and planning is the Duane Model.

The Duane Model is a graphical procedure providing visibility to the growth modeling process which otherwise would not be evident. It provides a means for making reliability projections, although it is unable to provide confidence bands for such projections, and with the proper data base it can serve as a tool in the structure and planning of reliability growth tests.

The Duane Model was documented in the paper "Learning Curve Approach to Reliability Monitoring," [18] showing empirical evidence that an approximate linear relationship existed between the logarithm of cumulated failure rate, $\lambda(T)$, and the logarithm of cumulative test time, T, for a number of different items undergoing what can be termed a TAAF process. See Fig. 11-11 for examples of the type of empirical relationship developed.

While each different item is seen to exhibit a different slope s and intercept on log $\lambda(T)$, each does indeed form a straight line when plotted on log–log paper. Since that time, similar plots have been noted for different types of items (see Refs. 19 and 20).

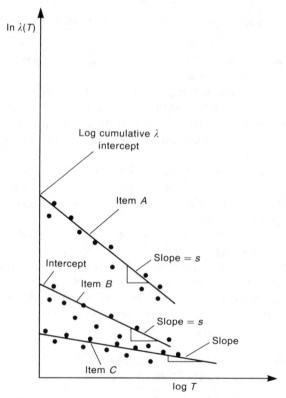

FIGURE 11-11.   Examples of a Duane type empirical relationship. T = cumulative operating hours; $\lambda(T)$ = cumulative failure rate = (total failures)/(cumulative operating hours).

The Duane type plots can be represented mathematically as (approximately)

$$\ln \lambda(T) = a - s \ln T \qquad (11\text{-}17)$$

where
$\lambda(T) = f/T =$ cumulative average failure rate over T
$f \quad =$ total failures occurring in time T
$T \quad =$ hours the item has been on test. If more than one item is on test, $T =$ total item hours where $T = \sum t_i$ and $t_i =$ hours on test for the ith item when the fth test failure occurs
$s \quad =$ the slope of the line plotted, where s is interpreted as the *growth rate*
$a \quad =$ intercept of the line on the $\ln \lambda(T)$ axis, where a is interpreted as the log of the initial value of $\lambda(T)$ at approximately the start of the growth test (actually at one hour into the growth test; when $\ln 1 = 0$)

The sense of the Duane plot is that reliability improvement will cease when the growth test terminates and the reliability attained at that point will be the reliability realized.

Taking antilogs transforming (11-17) to an exponential form,

$$\lambda(T) = e^a T^{-s} \qquad (11\text{-}18)$$

If we equate

$$\lambda(T) = \frac{f}{T} \qquad (11\text{-}19)$$

(11-19) may be expressed as

$$\frac{f}{T} = e^a T^{-s} \qquad (11\text{-}20)$$

or

$$f = e^a T^{1-s} \qquad (11\text{-}21)$$

Then

$\dfrac{f_{Tn} - f_{Tk}}{T_n - T_k} =$ the failure rate which would be apparent over the period of time between T_k and T_n

where

f_{Tn} = cumulative number of failures which would occur over T_n
f_{Tk} = cumulative number of failures which would occur over T_k
T_n = total growth test time (actual or projected)
T_k = a period of time less than T_n, $0 < T_k < T_n$

Under the assumptions of no negative growth the attained or projected failure rate $h(T_n)$ at test termination time T_n can be conservatively approximated as

$$h(T_n) \leqslant \frac{f_{Tn} - f_{Tk}}{T_n - T_k} = \frac{e^a[(T_n)^{1-s} - (T_k)^{1-s}]}{T_n - T_k} \tag{11-22}$$

For example, at $T_k = 0.9T_n$ (11-22) can be expressed as

$$h(T_n) < 10e^a(T_n)^{-s}[1 - (0.9)^{1-s}] \tag{11-23}$$

11.6.4.1 The Duane Model and Reliability Visibility Projection

If an item subjected to a reliability growth test takes on a plot to which a straight line can reasonably be plotted on log–log paper, as in Fig. 11-12, the values of the slope s and intercept a can be projected directly from the

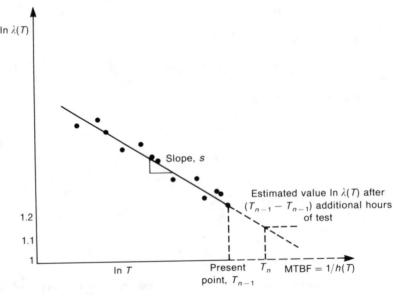

FIGURE 11-12.   Duane plot on log–log axes.

plot. The current attained value of failure rate $h(T)$ can be estimated and its future values can be extrapolated under the assumption that the growth process will continue its present trend.

Not all items, however, take on such a nice plot as shown in Fig. 11-12. Reasons for this may be attributed to the facts that:

1. Not all failures which occur will be design deficiency-related; some will occur as a result of defects in material, parts, or workmanship, which result in an apparent increasing failure rate function, at least early in the test. (The importance of taking such categories of failures into account was discussed earlier. Also plots that "look" approximately linear in log–log coordinates do not necessarily represent a growth situation.)
2. Not all deficiencies can be corrected immediately. Therefore, more failures due to a recognized deficiency may occur prior to implementation of the corrective action.
3. Not all items take on a form of reliability growth.

Some plots, for example, can take the form of Fig. 11-13, which shows the type of result which might occur as a result of (1) and (2), or a combination of the two. If a linear relationship does eventually make itself apparent, and if analysis shows it represents improvement due to design/material changes, a conservative estimate of growth projection could be based on the slope of the line produced.

Not all items will exhibit reliability growth. This is particularly true of a mature, proven design where there are no *cost-effective* design changes which would materially affect failure rate.

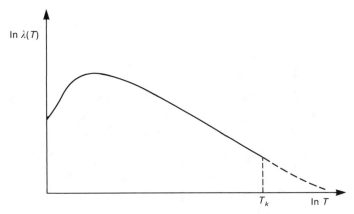

FIGURE 11-13. A Duane plot typical of early failures due to defects in materials and workmanship.

TABLE 11-6   **Examples of Growth Data for an Item Exhibiting a Constant Time Between Failures After 100 Hours**

Cumulative Hours	Number of Failures	Cumulative λ_c
100	5	0.0500
200	6	0.0303
300	7	0.0233
400	8	0.0200
500	9	0.0179
600	10	0.0167
700	11	0.0156
800	12	0.0149
900	13	0.0145
1000	14	0.0141

Some items will appear to have a plot indicative of a decreasing failure rate when the failure rate is actually constant with time. An illustration of this is provided in the RADC study report "Reliability Growth Prediction," [21] where an example is constructed in which an item undergoing growth test exhibits five failures due to infant mortality during the first 100 hours and then experiences *exactly* one failure per 100 operating hours for the next 900 hours of operation (see Table 11-6). The data is plotted in Fig. 11-14 and

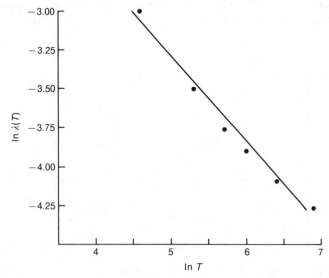

FIGURE 11-14.   A set of growth test data exhibiting what appears to be a pattern of growth.

gives the appearance of exhibiting growth over a 900 hour period when no growth has in fact occurred.

Even if care is taken in the analyses of the data to separate failures resulting from both causes, ambiguity with respect to whether or not growth has occurred can be present.

There are, however, relatively simple procedures which can be applied to resolve that situation. Assume that a reliability growth test is planned to run for a period of T_n unit hours and k items are to be put on test. The test will run until

$$T_n = \sum_i^k t_i \qquad (11\text{-}24)$$

where $t_i = $ the test hours associated with item i.

Whenever a failure occurs in any given unit, the failure is analyzed to find the cause, and corrective actions are determined. One of the following conditions/assumptions must apply:

1. Whenever a design change is identified, it is immediately implemented in all units.
2. The test is stopped periodically many times over T_n to implement design changes in all units.
3. In the event that changes cannot be implemented frequently, all corrective design changes identified will be considered 100% effective. All recurring failures due to the original design cause will not be counted. The validity of this assumption depends on the degree of confidence in the effectiveness of the design fix. In some cases, recurring failures are given some negotiated weight <1.

A determination is made of whether or not growth (negative or positive) is apparent on the basis of a test of hypothesis. In the simplest of terms, a *test of hypothesis* is based on the following rationale: If there is only a very small probability that an event will occur (i.e., an observed value of a parameter greater or smaller than a given value) if a given hypothesis is true, and if on the very first try that event occurs, we have reason to question our hypothesis.

In this case we want to make inferences about whether or not: (a) There is reason to believe that positive reliability growth is apparent. (b) There is reason to believe that negative reliability growth is apparent. (c) There is reason to believe that no reliability growth is apparent.

This can be accomplished with a simple test of hypothesis. *If there is no change in the failure rate of the items over T_n*, then an observed parameter Z will lie in range between $-H$ and H for 90%, 95%, etc., of the time. The particular percentage depends on the particular value of H chosen.

$$-H \leqslant Z \leqslant H \qquad (11\text{-}25)$$

If Z exceeds that range, then there is reason to believe that the failure rate has changed. The particular bound exceeded provides reason to believe that failure rate is undergoing positive or negative growth. If

$$Z \leqslant -H \qquad (11\text{-}26)$$

then a measurable degree of positive reliability growth is indicated. If

$$Z \geqslant H \qquad (11\text{-}27)$$

then a measurable degree of negative reliability growth is indicated.

The values of H are generated under the assumption that if failure rate remains constant over the test duration, each calculated value of $\lambda(T_j)$ will be normally distributed around the constant value of failure rate with a calculable standard deviation.

If H is assigned the following values, then the probability, P, that a value of observed Z will be within the range $-H$ to H is

$$H = 1.65, \qquad P = 90\%$$
$$H = 1.95, \qquad P = 95\%$$
$$H = 2.33, \qquad P = 98\%$$

As H increases, the more significant the happenstance of Z exceeding the boundaries of (11-25), and if it does, the more likely that growth has occurred.

The value of Z is calculated from the test times and number of failures observed. See Mil-Hdbk-189 for derivation.

$$Z = \frac{\left(\sum\limits_{j}^{F} \sum\limits_{i}^{k} t_{ji}/F \right) - (T_n/2)}{T_n(12n)^{-0.5}} \qquad (11\text{-}28)$$

where

t_{ji} = the accumulated test hours on item i when the jth test failure occurs
F = total failures occurring over the test period
k = number of items on test
T_n = number of item hours making up the test

In summary, the Duane Model can be used to estimate $h(T)$, both current and projected. However, the estimates are still approximations and have no confidence bound associated with them. Additionally, unless you are careful, the data may not represent what it appears to represent.

11.6.4.2   The Duane Model and Growth Test Planning

If reliability growth programs have been performed in the past by your organization on the same type of product, on similar types of products, under similar reliability program structures and policies, then the values of s and a (associated with (11-17)) calculated from previous programs may be used for scoping current and future growth programs. The information will aid in determining whether or not, and to what degree growth testing will pay off and how extensive a reliability growth test is generally required to meet a given failure-rate objective. For example, if the probable range of values of a and s could be defined, and if it were known that a linear plot of $\ln \lambda(T)$ vs. $\ln T$, would result for a given product, then

$$\ln \lambda(T) = a - s \ln T \tag{11-29}$$

could be exercised to determine the potential range of T prior to test start.

Knowledge of such parameters is required for growth test planning and projections.

Values of a and s could, of course, be assigned by subjective judgment. That would be a fallback position if no other guidance were available. The choice of the course of action to be taken in general will follow the order of the following preferences:

1. Choose values of a and s similar to those which resulted from growth test programs within your organization for the same type of product; having the same order of complexity; for application of similar technology; with the same program emphasis on reliability.
2. Based on growth tests performed within your organization on other programs which cover a number of different products, determine the range of values of a and s which resulted and use engineering judgment to narrow that range taking into account product type, complexity, technology employed, and reliability program makeup.
3. Based on data available in the literature, choose the range of values of a and s which may be best associated with your program.

Mil-Std-1635(EC) "Reliability Growth Testing," a Navy standard, dated March 1977, places the probable range of s as

$$0.3 < s < 0.6 \qquad (11\text{-}30)$$

"Reliability Growth Prediction"[21] indicates a considerable body of historical data which defines boundaries on s as

$$0.1 < s < 0.7 \qquad (11\text{-}31)$$

Its actual value within that range has been attributed by some to depend on the amount of control, rigor, and efficiency by which failures are discovered and corrected. The study "Reliability Growth Testing Effectiveness,"[17] defines a value of $s > 0.5$ as one reflecting a hard-hitting aggressive reliability program, and a value of $s < 0.1$ as reflecting a program designed to resolve only obvious design deficiencies. The paper "Reliability Growth in Real Life,"[20] provided general observations that s is larger for analog than digital items, for items under more severe test environments, and for less mature developments. Tables 11-7a and 11-7b provide *observed* reliability growth rates s extracted from Ref. 17.

The parameter a in equation (11-17), which is generally interpreted as the parameter representing the *log of the initial failure rate at the test start*, is a more ambiguous variable to define than s. It is generally defined by a backwards extrapolation because the first measurement of $\lambda(T)$ cannot be made until after the first failure occurs and more than one estimate of $\lambda(T)$ is required in order to perform the extrapolation.

Experience gained from previous programs on similar items provides the best available estimates for this parameter. One commonly used rule of thumb used to size a is to assume that it represents a failure rate ten times higher than the desired failure-rate objective (or that of the Mil-Hdbk-217 prediction) (see Ref. 21). Errors in the definition of this parameter can have significant effect on growth test planning.

11.6.5 Reliability Growth: Performance and Analysis—The AMSAA Method

The AMSAA method gets its name from the organization which originated it, the U.S. Army Material Systems Analysis Activity. The details of its mathematical deviation are included in Mil-Hdbk-189. The AMSAA method is a statistical procedure as opposed to a graphical one, and is more helpful

TABLE 11-7

Item	Observed s-value
(a) Reliability Growth Rates Observed from Electronic Equipment in Improvement Programs	
Airborne teletypewriter	−0.10
Airborne radar altimeter	−0.08
Airborne search radar	+0.01
Airborne computer recorder	+0.11
Airborne HF communications	+0.12
Airborne UHF communications	+0.13
Airborne navigation set	+0.14
Shipborne acquisition radar	+0.14
Shipborne data processor	+0.17
Airborne radio navigation	+0.19
Airborne sonobuoy receiver	+0.19
Airborne tactical data display (A)	+0.19
Airborne radar scan converter	+0.23
Airborne tactical data display (B)	+0.24
Airborne inertial navigation	+0.30
(b) Reliability Growth Rates Observed for Different Hardware Systems in Development Tests	
Gatling type AA gun	+0.40
Hydromechanical devices	+0.49
Pulse transmitter, radar	+0.35
Continuous-wave transmitter	+0.35
Aircraft generators	+0.39
Analog receivers	+0.49
Airborne radar	+0.48
Airborne radar (UK)	+0.43
Digital computer	+0.48
Jet engines	+0.35
High-power equipment (power supply, microwave amps)	+0.30
Satellite comm. terminal	+0.34
Modem (digital comm. terminal)	+0.29

Source: Ref. 17.

in tracking, assessing, and projecting failure rate and MTBF than for planning purposes.

The plots of linear relationships of values of $\log \lambda(T)$ vs. $\log T$, which we discussed earlier, can be interpreted in more than one way. Given the linear plot of $\ln \lambda(T)$ vs. $\ln T$,

$$\ln \lambda(T) = a - s \ln T \qquad (11\text{-}32)$$

recall from Chapters 1 and 2 that

$$\lambda(T) = \int_0^T \frac{h(t)\,dt}{T} \tag{11-33}$$

where
$h(t)$ = the hazard rate or the instantaneous failure rate of the item at time t
$1/h(t) = \theta(t)$ = a measure of the MTBF of the item at time t (11-34)

Transforming (11-32) to an exponential form,

$$\lambda(T) = e^a T^{-s} \tag{11-35}$$

and from (11-33),

$$e^a T^{-s} = \int_0^T \frac{h(t)\,dt}{T} \tag{11-36}$$

or, more generally,

$$e^a t^{1-s} = \int h(t)\,dt \tag{11-37}$$

or

$$\frac{d(e^a t^{1-s})}{dt} = h(t) \tag{11-38}$$

or

$$e^a (1 - s)t^{-s} = h(t) \tag{11-39}$$

e^a and $-s$ are constants. If we represent them as

$$e^a = \frac{1}{\alpha}, \qquad s = 1 - \beta$$

then (11-39) can be expressed as

$$h(t) = \frac{\beta}{\alpha} t^{\beta - 1} \tag{11-40}$$

which represents the hazard function for the Weibull distribution (see Chapter 2).

However, since during the growth process the system design hardware is subject to very significant changes, the maintenance of a Weibull failure distribution over the entire test period is not theoretically possible (see Ref. 19 and Chapters 2 and 9). It can be shown, however, that the linear plot will occur under a slightly altered set of conditions which are both logical and theoretically sound and will also provide some desirable computational capabilities (see Mil-Hdbk-186 for more mathematical details).

For our purposes it will be sufficient to outline the principles and assumptions involved in the AMSAA method. The AMSAA method assumes that

- A growth test is broken down into separate phases.
- Design changes which improve reliability are implemented on all items under test during the breaks between phases.
- During each phase hazard rate and hence *failure rate is constant*, and the number of failures occurring in each phase follow a nonhomogeneous Poisson distribution.
- After testing is complete the hazard, $h(t)$, will remain constant at the value given by the last phase, i.e., will become equivalent to failure rate $h(T)$.
- At the end of any phase after a total of T hours of test, the hazard (failure rate) can be evaluated as

$$h(T) = \frac{1}{\theta(T)} = \frac{\beta}{\alpha} T^{\beta-1} \qquad (11\text{-}41)$$

where
$h(T)$ = the hazard rate (rate of failure) that the item will exhibit after T hours
$\theta(T)$ = the MTBF that the item will exhibit after T hours of test
T = the total term of the test program when testing was completed

Relationships based on the assumptions outlined above have been developed for estimating values of β and α for use in relationship (11-41). And (11-41) can be used directly to assess $h(T)$ and $\theta(T)$ at test termination or at any projected point in time.

Let

$\hat{\beta}$ = the estimate of β generated by a particular set of growth test data
$\hat{\alpha}$ = the estimate of α generated by a particular set of growth test data

Estimating relationships for β and α then take the following forms for a time-truncated test which generates a significant quantity of failures (see

Mil-Hdbk-189, "Reliability Growth Management," for derivation detail):

$$\hat{\beta} = \frac{F}{\sum_{j=1}^{F} \ln(T_n/t_j)} \tag{11-42}$$

where
 F = total number of failures in total item test period T_n
 T_n = total item hours of the test
 t_j = cumulative number of hours to the jth failure

$$\hat{\alpha} = \frac{T_n^{\hat{\beta}}}{F} \tag{11-43}$$

and hence, for estimation purposes,

$$h(T) = \frac{1}{\theta(T)} = \frac{\hat{\beta}}{\hat{\alpha}} T^{\hat{\beta}-1} \tag{11-44}$$

Procedures and tables have been developed which allow for the structure of confidence bounds on (11-44), and for goodness-of-fit tests to provide confidence that the data resulting from the test is appropriate for application of the AMSAA model. These are summarized in Mil-Hdbk-781, "Reliability Test Methods, Plans and Environments," and detailed in Mil-Hdbk-186, "Reliability Growth Management," both of which are readily available (see Section 3.5).

11.7   ENVIRONMENTAL STRESS SCREENING: IMPROVING RELIABILITY BY REMOVING LATENT DEFECTS

Environmental stress screening (ESS) may be defined as a process or a series of processes involving the application of mechanical, thermal, or other stresses to items in order to precipitate latent defects to failure. It is a task which is usually implemented during the production phase of an engineering program, even though its planning origins are sometimes rooted in the full-scale engineering development phase.

We described reliability growth as a means for improving reliability through design improvement. The actual reliability attained even from the best of design efforts is affected by the defects in parts, materials, and workmanship or those defects otherwise introduced during the fabrication/

production process. Some such defects are immediately apparent during item inspection and performance tests before introduction to use operation. Other defects are latent (see Section 6.6) and will not be discovered until they precipitate to failure during operational use. Examples of such defects include imperfect bonds and other defects in parts and devices, soldering defects, broken strands of wire, improper crimps, etching defects, poor connector contacts, etc. It is to these defects that ESS is directed. To put things in proper perspective, however:

1. Not all defects result in failure over service life. When and whether a failure results depends on the exact nature of the defect, the stresses imposed on the defect, and its failure mechanisms. A part with a given defect which is subjected to very low operational and application stresses might very well survive the entire service life period. The same part, if subjected to significantly higher operational or application stresses, could fail during early service life.
2. Not all latent defects have their failure mechanisms accelerated by the same set of screens. Hence, a combination of screens as opposed to a single screen must be considered.
3. It is only those latent defects which can be precipitated to failure by the particular screens/treatments imposed in a factory production environment which are candidates for ESS.
4. While not structured to specifically discern design or part/component specification defects, analysis of failures generated by the ESS process can pinpoint some such deficiencies which were overlooked or escaped notice through other tasks.

While ESS is most sensitive to the elimination of those failures which tend to dominate the reliability of items during early life, studies have shown that the results of ESS can go beyond "early" life. For example, in a 1981 survey by The Institute of Environmental Sciences, "Environmental Stress Screening Guidelines," evidence indicated that reliability could be improved by 25–90% through ESS application. Another study, "Improved Operational Readiness Through Environmental Stress Screening," [22] used field data to assess the difference between the mean time between removals of five different types of units. Quantities of each type were subjected to ESS in the contractor's plant, others were not. The findings, expressed in terms of mean flying hours between removal (MFHBR), are shown in Table 11-8 (extracted in part from the report).

It would be most desirable to be able to structure a test or a combination of tests/screens which could identify or precipitate a failure for virtually every defect of every magnitude within an item. The reality is however, that: (a)

TABLE 11-8   Reliability Improvement Through ESS

Equipment	Non-ESS MFHBR	ESS MFHBR
A	32	70
B	66	127
C	83	149
D	92	278
E	570	1110

Source: Ref. 22.

Tests and screens sensitive to all types and magnitudes of defects do not exist or are at present beyond the state of the art. (b) Tests and screens structured to identify given defects are not 100% effective; only a proportion of the type of defects sensitive to the screen are usually identified. (c) In some cases tests and screens associated with certain defects are not cost-effective to run; for example, 100% inspection and test of each part.

The types of screens/tests most often associated with ESS include constant-temperature, cycle temperature, random vibration, swept sine vibration, on and off cycling, high-temperature burn in, and cold soaks. Each can be varied by increasing respective stress levels and stress duration.

11.7.1   Clues to When An ESS Program Will Pay Off

The payoffs from an ESS program are a decreased number of failures and lower support costs. However, the cost of implementing an ESS program, including test facilities, engineering and technical manpower, failure analysis, tests needed to isolated failures in assemblies and units, record keeping, etc., can be significant. While most items will benefit from application of ESS procedures, the payoff resulting must be greater than the cost incurred. It is important, therefore, to be able to make a judgment about potential payoff for an ESS program as early in the development phase as possible.

It would be helpful, then, to identify those general factors which tend to affect the quantity of defects in a given item and hence to recognize situations where ESS will provide the greatest general payoff. The following factors to that end are extracted, in part, from the RADC technical report "RADC Guide to Environmental Stress Screening,"[23] which describes the rationale and procedures associated with ESS programs.

1. *Complexity.* The quantity and type of parts and interconnections used in the product affects defect density. Increased complexity creates more opportunities for defects.

2. *Part quality level/grade.* The quality levels of parts are established by part screening requirements. The number of defects which remain in a lot of screened parts is determined by the type and extent of screening and testing to which the parts are subjected under Mil-Std and other screening requirements. (See Sections 3.5.2, 6.2, and 6.3.)

3. *Field stress environment.* The stress conditions to which the equipment will be exposed in the field environment will affect the proportion of defects which will precipitate to failure. A defect may be precipitated to early failure in a harsh field operating environment, but may survive product life in a benign field environment. Hence, in general, the harsher the field environment, the more need for ESS.

4. *Process maturity.* Production requires time to identify and correct planning and process problems, train personnel, and establish vendor and process controls. Maturity is dependent on both volume and how long production has been taking place; on how long the particular production processes involved have been utilized. Hence, the need for ESS is inversely related to the degree of process maturity.

5. *Packaging density.* Electronic assemblies with high part and wiring density are more susceptible to process, workmanship, and temperature-induced defects due to smaller error margins, increased rework difficulty, and thermal control problems.

6. *Past experience.* Prior screening experience on similar items can provide clues regarding what can be expected and which screens can be employed. If screening programs on similar items, using the same general manufacturing processes under the same general manufacturing policies, have yielded a high defect fallout in the past, they are likely to have the same effect on the current development. If past programs involving the same or modified versions of an item, its components, or its parts, have experienced a high incidence of failures attributed to latent defects, expect the same to occur with the current development (unless changes in parts, processes, or policies have been made and verified). Consider introducing other screens to reduce them.

11.7.2 The Mechanics of an ESS Program

An ESS program requires that decisions be made concerning (1) The level(s) at which screening will be carried out, e.g., the part level, assembly level, unit level, equipment level; the facilities required to apply the screens at each level; and the instrumentation and test regimen required to support test performance. (2) The nature of the screens to be applied at each level.

11.7.2.1   Screening Levels and Regimens

The lowest level screen possible is at the part level. In Section 6.6 we described the impact of latent part defects on reliability, developed relationships which quantified the effect, and related the incidence of latent defects in a product to the quality of parts procured. In general, as the quality of parts procured increases, the tests, screens, and burn-in that are an *integral part of the general quality process* will decrease the latent defects associated with manufactured lots of such parts.

For some reliability-critical parts, the developer may require additional burn-in and test on each lot of parts delivered. In some instances such actions may be performed on a sampling as opposed to a 100% basis. It may be performed by the vendor, a third-party test facility, or within the developer's organization.

ESS is, however, more generally focused at the assembly level (e.g., a printed circuit board), the unit level (an item comprised of assemblies, e.g., a shop-replaceable unit), or the equipment or system level (the item under development.)

Many different parts go into a given assembly; many different assemblies go into a given unit; many different units go into a given equipment or system. An ESS program usually employs screens at each level of indenture to cull out as many defects as possible before the item is combined with other items as part of the next higher level of indenture. This is done for two primary reasons.

1. To reduce the proportion of latent defects in the items formed at the next higher level of indenture. We saw in Chapter 6 (see in particular Section 6.6, which discusses latent defects) that if

p = the proportion of parts in a given population having latent defects
n = the number of parts randomly chosen from that population to form an assembly

then the proportion $p(x)$ of assemblies produced which will contain x parts which have latent defects can be expressed as

$$p(x) = \frac{n!}{(n-x)!\,x!} p^x (1-p)^{n-x} \qquad (11\text{-}45)$$

and Chart 6-5 showed that if 0.005 of the population of parts has latent defects and if each assembly contains 100 parts, then almost 40% of the assemblies produced would contain one or more defective parts. If 0.02 of the population of parts has latent defects, then almost 87% of the assemblies

TABLE 11-9   Average Rework Costs for Different
                       Screening Levels

Parts/printed circuit board wiring	$1 −5
Assemblies	$20 −50
Units	$250−500
Equipment/system	$500−1000

produced would contain one or more defective parts. Hence, a relatively small increase or decrease in defective parts, assemblies, or units will have significant impact on the number of defects found at the next level of indenture. Also note from Chart 6-5 that a significant number of assemblies will contain more than one defect. Multiple defects in an assembly or higher-level unit mean either multiple reworks on the same unit during its screening process or multiple failures which occur during the same screen, making failure isolation difficult and ambiguous.

2. Going hand in hand with the above is the fact that rework costs, which include the time and resources to isolate the failure, escalate with respect to the indenture level. Average rework costs associated with each general level of screening are shown in Table 11-9 (see Ref. 23). These are presented more as guides for sizing the general magnitudes of costs than as indicators of actual costs.

As the level at which screening is performed increases, the size or quantity of test resources also increases: temperature chambers, vibration tables, mounting fixtures, controls, test instrumentation, test fixtures, etc.
No matter what level of screening is chosen, the test regimen to be followed will be similar:

1. The item is tested prior to the stress screen to assure that no failure exists prior to initiation of the screen.

2. The item is tested after completion of the stress screen using the same test procedures applied before the screen to determine whether or not failure has occurred as a result of the screen.

3. The stress screen must represent a powered use of the item under a stress representative of normal operating conditions. Specific guidelines associated with the "severity" of a stress test will be addressed in more detail in a following section.

4. Failures are to be analyzed.

5. Data is collected as production maturity increases to determine defect rates associated with each screen, sensitivity of the item to various screens, and screening strategies.

6. Results are closely monitored to ensure both technical and cost effectiveness. Changes to screens are made as required.

11.7.2.2   The Nature of Screens,
Their Applications, and Effectiveness

The screens most commonly employed take the form of application of thermal or vibration stresses over time. In general, the types of defects detectable from each of those two kinds of test (see Ref. 23) are shown in Chart 11-6. Generally, vibration screens are considered more effective for precipitating workmanship defects; thermal screens are considered more effective for electronic part defects.

Vibration and thermal screens can be further broken down into temperature-cycle screens, constant-temperature screens, random vibration screens, and swept sine vibration screens. Chart 11-7 (see Ref. 23) provides general

CHART 11-6 Assembly-Level Defect Types Precipitated by Thermal and Vibration Screens

Defect Type Detected	Thermal Screen	Vibration Screen
Defective part	×	×
Broken part	×	×
Improperly installed part	×	×
Solder connection	×	×
PCB etch, shorts and opens	×	×
Loose contact		×
Wire insulation	×	
Loose wire termination	×	×
Improper crimp or mating	×	
Contamination	×	
Debris		×
Loose hardware		×
Chafed, pinched wires		×
Parameter drift	×	
Hermetic seal failure	×	
Adjacent boards/parts shorting		×

Source: Ref. 23.

information on the efficacy of various types of screens to various levels of assembly.

The screens chosen are not ordinarily outside the maximum temperature or vibration boundaries associated with a given part, assembly, unit, or equipment. Both thermal and vibration screens are generally applied within the following boundaries:

- The maximum temperature and vibration to which a given item is exposed, which is equal to or less than the maximum operating temperature and vibration rating of any part or material making up the item.
- The minimum temperature to which a given item is exposed will be equal to or greater than the minimum operating temperature rating of any part or material making up the item.

Depending on its nature, the following must also be established/defined for each screen:

- The temperature and vibration range, the difference between the minimum and maximum temperature and vibration level.
- The period of time during which an item will be maintained at a given minimum or maximum temperature and vibration level.
- The average rate of change with time between minimum and maximum operating temperature and vibration levels or vice versa.
- The number of transitions between temperature extremes.
- The duration of the screen in terms of time.

The nature of the screens performed should be tailored to the particular characteristics of the item being developed and the manufacturing processes applied. However, investigations, surveys, and studies have been performed which provide general guidance as to the efficacy of various thermal and vibration screens. In addition, standards and guides have been developed recommending given screening tests and procedures. These include:

- Nav Mat-P-9492, "Navy Manufacturing Screening Program," 1979
- Mil-Std-2164 (EC), "Environmental Stress Screening Process for Electronic Equipment," 1985
- "Environmental Stress Screening Guidelines," The Institute of Environmental Sciences, 1981
- "Environmental Burn-in Effectiveness," AFWAL-TR-80-3086
- "Stress Screening of Electronic Hardware," RADC-TR-82-87
- "Environmental Stress Screening," RADC-TR-86-149
- "Improved Operational Readiness Through Environmental Stress Screening," RADC-TR-87-225
- "RADC Guide to Environmental Stress Screening," RADC-TR-86-138

CHART 11-7 Guidelines for Initial Screen Selection and Placement

| Level | Selection[a] | | | | Placement | |
| of | | | | | Advantages | Disadvantages |
Assembly	Temp. Cycle	Const. Temp.	Rand. Vib.	S.S. Vib.		
Assembly	$E^{(1)}$	$M^{(2)}$	$M^{(3)}$	N	• Cost per flaw precipitated is lowest (unpowered screens) • Small size permits batch screening • Low thermal mass allows high rates of temperature change • Temperature range greater than operating range allowable	• Test detection efficiency is relatively low • Test equipment cost for powered screens is high

(1) Particularly if power is applied and performance is monitored at temperature extremes.

(2) Effective where assemblies contain complex devices (RAMs, microprocessors, hybrids).

(3) Effectiveness highly dependent on assembly structure. Not effective for small, stiff PWAs.

				Advantages	Disadvantages
Unit	E	M	E	M	• Relatively easy to power and monitor performance during screen • Higher test detection efficiency than assembly level • Assembly interconnections (e.g., wiring backplane) are screened	• Thermal mass precludes high rates of change, or requires costly facilities • Cost per flaw significantly higher than assembly level • Temperature range reduced from assembly level
System	E	M	E	M	• All potential sources of flaws are screened • Unit interoperability flaws detected • High test detection efficiency	• Difficult and costly to test at temperature extremes • Mass precludes use of effective vibration screens, or makes use costly • Cost per flaw is highest

[a] E = effective; M = marginally effective; N = not effective.

Source: Ref. 23.

As an aid in the preliminary planning and choice of screens, empirical relationships have been generated based upon estimates derived from a combination of actual screening program data, experiments and published data. The relationships can generally be used (a) in the absence of other, more specific or precise data concerning the effects of screens on the types of latent defects particular to a given item and its fabrication process; (b) in the initial planning of the structure of a screening program; (c) in the fine tuning of specific screening stresses. They provide estimates of *screening strengths*, SS, a measure of the proportion of latent defects which will be precipitated to failure by a given screen.

Measures of SS for given screens were first developed in the RADC report "Stress Screening of Electronic Hardware," [24] and later refined in the RADC report "Environmental Stress Screening." [25] Screening strength relationships were generated for random vibration screens, swept sine vibration screens, single-frequency vibration screens, cycled thermal screens, and constant thermal screens. (Bear in mind that these are empirical relationships based on data available and subject to assumptions during their development. As such, they should be used with a degree of caution, or when no better information is available.)

The relationships developed took the following forms. See Refs. 24 and 25 for their derivation.

1. *Random vibration screening strength, SS_{rv}.*

$$SS_{rv} = 1 - \exp[-0.0046(G_r)^{1.71}(t)] \tag{11-46}$$

where

G_r = the rms value of the applied power (power spectral density, g^2/Hz) over the vibration frequency spectrum.

t = the duration of the applied vibration excitation, in minutes.

2. *Swept sine vibration screening strength, SS_{ssv}.*

$$SS_{ssv} = 1 - [-0.000727(G_a)^{0.863}(t)] \tag{11-47}$$

where

G_a = the constant acceleration in terms of g's applied to the equipment being screened throughout the frequency range above 40 Hz. The value of G_a to be applied below 40 Hz may be less.

t = time in minutes.

3. *Single-frequency vibration screening strengths, SS_{sfv}.*

$$SS_{sfv} = 1 - \exp[-0.00047(G)^{0.490}(t)] \qquad (11\text{-}48)$$

where
G = g-level
t = time, in minutes

4. *Cycled thermal screens screening strengths, SS_{tc}.*

$$SS_{tc} = 1 - \exp\{-0.0017(R + 0.6)^{0.6}[\ln(e + DT]^3(N_{cyc})\} \qquad (11\text{-}49)$$

where
R = range. This is the difference between the maximum and minimum applied external (chamber) temperature or inlet cooling air temperature for independently cooled items, ($T_{max} - T_{min}$). Temperatures are in °C.
DT = temperature rate of change. This is the average value of the temperature rate of change of the item being screened as it transitions between the temperature extremes. DT is in °C/min.

$$DT = \frac{1}{2}\left(\frac{T_{max} - T_{min}}{t_1} + \frac{T_{max} - T_{min}}{t_2}\right)$$

where
t_1 = the transition time from T_{min} to T_{max} (in minutes)
t_2 = the transition time from T_{max} to T_{min} (in minutes)
N_{cyc} = number of temperature cycles

5. *Constant thermal stress screen screening strength, SS_{ct}.*

$$SS_{ct} = 1 - \exp[-0.0017(R + 0.6)^{0.6}(t)] \qquad (11\text{-}50)$$

where
R = the temperature range, defined as the absolute value of the difference between the screening temperature (in °C) and 25°C.
t = the screening time in hours

References
1. Mil-Std-1629. "Failure Modes and Effects Analysis."
2. Mil-Hdbk-338. "Reliability Design Handbook."
3. Coutinho, J. (1974). *Introduction to Systems Assurance.* U.S. Army Material Systems Analysis Agency Special Publication SP-9.

4. Dussault, H. (1983). "The Evolution and Practical Application of Failure Modes and Effects Analyses." RADC Technical Report RADC-TR-83-72. Available through the Defense Documentation Center. Report AD-A131358.

5. Barbour, G. (1977). "Failure Modes and Effects Analysis by Matrix Methods." Proceedings, 1977 Annual Reliability and Maintainability Symposium, IEEE, Piscataway, NJ.

6. Schneeweiss, W. (1985). "Fault Tree Analysis Using A Binary Decision Tree." *IEEE Transactions on Reliability*, Dec.

7. Rushdi, A. (1985). "Uncertainty Analysis of Fault Tree Outputs." *IEEE Transactions on Reliability*, Dec.

8. Helman, P. and Rosenthal, A. (1989). "A Decomposition Scheme for the Analysis of Fault Trees and Other Combinatorial Circuits." *IEEE Transactions on Reliability*, Aug.

9. Jackson, T. (1986). "Integration of Sneak Circuit Analysis With FMEA." Proceedings, 1986 Annual Reliability and Maintainability Symposium, IEEE, Piscataway, NJ.

10. Boeing Aerospace Company (1982). "Sneak Analysis Application Guidelines." RADC Technical Report RADC-TR-82-179. Available from the Defense Documentation Center. Report AD-A118479L.

11. SoHaR, Inc. (1989). "Sneak Circuit Analysis for the Common Man." RADC Report RADC-TR-89-223. Available from the Defense Documentation Center. Report AD-A215275.

12. Hughes Aircraft Company (1982). "RADC Thermal Guide for Reliability Engineers." RADC Report RADC-TR-82-172. Available from the Defense Documentation Center. Report AD-A118839.

13. Boeing Aerospace Company (1982). "Reliability Parts Derating Guidelines." RADC Report RADC-TR-82-177. Available from the Defense Documentation Center. Report AD-A120367.

14. Martin Marietta Corp. (1984). "Reliability Derating Procedures." RADC Technical Report RADC-TR-254. Available from the Defense Documentation Center. Report AD-A153744.

15. Morris, S. et al. (1988). "RADC Reliability Engineer's Toolkit." Available from the Reliability Analysis Center, P.O. Box 4700 Rome, NY 13440-8200.

16. Hughes Aircraft Company (1975). "Reliability Growth Studies." RADC Technical Report RADC-TR-75-253. Available from the Defense Documentation Center. Report AD-A023926.

17. MacDiarmid, P. and Morris, S. (1984). "Reliability Growth Testing Effectiveness." RADC Technical Report RADC-TR-84-20. Available from the Defense Documentation Center. Report AD-A141232.

18. Duane, J. (1964). "Learning Curve Approach to Reliability Monitoring." *IEEE Transactions Aerospace*, Vol. 2.

19. Crow, L. (1975). "On Tracking Reliability Growth." Proceedings, 1975 Annual Reliability Symposium, IEEE, Piscataway, NJ.

20. Codier, E. (1968). "Reliability Growth in Real Life." Proceedings, 1968 Annual Reliability Symposium, IEEE, Piscataway, NJ.

21. The Analytical Sciences Corp. (1986). "Reliability Growth Prediction." RADC Technical Report RADC-TR-86-148. Available from the Defense Documentation Center. Report AD-A176128.
22. Grumman Corporation (1987). "Improved Readiness Through Environmental Stress Screening." RADC Technical Report RADC-TR-87-225. Available from the Defense Documentation Center. Report AD-A193788.
23. Fiorentino, E. (1986). "RADC Guide to Environmental Stress Screening." RADC Technical Report RADC-TR-86-138. Available from the Defense Documentation Center. Report AD-A174333.
24. Hughes Aircraft Company (1982). "Stress Screening of Electronic Hardware." RADC Technical Report RADC-TR-82-87. Available from the Defense Documentation Center. Report AD-A118261.
25. Hughes Aircraft Company (1986). "Environmental Stress Screening." RADC Technical Report RADC-TR-86-149. Available from the Defense Documentation Center, Report AD-A176847.

12

Reliability Measurement and Demonstration

Reliability measurement and demonstration are two separate but related areas. One of the problems most often encountered is the fact that often, and erroneously, the concepts are used interchangeably and indiscriminately. For the purposes of this discussion we will define *reliability measurement* as an estimate of reliability characteristics under given conditions, based on available test or operational data. If desired, a confidence bound may be associated with the estimate. More will be said about confidence bounds as we proceed. *Reliability demonstration* is a test performed under given conditions on an item to determine whether or not a given reliability objective has been achieved. Consumer and producer risks are commonly associated with such tests. More will be said about these as we proceed.

The former tells program engineers and managers how well they have succeeded thus far, or how well a given item is performing. The latter determines acceptance or rejection of an item or its design.

12.1 RELIABILITY MEASUREMENT AND ESTIMATION

Estimators of the reliability of an item or component are provided to establish a baseline for initial (conceptual estimate based on the performance of existing similar items) reliability specification and prediction in a new development program; to determine the reliability of an off-the-shelf item (part, component, equipment, etc.) to be used in a development program; to determine the reliability performance of a particular item in field or test use; and to determine the reliability performance of an item under development or under test prior to demonstration. The measures discussed

482

in all cases will be those which are based on available information on operating time, observations of number of failures occurring in that period, and the quantities of items on test or in use.

Reliability measurements are commonly performed on items of virtually every type, "single-shot" items such as ordnance and missiles, repairable and non-repairable parts, components, equipments, and systems. The nature of the measurements and the procedures followed on a case-to-case basis generally depend on the characteristics of the item. While there are many different statistical estimation procedures available, the discussions which follow will concentrate on those which are most commonly used and easy to grasp.

12.1.1   Confidence Bounds and Intervals: What They Are

Confidence bounds and intervals are used in the reliability measurement process to express reliability estimates of a given parameter in terms of a *range* of values. They express the probability that the range of values, so structured, will include the actual value of the parameter. A confidence bound expresses the probability P that the actual value of a parameter is included in a range of values greater than a specific estimate (the estimated value is referred to as the lower confidence bound), or a range of values less than a specific estimate (that estimate value is referred to as the upper confidence bound). A *confidence interval* is made up of a lower and upper confidence bound and expresses the probability P that the actual value of the parameter is included in the range between the upper and lower bounds. Bear in mind that we are *not* saying, and we cannot say or infer that the probability is equal to P that the value of the parameter is actually less than or greater than a particular bound or that the probability is P that the value of the parameter is between two bounds. What is being said, and this is the proper interpretation of a confidence bound or interval, is that the probability is equal to P that the bounds *we define* will contain the actual value of the parameter.

For example, if we were to: (a) take many, many samples of the same size (where, say, each sample contained the same quantity of items) from the same population, (b) expose each to the same test, and (c) calculate a given reliability parameter for each *sample* and a *90%* confidence bound (upper or lower) for each sample based on the calculation, then approximately *90% of all confidence bounds* we generate will include the true value of the parameter. If we were to calculate from each of the same samples an 80% *confidence interval*, then approximately 80% of the confidence intervals would include the true value of the parameter.

12.1.1.1 Confidence Bounds and Intervals: Why Do We Use Them?

If it were desired to estimate a parameter (i.e., a mean, a proportion, a percentile, etc.), the most straightforward means that would occur to most people would be to take a sample of such items, say n in number, and subject the sample to trial use. For example, if we were interested in estimating the proportion of items in a population, p, which could be successfully used, for example, rockets, we could acquire n items, subject them all to firing and determine, r, the number of failures which occur. Then the quantity

$$\frac{n-r}{n} = \text{the proportion of the sample which performed successfully}$$

and $(n-r)/n$ would represent an estimate of p. It would be tempting to take that action and base a conclusion on it; however, it could also be misleading.

Suppose, for example, that a large population of items existed and p represented the proportion of the population which was actually good. If samples of size n from the large population were collected and each sample of size n was tested, a value of $(n-r)/n_i$ for each sample would result:

$$(n-r/n)_i = \text{ratio of items that successfully perform to total number of items in sample } i$$

Such values would vary from sample to sample. Some sample values would be greater than p. Some would be less than p. As a result, a single point estimate might represent a significant variation from the actual value of p.

For any observed value of $(n-r)/n$, however, upper and lower limits on the probable values of p capable of resulting in that value of $(n-r)/n$ can be structured. Such limits form the basis for confidence bounds and intervals.

12.1.2 Reliability Measurement of a "Single-Shot" or Go–No Go Operation

This category of measurement can include missiles, rockets, ordnance, or any item subject to go–no go operation. For such items the most commonly used reliability measures are the fraction or proportion of items which perform successfully equals p, or alternately, the probability equals p that the item will perform successfully.

Whenever such measures as proportion, fraction or probability must be estimated, the binomial distribution (discussed in Chapter 2) is commonly used to provide estimates of such parameters. The binomial distribution directly defines:

$P(r)$, the probability that r out of n trials, tests, firings, etc., will be failures, if p is of a certain value, or

$P(s)$, the probability that s out of n trials will be successes, if p is of a certain value, where $s = (n - r)$.

In addition it may be utilized *to estimate* p as a result of observing a given number of successes, s, or failures, r, over a given number of trials, n. It is this latter application which will be of most interest. In order to best get a feel for its use as an estimator, however, a knowledge of its application in assessing $P(r)$ and $P(s)$ is needed.

For example, if it is desired to determine the probability that r out of n missile firings will be failures, $P(r)$, when the probability of a successful fire for each missile is p, $P(r)$ can be expressed as:

$$P(r) = \frac{n!}{(n - r)!r!} (1 - p)^r p^{n-r} \qquad (12\text{-}1)$$

The probability that s out of n missile firings, $P(s)$, will be successes, when the probability of a successful fire is p is similarly expressed as

$$P(s) = \frac{n}{(n - s)!s!} p^s (1 - p)^{n-s} \qquad (12\text{-}2)$$

where

n = number of items tested
r = number of items that fail
s = number of items that perform successfully
$P(r)$, $P(s)$, and p are defined as above

As will be seen, either (12-1) or (12-2) can be used to provide estimates of the p parameter. The discussions which follow will be framed around (12-1).

12.1.2.1 *Reliability Measurement When Sample Sizes Are Small*

When the sample size upon which the estimate of p is to be based is small, the following exact procedure should be followed.

The cumulative form of the binomial distribution is employed, which can be expressed alternately as

$$P[(r > r_0)/p] = \sum_{r=r_0+1}^{n} \frac{n!}{(n - r)!r!} (1 - p)^r p^{n-r} \qquad (12\text{-}3)$$

or

$$P[(r > r_0)/p] = 1 - \sum_{r=0}^{r_0} \frac{n!}{(n-r)!r!} (1-p)^r p^{n-r} \qquad (12\text{-}4)$$

where

$P[(r > r_0)/p]$ = probability of experience more than r_0 failures if the actual proportion of items in the *population* that will perform successfuly is p

r_0 = number of failures experienced

$(1-p)$ = actual proportion of the items which are defective

p = actual proportion of the items which will perform satisfactorily (not defective)

n = the number of items in the sample

Equation (12-3) facilitates the computation when the number of items that perform successfully are smaller than the number that fail. Equation (12-4) facilitates computation when the opposite situation is true.

The mechanics for determining a lower-bound value for p (designated as p_a), to a given degree of confidence, P_A, based on the results from a single sample of size n, follows directly from the application of (12-3):

$$P[(r > r_0)/p_a] = P_A = \sum_{r=r_0+1}^{n} \frac{n!}{(n-r)!r!} (1-p_a)^r p_a^{n-r} \qquad (12\text{-}5)$$

where in essence a probability value P_A is assigned (say, $P_A = 0.9$), which becomes the confidence value; the failures experienced by the sample is equal to r_0; the number of items in the sample is taken as n. P_A = the degree of confidence desired = the probability that *more than* r_0 failures would be expected to occur in any sample if the actual proportion of good items in the population was as low (as bad) as p_a.

Therefore, the values of P_A (representing the confidence bound chosen), n, and r_0 associated with the sample are inserted in (12-5) and p_a is ascertained by trial and error. For example, if $P_A = 0.90$ was chosen, and if the actual proportion of *good items in the population* was as low (bad) as p_a, a *single test result* showing as small a number of failures of r_0 or less would be expected to happen only 10% of the time.

Hence p_a is interpreted as the P_A lower confidence bound for p, i.e., the probability is P_A that the actual value of p will be included in the range

$$p_a \leqslant p \leqslant 1 \qquad (12\text{-}6a)$$

Note that since (12-5) is a discrete relationship, in many instances *no* value of p_a for a given value of r_0 will yield exactly P_A. The usual course of action under those circumstances is to choose the largest value of p_a such that

$$\sum_{r=r_0+1}^{n} \frac{n!}{(n-r)!r!}(1-p_a)^r p_a^{n-r} \geqslant P_A \tag{12-7}$$

This provides for a confidence of at least P_A.

Similarly, a P_B upper confidence bound for p (designated as p_b), can be determined from the cumulative form of the binomial distribution and expressed as

$$P[(r < r_0)/p_b] = P_B = \sum_{r=0}^{r_0-1} \frac{n!}{(n-r)!r!}(1-p_b)^r (p_b)^{n-r} \tag{12-8}$$

or

$$P[(r < r_0)/p_b] = P_B = 1 - \sum_{r=r_0}^{n} \frac{n!}{(n-r)!r!}(1-p_b)^r (p_b)^{n-r} \tag{12-9}$$

where P_B = the degree of confidence desired = the probability that fewer than r_0 failures will occur in any sample if the actual proportion of good items in the population was as high (as good) as p_b; p_b = an upper-bound quality level.

Therefore, a value of P_B is chosen, the values of n and r_0 associated with the sample are inserted into (12-8) or (12-9) and the upper-bound value p_b is ascertained. Note that since (12-9) is a discrete relationship, in many instances *no* value of p_b for a given value of r_0 will yield exactly P_B. The usual course of action under those circumstances is to choose the *smallest value of* p_b such that

$$\sum_{r=0}^{r_0-1} \frac{n!}{(n-r)!r!}(1-p_b)^r p_b^{n-r} \geqslant P_B$$

For example, if $P_B = 0.90$ was chosen, a sampling result showing a number of failures *equal to or greater than* r_0 would be expected to happen only 10% of the time if the actual proportion of good items in the population was as high as p_b.

Hence p_b is interpreted as the P_B upper confidence bound for p, i.e., the probability is P_B that the actual value p will be included in the range

$$0 \leqslant p \leqslant p_b \tag{12-6b}$$

The confidence interval, P_I, is defined as a function of the upper and lower confidence bounds:

$$P_I = P_A - (1 - P_B) \tag{12-10}$$

where the range for p in that interval is given as $p_a \leqslant p \leqslant p_b$. Commonly, when generating confidence intervals, P_A is chosen equal to P_B.

Charts are available in many statistics texts which directly provide information on such confidence bounds and intervals. They are usually contained in sections identified as confidence belts for proportion, confidence intervals for the binomial distribution, or confidence intervals for proportion. See Refs. 1 to 4 for examples of such charts and tables.

Example application   A sample of 22 rockets are test-fired; 19 fire successfully, 3 fail to perform satisfactorily. Define the 90% upper and lower confidence bounds and the 80% confidence interval (i.e., probability is 0.80 that the true value of p will be contained between p_a and p_b) for p.

$$P_A = P_B = 0.90$$

A value for p_a is determined given n, r_0, and P_A such that

$$P[(r > r_0)/p_a] = \sum_{r=r_0+1}^{n} \frac{n!}{(n-r)!r!}(1-p_a)^r p_a^{n-r} \geqslant 0.90$$

From a table of the cumulative binomial distribution, for $n = 22$, $r_0 = 3$, the value which comes closest to $P_A = 0.9$, is 0.905, and that corresponds to $p_a \sim 0.72$.

$$P[(r > r_0)/p_a] = \sum_{r=4}^{22} \frac{n!}{(n-r)!r!}(1-p_a)^r p_a^{n-r} = 0.905$$

where $p_a = 0.72$

Hence, the probability is equal to 0.905 that we will be correct in saying the true value of p will be contained in an interval bounded by 0.72 and 1 and hence 0.72 is considered a lower bound.

With respect to the upper bound, the procedure is much the same.

$$P[(r < r_0)/p_b] = \sum_{r=0}^{r_0-1} \frac{n!}{(n-r)!r!}(1-p_b)^r p_b^{n-r} \geqslant 0.90$$

From a table of the cumulative binomial distribution, if $n = 22$, $r_0 = 3$, the value which comes closest to $P_B = 0.9$ is 0.905 and that corresponds to a value of $p_b \sim 0.95$.

$$P[(r < r_0)/p_b] = \sum_{r=0}^{2} \frac{n!}{(n-r)!r!}(1 - p_b)^r p_b^{n-r} = 0.905$$

where $p_b = 0.95$

Hence, 0.905 of the time we will be correct in saying that the true value of p will be contained in an interval bounded by 0 and 0.95 and hence 0.95 is considered an upper bound.

Since $P_A - (1 - P_B)$, the probability that we will be correct in saying that p will be between p_a and p_b, is approximately 0.80. Hence the probability is approximately 0.80 that we will be correct in saying that the true value of p will be between 0.72 and 0.95.

12.1.2.2   Reliability Measures When Sample Sizes Are Large

Relationships (12-5) and (12-7) can be utilized to provide accurate confidence bounds and intervals for all sample sizes, but when the sample size is very large they become cumbersome to apply without recourse to a computer. In such cases, when estimates are required, approximation procedures can be applied to define such confidence measures. Two criteria establish a general rule of thumb for the reasonable application of the approximation procedures which follow: (1) a minimum sample size of 30 and (2) occurrence of *more* than five failures and five successes.

The approximation is based on the fact that when a sample size, n, is large, the value of p_0, the proportion of items in a sample which perform successfully, where $p_0 = (n - r_0)/n$, will be normally distributed over all samples of size n taken from the same population. The distribution in question will have a mean of p and a standard deviation of $[p(1 - p)/n]^{1/2}$, where p equals the actual proportion of good items in the *population*.

Many elementary statistics texts show the basis for this. An especially clear, detailed discussion can be found in Ref. 5. A rudimentary explanation can be found in Appendix A19.

A lower P_A confidence bound, an upper P_B confidence bound and a $[P_A - (1 - P_B)]$ confidence interval may be determined, respectively, through use of the following relationships,

$$p_a + K_{PA}[p_a(1 - p_a)/n]^{1/2} \geqslant p_0 \qquad (12\text{-}11)$$

which says that P_A of the time a value of p_0 less than or equal to the left side of the inequality would result if the actual proportion of items which perform successfully is as bad as p_a, where the smallest value of p_a (as defined previously) capable of satisfying the inequality is selected.

$$p_b - K_{PB}[p_b(1 - p_b)/n]^{1/2} \leq p_0 \qquad (12\text{-}12)$$

which says that P_B of the time a value of p_0 greater than or equal to the left side of the inequality would result if the actual proportion of items which perform successfully is as good as p_b, where the largest value of p_b (as defined previously) capable of satisfying the inequality is selected and a $[P_A - (1 - P_B)]$ confidence interval may be defined as

$$p_a \leq p \leq p_b$$

where

$p_0 = [n - r_0]/n$, calculated from the sample
$K_{PA} = $ coefficient of the normal distribution such that

$$\int_{-\infty}^{K_{PA}} \phi(u) \, du = P_A$$

$K_{PB} = $ coefficient of the normal distribution such that

$$\int_{-\infty}^{K_{PB}} \phi(u) \, du = P_B$$

$\phi(u) = $ the normal density function
n, P_a, and P_b are as defined previously

Example   A sample of 500 bullets from long-term storage of ammunition was taken to determine whether or not items from that store would still be effective for use. All were fired: 475 performed successfully; 25 did not. Compute the lower 0.90 confidence bound, the upper 0.90 confidence bound and the 0.80 confidence interval.

$$p_a + K_{PA}[p_a(1 - p_a)/n]^{1/2} \geq p_0$$

In this case, $K_{PA} = 1.28$, $p_0 = 475/500 = 0.95$, and a value of $p_a \sim 0.94$ will fit the constraints for the upper bound.

$$p_b - K_{PB}[p_b(1 - p_b)/n]^{1/2} \leq p_0$$

and a value of $p_b \sim 0.96$ will fit the constraints.

The 0.80 confidence level interval $[P_A - (1 - P_B)]$ for p is

$$0.94 \leqslant p \leqslant 0.96$$

12.1.3   Measuring the Probability of Successful Use When the Reliability of an Item is Known

In the last section our concern was restricted to the reliability of one-shot items based on knowledge of direct quality measures such as the proportion of items which perform successfully. Here it is assumed that a reliability, as opposed to a proportion of items which perform successfully, can be associated with an item or item population. Applicable to both single shot and multiusage items the concern is with (1) the proportion of missions, flights, or jobs of a given type performed successfully, and (2) the probability that a given task will be successfully performed.

The mechanics of treatment is identical to that in the last section. That is to say, equations (12-5), (12-7), and (12-8) to (12-12), apply in the same way for this measure as they did for the last. The primary difference is in the way p is defined. p is defined in this instance as equivalent to reliability $R(T)$:

$$p = R(T)$$

However, both of the following are assumed.

1. Mission or job time is essentially the same from use to use because otherwise p would vary from mission to mission and the binomial distribution as defined would not hold.
2. Either:
   a. All items are the same age at the start of a given mission and the objective is to structure a confidence bound (interval) around the proportion of items *which will successfully perform the current mission*. In that case, if an *approximation* of the distribution of times to failure for the item could be made p could be taken equal to $R(T_2 - T_1)$ where $T_2 = $ the operational age of each item after the mission, $T_1 = $ the operational age of each item prior to the mission.
   b. Time to failure for each item follows an exponential distribution.

Once upper and lower confidence bounds p_a and p_b have been established using the procedures in the last section,

$$p_a = R(T)_a = \text{a lower bound on } R(T)$$

$$p_b = R(T)_b = \text{an upper bound on } R(T)$$

If (2.a) or (2.b) holds, and one of them must in order for the procedure to be followed, confidence bounds on $R(T)$ can be converted to confidence bounds on mean time to failure.

This will be illustrated for the case of time to failure following an exponential distribution. If

$$p = R(T) = e^{-\lambda T} = e^{-T/\theta} \qquad (12\text{-}13)$$

where

λ = the failure rate of an item
$1/\lambda = \theta$ = mean time to failure for the item
T = the mission time

Then the resulting values of p_a and p_b are

$$p_a = R(T)_a = \exp(-\lambda_a T) = \exp(-T/\theta_a) \qquad (12\text{-}14)$$

$$p_b = R(T)_b = \exp(-\lambda_b T) = \exp(-T/\theta_b) \qquad (12\text{-}15)$$

Since T is fixed a P_A confidence bound relative to p_a is equivalent to a P_A confidence bound relative to θ_a, and a P_B confidence bound relative to p_b is equivalent to a P_B confidence bound relative to a θ_b.

Example   A type of aircraft carries a subsystem which, while not flight critical, affects the economy of operation of the aircraft. It is known that subsystem failures follow an exponential distribution. The aircraft has a typical mission of 4 hours. To date, aircraft have flown a total of 420 missions and experienced 50 failures of the subsystem. Compute the lower 0.90 confidence bound, the upper 0.90 confidence bound, and the 0.80 confidence interval for its reliability, $R(T) = p$. What can we say about its mean time to failure?

$$p_a + K_{PA}[p_a(1 - p_a)/n]^{1/2} \geq p_0$$

$$p_b - K_{PB}[p_b(1 - p_b)/n]^{1/2} \leq p_0$$

Since $K_{PA} = 1.28$, $K_{PB} = 1.28$, $p_0 = 370/420 = 0.881$, the approximate values of p_a and p_b which satisfy the equation constraints are $p_a \sim 0.86$, $p_b \sim 0.90$.

The 0.80 confidence interval $[P_A - (1 - P_B)]$ relative to p is

$$0.86 \leqslant p \leqslant 0.90$$

The lower 0.90 confidence bound on θ, i.e., θ_a, can be found from the relationship

$$\exp(-\lambda_a T) = \exp(-T/\theta_a) = p_a$$

Since $p_a = 0.86$ and $T = 4$,

$$0.86 = \exp(-4\lambda_a) = \exp(-4/\theta_a)$$

Taking logs

$$-0.150 = -4\lambda_a = -4/\theta_a$$

$\theta_a = 26.67 =$ lower 0.90 confidence bound
   The upper 0.90 confidence bound on θ, i.e., θ_b, is

$$\exp(-\lambda_b T) = \exp(-T/\theta_b) = p_b$$

Since $p_b = 0.90$ and $T = 4$,

$$0.90 = \exp(-4\lambda_b) = \exp(-4/\theta_b)$$

Taking logs

$$-0.104 = -4\lambda_b = -4/\theta_b$$

$\theta_b = 38.46 =$ upper 0.9 confidence bound.
The 0.80 confidence interval on θ, is $P_A - (1 - P_B)$,

$$\theta_a \leqslant \theta \leqslant \theta_b$$

12.1.4   Measuring the Mean Time to Failure for
Items Having Exponential Distributions of Failure

The mean time to failure (MTTF) or mean time between failures (MTBF) (both terms may be considered equivalent when the distribution of time to failure is exponential), denoted by θ, is estimated by taking the ratio of operating time (T) over a given observation or test period to the number of failures (r) which occur during that period:

$$\hat{\theta} = \frac{T}{r}$$

In particular, when a measurement of θ is made from available data, what results is an *estimate* of the parameter (sometimes called a point estimate). $\hat{\theta}$, like $(n - r_0)/n$ in the last section, is a point estimate. As we did for $(n - r_0)/n$, and for the same reasons, we structure confidence bounds and intervals based on distribution of the estimate.

Because of the assumption of an exponential distribution of times to failure, advantage can be taken of certain properties of that distribution in the structure of such bounds and intervals. In particular these involve the ways that operating time T can be interpreted and in the distribution of the observed values of $\hat{\theta}$ which results.

12.1.4.1   Measuring the Mean Time to Failure:
Defining Operating Time, T
It is at this point that advantage is taken of the unique characteristics of the exponential distribution. In particular, advantage is taken of the fact that failure rate will remain constant throughout an item's life time. This means that an item has the same failure rate over any T-hour operating interval over its service life as it had during the first T hours of its service life. For example, an item which is brand new has the same reliability as an item which is several years old. An item which has failed and has been repaired any number of times has the same reliability as an item which is brand new. This characteristic also leads to the definition of what will be called "equivalents":

1. Operating a single item for T hours is the same reliability-wise as operating n identical items for T/n hours each. This can be shown simply. Since reliability is expressed as

$$R(T) = e^{-\lambda T}$$

a single item which has a failure rate of λ and which operates for T hours has a reliability equal to

$$R(T) = e^{-\lambda T} \tag{12-16}$$

A quantity of $n = T$ items, each of which has a failure rate of λ, and where each operates for 1 hour, has a reliability equal to

$$R(T) = \prod_{1}^{n=T} e^{-\lambda} = (e^{-\lambda})^T = e^{-\lambda T} \tag{12-17}$$

More generally, a quantity of n items, each of which has a failure rate of λ, operating for T/n hours each, has a reliability equal to

$$R(T) = \prod_{1}^{n} e^{-\lambda[T/n]} = (e^{-\lambda T/n})^n = e^{-\lambda T} \tag{12-18}$$

2. Operating a single item over a time T and acquiring r failures is equivalent to operating n items for T/n hours each and acquiring r failures. More generally, it is equivalent to operating a group of n items for a cumulative total of T item operating hours and acquiring r failures without regard to how many hours are on each.

$$T_{\text{cum}} = \sum_{i}^{n} T_i = T \tag{12-19}$$

where

T_{cum} = cumulative item hours (over which time r failures occurred)
T_i = hours of operation on item i

Hence, for this distribution, what is important in reliability measurement in general is the *total number of item hours* acquired and the number of failures which occur, not the number of items which are considered. The number of items on test, however, can be used as a test time schedule multiplier. For example, one item on test for a total of 500 hours is equivalent to such alternatives as two items on test for 250 hours each; four items on test for 125 hours each; ten items on test for 50 hours each; or fifteen items on test for 20 hours each and five items on test for 40 hours each.

The last case exemplifies the situation where test time data is available on a number of different groups of identical equipments. Assume that

N_1 items have been *operating* T_1 hours each and have acquired a total of r_1 failures.

N_2 items have been *operating* for T_2 hours each and have acquired a total of r_2 failures.

N_m items have been *operating* for T_m hours each and have acquired a total of r_m failures.

Total test time, T, for the above situation and the total number of failure, r, over T can be computed for measurement purposes as

$$T_{\text{cum}} = T = \sum_{j}^{G} N_j T_j \tag{12-20}$$

$$r = \sum_{j}^{G} r_{Tj} \tag{12-21}$$

where
   G = number of different groups of equipments
   T_j = number of operating hours on each item in group j
   r_{Tj} = total of failures of items in group j over T_j
   N_j = number of items in group j

Further generalizing, we can define T and r as

$$T = \sum_1^n T_i \tag{12-22}$$

$$r = \sum_1^n r_{Ti}$$

where
   n = total number of items operating over the observation/test period
   T_i = total operating hours on item i
   r_{Ti} = total failures of item i over T_i, where $r_{Ti} \geqslant 0$

Equation (12-22) can be used to define T whether or not items are repaired/replaced or discarded upon failure as long as the definitions for n, T_i, and r_{Ti} are adhered to.

12.1.4.2 Measuring the Mean Time Between Failure: Using Distribution Characteristics to Structure Confidence Bounds and Intervals

If the times to failure of a given item follow an exponential distribution, then the measurement/estimate of mean time between failure, θ, calculated from available data,

$$\hat{\theta} = \frac{T}{r} = \sum_{i=1}^n \frac{T_i}{r} \tag{12-23}$$

will taken on a gamma distribution (see Ref. 6) as shown previously in (2-89), where

$$f(\hat{\theta}) = \frac{\alpha}{\Gamma(r)} r(r\alpha\hat{\theta})^{r-1} e^{-r\alpha\hat{\theta}} \tag{12-24}$$

Rewriting (12-24) with $\alpha = 1/\theta$, $r_0 =$ the number of failures observed, and making a few minor manipulations,

$$f(\hat{\theta}) = \frac{1}{\Gamma(r_0)} \left(\frac{r_0}{\theta}\right)^{r_0} \hat{\theta}^{r_0-1} e^{-r_0 \hat{\theta}/\theta} \qquad (12\text{-}25)$$

The cumulative form of (12-25) is

$$P[(\hat{\theta} \geqslant (T/r_0)/\theta] = 1 - \int_0^{T/r_0} f(\hat{\theta})\, d\hat{\theta} \qquad (12\text{-}26)$$

where $P[(\hat{\theta} \geqslant (T/r_0)/\theta] =$ probability that any sample estimate of mean time to failure, $\hat{\theta}$, will be greater than or equal to (T/r_0) if the actual value of the MTTF was θ.

In most situations, however, the gamma distribution does not have to be used directly for reliability measurement purposes; instead, other distributions with simpler forms directly *derivable from* the gamma are applied. Their use is in fact much more common than the application of the gamma, and is justified through their relationship to that distribution.

Recall from Section 2.4 that the gamma distribution can be transformed into the Poisson distribution where r is an integer, and it is an integer in this case. As a result, from Section 2.4 we can equate $\int f(\hat{\theta})\, d\hat{\theta}$ in (12-26) to

$$\int_0^{T/r} f(\hat{\theta})\, d\hat{\theta} = 1 - \sum_{k=0}^{r-1} \frac{(T/\theta)^k e^{-T/\theta}}{k!} \qquad (12\text{-}27)$$

where $T = r\hat{\theta}$.

As a result, the relationship of (12-27) can be used to assess the probability that any sample estimate of MTBF observed will be less than or equal to T/r given the actual MTBF $= \theta$.

Tables of the Poisson distribution are readily available in most elementary statistics texts or books of tables. As was discussed in Section 2.4, the gamma distribution (12-26) can also be related to the chi-square (χ^2) distribution, which, while not quite as intuitive a form as (12-27) is, nevertheless, simpler than (12-26) and sometimes easier to work with than (12-27).

In that instance, advantage is taken of the fact that when the quantity $2r\hat{\theta}/\theta = 2T/\theta$ is taken equal to a variable u (note that we equate $r\hat{\theta}$ equal to T), then u is chi-square-distributed and the distribution shown in (12-24) can be expressed as

$$f(u) = \frac{1}{2^r \Gamma(r)} \int_0^{2T/\theta} u^{n-1} e^{-u/2}\, du \qquad (12\text{-}28)$$

As a result, the probability of observing values of $2T/\theta$ greater or less than any chi-square value can be calculated. Where u is defined as the chi-square (χ^2) variable in question using a rather common notation we can express:

- the probability of observing a specific value of chi-square less than or equal to some observed value $(2T)/\theta$ as

$$P\left[\chi^2_{(1-p)} \leqslant \frac{2T}{\theta} \right] = 1 - p \qquad (12\text{-}29a)$$

for a given value θ.
- the probability of observing a specific value of chi-square greater than or equal to some observed value $(2T)/\theta$ as

$$P\left[\chi^2_p \geqslant \frac{2T}{\theta} \right] = p \qquad (12\text{-}29b)$$

for a given value θ, where χ^2_k = a value of chi-square that will be exceeded $k\%$ of the time (available from any table of the chi-square distribution).

Hence, either the Poisson or the chi-square distribution may be applied to determine reliability estimates.

For the *Poisson distribution*, we proceed as in Section 12.1.2.1. A lower confidence bound P_A is chosen such that

$$1 - \sum_{k=0}^{r_0} \frac{(T/\theta_a)^k \exp(-T/\theta_a)}{k!} \geqslant P_A \qquad (12\text{-}30)$$

(i.e., if θ is as poor as θ_a, only $(1 - P_A)$ of the time could we expect to observe r_0 or fewer failures in time T). θ_a will constitute the lower P_A confidence bound for θ.

As before, since (12-30) is a discrete relationship, in many instances no value of θ_a for a given value of T and r_0 will yield exactly P_A. The usual course of action under those circumstances is to choose the largest value of θ_a that will yield a value of (12-30) equal to or greater than P_A. For this reason the Poisson approach to the calculations of confidence bounds will provide a minimum of a P_A confidence bound. (The same characteristic applies to the calculation of the upper confidence bound which will be described next.)

A P_B upper confidence bound will be chosen such that

$$\sum_{k=0}^{r_0-1} \frac{(T/\theta_b)^k \exp(-T/\theta_b)}{k!} \geq P_B \tag{12-31}$$

(i.e., if θ is as good as θ_b, only $(1 - P_b)$ of the time could we expect to observe r_0 or more failures in T). The smallest value of θ_b which will satisfy the above inequality will constitute the P_B upper confidence bound.

The $P_A - (1 - P_B)$ confidence interval on θ corresponds to

$$\theta_a \leq \theta \leq \theta_b \tag{12-32}$$

Using the chi-square distribution, both the upper and lower confidence bounds are determined through use of a table of the chi-square distribution. The chi-square values are a function of degrees of freedom, DF. DF are related to the number of failures observed:

- For upper-bound values of MTBF, DF is always $= 2r_0$ for observations which terminate either at a fixed time or observations which terminate after a fixed number of failures have occurred.
- For lower bound values of MTBF, DF $= 2r_0$ for observations which terminate after a fixed number of failures have occurred; DF $= 2(r_0 + 1)$ for observations which terminate after a fixed time. (See Refs. 6 and 7.)

In order to determine the lower and upper bounds for MTBF associated with values of P_A and P_B, the values of $(\chi_{(1-P_A)})^2$ associated with $2r + 2$ degrees of freedom [denoted as $(\chi_{(1-P_A)})^2(2r + 2)$] and $(x_{P_B})^2$ associated with $2r$ degrees of freedom [denoted as $(\chi_{P_B})^2(2r)$] are found and inserted below. Equation (12-29a) can be expressed as

$$\theta_a \geq \frac{2T}{[(\chi_{(1-P_A)}^2(2r + 2)]} \tag{12-33}$$

which means that only $1 - P_A$ of the time would a value of MTBF, θ_a, as low or lower than $2T/[\chi_{(1-P_A)}^2(2r + 2)]$ yield a result as good as the observed result.

Equation (12-29b) can be expressed as

$$\theta_b \leq \frac{2T}{[\chi_{P_B}^2(2r)]} \tag{12-34}$$

which means that only $1 - P_B$ of the time would a value of MTBF, θ_b, as high or higher than $2T/[\chi_{P_B}^2(2r)]$ yield a result as bad as the observed result.

The $P_A - (1 - P_B)$ confidence interval corresponds to

$$\frac{2T}{[\chi^2_{(1 - P_A)}(2r + 2)]} < \theta < \frac{2T}{[\chi^2_{P_B}(2r)]} \tag{12-35}$$

(Note that lower and upper bounds have associated with them $(2r + 2)$ and $2r$ degrees of freedom, respectively.)

Example   Data from the field indicates that a particular group of equipments operating under a specific set of conditions has accumulated a total of 3000 equipment operating hours and has suffered 8 failures. Calculate the lower 0.90 confidence bound on θ; the upper 0.90 confidence bound on θ; the 0.80 confidence interval on θ. Use both the Poisson distribution and the chi-square distribution to determine each bound.

Using the Poisson, the lower and upper bounds are calculated using, respectively,

$$1 - \sum_{k=0}^{r_0} \frac{(T/\theta_a)^k \exp(-T/\theta_a)}{k!} \geqslant P_A$$

$$\sum_{k=0}^{r_0 - 1} \frac{(T/\theta_b)^k \exp(-T/\theta_b)}{k!} \geqslant P_B$$

Using a table of Poisson distribution where (T/θ_a) is taken as the value of the exponent, we find that the largest value of θ that will yield a value of probability $\geqslant P_A$ is

$$\theta_a/T = 13; \qquad \theta_a = T/13 \qquad \text{or} \quad 3000/13 \sim 231$$

(It happens in this case that for $\theta_a = 231$, the Poisson expression exactly equals 0.9.) Similarly, the smallest value of θ_b that will yield a value of probability $\geqslant P_B$ is

$$\theta_b/T = 4.6; \qquad \theta_b = T/4.6 \qquad \text{or} \quad 3000/4.6 \sim 652$$

(It happens in this case that for $\theta_b = 652$ the Poisson expression takes on a value of 0.905. As a consequence, we are actually defining the 0.905 upper confidence bound.)

The *better than* 0.80 confidence interval for θ is

$$231 < \theta < 652$$

Using the chi-square distribution the lower and upper bounds are calculated using, respectively,

$$\theta_a > \frac{2T}{[\chi^2_{(i-P_A)}(2r + 2)]}$$

$$\theta_b < \frac{2T}{[\chi^2_{P_B}(2r)]}$$

Using a table of the chi-square distribution, $[\chi^2_{(i-P_A)}(2r + 2)]$ when $P_A = 0.9$ and $r = 8$ translates into $\chi^2(18) = 25.99$.

Similarly $[\chi^2_{P_B}(2r)]$ when $P_B = 0.9$ and $r = 8$ translates into $\chi^2(16) = 9.31$.

$$\theta_a > 6000/25.99 \sim 231 = \theta_a$$

$$\theta < 6000/9.31 \sim 644 = \theta_b$$

(Note the discrepancy between the values of θ_b calculated using the Poisson distribution, which provides a *minimum* of a 0.90 confidence bound, and the *exact* 0.90 confidence bound generated using the chi-square distribution.)

The 0.80 confidence interval for θ is

$$231 < \theta < 644$$

12.1.5 Measuring the Mean Time to Failure for Items Having Nonconstant Hazards

Measuring the mean time to failure of failure distributions which have one unknown parameter, such as *mean time to failure*, is quite tractable. Estimation of MTTF for items which have two or more unknown parameters is much more complicated statistically and beyond the scope of this text. For an in-depth treatment of the subject, see Ref. 7.

There are methods available, however, to handle special cases. As was done previously, for example purposes assume a Weibull distribution of time to failure with a known shape parameter, β, but an unknown scale parameter, α. In this case assume that we are dealing with data on a population of nonrepairable components, the failure times of which take on a Weibull distribution.

Recalling the discussion of the Weibull distribution in Section 2.1.2, an item with such a distribution of failures has the following reliability function,

$R(T)$, and mean time to failure, ϕ:

$$R(T) = \exp\left[-\left(\frac{T}{\alpha}\right)^{\beta} \right]$$ (12-36)

$$\text{MTTF} = \phi = \alpha\Gamma\left(1 + \frac{1}{\beta} \right)$$ (12-37)

where
  T = interval of operation, measured from the time the item was first energized
  α = a scale parameter
  $$\alpha = \frac{\phi}{\Gamma[1 + (1/\beta)]}$$
  β = a shape parameter
  ϕ = MTTF

For purposes of measurement, as in the previous sections, both an upper and a lower confidence bound can be associated with MTTF, or ϕ.

One way to make a measurement of the reliability of an item which does not follow an exponential distribution of time to failure is to perform a series of transformations and substitutions on its reliability function to try to transform it to an exponential form. After such steps, the same techniques used for measurement/estimation under the assumption of an exponential distribution of time to failure can be applied. The resultant procedure to be used in determining such bounds for a Weibull distribution is described below. See Appendix A21 for the detail and logic of its development.

The measurement is treated as it were from an exponential distribution (see Section 12.1.4), where observed operating test time is transformed to another timescale. In particular, "operating time," T_x, is measured in terms of

$$T_x = T^{\beta}$$ (12-38)

where T and β are defined as before.

Under that transformation, (12-36) takes the form

$$R(T_x) = R(T^{\beta}) = \exp(-T^{\beta}/\alpha_x) = \exp(-T_x/\alpha_x)$$ (12-39)

where $\alpha_x = \alpha^{\beta}$, a constant. Equation (12-39) is an exponential form, where α_x is treated as if it were MTBF, or θ.

Confidence bounds can then be associated with α_x as they were with θ previously. For example, if n items are put on test at the same time and r

of the items fail at times $T_1, T_2, \ldots, T_r$ and the remaining items survive, until test termination point T_m a particular point estimate of MTTF, in terms of the new timescale T_x, would be computed as

$$\hat{\theta}_x = \frac{\sum\limits_{i=1}^{r} (T_i)^\beta + (n - r)T_m^\beta}{r} \tag{12-40}$$

where $\hat{\theta}_x$ = a measurement of MTTF using the new timescale, T_x.

$\hat{\theta}_x$ can then be treated as T/r_0 was treated in the last section and upper and lower confidence bounds $\theta_{ax} \equiv \alpha_{ax}$ and $\theta_{bx} \equiv \alpha_{bx}$ can be developed *based on the* T_x time transformation.

α_{ax} and α_{bx} values can then be retransformed to values of α_a and α_b, since

$$\alpha_x = \alpha^\beta$$

and transformed to upper and lower bounds on MTTF, since

$$\text{MTTF} = \phi = \alpha\Gamma\left(1 + \frac{1}{\beta}\right)$$

so that upper and lower bounds on MTTF, ϕ_a and ϕ_b, can be generated.
A $[P_A - (1 - P_B)]$ confidence interval is of the form

$$\phi_a < \theta < \phi_b$$

12.2 RELIABILITY, QUALIFICATION, DEMONSTRATION, AND TEST

While reliability measurement and estimation is used in the design/development phase to determine a baseline for design and assess progress toward development objectives, it is seldom a program requirement. Reliability demonstration, test, qualification—call it what you will—on the other hand, usually is a program requirement.

Prior to implementing a formal demonstration/qualification of any type, especially when dealing with a new design or product, it is logically prudent to take whatever practical steps are possible to protect against a reject decision. Some examples of simple steps that can be taken in that direction are:

- Reviewing the results of the final reliability prediction to make sure the prediction is close to the upper MTBF value, which will provide a high probability of passing the test. (We will discuss upper and lower MTBF values in more detail later in the next section.)

- Performance of a reliability growth test prior to the demonstration test.
- Running burn-in tests on the items to be demonstrated prior to actual start of the test.

There are various types of demonstrations/tests/qualifications that relate to reliability; from qualifications of parts and components to ensure suitability for inclusion in a development program, to demonstrations that a given design or product has met a given reliability requirement, to demonstration that the quality inherent in a design can be maintained through the production process.

Each type of qualification/demonstration requires that samples of the item developed or produced be subjected to test under a specified set of conditions. Failures are counted against test time and a decision to accept or reject the item or groups of items is made based on the data generated.

A number of government standards/handbooks are available which describe such demonstration tests in depth. The three which pertain most closely to the topics to be covered in this section are:

Mil-Std-781, "Reliability Testing for Engineering Development, Qualification and Production"

Mil-Hdbk-781, "Reliability Test Methods, Plans and Environments for Engineering Development, Qualification and Production"

Mil-Hdbk-108, "Quality Control and Reliability—Sampling Procedures and Tables for Life and Reliability Testing"

12.2.1 Consumer and Producer Risks: What They Are

Tests are probabilistic in nature, and in the same way that a coin can be tossed five times and result in five heads in a row, an item with a better than required level of reliability can fail a demonstration test. Similarly, an item with a much worse reliability than required can pass a demonstration test. Such occurrences indicate risks. Two kinds of risks are associated with every test. A *consumer risk*, usually denoted as β, which defines the probability that an accept decision will be made when the reliability/quality of the item or group of items is less than required. The risk is structured in such a way that it *decreases* as the actual reliability of the product gets worse. The second kind of risk is the *producer risk*, usually denoted as α, which defines the probability that a decision will be made to not accept (in other words, reject) an item or group of items when the reliability/quality of the items is as good as or better than that required. This risk is structured in such a way that it *decreases* as the magnitude of the actual reliability increases.

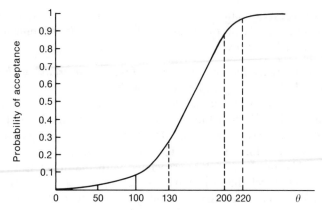

FIGURE 12-1.   An operating characteristic (OC) curve showing how risk changes with MTTF (= θ).

A graphical representation of both risks, called an operating characteristic (OC) curve, is shown in Fig. 12-1. The risks are presented in terms of probability of acceptance. In this particular case, the probability of acceptance is only 10% if the MTTF is 100 hours; it is only 5% if the MTTF is 50 hours. If the MTTF is 130 hours, the probability of acceptance would increase to 30%; if the MTTF is 200 hours, the probability of acceptance increases to 90%; if it increased to 220 hours, the probability of acceptance is 95%.

12.2.2 Structuring a Demonstration Plan to Your Needs

Choosing and structuring a demonstration test demands consideration of a number of technical factors and cognizance of a number of cautions, such as:

1. *Addressing the minimum reliability requirement,* sometimes referred to as a lower test MTBF (or MTTF). This is the *contractual* minimum requirement.

2. *Definition of a reliability design objective value,* sometimes referred to as an upper test MTBF. This value is greater than the lower test MTBF and represents the value of reliability believed to be cost effectively attainable through use of state-of-the-art technology and materials.

(In some instances both of the above may be defined, at least indirectly, by the customer; for example, by the customer defining *both* a minimum MTBF requirement *and* a specific test plan which is based on a given ratio of upper to lower test MTBF (defined as the test discrimination ratio). In other instances the developer may be free to define an upper test value consistent with best estimates of what is believed can be achieved; for example, the design target goals or the final results of the apportionment process described earlier in the text.)

3. *Making provision for, or being aware of the consumer and producer risks* associated with the test limits chosen and with what the actual reliability is thought to be. The consumer risk is commonly associated with the lower test reliability value. If the actual reliability is equal to the lower test value and if the consumer risk is equal to 10%, then the probability that an item having the absolute minimum acceptable reliability will pass the demonstration test is 10%. The producer risk is associated with the upper test reliability value. If the actual reliability is equal to the upper test reliability, and the producer risk is 10%, then the probability is only 10% that the item will fail the demonstration test. The consumer risk is almost always defined by the customer. In some instances, both consumer and producer risks are established by the customer. When this occurs, it is common practice to levy risks of the same magnitude on both consumer and producer. In many instances, it is left to the producer to select a producer risk. In any case:

1. If the actual value of item reliability is less than the lower limit, the probability of *acceptance* is less than the established risk value.
2. If the actual reliability of the item is greater than the upper limit, the probability of *test failure* is less than the established risk value.
3. If the actual reliability of the item is between the upper and lowest test limits, the probability of item acceptance will be greater than the established consumer risk, but less than (1 − established producer risk).
4. It is always a good idea to use or develop an OC curve to determine the probability of item acceptance at intermediate values of MTBF and values of MTBF greater than the upper limit. By doing this, a feel can be gotten concerning probability of acceptance over a wide range of MTBFs. This is particularly valuable if uncertainty exists about the actual value of MTBF produced.

4. *Taking into account the effects of test time on consumer and producer risk.* In general, the lower both risk levels are for the same test, the more test time or test items required. Relatively shorter-time tests can be structured by making one risk substantially greater than the other. Also be aware that the shape of the OC curve in general can be changed by changing test time.

5. *Defining an acceptable test duration.* Test duration is a critical consideration in test planning. Reliability testing impacts schedule and requires test resources and the expenditure of manpower. Risks are a function of test duration and/or number of items put on test. For items having exponential distributions of time to failure, more items on test will result in a shorter test duration at the same risk levels as a longer duration test. In the event that insufficient test time, T, is available to meet both risks, consideration should be given to:

1. Negotiating tradeoffs between both risks consistent with constraints on T.
2. Considering weighting demonstration results with growth test or other life-test data.
3. Considering means to increase the reliability of the item. This will put the item at another place on the OC curve and increase the probability of acceptance in any test plan.

6. *Determining the environmental conditions under which the test is to be run.* The mission/use profile discussed earlier is used to determine the makeup of the environmental specifications of the test. Tests should be performed under the combined effects of vibration, humidity, and other appropriate test conditions expected to be encountered during item use and operation. Guides for the identification and selection of such environments can be found in Mil-Hdbk-781, "Reliability Test Methods, Plans and Environments for Engineering Development, Qualification and Production."

7. *Choosing a test plan framework.* There are two basic catagories of test plans from which to make a choice:

1. A fixed-time test, or fixed number of samples test. In this case, items are put on test for a fixed period of time. Risks are chosen and associated with upper and lower test MTBF values. A maximum allowable number of failures is computed consistent with test time and risks chosen. If the fixed period of time passes with *no more than the maximum* number of failures occurring, the item is accepted. If at any time during the test more than the maximum number of failures occur, *the item is rejected.* Hence, this procedure defines a firm fixed test time for *accept decisions* and criteria for rejection prior to that fixed time. Since it provides a maximum test time, it provides for a minimum degree of uncertainty with respect to test time planning.
2. A time-truncated sequential test. In this case items are put on test and acceptance/rejection boundaries which are both risk- and "dynamic" test time-dependent (i.e., the accept/reject criteria change as the test progresses) are established. A maximum test time (time of truncation) is usually defined, at that time an accept/reject decision is forced. Prior to that time however, either an *accept* or *reject* decision can be made if an accept/reject boundary has been reached. Hence, the test procedure

provides not only a maximum test time (for planning purposes), but opportunity for either acceptance or rejection prior to that time. A sequential test will on the average foster an accept/reject decision in a shorter period of time than a fixed-time test (see Ref. 7).

8. When choosing a particular test plan, pay attention to the number of failures permissible before an accept/reject decision is made. Generally it is not prudent for the producer to choose a test plan which will result in a rejection with very few failures no matter what the producer risk indicates. From Murphy's Law (while not necessarily a statistical basis), it is all too easy for a few defects in an item to slip by even the most careful screens and examinations and result in failures during early portions of operation of a new system. The more failures permissible prior to reject, the more of a buffer for such incidents.

12.2.3   Qualification Tests on Parts or Components: Acceptable Quality Level (AQL) and Lot Tolerance Percent Defective (LTPD)

When high-reliability parts or components are required in order to meet a stringent reliability requirement, there is a need to provide some degree of confidence that population lots of such parts provided by vendors meet the quality requirements needed. As a consequence, tests are performed on such lots, usually by the vendor but in some cases by the buyer, to determine whether or not quality is being maintained. Tests can be performed on reliability attributes *directly*, taking into account failure rates observed, or *indirectly*, when it is assumed that proportion of defects (defined by the results of a given series of tests or screens) in a population is proportional to the failure rate of the population of parts.

For the former, reliability is defined in terms of $R(T)$,

$$R(T) = e^{-\lambda T} \qquad (12\text{-}41)$$

where
$\lambda =$ the failure rate of the item
$T =$ a period of test for each item

For the latter, we define reliability in terms of proportion of items, p, *capable of passing a series of tests and screens.* Define

AQL = an "acceptable" or desirable (upper test value) value of
$R(T) = R(T)_g$ (or $p = p_g$)

LTPD = an "unacceptable" (lower test value) value of
$R(T) = R(T)_u$ (or $p = p_u$)

We want to structure a test which will provide risks (expressed in terms of probability) (a) equal to or less than β (consumer risk) of accepting or shipping a part lot which is as bad as or worse than $R(T)_u$ (or p_u) or (b) equal to or less than α (producer risk) of accepting or shipping a part lot which is as good as or better than $R(T)_g$. (Alternately, producer risk is often described in terms of $(1 - \alpha)$, the probability of acceptance if upper test values are achieved.)

As previously, the cumulative binomial distribution can be applied to express the problem in terms of p, (or $R(T)$ substituted for p).

$$P[(r \leqslant r_0)/p_u] = \sum_{r=0}^{r_0} \left(\frac{n!}{(n-r)!r!} \right)(1 - p_u)^r p_u^{n-r} \leqslant \beta \qquad (12\text{-}42)$$

$$P[(r \leqslant r_0)/p_g] = \sum_{r=0}^{r_0} \left(\frac{n!}{(n-r)!r!} \right)(1 - p_g)^r p_g^{n-r} \geqslant (1 - \alpha) \qquad (12\text{-}43)$$

where (12-42) expresses that the probability of observing r_0 or fewer failures is less than or equal to β if the reliability, p, is equal to p_u. Stated another way: If p is as bad as p_u, r_0 or fewer failures would be expected to be observed only β (or less) of the time.

If a test plan was established where r_0 or fewer failures out of a sample of n would be grounds for acceptance, then we could express (12-42) above as: *If p is as bad as p_u, we would accept the item only β (or less) of the time,* a clear statement of the consumer risk.

Similarly, under the same grounds for acceptance (i.e., r_0 or fewer failures out of a sample of n), we could interpret (12-43) as saying: *If p is as good as p_g, we would accept the item $1 - \alpha$ (or more) of the time,* a clear statement of the producer risk.

It follows, then, that if one wants to structure a qualification test to meet specific α, β, p_g, and p_u needs, all that is required is to determine a combination n and r_0 which satisfies *both* (12-42) and (12-43). For any values of β, α, p_a, and p_u chosen, many values of n and r_0 will satisfy the inequalities, but there will be one combination of r_0 and n which will come closer to the actual values of α and β than others for given values of p_u and p_g. Such values can be determined in many different ways:

1. Trial-and-error using (12-42) and (12-43).
2. Use of tables of the binomial distribution in lieu of computation for sample sizes less than 50.
3. Use of approximations to the normal distribution for large sample sizes.
4. Use of ad-hoc tables prepared specifically for this purpose, for example:
   a. *Sampling Inspection Tables*, Dodge and Romig, Wiley, 1956

b. Mil-Std-108, "Sampling Procedures and Tables for Life and Reliability Testing"
c. Mil-Std-105, "Sampling Procedures for Inspection of Attributes"

The ideal objective of any qualification test is to provide low risk to both parties. However, what if the value of n required is too large to be afforded? In that case there would be no alternative other than adhering to the most "important" risk and determining whether or not the second risk associated with the maximum allowable value of n was low enough to take the chance. The same procedures/resources described and/or listed previously will similarly define that risk.

For example, if a consumer risk $\leq \beta$ was the most important condition of the procurement, the *different sets* of sample sizes, n, and number of failures, r_0, capable of meeting that need, up to the maximum value of n which could be economically considered, would be identified. Recall that there are many combinations of n and r_0 that can meet any single value of β (or α). Each set would then be inserted into (12-43) to see what producer risks were tendered.

The development of accept/reject and sample size criteria will be illustrated using a table of the binomial distribution. Assume that a high proportion of items in a population capable of passing a given test is indicative of a population which will exhibit high reliability. Since it is impossible to test each item in the population separately, a sample from the population will be tested and inferences about the population will be made based on data from the sample. Specifically, take the case where $p_g \geq 0.95$ is indicative of a population which will provide a very low value of failure rate and a $p_u \leq 0.85$, on the other hand, is indicative of a population which will provide an unacceptably high value of failure rate.

Values of consumer risk, $\beta = 0.2$, and producer risk, $\alpha = 0.2$, are selected. What are the accept/reject criteria that should be used to provide those risks for $p_u = 0.85$, $p_g = 0.95$? We peruse a table of the cumulative binomial distribution starting at a very small value of n, say $n = 5$. For $n = 5$, we determine the probability of zero, one or fewer, two or fewer, etc., to five or fewer failures when $p = 0.85$, and $p = 0.95$. What we are looking for is any value (maximum number of failures) which will provide both a probability value of 0.20 (i.e., β), when $p = 0.95$ and a probability value of 0.80 (i.e., $[1 - \alpha]$), when $p = 0.85$ for the same value of n. That maximum number of failures will be equal to r_0. If we cannot come suitably close to such values for $n = 5$, we look at $n = 6$, $n = 7$, and so on.

We start at $n = 5$ and see that when $p = 0.85$ and $p = 0.95$ we come nowhere close to the values of β (i.e., 0.20) and $[1 - \alpha]$ (i.e., 0.80) for any possible maximum value of failures. So we try $n = 6$, and so on. At $n = 15$

we find that for 1 or less failures, $(1 - \alpha) = 0.83$ when $p = 0.95$, and $\beta = 0.32$ when $p = 0.85$—closer to our objective than before but not close enough. At $n = 26$ we find that for 2 or fewer failures, $(1 - \alpha) = 0.86$ when $p = 0.95$, and $\beta = 0.23$ when $p = 0.85$—still closer to our objective. At $n = 28$ we find that for 2 or fewer failures, $(1 - \alpha) = 0.84$ when $p = 0.95$, and $\beta = 0.19$ when $p = 0.85$. In general, these values come closer than any other to satisfying the inqualities.

As a result, the qualification procedure will be to test 28 items, accept delivery if two or fewer fail, and reject delivery if more than two fail. That qualification test has associated with it a consumer risk of 0.19 and a producer risk of 0.16.

12.2.4   Demonstrating a Development Requirement for Items Having An Exponential Distribution Failure

The objective of demonstrating reliability at the end of the development phase is to provide a degree of proof that the design is capable of meeting the reliability requirement imposed. This is, perhaps, the most critical reliability demonstration of all to both the customer and developer.

To the customer, corrections of design-based reliability problems once design is accepted and released to production are, as a rule, extremely expensive and difficult to manage. Often, after design acceptance, if problems with the design surface no clear-cut trail of responsibility to the developer exists. With respect to the developer, failing a reliability demonstration has not only direct economic consequences but legal and contractual implications as well. To the consternation of both, prototypes available for testing are usually relatively few, and demonstration test time is under the constraints of the program schedule. These are the reasons both developer and customer must rely not only on demonstration results but on a well-founded reliability program plan preceeding demonstration as well.

Demonstration is generally implemented using MTBF measures. Two general classes of demonstration tests are used for this purpose: fixed-time tests and sequential tests. We will discuss both. One note of warning should be heeded, however, concerning the application of such demonstration tests. Fault-tolerant systems composed of redundant units, even when each unit has an exponential distribution of failures, do not follow an exponential distribution of failure. Hence, only the last of the procedures in this chapter and none of the procedures in the current version of Mil-Hdbk-781, "Reliability Test Methods, Plans and Environments for Engineering Development, Qualification and Production," are applicable to reliability demonstration of fault-tolerant systems.

12.2.4.1 Demonstrating a Development Requirement: The Fixed-Time Test

A fixed-time reliability demonstration test is characterized by a fixed value of *item test hours*, or time, and a maximum allowable number of failures, r_0, during the test period.

The fixed operating time, T, over which the test is run is not necessarily the same as clock time. As was discussed in Section 12.1.4.1, operating time for an item having an exponential distribution of failure times is calculated as

$$T = \sum_{i}^{n} T_i \qquad (12\text{-}44)$$

where

T = total item test hours
T_i = number of operating hours on item i
n = number of items under test

As a consequence, if three items were put on test as shown in Fig. 12-2 and run until a total of $T = 80$ operating hours accrued, only 35 hours of *clock* time would have passed. Items during that 35 hours may be taken off test to make repairs when failure occurs (for example, items 1 and 3) or removed permanently to be used for other purposes (for example, item 2). All that is required is that a fixed number of operating hours, T, be accrued.

While clock time impacts schedule and is a concern for both customer and developer, the characteristics of the test are always described in terms of item operating hours T as opposed to clock time.

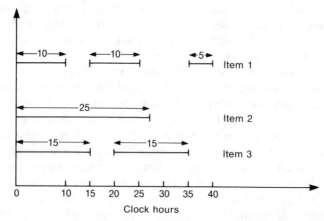

FIGURE 12-2.   Accumulating 80 operating hours over a period of 35 clock hours.

As in Section 12.2.3, a consumer risk β is associated with the lower test bound reliability, in this case θ_u. This indicates that if the actual MTBF is as bad as θ_u, the probability is only β that the item will be accepted. Similarly, a producer risk α is associated with the upper test bound reliability, in this case θ_g. This indicates that if the MTBF is as good as θ_g, the probability is only α that the item will be rejected. (Alternatively, we could say that the probability is $(1 - \alpha)$ that the item will be accepted.) Naturally, if the reliability of the product is worse than θ_u, the probability of acceptance will be even lower; and if the reliability of the product is better than θ_g, the probability of rejection will be less than α.

The maximum number of failures, r_0, allowable and the required value of T are determined by taking into account the consumer and producer risks associated with the test. The test is terminated when either T is reached or $(r_0 + 1)$ failures occur prior to T, in which case the item is rejected at the time $(r_0 + 1)$ failures occur.

This can be interpreted graphically as in Fig. 12-3a. In this case the number of failures which occur as T increases is plotted and an accept decision occurs only if and when the vertical accept line is reached. A rejection occurs if the plot reaches the horizontal reject line (i.e., $(r_0 + 1)$ failures) prior to reaching the accept line.

As the test progresses, a plot similar to Fig. 12-3b results and can provide the user with easy-to-interpret insights relative to test status. This particular plot indicates that the test is still continuing satisfactorily, and barring any unforeseen problems, it looks as if it should end at an accept decision. In the event that the test looks as if it might end at a reject boundary (for example, if at time T_1, r_0 failures had already occurred), it might be prudent to terminate the test prior to formal rejection and consider design changes and modifications.

One of the easiest means to get a feel for the logic of this demonstration test is through use of the Poisson distribution, which can be used for test structure and analysis.

As discussed in Chapter 2, the Poisson distribution tells a story. It defines the probability $P(r)$ that r failures will occur when a are expected:

$$P(r) = \frac{a^r e^{-a}}{r!} \tag{12-45}$$

Since, as we discussed, in Chapters 1 and 2,

$$\lambda T = \frac{T}{\theta} = a = \text{expected number of failures}$$

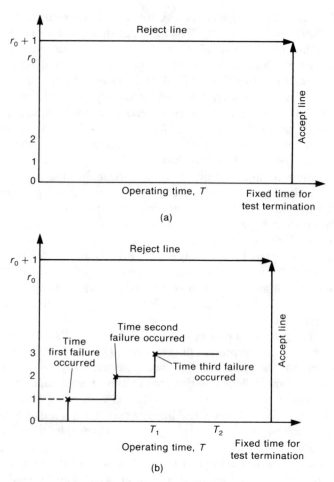

FIGURE 12-3. (a) Graphical representation of a fixed-time test. (b) Resulting plot of failure versus accept/reject boundaries to time T_2.

where
  λ = failure rate of an item
  T = operating hours of item operation
  θ = MTBF

(12-45) may be expressed as

$$P(r) = \frac{(T/\theta)^r e^{-T/\theta}}{r!} \qquad (12\text{-}46)$$

The cumulative form of the Poisson distribution, the relationship which can be utilized for both the analysis and structure of reliability-fixed time test plans, can be expressed as

$$P[(r \leqslant r_0)/\theta] = \sum_{r=0}^{r_0} \frac{(T/\theta)^r e^{-T/\theta}}{r!} \qquad (12\text{-}47)$$

where $P[(r \leqslant r_0)/\theta]$ = probability that r_0 or fewer failures will occur over operating time T given the MTBF = θ.

If we were to establish a test plan in which r_0 or fewer failures in operating time, T, would be grounds for acceptance, then the above could be expressed as: If the item had an MTBF = θ, the probability that it would be accepted would be equal to $P[(r \leqslant r_0)/\theta]$. As a result (12-47) *allows its user to calculate the probability of acceptance of an item having a given MTBF, θ, for any combination of maximum allowable number of failures, r_0, and test time, T.* It is a means through which the OC curve for any given test plan may be developed.

Aside from allowing for the analysis of test plans, (12-47) also allows for the development of test plans to meet given consumer and producer risks and for the development of the "best" test plan consistent with time constraints.

With respect to the former, suppose that a test with maximum values for consumer risk of β, tied to a lower test value MTBF of θ_u, and a producer risk of α, tied to an upper test value MTBF of θ_g, was desired. The equations necessary to structure a fixed-time test to any levels of risk, α and β, under the framework of (12-47) take the form

$$P[(r \leqslant r_0)/\theta_u] = \sum_{r=0}^{r_0} \frac{(T/\theta_u)^r \exp(-T/\theta_u)}{r!} \leqslant \beta \qquad (12\text{-}48)$$

$$P[(r \leqslant r_0)/\theta_g] = \sum_{r=0}^{r_0} \frac{(T/\theta_g)^r \exp(-T/\theta_g)}{r!} \geqslant (1 - \alpha) \qquad (12\text{-}49)$$

Note that the inequality signs are included because a Poisson distribution takes on only integer values of r and, for a given value of (T/θ), no value of r might exist which will yield an exact value of β or $(1 - \alpha)$.

Equation (12-48) makes the following equivalent statements:

1. The probability of observing r_0 or fewer failures in test time T is less than or equal to β if θ is equal to θ_u.
2. If θ is as bad as θ_u, r_0 or fewer failures would occur no more than β of the time in test time T.

3. If θ were as bad as θ_u we would accept the item only β or less of the time (if the test time was T and the maximum allowable number of failures was r_0),

and (3) defines a clear statement of consumer risk.

Similarly, (12-49) can be interpreted as saying that if θ is as good as θ_g, we would accept the item $(1 - \alpha)$ or more of time (if the test time was T and the maximum allowable number of failures was r_0)—a clear statement of producer risk.

It then follows that any combination of T and r_0 which would satisfy *both* (12-48) and (12-49) for given values of θ_u, θ_g, β, and α, would express the accept/reject requirement for such a test.

If test time, T, is not a problem, the values of T and r_0 which come closest to β and $(1 - \alpha)$ under their inequality constraints are chosen.

If maximum allowable test time is limited to a value of T, then adherence to a desired set of risks may not be possible. The value of r_0 chosen in that case is one that allows the left portions of (12-48) and (12-49), shown below, to result in values of risk which are deemed most acceptable. This may result in one risk at the originally desired level and the other at a less desirable level, or in a small to moderate compromise to both risks.

$$P[(r \leqslant r_0)/\theta_u] = \sum_{r=0}^{r_0} \frac{(T/\theta_u)^r \exp(-T/\theta_u)}{r!} \qquad (12\text{-}50)$$

which is the consumer risk associated with a fixed-time test of duration T, maximum number of failures r_0, given θ_u as a lower test limit.

$$P[(r \leqslant r_0)/\theta_g] = \sum_{r=0}^{r_0} \frac{(T/\theta_g)^r \exp(-T/\theta_g)}{r!} \qquad (12\text{-}51)$$

which is $(1 - \text{producer risk})$ associated with a fixed time test of duration T, maximum number of failures r_0, given θ_g as an upper test limit.

Equations (12-50) and (12-51) may be used in general to evaluate the consumer and producer risks associated with any fixed time test.

While (12-48) and (12-49) provide an excellent degree of visibility concerning the logic of the demonstration test, they can be rather cumbersome to work with in determining required r_0 and T parameters. However, based on the fact that there exists a relationship between the Poisson and the chi-square distribution, (12-48) and (12-49) can be expressed in terms of the chi-square distribution. This provides advantages in more easily determining required r_0 and T values. These can be expressed as:

$$[\chi_\beta^2(2r_0 + 2)] \geqslant \frac{2T}{\theta_u} \qquad (12\text{-}52)$$

$$[\chi_{1-\alpha}^2(2r_0 + 2)] \leqslant \frac{2T}{\theta_g} \qquad (12\text{-}53)$$

[Note that r_0 is defined as the maximum number of failures allowable in order to make an *accept* decision. In a number of documents, test plans are structured based on the minimum number of failures, r_r, required in order to make a *reject* decision. In that case in both (12-52) and (12-53) replaces $(2r_0 + 2)$ with $(2r_r)$.]

Where, as before,

χ_k^2 = a value of chi-square that will be exceeded $k\%$ of the time (available from any table of the chi-square distribution)

α, β, T, θ_g, θ_u are defined as before,

and where the values of r_0 (the maximum allowable number of failures for an accept decision) and T that satisfy both relationships define the test characteristics.

Equations (12-52) and (12-53) can be used in place of (12-48) and (12-49) to structure and analyze reliability demonstration plans. An example of their use is shown next.

Structuring Reliability Demonstrations To Meet a
Given Risk

Assume θ_g and θ_u are established and that the consumer risk is defined by the customer as a requirement that must be met, but the developer is free to establish producer risk. Since the acceptable value of consumer risk is β, we replace the inequality of (12-52) with an equality.

This will provide the customer with a test plan having a needed value of β and will also provide the developer with the opportunity for a lower value of α than the target (the reverse action, could be taken also, providing the developer with a needed value of α, by replacing the inequality in (12-53) with an equality). This allows a ratio of (12-52) (with inequality replaced by an equality) to (12-53) to be made, with the result

$$\frac{[\chi_\beta^2(2r_0 + 2)]}{[\chi_{1-\alpha}^2(2r_0 + 2)]} \geqslant \frac{\theta_g}{\theta_u} \qquad (12\text{-}54)$$

which can be employed to quickly determine the value of r_0 and T which will satisfy the α and β risks.

We proceed as follows. Since θ_g and θ_u are known, a table of chi-square values can be used (an abbreviated table is illustrated in Table 12-1) as a basis for the search. We concentrate on the columns associated with probability equal to β and a target value $(1 - \alpha)$ and search for the value of degrees of freedom, F, that will provide the smallest value of $[\chi_\beta^2(F)]/[\chi_{1-\alpha}^2(F)]$ such that

$$\frac{[\chi_\beta^2(F)]}{[\chi_{1-\alpha}^2(F)]} \geq \frac{\theta_g}{\theta_u} \tag{12-55}$$

Since $F = (2r_0 + 2)$, r_0 may be determined,

$$r_0 = \frac{F - 2}{2} \tag{12-56}$$

where F is degrees of freedom read from the chi-square table.

The value of $\chi_\beta^2(F)$ found appropriate to (12-55) is used to compute the required value of T. T is computed through (12-52), replacing the inequality with an equal sign such that

$$T = \frac{\theta_u \chi_\beta^2(2r_0 + 2)}{2} = \frac{\theta_u \chi_\beta^2(F)}{2} \tag{12-57}$$

TABLE 12-1   Values of the Chi-Square Distribution (The Probability is k of Observing a Value of Chi-Square Greater than $(\chi_k)^2$)

F	$(\chi_{0.90})^2$	$(\chi_{0.80})^2$	$(\chi_{0.70})^2$	$(\chi_{0.50})^2$	$(\chi_{0.30})^2$	$(\chi_{0.20})^2$	$(\chi_{0.10})^2$
1	0.016	0.064	0.150	0.460	1.07	1.64	2.71
2	0.210	0.450	0.712	1.39	2.41	3.22	4.61
3	0.580	1.01	1.42	2.37	3.67	4.64	6.25
4	1.06	1.65	2.20	3.36	4.88	5.97	7.78
5	1.61	2.34	3.00	4.35	6.06	7.29	9.24
6	2.20	3.07	3.83	5.35	7.23	8.56	10.65
7	2.83	3.82	4.67	6.35	8.38	9.80	12.02
8	3.49	4.59	5.53	7.34	9.52	11.03	13.36
9	4.17	5.38	6.39	8.34	10.66	12.24	14.68
10	4.87	6.18	7.27	9.34	11.78	13.44	15.99
11	5.58	6.99	8.15	10.34	12.9	14.63	17.28
12	6.30	7.81	9.03	11.34	14.01	15.81	18.55

Example   Assume that a consumer risk of 0.30 and a producer risk target of 0.30 are required/desired. What test time, T, and what maximum allowable number of failures over T will be required to meet demonstration requirements if $\theta_g = 2.2\theta_u$.

Concentrating on the $\chi^2_{0.30}$ and $\chi^2_{0.70}$ columns of Table 12-1, we proceed until the smallest ratio appears such that

$$\frac{\chi^2_{0.30}}{\chi^2_{0.70}} \geqslant 2.2$$

F-value	Ratio
$F = 2$	3.38
$F = 3$	2.58
$F = 4$	2.22
$F = 5$	2.02
$F = 6$	1.89

As a consequence, $F = 4$ is chosen, which results in a value of $r_0 = 1$.
For the example, $\chi^2_{0.30}(4) = 4.88$, so that

$$T = \frac{\theta_u(4.88)}{2} = 2.44\theta_u$$

or 2.44 multiples of θ_u.

Thus, the test plan that will yield $\beta = 0.3$ will require a test time of $2.44\theta_u$, and a maximum number of failures $r_0 = 1$. The value of producer risk in this case will be very close to 0.30

Chart 12-1 and Fig. 12-4, extracted from Mil-Std-781, "Reliability Test Methods, Plans and Environments for Engineers," shows the structure of various fixed-time test plans for given discrimination ratios, i.e., θ_g/θ_u, and an example of the OC curve for a sample test plan in the document. However, for those test plans no attempt was made to fix either α or β.

Example Application   Prototypes of an equipment have been fabricated. The MTBF requirement is $\theta_u = 200$ hours. Based on an analysis of the design, it is believed that an MTBF of $\theta_g = 400$ hours has been attained. The maximum consumer risk which must be represented by the test structure is given as 0.10. The developer is free to establish his own producer risk, but test time costs money and he wants to assess the trades between risk and test time and make whatever adjustments between the two are deemed desirable. If a small increase in test time will accrue significant advantages, the developer wants to know about it. Likewise, if a significant increase in

CHART 12-1 Various Fixed-Time Test Plans

Test Plan	True risks (%)		Discrimination Ratio (d), θ_g/θ_u	Test Duration = T (Multiples of θ_u)	Number of Failures	
	α	β			Reject (Equal or More)	Accept (Equal or Less)
					$r_0 + 1$	r_0
IX-D	12.0	9.9	1.5	45.0	37	36
X-D	10.9	21.4	1.5	29.9	26	25
XI-D	19.7	19.6	1.5	21.5	18	17
XII-D	9.6	10.6	2.0	18.8	14	13
XIII-D	9.8	20.9	2.0	12.4	10	9
XIV-D	19.9	21.0	2.0	7.8	6	5
XV-D	9.4	9.9	3.0	9.3	6	5
XVI-D	10.9	21.3	3.0	5.4	4	3
XVII-D	17.5	19.7	3.0	4.3	3	2

Source: Mil-Hdbk-781.

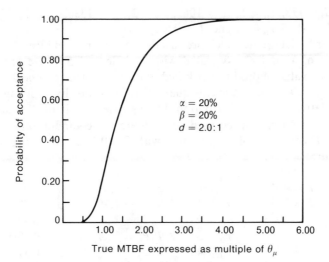

FIGURE 12-4.   OC curve of test plan 14D of Mil-Hdbk-781.

test time will buy insignificant decrease in risk, the developer will want to
know that also.

The identification and analysis of alternate fixed-time test plans which
will meet the customer's requirement and the developer's needs starts with
a listing of potential values of r_0, from 0 to some target number which can
be increased, if necessary, as the analysis proceeds. In this case, values of r_0
are chosen from 0 to 6. Since the chi-square distribution "works" as a function
of degrees of freedom, F, and since $F = (2r_0 + 2)$, the values of concern for
F are 2, 4, 6, 8, 10, 12, 14.

A table of the chi-square distributions is entered at the $\chi^2_{0.10}$ column, the
value of χ^2 is noted for each value of F; for example

$$
\begin{array}{ll}
F = 2 & \chi^2(2) = 4.61 \\
F = 4 & \chi^2(4) = 7.78 \\
F = 6 & \chi^2(6) = 10.65 \\
F = 8 & \chi^2(8) = 13.36 \\
F = 10 & \chi^2(10) = 15.99 \\
F = 12 & \chi^2(12) = 18.55 \\
F = 14 & \chi^2(14) = 21.06
\end{array}
$$

Applying (12-52) with $\theta_u = 200$, $\beta = 0.1$, and the inequality replaced with
an equal sign, as we did before, results in identifying a number of fixed-time
tests which yield a consumer risk of exactly 0.10: test for $T = 461$ hours with

$r_0 = 0$; $T = 778$, $r_0 = 1$; $T = 1065$, $r_0 = 2$; $T = 1336$, $r_0 = 3$; $T = 1599$, $r_0 = 4$; $T = 1855$, $r_0 = 5$; and $T = 2106$, $r_0 = 6$.

In order to determine which producer risks are tendered by each test plan, the values of T and r_0 associated with each test plan above are used to evaluate the value of producer risk tied to $\theta_g = 400$. That can be done either as before from a table of the chi-square distribution or by applying (12-51), the cumulative Poisson distribution. In this case a table of the cumulative Poisson distribution will be used to facilitate the computations. For each test time, T, and value of r_0 consistent with $\beta = 0.1$, the following producer risks result:

$T = 461$	$r_0 = 0$	$T/\theta_g = 1.5$	$\alpha \sim 0.78$
$T = 778$	$r_0 = 1$	$T/\theta_g = 1.95$	$\alpha \sim 0.58$
$T = 1065$	$r_0 = 2$	$T/\theta_g = 2.66$	$\alpha \sim 0.49$
$T = 1336$	$r_0 = 3$	$T/\theta_g = 3.34$	$\alpha \sim 0.43$
$T = 1599$	$r_0 = 4$	$T/\theta_g = 4$	$\alpha \sim 0.37$
$T = 1855$	$r_0 = 5$	$T/\theta_g = 4.64$	$\alpha \sim 0.32$
$T = 2106$	$r_0 = 6$	$T/\theta_g = 5.27$	$\alpha \sim 0.28$

If test time and the value of r_0 could be increased further, reduction of producer risk would result. Hence, the risk to be taken can be balanced by test time until an acceptable mix results.

12.2.4.2 Demonstrating a Development Requirement:
The Sequential Test

Sequential test plans in the area of quality control were first introduced in the early 1940s. In the 1950s they were applied to reliability demonstration. Sequential tests take on the graphical representation as in Fig. 12-5. In this case, the number of failures which occur as T increases are plotted and an accept decision occurs the first time the acceptance line is reached; a reject decision occurs the first time the rejection line is reached. Such tests, on the average, require less time to reach an accept/reject decision than fixed-time tests, hence their popularity for reliability tests.

The mathematical proof for this savings is beyond the scope of this text, but an intuitive feel for the reason may be gleaned by comparing Fig. 12-3a with Fig. 12-5. Note that the region associated with "continue test" for the fixed-time test is the rectangle associated with the accept/reject boundaries. For the sequential test the continue test region is restricted to the area between the accept/reject lines. In addition, it is clear from review of both figures that, for the sequential test, either a reject or an accept decision can be made prior to termination time. For a fixed-time test, *only* a reject decision can be made prior to the termination time.

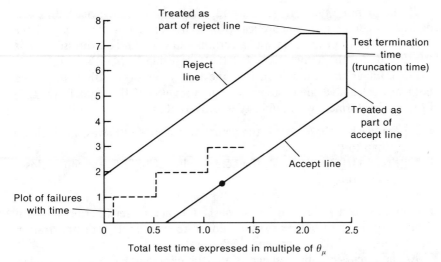

FIGURE 12-5.   Graphical representation of a sequential test.   [From Mil-Hdbk-781.]

The Poisson distribution is used as the basis of structure for the sequential test, and, as before, an upper test value of MTBF, θ_g, a lower test value of MTBF, θ_u, a consumer risk, β, and a producer risk, α, are associated with the test. The test logic works on a basis of a ratio and therefore is known as the Probability Ratio Sequential Test (PRST).

We define

$$P(r/\theta) = \frac{(T/\theta)^r e^{-T/\theta}}{r!} \tag{12-58}$$

where we interpret $P(r/\theta)$ as the probability of realizing r failures in T hours of operation when the MTBF of an item is θ, and specifically

$$P(r/\theta_u) = \frac{(T/\theta_u)^r \exp(-T/\theta_u)}{r!} \tag{12-59}$$

$$P(r/\theta_g) = \frac{(T/\theta_g)^r \exp(-T/\theta_g)}{r!} \tag{12-60}$$

where

$P(r/\theta_u) =$ the probability of realizing r failures in T hours of operation when $\theta = \theta_u$

$P(r/\theta_g) =$ the probability of realizing r failures in T hours of operation when $\theta = \theta_g$

If one or more items is on test and r failures are experienced, and it is desired to compute the theoretical probability of getting that result if the actual reliability of the item was a certain value, it can be expressed (a) as $P(r/\theta_u)$, under the assumption that the true value of MTBF is θ_u; (b) as $P(r/\theta_g)$, under the assumption that the true value of MTBF is θ_g. When both are calculated after observing r failures and if the ratio $[P(r/\theta_u)]/[P(r/\theta_g)]$ is formed, we would say intuitively that:

- If $[P(r/\theta_u)]/[P(r/\theta_g)] \gg 1$, the actual value of MTBF was much closer to θ_u than to θ_g.
- If $[P(r/\theta_u)]/[P(r/\theta_g)] \ll 1$, the actual value of MTBF was much closer to θ_g than to θ_u.

If limits for such ratios can be established as accept/reject criteria, and if defined consumer/producer risks are linked to such limits, a demonstration test results. The PRST represents such a test.

As the discussion above suggests and the name implies, the PRST takes the form of the ratio of (12-59) to (12-60):

$$P_r = \frac{P(r/\theta_u)}{P(r/\theta_g)} = \frac{(T/\theta_u)^r e^{-T/\theta_u}}{(T/\theta_g)^r e^{-T/\theta_g}} = \left(\frac{\theta_g}{\theta_u}\right)^r \exp\left[-T\left(\frac{1}{\theta_u} - \frac{1}{\theta_g}\right)\right] \qquad (12\text{-}61)$$

As the test progresses, the value of P_r changes as r and T increase. As each new failure occurs, the number of failures, r, is incremented by one; the current value of operating test time, T, is inserted in (12-61), and P_r is examined to determine whether it is between two limits, say, A and B:

$$B < P_r < A \qquad (12\text{-}62)$$

where
A = one accept/reject limit associated with a given set of risks
B = the second accept/reject limit associated with a given set of risks

If P_r is contained in the inequality bounded by A and B, the test continues. When one or the other inequality is violated, an accept/reject decision is made.

Exact methods to structure such accept/reject criteria in accord with defined risks and maximum test time are complex:

- An analytical method and a computer program capable of defining characteristics of such tests is described in Ref. 8.
- An iterative method which will provide exact results is described in Ref. 9. A computer program could be developed relatively easily using the procedure described.

- Monte Carlo simulation programs can be relatively easily developed to evaluate the risks associated with any arbitrary set of accept/reject criteria and maximum test time. Many exist, for example, at RADC.
- Approximations may be employed to provide approximate accept/reject criteria and maximum values of test time consistent with a set of consumer/producer risks.
- Mil-Hdbk-781, "Reliability Test Methods, Plans and Environments for Engineering Development, Qualification and Production," contains a number of the most commonly used PRSTs.

An approximate relatively simple procedure for structuring such a test will be illustrated. As described earlier, suppose the ratio

$$\frac{P(r/\theta_u)}{P(r/\theta_g)} = P_r$$

is formed. Limits A and B on the value of P_r must be defined such that if:

$P_r \geqslant A$	we can reject the item with some confidence that θ is much closer to θ_u than θ_g
$P_r \leqslant B$	we can accept the item with some confidence that θ is much closer to θ_g than to θ_u

Based on the results documented in Ref. 10, A and B can be approximated, related to α and β, and related to test time as follows: A and B can be expressed as

$$A \sim \frac{1 - \beta}{\alpha}, \qquad B \sim \frac{\beta}{1 - \alpha}$$

where β and α represent the consumer and producer risk as defined in the last section. As long as

$$\frac{\beta}{1 - \alpha} < P_r < \frac{1 - \beta}{\alpha} \tag{12-63}$$

the demonstration continues with no decision made. As soon as $P_r \geqslant (1 - \beta)/\alpha$ or $P_r \leqslant \beta/(1 - \alpha)$, the test ends with a decision to accept or reject as described above.

We can express (12-63) in a much more visible form by expressing it as

$$\frac{\beta}{1 - \alpha} > \left(\frac{\theta_g}{\theta_u}\right)^r \exp\left[-T\left(\frac{1}{\theta_u} - \frac{1}{\theta_g}\right)\right] < \frac{1 - \beta}{\alpha} \tag{12-64}$$

Taking logarithms and manipulating the results, (12-64) can be expressed as

$$\frac{-\ln[\beta/(1-\alpha)] + r\ln(\theta_g/\theta_u)}{(1/\theta_u) - (1/\theta_g)} > T > \frac{-\ln[(1-\beta)/\alpha] + r\ln(\theta_g/\theta_u)}{(1/\theta_u) - (1/\theta_g)}$$

(12-65)

Simplifying by using the discrimination ratio, d, where $d = \theta_g/\theta_u$, after some minor manipulation (12-65) takes the form

$$\frac{-\ln[\beta/(1-\alpha)] + r\ln d}{(1/\theta_u)[1 - (1/d)]} > T > \frac{-\ln[(1-\beta)/\alpha] + r\ln d}{(1/\theta_u)[1 - (1/d)]}$$

(12-66)

where we accept when T equals (or exceeds) the left-hand term, and reject when T equals (or is less than) the right-hand term.

The minimum conditions for acceptance can be described as

$$T = \frac{rd\theta_u \ln d}{d-1} - \frac{\theta_u d \ln[\beta(1-\alpha)]}{d-1}$$

(12-67)

which is the equation for a straight line, the accept line (see Fig. 12.5).

Similarly, the minimum conditions for rejection can be described as

$$T = \frac{rd\theta_u \ln d}{d-1} - \frac{d\theta_u \ln[(1-\beta)/\alpha]}{d-1}$$

(12-68)

which is also the equation for a straight line, the reject line (see Fig. 12-5).

By choosing a range of values of r, representing failures, from 0 to n and inserting these one by one into (12-67) and (12-68), the values of T associated with accept and reject decisions for any given test can be defined and plotted, forming the accept/reject lines (decision boundaries). For example, for Fig. 12-5:

- We would reject the item at *any* of the following points in time:
  If two failures occurred prior to approximately $0.12\theta_u$
  If three failures occurred prior to approximately $0.5\theta_u$
  If four failures occurred prior to approximately $0.85\theta_u$
  If five failures occurred prior to approximately $1.15\theta_u$
  etc.
- We would accept the item at *any* of the following points in time:
  If no failures occurred prior to approximately $0.6\theta_u$
  If only one failure occurred prior to approximately $1.05\theta_u$
  If only two failures occurred prior to approximately $1.45\theta_u$
  If only three failures occurred prior to approximately $1.75\theta_u$
  etc.

Equations (12-67) and (12-68) provide the capability to develop approximate sequential test procedures for any combination of consumer and producer risk, and for any discrimination ratio.

Theoretically speaking, the approximate α and β values assigned relate only to tests which are not truncated (tests that continue in time until an accept or reject decision is made). Practically speaking, a maximum test time must be assigned. It is to be expected, then, that the values of α and β will change as test time is limited. Empirical evidence and theoretical studies (see Ref. 11) suggest that the degree of consumer and producer risk change due to truncation is a function of the magnitude of discrimination ratio (the larger the discrimination ratio, the less the degree of change in risk for a given truncation time) and the values of the risks themselves (the higher the risk, the lower the degree of change for a given truncation time). Consequently, as a general rule of thumb, risks will not be significantly impacted if $\alpha,\beta \geqslant 0.20$ and $d \geqslant 2$ if truncation time is $10\theta_u$ or more. Mil-Hdbk-781 provides examples of a number of different truncated tests and their actual risks. See Refs. 8 and 9 for detailed procedures available to evaluate such risks.

The capability to generate the OC curve for a sequential test is more limited than for any other of the tests discussed previously. However, an approximate mapping of the OC curve (see Refs. 10 and 11), i.e., the probability of acceptance, P_a, can be made when:

$$\theta = 0 \qquad\qquad P_a = 0$$
$$\theta = \theta_u \qquad\qquad P_a = \beta$$
$$\theta = \frac{\ln d}{1/\theta_u(1 - 1/d)} \qquad P_a = \frac{\ln[(1 - \beta)/\alpha]}{\ln[(1 - \beta)/\alpha] - \ln[\beta/(1 - \alpha)]}$$
$$\theta = \theta_g \qquad\qquad P_a = (1 - \alpha)$$
$$\theta = \infty \qquad\qquad P_a = 1$$

In order to illustrate the development of a sequential test and its approximate OC curve, suppose we were asked to develop a sequential test such that $\alpha = \beta = 0.2$ and $\theta_g/\theta_u = d = 2$. The equation for the accept line would take the form

$$T = \frac{r\theta_u d \ln d}{d - 1} - \frac{\theta_u d \ln[\beta/(1 - \alpha)]}{d - 1}$$

Expressing time, T, in terms of multiples of θ_u results in the above expression taking the form

$$\frac{T}{\theta_u} = \frac{-rd \ln d}{d - 1} - \frac{d \ln[\beta/(1 - \alpha)]}{d - 1} \qquad\qquad (12\text{-}69)$$

Substituting values for d, β, and α in (12-69),

$$\frac{T}{\theta_u} = 1.39r + 2.77 = \text{accept line} \qquad (12\text{-}70)$$

Doing the same for the reject line (12-68) results in

$$\frac{T}{\theta_u} = \frac{rd \ln d}{d - 1} - \frac{d \ln[(1 - \beta)/\alpha]}{d - 1} \qquad (12\text{-}71)$$

After substitutions are made,

$$\frac{T}{\theta_u} = 1.39r - 2.77 = \text{reject line} \qquad (12\text{-}72)$$

Figure 12-6 represents a plot of the above accept/reject lines compared to the actual accept/reject lines computed by a more exact method, as they

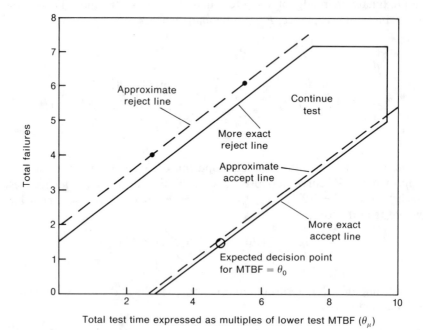

FIGURE 12-6.   Approximation of accept/reject lines versus actual lines.   [From Mil-Hdbk-781.]

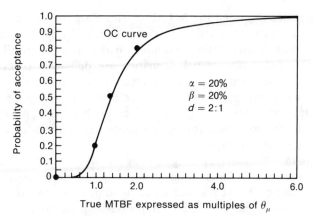

FIGURE 12-7.   Three-point approximation to OC curve (infinity not shown) versus actual OC curve.   [From Mil-Hdbk-781.]

appear in Mil-Hdbk-781. Plotting an approximation for the OC curve:

$\theta = 0$ $\qquad\qquad\qquad\qquad\qquad$ $P_a = 0$

$\theta = \theta_u$ $\qquad\qquad\qquad\qquad\qquad$ $P_a = 0.20$

$\theta = \dfrac{\ln d}{(1/\theta_u)[1 - (1/d)]} = 1.38\theta_u$ $\qquad$ $P_a = \dfrac{\ln[(1 - \beta)/\alpha]}{[\ln[(1 - \beta)/\alpha] - \ln[\beta/(1 - \alpha)]}$

$\theta = \theta_a$ $\qquad\qquad\qquad\qquad\qquad$ $P_a = 0.2$

$\theta = \infty$ $\qquad\qquad\qquad\qquad\qquad$ $P_a = 1$

Figure 12-7 shows these points as dots superimposed on a more exact OC curve extracted from Mil-Hdbk-781.

12.2.5   Demonstrating a Production Requirement for Items Having Exponential Distribution of Failure

The objective of demonstrating reliability during the production phase of development is to provide assurance that the reliability of the inherent design has not been degraded by the materials used and the fabrication processes applied during production. The prototypes or pilot production models of the item which underwent developmental test are not necessarily representative of the full-scale production models of the item to be produced. In addition, during production, reliability/quality is subject to variation depending on the degree of control exercised with respect to the quality of the parts and material used; the manufacturing process chosen and its calibration; and the training and motivation of manufacturing personnel.

The types of demonstration tests applicable are the same as used in the development/design, i.e., fixed-time or sequential tests. The philosophies and circumstances which guide risk selection are different, however.

In some instances, when the MTBF of an item is high and production quantities are small or schedule is tight, sufficient time to provide the producer/consumer risks desirable is not available. In virtually all instances the consumer and producer risk scenarios are different. A development reliability demonstration test is considered to be a one-time event with potential dire consequences to the customer, if an item with poor reliability passes the test, or to the developer, if an item with either poor or superior reliability fails the test. As a result, low consumer and producer risks are prudent for tests of that type.

Assuming that the inherent design of the product is capable of providing the reliability required, a functioning quality program encompassing material and process control, FRACAS, and ESS acts to maintain the reliability of a product. It sends signals of potential problems prior to production lot demonstration and allows adjustments to reliability-sensitive processes to be made as production continues. As a result, in its presence a reliability demonstration becomes more of a type of safety net for the long-term effectiveness of such a program; a validation of its effectiveness as production continues; and a guard against potential significant slip-ups and failings in the quality program which can create long-term effects.

Demonstration in this phase is performed basically on a lot-to-lot basis to determine the existence of large deviations in quality in any single lot and to flag any significant deviations which continue over a number of lots. As a result, consideration is usually given to allowing the use of higher-risk reliability demonstrations on a lot basis.

For the customer, acceptance of a relatively few lots of non-safety-critical items with reliability discrepancies that can be corrected the first time the item component is returned for repair is usually less than catastrophic. Assume the following: that problems in production reliability are not recognized until a lot fails its demonstration test, that a consumer risk β is associated with the test, and that the reliability of the items making up the lot dropped to the θ_u limit due to undetected changes in the manufacturing process. The probability, P_c, of detecting such changes, *just* due to the demonstration test, within m lots of its "drop" can be estimated:

$$P_c = 1 - \beta^m$$

Hence, if $\beta = 0.30$, the probability that it will be recognized within two lots is $P_c = 0.91$; the probability that it will be recognized within three lots is $P_c = 0.97$.

Consequently, under the conditions outlined above, the customer can usually tolerate more moderate risks during production reliability tests than during a development test. Generally, at the start, the consumer risk associated with production tests should be no greater than that used during development tests. Those risks should persist until reliability production control has been established. At that time, consumer risk can be increased to a moderate level and held at that level until evidence appears of potential loss of control of the reliability production process. When that occurs, consumer test risk is reduced until control has been reestablished.

The producer's knowledge and control of reliability status provides visibility relative to the actual value of θ being produced. Unfortunately, a value of $\theta = \theta_g$ (the value the producer feels the item has, and saw use as an upper test limit) will still result in rejection of α of the lots tested. As a result, while the consequences of lot rejection are less critical to the producer than a rejection of a developmental product (it may mean perhaps, performance of additional tests and screens, performing rework, and retests of a small proportion of lots), increase in risk still translates into additional producer effort. For this reason, while increases in individual test consumer risk may be practically afforded, increases in producer risk in general cannot.

The "All Equipment Production Reliability Test Plans" described in Mil-Hdbk-781 have such general consumer risk characteristics when applied to item lots.

The tests are sequential in nature, are carried out in a unique procedural fashion, and have risks which *vary* significantly in accord with maximum allowable test time.

In the test plan, failures versus test time are plotted as before for sequential tests; however, in this case, *no* accept decision is made if the plot crosses the accept boundary. The plot continues beyond the boundary until a failure occurs. If at that time the plot is *one or more* failure intervals below the accept line, the plot is brought up vertically to the accept boundary line. If at that time the plot is *less* than one failure interval below the accept line, the plot is brought vertically up one failure interval, crossing the accept line. The plot continues until the maximum item test time is reached. If no reject decision is reached up to that time, the lot is accepted. If a reject decision is made at any time before the maximum test time is reached, the lot is rejected. Figure 12-8 shows a typical plot of this nature.

Figures 12-9a and b show accept/reject boundaries and the operating characteristic curves for one test plan. Note that the OC curve varies as a function of maximum test time.

12.2.6 Reliability Demonstration for Items Having Nonconstant Hazards

Demonstrating the mean time to failure of items having failure distributions of one unknown parameter such as λ (or $1/\lambda = \theta$) is relatively uncomplicated.

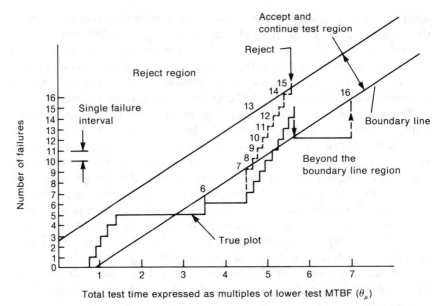

FIGURE 12-8.   The special plotting criteria for the "All Equipment Reliability Test." [From Mil-Hdbk-781.]

When the distribution of time to failure is known and has more than one parameter (i.e., the Weibull, chi-square, gamma or log-normal distributions), the process, like the measurement process discussed earlier, becomes significantly more complicated. While it is possible to structure a demonstration plan taking into account more than one parameter, through either analytical or nonparametric means, such plans are beyond the scope of this text (see Ref. 12 for procedures which can be applied to repairable items in general). We will limit discussion to demonstration procedures applicable when all but one parameter is known or assumed.

Demonstrations of mean time to failure for redundant systems also fall into this category. Even when all components of the system have exponential distributions of time to failure, the time to failure of the system will not follow an exponential distribution of time to failure.

12.2.6.1   Demonstrating Mean Time to Failure When the Item Has a Weibull Distribution of Time to Failure

As before, the Weibull distribution is used as an example of a general distribution which can result in an increasing or decreasing hazard (failure rate) with time. The purpose is to structure a test for the MTTF of an item when it is known that the item exhibits a distribution of failures that is

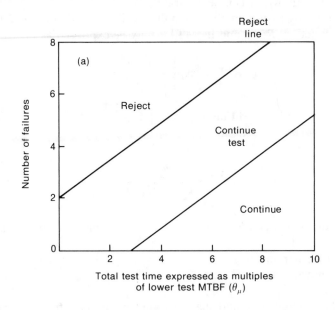

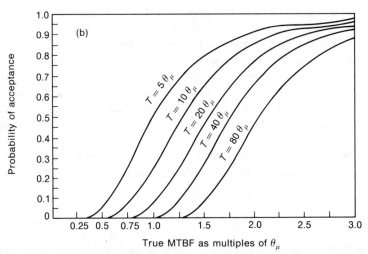

FIGURE 12-9. (a) Accept/reject boundaries for an "All Equipment Production Reliability Test." (b) Operating characteristic curves for the test plan in (a) as a function of maximum test time, T. [From Mil-Hdbk-781.]

approximately Weibull in character and when the shape parameter of the distribution is known or can be accurately estimated.

From our discussion of the Weibull distribution in Section 2.1.2, its reliability function, $R(T)$, and its MTTF may be expressed as

$$R(T) = \exp\left[-\left(\frac{T}{\dot{\alpha}}\right)^{\dot{\beta}} \right] \tag{12-73}$$

$$\text{MTTF} = \phi = \dot{\alpha}\Gamma\left[1 + \frac{1}{\dot{\beta}} \right] \tag{12-74}$$

or

$$\dot{\alpha} = \frac{\phi}{\Gamma[1 + (1/\dot{\beta})]}$$

where

$T =$ interval of operation measured from the time the item was first energized

$\dot{\alpha} =$ a scale parameter

$\dot{\beta} =$ a known shape parameter

(*Note:* The Weibull scale and shape parameters are denoted as $\dot{\alpha}$ and $\dot{\beta}$ in order to distinguish those parameters from the producer and consumer risk characteristics, which are traditionally defined as α and β.)

One of the simplest means to implement a demonstration based on this type of distribution is to perform a series of transformations and substitutions such that the same techniques used for demonstration under the assumption of an exponential distribution of times to failure can be used. The resultant procedure to be used in performing the demonstrations is described below. It has the same basis as the procedure employed in Section 12.1.5 in measuring MTTF.

The demonstration is performed as if a new system for measuring and expressing time existed, where T units of time in our frame of reference is equivalent to

$$T^{\beta} = T_x \tag{12-75}$$

units of time in the new system. For example, in Fig. 12.3, the x-axis would read $T^{\dot{\beta}}$ instead of T and a demonstration test scheduled to run for 300 hours would be considered as a test running for $(300)^{\dot{\beta}}$ hours. Under that transformation, (12-73) would take the form

$$R(T_x) = R(T^{\dot{\beta}}) = \exp(-T^{\dot{\beta}}/\dot{\alpha}_x) = \exp(-T_x/\dot{\alpha}_x) \tag{12-76}$$

where

$\dot{\alpha}_x = \dot{\alpha}^{\dot{\beta}}$ in the new time system, a constant

T = interval of operation measured from the time the item was first energized in our usual time system

$T_x = T^{\dot{\beta}}$ = interval of operation measured from the time the item was first energized in the new time system.

Since (12-76) is an exponential form, $\dot{\alpha}_x$ can be treated as if $\dot{\alpha}_x = \theta$, MTBF, where the demonstration test in the new time system is between an upper value of $\dot{\alpha}_{gx}$

$$\dot{\alpha}_{gx} = (\phi_g / \Gamma [1 + (1/\dot{\beta})])^{\dot{\beta}} \qquad (12\text{-}77)$$

and a lower value of $\dot{\alpha}_{ux}$

$$\dot{\alpha}_{ux} = (\phi_u / \Gamma [1 + (1/\dot{\beta})])^{\dot{\beta}} \qquad (12\text{-}78)$$

and with a discrimination ratio d_x

$$d_x = \left(\frac{\phi_g}{\phi_u} \right)^{\dot{\beta}} \qquad (12\text{-}79)$$

(but where the actual discrimination ratio between ϕ_g and $\phi_u = \phi_g / \phi_u = d$).

Hence, from knowledge of the needed values of ϕ_u and ϕ_g, the discrimination ratio d, the scale parameter $\dot{\alpha}$, the shape parameter $\dot{\beta}$, values of $\dot{\alpha}_{gx}$, $\dot{\alpha}_{ux}$ and d_x can be generated.

Substituting

$$T_x \text{ for } T$$
$$\dot{\alpha}_{gx} \text{ for } \dot{\alpha}_g$$
$$\dot{\alpha}_{ux} \text{ for } \dot{\alpha}_u$$
$$d_x \text{ for } d$$

the demonstration procedures of the previous sections can be applied.

12.2.6.2 Demonstrating Mean Time to Failure For Redundant or Fault-Tolerant Systems

Redundant systems and subsystems, even when made up of components with exponential distributions of time to failure, will not have an exponential distribution of time to failure. As a result, the procedures described in the preceding sections will not apply. Reliability demonstration for redundant configurations requires the use of special techniques. These range from generating a single specific statistical test plan keyed to the configuration and specifics of the subsystem, to the application of individual standard single test plans to each component in the redundant configuration, to application of a single standard test to the subsystem based on a "single-thread" concept. All have their advantages and disadvantages. Discussion will be restricted to those configurations made up of components which exhibit exponential distributions of repair time.

Developing Unique Test Plans for Redundant Configurations

Every reliability test plan is composed of a series of points in time at which an accept/reject decision can be made based on the number of failures observed to that time. With knowledge of the distribution of time to failure of an item, test plans can be structured in advance which will always yield the same consumer and producer risks for a given discrimination ratio, d.

If a test plan which was generated in accord with one distribution of time to failure were to be applied to an item which takes on a different distribution of time to failure, different consumer and producer risks for that same discrimination ratio would result. Indeed, a different OC curve would result.

The makeup of all but the most simple redundant systems is unique. The number of redundant components in each subsystem, the minimum number of components necessary to perform the subsystem function, the number of different subsystems comprising the system, and the relative value of the reliability of each component; each singly and in combination can affect the distribution of failure time of a subsystem. All of these militate against the availability or existence of an appropriate standard subsystem test plan to fit the system to be tested.

However, a system test plan of that type, one capable of directly demonstrating the reliability of the system as a whole, is the most desirable. It is not only the most realistic, but additionally it allows for consideration of the effects of compensation. In other words, if one subsystem turns out to be less reliable than expected (or planned for), but one subsystem turns out to be more reliable than expected, the test plan is sensitive to their joint impacts on the reliability of the system.

The issue, then, is the structure/determination of a test plan which, when applied to the unique system, will provide the risks needed.

One practical approach to the problem which has been employed involves the evaluation of the risks associated with any given test plan. Recall that any series of points in time where accept/reject decisions are made based on a defined number of failures observed to those times is a test plan. The approach involves the use of Monte Carlo simulation to analyze/synthesize such plans (see Ref. 13). Monte Carlo simulation involves performing a large number of experiments under a given set of assumptions and then using the results of such experiments to develop the OC curve for a given system. A large enough number of experiments can be selected practically to provide an answer to an acceptable level of accuracy. One software program for this purpose is available at the Rome Laboratory, Griffiss AFB (U.S.A.F.) New York. Example applications and results are shown in the above-cited reference.

Demonstration by Parts

This procedure involves the derivation of individual component values of θ_{ui} and θ_{gi} from the θ_u and θ_g values associated with a redundant subsystem configuration, and performing individual reliability demonstrations on each component (type).

When all components making up the configuration are identical, the procedure is maximally effective, taking into account reliability compensating factors, and produces results which are analytically derived. The procedures in Chapter 10 provide the tools necessary for the structure of the test. The models in the chapter which relate the MTTF/MTBF of the configuration to the component makeup and redundancy strategy are employed. For example, for a full-on redundant subsystem,

$$\text{MTTF}_i = \text{MTFF}_i = \theta_i = \frac{1}{\lambda} \sum_{k=d}^{n} \frac{1}{k} = \theta \sum_{k=d}^{n} \frac{1}{k} \qquad (12\text{-}80)$$

where
  $1/\lambda = \theta = \text{MTTF/MTFF of a component}$
  d = number of operating components necessary for the subsystem to function
  n = number of components making up the subsystem

Assume that

  θ_u = minimum acceptable MTTF for the subsystem
    = the lower test reliability

  θ_g = the design objective MTTF for the subsystem
    = the upper test reliability

The lower and upper test MTTF values, θ_{ui} and θ_{gi}, for each component can be derived by setting (12-80) equal to θ_u then to θ_g and solving for θ_{ui}

and θ_{gi}, respectively:

$$\theta_u = \theta_{ui} \sum_{k=d}^{n} \frac{1}{k} \tag{12-81}$$

$$\theta_{ui} = \frac{\theta_u}{\sum_{k=d}^{n} (1/k)} \tag{12-82}$$

$$\theta_g = \theta_{gi} \sum_{k=d}^{n} \frac{1}{k} \tag{12-83}$$

$$\theta_{gi} = \frac{\theta_g}{\sum_{k=d}^{n} (1/k)} \tag{12-84}$$

where θ_{gi}, θ_{ui} = upper and lower levels of test MTBF associated with each component.

If the components have exponential distributions of failure, their reliability can be demonstrated using the techniques described in the previous sections, using θ_{ui} and θ_{gi} as lower and upper test MTBF values and a discrimination ratio of θ_{gi}/θ_{ui}. *The risks associated with the demonstration of a component of a given type will be equivalent to those associated with the demonstration of a subsystem made up of such components.* The fact that more than one component is available for test will serve to reduce test time as described in Section 12.1.4.1.

When the components making up a redundant configuration are not all identical, for example, a system made up of n different subsystems, some redundant, some not redundant, more complex translation and test procedures must be applied. The objectives are:

1. To determine those minimum acceptable values of component MTBF which would yield the required value of θ_u.
2. To determine those upper test values of component MTBF which will yield the desired value of θ_g.
3. To develop test plans for each different component which will yield the *system* risk levels desired and perform independent tests on each component.

Such objectives are more easily stated than achieved. Thus far, only approximations and worst-case approaches to such needs have been suggested and those are quite complex. Some approaches to the problem are contained in Refs. 7 and 14.

Conservative consumer risk consequences at the subsystem level (less risk than that defined) always result if values of θ_{ui} which will yield a *system level* value of θ_u are identified and then independently demonstrated to the stated level of consumer risk.

However, this test procedure will not take into account reliability compensation among components. Hence, there will be a greater probability that a subsystem that meets requirements will be rejected, i.e., the fact that a reliability shortfall in one subsystem can be compensated for by a better than planned reliability in another subsystem will not be taken into account.

Single-Thread Demonstration Procedures

Single-thread demonstration procedures involve treating the redundant system as if all components were connected in series:

$$\theta_t = \frac{1}{\sum\limits_{j=1}^{n} \lambda_j} \qquad (12\text{-}85)$$

where
θ_t = single-thread MTBF
λ_j = failure rate of component j in the system or subsystem
n = number of components in the system or subsystem

Values of θ_{tg} and θ_{tu} are selected representing upper and lower test values of single thread MTBF, where

θ_{tu} = the value consistent with the lower test bound for a logistics reliability requirement or the value which results if all components in the subsystem take on assigned minimum, or worst case reliability values which will still yield an acceptable value of subsystem reliability, θ_u.

θ_{tg} = the value consistent with the upper test bound for a logistics reliability requirement or the value which results if all components take on their final reliability prediction values

A demonstration is then performed consistent with selected risks α and β assuming θ_t represents the MTBF of an item: (1) assuming an exponential distribution of time to failure for each component and by extension for the series combination of components; (2) applying the procedures described before for nonredundant items.

In essence, it is a demonstration of logistics reliability (see Chapter 1). In this case, θ can be directly defined as a *logistics reliability requirement* which must be demonstrated in *addition* to any mission/use-related reliability requirement.

When used to demonstrate logistics reliability θ_t it provides for an accurate test. When used to demonstrate mission/use reliability, the procedure, at best, is an approximate one. It *infers* compliance with a mission/use requirement rather than demonstrating it. The procedure demonstrates reliability prediction accuracy. Assuming there is consistency of accuracy among all predictions, it is inferred that the minimum values of reliability from which θ_{tu} was developed have been attained or bettered. However, it is relatively simple to run and accounts for some degree of reliability compensation among components.

References
1. Hahn, G. and Shapiro, S. (1967). *Statistical Models in Engineering*. Wiley, New York.
2. Dixon, W. and Massey, F., Jr. (1957). *Introduction to Statistical Analysis*. McGraw-Hill, New York.
3. Bulfinch, A. (1962). *Reliability Handbook*. Quality Assurance Division Picatinny Arsenal. Dover, New Jersey, Report ORDBB-NR126.
4. Air Training Command, Chanute Technical Training Center (1968). *Statistical Tables*. 4OST2895G-1-102B-2.
5. Bhattacharya, G. and Johnson, R. (1977). *Statistical Concepts and Methods*. Wiley, New York.
6. Epstein, B. (1960). "Estimation From Life Test Data." *Technometrics*, Nov.
7. Mann, N., Schafer, R., and Singpurwalla, N. (1974). *Methods for Statistical Analysis of Reliability and Life Data*. Wiley, New York.
8. Goel, A. and Klion, J. (1978). "Analysis of Truncated SPRT: Exponential Case." Proceedings, 1978 Annual Reliability and Maintainability Symposium, IEEE, Piscataway, NJ.
9. Sumerlin, W. (1972). "Confidence Calculations for Mil-Std-781." Proceedings, 1972 Annual Reliability and Maintainability Symposium, IEEE, Piscataway, NJ.
10. Epstein, B. (1960). "Statistical Life Test Acceptance Procedures." *Technometrics*, Nov.
11. Epstein, B. (1954). "Truncated Life Tests in the Exponential Case." *Annals of Mathematical Statistics*, **25**.
12. Ascher, H. and Feingold, H. (1984). *Repairable Systems Reliability*. Marcel Dekker, New York.
13. Bivens, G. et al. (1987). "Reliability Demonstration Technique for Fault Tolerant Systems." Proceedings, 1987 Annual Reliability and Maintainability Symposium, IEEE, Piscataway, NJ.
14. Reliability Analysis Center (1984). "Revision of Mil-Std-781C." Report, prepared for Naval Electronics System Command by Reliability Analysis Center.

Appendix

A0 A GENERAL RELIABILITY EXPRESSION

Given a population of items which has *any* failure density function (failure distribution), $f(t)$:

- The proportion of the initial population which will fail between the time the items *were first put in operation* (time $T_1 = 0$) and any selected time T_2 is

$$\int_0^{T_2} f(t)\, dt = \text{proportion of initial items failing in the interval}$$
$$T = (T_2 - T_1) = T_2 \tag{A0-1}$$

$$= \text{probability a brand new item selected from the population and put into operation at time 0 will fail by time } T = T_2.$$

- The proportion of the initial population which will fail between *any time T_1 and another time T_2 ($T_2 > T_1$)* is

$$\int_{T_1}^{T_2} f(t)\, dt = \text{proportion of } \textit{initial} \text{ items failing in the interval } T \text{ between } T_1$$
$$\text{and } T_2 \tag{A0-2}$$

$$= \text{probability an item selected from the population at time 0 and put into operation will first fail between } T_1 \text{ and } T_2$$

Of course, as time approaches infinity, 100% of all of the initial items will fail:

$$\int_0^{\infty} f(t)\, dt = 1 \tag{A0-3}$$

$$= \text{proportion of initial items failing in the time interval 0 to } \infty.$$

From (A0-1),

$$R(T) = 1 - \int_0^T f(t)\, dt \tag{A0-4}$$

where

$R(T) =$ reliability, the proportion of initial items which survive the operating time period 0 to T

$=$ probability a brand new item selected at random from the population at time 0 will not fail by time T

$=$ the probability that an item will perform satisfactorily over an interval of time, T, given that it is operable at time 0.

However, suppose you are at time T_1 and you want to determine the *proportion of your current population* (the items which have survived T_1 hours of operation) which will fail between T_1 and T_2. If n is the number of items in the original population,

$$\left(1 - \int_0^{T_1} f(t)\, dt\right) n = \text{the number of items that are in the current population}$$
(that is to say, the number of items which have survived T_1 hours of operation)

$$\left(\int_{T_1}^{T_2} f(t)\, dt\right) n = \text{the number of items which will fail between } T_1 \text{ and } T_2, \text{ then}$$

$$\frac{\left(\displaystyle\int_{T_1}^{T_2} f(t)\, dt\right) n}{\left(1 - \displaystyle\int_0^{T_1} f(t)\, dt\right) n} = \frac{\displaystyle\int_{T_1}^{T_2} f(t)\, dt}{1 - \displaystyle\int_0^{T_1} f(t)\, dt} = \frac{\displaystyle\int_{T_1}^{T_2} f(t)\, dt}{\displaystyle\int_{T_1}^{\infty} f(t)\, dt} \tag{A0-5}$$

$$= G(T_2 - T_1)$$

where

$G(T_2 - T_1) = G(T) =$ the proportion of the items functioning at T_1 which will fail between T_1 and T_2 (i.e., the proportion of the *current* population which will fail in the interval T_1 to T_2)

$=$ the probability that a particular item which is in a nonfailed state after operating for time T_1 will fail in the interval of time $T = T_2 - T_1$.

Reliability $R(T_2 - T_1)$ over the specific time interval $T = T_2 - T_1$ is

$$R(T) = R(T_2 - T_1) = 1 - G(T_2 - T_1) = 1 - \frac{\displaystyle\int_{T_1}^{T_2} f(t)\,dt}{1 - \displaystyle\int_{0}^{T_1} f(t)\,dt} \qquad (A0\text{-}6)$$

where, following the pattern established previously, $R(T) = R(T_2 - T_1) =$ the proportion of items functioning at time T_1 which will not fail between T_1 and $T_2 =$ probability that an item will perform satisfactorily over an interval of time $T = (T_2 - T_1)$ given that the item has acquired an operational history equal to T_1 and is operable at time T_1.

In order to establish a more workable form for (A0-6), first note that (A0-3) can be expressed as

$$\int_0^\infty f(t)\,dt = \int_0^{T_1} f(t)\,dt + \int_{T_1}^{T_2} f(t)\,dt + \int_{T_2}^\infty f(t)\,dt = 1 \qquad (A0\text{-}7)$$

and from (A0.4),

$$\int_0^{T_2} f(t)\,dt = 1 - R(T_2 - 0) = 1 - R(T_2)$$

where

$$\int_{T_2}^\infty f(t)\,dt = R(T_2)$$

and

$$\int_0^{T_1} f(t)\,dt = 1 - R(T_1 - 0) = 1 - R(T_1)$$

Equation (A0-7) can then be expressed as

$$1 - R(T_1) + \int_{T_1}^{T_2} f(t)\,dt + R(T_2) = 1$$

Rearranging,

$$\int_{T_1}^{T_2} f(t)\, dt = R(T_1) - R(T_2)$$

Substituting into (A0-6),

$$R(T) = R(T_2 - T_1) = 1 - \frac{R(T_1) - R(T_2)}{R(T_1)}$$

$$= 1 - \left(1 - \frac{R(T_2)}{R(T_1)}\right) = \frac{R(T_2)}{R(T_1)} \qquad \text{(A0-8)}$$

where $R(T) = R(T_2 - T_1) =$ the probability that an item will perform satisfactorily over the interval of time $T = (T_2 - T_1)$ given that the item has acquired an operational history equal to T_1 and is operable at T_1.

Equation (A0-8) represents the most general form for the expression of reliability and holds for any distribution of time to failure.

A1   AN EXPRESSION FOR RELIABILITY WHEN TIME TO FAILURE FOLLOWS AN EXPONENTIAL DISTRIBUTION

Given that $f(t)$ represents an exponential failure probability density function

$$f(t) = \lambda e^{-\lambda t}$$

from (A0-5), since $1 - \int_0^{T_1} f(t)\, dt = \int_{T_1}^{\infty} f(t)\, dt$,

$$G(T_2 - T_1) = \frac{\displaystyle\int_{T_1}^{T_2} f(t)\, dt}{\displaystyle\int_{T_1}^{\infty} f(t)\, dt} = \frac{\displaystyle\int_{T_1}^{T_2} \lambda e^{-\lambda t}\, dt}{\displaystyle\int_{T_1}^{\infty} \lambda e^{-\lambda t}\, dt}$$

$$= 1 - \exp[-\lambda(T_2 - T_1)]$$

and

$$R(T_2 - T_1) = \exp[-\lambda(T_2 - T_1)]$$

Hence, for *any given length of operating time*, T, where $T = (T_2 - T_1)$ (say, for example, $T_2 - T_1 = 100$ hours), the reliability will be the same *regardless of which values of T_1 or T_2 are chosen*. This is a characteristic unique to the exponential distribution. A 100-hour period of operating time when the item is almost brand new will have the same probability of failure as a 100-hour period of operating time when the item has been in operation for 1000 hours.

A2   AN EXPRESSION FOR RELIABILITY WHEN TIME TO FAILURE FOLLOWS A WEIBULL DISTRIBUTION

Given that $f(t)$ represents a Weibull failure probability density function

$$f(t) = \frac{\beta}{\alpha}\left(\frac{t}{\alpha}\right)^{\beta-1}\exp\left[-\left(\frac{t}{\alpha}\right)^{\beta}\right]$$

from (A0-5),

$$G(T_2 - T_1) = \frac{\displaystyle\int_{T_1}^{T_2} f(t)\,dt}{\displaystyle\int_{T_1}^{\infty} f(t)\,dt} = \frac{(\beta/\alpha^{\beta})\displaystyle\int_{T_1}^{T_2} t^{\beta-1}e^{-(t/\alpha)^{\beta}}\,dt}{(\beta/\alpha^{\beta})\displaystyle\int_{T_1}^{\infty} t^{\beta-1}e^{-(t/\alpha)^{\beta}}\,dt}$$

$$= \frac{e^{-(T_1/\alpha)^{\beta}} - e^{-(T_2/\alpha)^{\beta}}}{e^{-(T_1/\alpha)^{\beta}}} = 1 - \exp\left[-\left(\frac{1}{\alpha}\right)^{\beta}(T_2^{\beta} - T_1^{\beta})\right]$$

$$R(T_2 - T_1) = 1 - G(T_2 - T_1) = \exp\left[-\left(\frac{1}{\alpha}\right)^{\beta}(T_2^{\beta} - T_1^{\beta})\right] \qquad \text{(A2-1)}$$

Hence, for any given length of operating time, T, where $T = (T_2 - T_1)$ (say, for example, $T_2 - T_1 = 100$ hours), the reliability will vary depending on the values of T_1 and T_2 chosen. A 100-hour period of operating time when the item is almost brand new will have a markedly different reliability than when the item has been in operation for 1000 hours.

A3   THE GAMMA DISTRIBUTION AND THE POISSON DISTRIBUTION

The gamma distribution can be expressed in terms of a Poisson distribution for integer values of β:

$$\int_{T}^{\infty} \frac{\alpha}{\Gamma(\beta)}(\alpha t)^{\beta-1}e^{-\alpha t}\,dt = \sum_{k=0}^{\beta-1} \frac{(\alpha t)^{k}(e^{-\alpha t})}{k!}$$

(See any table of integrals or integrate by parts.)

From (A0-8) then,

$$R(T_2 - T_1) = e^{-\alpha(T_2 - T_1)} \left[\frac{\sum\limits_{k=0}^{\beta-1} (\alpha T_2)^k / k!}{\sum\limits_{k=0}^{\beta-1} (\alpha T_1)^k / k!} \right] \qquad (A3\text{-}1)$$

A4   THE DERIVATION OF MEAN TIME TO FIRST FAILURE (MTFF) AND MEAN TIME TO FAILURE (MTTF)

The derivations of mean time to first failure (MTFF) (a term mostly applied to repairable items) and mean time to failure (MTTF) (a term mostly applied to nonrepairable items) have the same basic formulation. The formulation holds with respect to every failure distribution, whether or not the item in question is redundant (fault-tolerant) or in a series connected configuration, a single repairable item or a population of items. For convenience, let us designate both (MTFF) and (MTTF) as M. Further, let $R(T) = R(T_2 - 0) = $ the probability the item will perform satisfactorily over an interval of time, T, between time 0 and T_2. We assume the item is brand new and operable at time $T_1 = 0$, so that $T = T_2$.

From (A0-4),

$$R(T) = 1 - \int_0^T f(t)\, dt$$

or, more generally,

$$R(t) + C = 1 - \int f(t)\, dt \qquad (A4\text{-}1)$$

where C = any constant and, of course, from elementary calculus,

$$f(t) = -\frac{dR(t)}{dt} \qquad (A4\text{-}2)$$

The mean, M, associated with any distribution, whether it be related to the salaries of engineers or with time to failure, is

$$M = \int x f(x)\, dx \qquad (A4\text{-}3)$$

where

x = a value of x—time to failure, a salary.

$f(x)\,dx \sim$ probability that the value x, will be achieved.

The integral sign denotes a summation over all possible values of x. Hence, for reliability purposes,

$$M = \int_0^\infty tf(t)\,dt \qquad (A4\text{-}4)$$

= the mean or expected value of time to failure measured from the time the equipment is brand new, i.e. measured from time 0

Since we measure the mean, starting from time 0, the mean generated will not only be the mean time to failure (associated with a population of nonrepairable items) but the mean time to first failure (associated with repairable items).

Substituting (A4-2) in (A4-4),

$$M = -\int_0^\infty t\,\frac{dR(t)}{dt}\,dt \qquad (A4\text{-}5)$$

Integrating (A4-5) by parts,

$$M = -tR(t)\Big|_0^\infty + \int_0^\infty R(t)\,dt \qquad (A4\text{-}6)$$

and evaluating the results by limit theorems, the first term drops out of (A4-6) and

$$M = \int_0^\infty R(t)\,dt \qquad (A4\text{-}7)$$

A5 OTHER MEASURES OF THE MEAN USED IN RELIABILITY EVALUATION

The mean measure just discussed defined MTFF and MTTF assuming an item or population of items started operation at time 0 and continued operation to time approaching ∞. In some instances, MTFF or MTTF must be evaluated for an item or a population of items which have the following characteristics of use:

1. Starting operation at time 0 (when the item is brand new) but operating only to a finite time, T_m where it is desired to determine the average uninterrupted operating time, AUL, or average satisfactory operating hours per item experienced over T_m.
2. Starting operation after acquiring a prior history of use, T_1, and continuing to operate until failure of the item or of the population of items where it is desired to determine the average time to the next failure, or mean residual life, MRL, measured from T_1.

Evaluating AUL means defining the value of a mean that describes the AUL estimate, ϕ, where

$$\phi = \frac{\sum_{i=1}^{r} t_i + (n-r)T_m}{r} = \sum_{i=1}^{r} \frac{t_i}{r} + \frac{(n-r)T_m}{r} \tag{A5-1}$$

where

t_i = the time to failure of the ith item during T_m
r = the number of failures during T_m
n = the population of items undertaking a mission of duration T_m
$(n-r)/n$ = the proportion of items surviving T_m without failing

The final form for (A5-1) is made up of two terms, one representing the average operating hours for all the items which failed, i.e., $\sum_i^r t_i/r$, the other representing the product of the proportion of survivors and T_m.

The theoretical expression which will yield the average operating hours for all the items which fail during T_m is

$$\int_0^{T_m} tf(t)\,dt \tag{A5-2}$$

where
$f(t)$ = the probability density function
T_m = duration of the mission or use in question

The theoretical expression which will yield the product of the proportion of survivors and T_m is

$$R(T_m)T_m \tag{A5-3}$$

Adding (A5-2) and (A5-3) results in

$$\text{AUL} = \int_0^{T_m} tf(t)\, dt + R(T_m) T_m \qquad \text{(A5-4)}$$

Solving (A5-2) by integration by parts yields

$$\int_0^{T_m} tf(t)\, dt = \int_0^{T_m} R(t)\, dt - R(T_m) T_m$$

so that

$$\text{AUL} = \int_0^{T_m} R(t)\, dt \qquad \text{(A5-5)}$$

Evaluation of mean residual life (MRL) or mean time to the next failure is required when there is a need to assess the mean time to failure associated with the current survivors of a population starting from the time T_1, i.e., when it is required to evaluate the mean time to failure of an item measured *from* the time after it has acquired an operational history of T_1 hours. MRL is a general form of MTFF (and of MTTF).

The general basis for deriving a relationship that can be used to assess MRL is the same as that used in Section A0

$$\text{MRL} = \int_{T_1}^{\infty} xf(x)\, dx \qquad \text{(A5-6)}$$

where
$f(x)$ = a probability density function
x = time to failure based on some reference point (i.e., T_1)
$f(x)\, dx \sim$ probability that a failure will occur during dx
T_1 = acquired operational history on the item at the time MRL is to be assessed (MRL is to be measured *from* T_1).

In order to develop an expression for MRL: 1. the basic failure density function, $f(x)$, associated with the item must be normalized to account for only those failures which occur after T_1; 2. measures of time to failure must be adjusted to use T_1 rather than 0 as a base.

The normalized failure density, $f(t)_n$, takes the form

$$f(x) = f(t)_n = \frac{f(t)}{R(T_1)} \qquad \text{(A5-7)}$$

where $f(t) =$ the failure density associated with the item.

The value of x is normalized to take T_1 as a base,

$$x = (t - T_1) \tag{A5-8}$$

and

$$\mathrm{MRL} = \int_{T_1}^{\infty} (t - T_1) f(t)_n \, dt = \int_{T_1}^{\infty} (t - T_1) \frac{f(t)}{R(T_1)} \, dt$$

$$= \int_{T_1}^{\infty} t \frac{f(t)}{R(T_1)} \, dt - T_1 \int_{T_1}^{\infty} \frac{f(t)}{R(T_1)} \, dt$$

Since

$$T_1 \int_{T_1}^{\infty} \frac{f(t)}{R(T_1)} \, dt = \frac{T_1}{R(T_1)} \int_{T_1}^{\infty} f(t) \, dt$$

And since $\int_{T_1}^{\infty} f(t) \, dt = R(T_1)$

$$\mathrm{MRL} = \int_{T_1}^{\infty} t \frac{f(t)}{R(T_1)} \, dt - T_1 \tag{A5-9}$$

Since from (A4-2), $f(t) = -[dR(t)/dt]$, the integral part of (A5-9) can be expressed as

$$-\left(\frac{1}{R(T_1)} \right) \int_{T_1}^{\infty} t \frac{dR(t)}{dt} \, dt$$

Integrating the above by parts yields

$$\frac{1}{R(T_1)} \left[\left(-tR(t) \Big|_{T_1}^{\infty} \right) + \int_{T_1}^{\infty} R(t) \, dt \right] = T_1 + \frac{1}{R(T_1)} \int_{T_1}^{\infty} R(t) \, dt \tag{A5-10}$$

Substituting in (A5-9)

$$\mathrm{MRL} = \frac{1}{R(T_1)} \int_{T_1}^{\infty} R(t) \, dt \tag{A5-11}$$

Note that when $T_1 = 0$,

$$\mathrm{MRL} = \int_{0}^{\infty} R(t) \, dt = M = \mathrm{MTFF}$$

MRL for distributions exhibiting nonconstant hazard/failure rates is more difficult to solve in closed form than it appears. Closed-form solutions for special cases, however, can be found. For example, if the item had a Weibull failure distribution, $R(t)$ in (A5-11) would take the form

$$R(t) = \exp\left[-\left(\frac{t}{\alpha}\right)^{\beta} \right]$$

where α and β are scale and shape parameters associated with the distribution.

$$MRL = \frac{1}{R(T_1)} \int_{T_1}^{\infty} R(t) \, dt = \frac{1}{R(T_1)} \int_{T_1}^{\infty} \exp\left[-\left(\frac{t}{\alpha}\right)^{\beta} \right] dt \quad (A5\text{-}12)$$

We will change the variable of integration by substitution. Let

$$\left(\frac{t}{\alpha}\right)^{\beta} = z, \qquad t^{\beta} = \alpha^{\beta} z, \qquad t = \alpha(z)^{1/\beta}, \qquad \beta t^{\beta-1} \, dt = \alpha^{\beta} \, dz$$

$$dt = \frac{\alpha^{\beta}}{\beta} \frac{1}{t^{\beta-1}} \, dz = \frac{\alpha^{\beta}}{\beta} \frac{1}{\alpha^{\beta-1} z^{(\beta-1)/\beta}} \, dz = \frac{\alpha}{\beta} z^{(1/\beta-1)} \, dz$$

Equation (A5-12) can then be expressed as

$$MRL = \frac{1}{R(T_1)} \int_{z_1 = (T_1/\alpha)^{\beta}}^{\infty} \left(\frac{\alpha}{\beta}\right) z^{(1/\beta-1)} e^{-z} \, dz \quad (A5\text{-}13)$$

where the range of integration goes from $z_1 = (T_1/\alpha)^{\beta}$ to ∞, because of the change of the variable of integration. In the special case when $1/\beta$ is an integer, n, (A5-13) takes the form

$$MRL = A \int_{z_1}^{\infty} (z^{n-1}) e^{-z} \, dz \quad (A5\text{-}14)$$

$A = $ a constant $= \alpha/\beta R(T_1)$, which can be found in almost any table of integrals. As a consequence

$$MRL = -Ae^{-z}[z^{n-1} + (n-1)z^{n-2} + (n-1)(n-2)z^{n-3} + \cdots$$
$$+ (n-1)!] \Big|_{(T_1/\alpha)^{\beta}}^{\infty}$$

$$= \frac{\alpha}{\beta}[z^{n-1} + (n-1)z^{n-2} + (n-1)(n-2)z^{n-3} + \cdots + (n-1)!] \quad (A5\text{-}15]$$

where $z = (T_1/\alpha)^{\beta}$.

A6 EVALUATING MTTF OF A WEIBULL TIME TO FAILURE DISTRIBUTION WHEN $\lambda(T_p)$ IS USED AS A SURROGATE FOR α

$$\text{MTTF} = \int_0^\infty R_e(t)\, dt$$

$$R_e(t) = \exp\left[-\left(\frac{\lambda(T_p)t^\beta}{T_p^{\beta-1}}\right)\right]$$

where
T_p = a period of service life.

Let

$$\frac{\lambda(T_p)}{T_p^{\beta-1}} = A$$

$$\text{MTTF} = \int_0^\infty \exp(-At^\beta)\, dt$$

Let

$$t^\beta = x, \qquad t = x^{1/\beta}$$

$$\beta t^{\beta-1}\, dt = dx, \qquad dt = dx(\beta x^{(\beta-1)/\beta})^{-1}$$

$$\text{MTTF} = \frac{1}{\beta}\int_0^\infty x^{[(1/\beta)-1]} \exp(-Ax)\, dx$$

which is an integral included in most tables of integrals.

$$\int_0^\infty x^n e^{-ax}\, dx = \frac{\Gamma(n+1)}{a^{(n+1)}}$$

$$\text{MTTF} = \frac{1}{\beta}\Gamma(1/\beta)\bigg/\left(\frac{\lambda(T_p)}{T_p^{\beta-1}}\right)^{1/\beta} = \frac{(T_p)^{[1-(1/\beta)]}\Gamma(1/\beta)}{\beta\lambda(T_p)^{1/\beta}}$$

$$= \frac{(T_p)^{[1-(1/\beta)]}\Gamma(1+(1/\beta))}{\lambda(T_p)^{1/\beta}} \tag{A6-1}$$

A7   DETERMINING THE DISTRIBUTION
OF AN ESTIMATOR OF MTTF

Reference is made to Section 2.3.3 where the distribution of $T = \sum_{i=1}^{N} t_{i/r}$ was described.

Let

$$\hat{\theta} = \frac{T}{r} = \frac{\sum_{i=1}^{r} t_{i/r}}{r} \tag{A7-1}$$

r = number of failures

$t_{i/r}$ = the total operating hours on an item at the time of the rth failure.
　　If unit i failed at time t_i prior to the rth failure, then $t_{i/r}$ will equal t_i

t_i = time to failure of the ith component

$\hat{\theta}$ = average computed time to failure of r components

We know the distribution for T and it is desired to derive a distribution function $g(\hat{\theta})$, which would describe the distribution of the average computed time to failure $\hat{\theta}$. Since

$$\hat{\theta} = \frac{T}{r}$$

a simple transformation of variables can be performed for that purpose.

In general, if a probability distribution of the variable x, $f(x)$, exists and we want to develop a probability distribution for some variable y, where

$$y = ax$$

then the distribution of y, $g(y)$, is

$$g(y) = \left| \frac{1}{a} \right| f(y/a) \tag{A7-2}$$

where $f(y/a)$ is the function $f(x)$ with each x replaced by (y/a).

As a consequence,

$$g(\hat{\theta}) = |r| f(r\hat{\theta}) \tag{A7-3}$$

A8 EVALUATING THE FRACTION OF DEFECTIVE UNITS WHICH RESULT FROM THE USE OF PART POPULATIONS WITH VARIOUS LEVELS OF DEFECTS

Consider a situation where we have a population of parts, M, containing D defective parts. N of such parts are chosen randomly from the population to make up a unit. If one or more defective parts are included in the unit, the unit is considered defective. The objective is to form S units from the parts available, where

$$\frac{M}{N} = S$$

What is the proportion of defect-free units that can be expected?

To start, consider the fact that each unit formed has the same probability of being defect-free. To show this, it is only necessary to consider two white balls and one black ball in a hat. Three individuals choose a ball from the hat sequentially but do not look at it. After all have chosen, what is the probability for each that he has picked a black ball? Does the fact that a person picked first increase or decrease the probability that he will have the black ball?

Actually, all three individuals have the same probability of having picked the black ball. Our case is identical since it is assumed that no unit will be tested to ascertain its status until all units are assembled.

As a consequence, if the probability of the first element being defect-free is calculated, the probability of any element being defect-free is calculated.

$$\binom{M}{N} = \frac{M!}{(M - N)!N!} \tag{A8-1}$$

is a combinatorial expression denoting the number of different combinations of N that can be extracted from a total population of M items. Take, for example, six items making up a population A_1, A_2, A_3, A_4, A_5, A_6. What are the number of different ways that two items can be chosen (where two will make up a unit)?

$$
\begin{array}{lllll}
A_1 A_2 & A_2 A_3 & A_3 A_4 & A_4 A_5 & A_5 A_6 \\
A_1 A_3 & A_2 A_4 & A_3 A_5 & A_4 A_6 & \\
A_1 A_4 & A_2 A_5 & A_3 A_6 & & \\
A_1 A_5 & A_2 A_6 & & & \\
A_1 A_6 & & & &
\end{array}
$$

From (A8-1), this number is

$$\frac{6!}{4!\,2!} = 15$$

Hence, the first unit formed will be made up of one of 15 possible combinations of parts, each equally likely to be picked.

Let us imagine that two of the parts in the population are defective. It does not matter which two, the result will be the same. Assume, then, that A_5 and A_6 are defective, so that out of the 15 possible different combinations depicted above (each equally possible) only 6 represent combinations which are defect-free. As a result, only 6 defect-free units are formed.

The number of defect-free combinations possible can be calculated directly from the combinatorial expression

$$\binom{M - D}{N} = \frac{(M - D)!}{(M - D - N)!\,N!} \tag{A8-2}$$

where D = the number of defective parts in the population. With $M = 6$, $D = 2$, $N = 2$, the above reduces to the number observed in the illustration,

$$\frac{4!}{2!\,2!} = 6$$

The probability of a unit being defect-free, P_{df} (the proportion of defect-free units produced), is equal to the ratio of the *number* of defect-free part combinations of N parts possible to the total number of all combinations (defect-free and otherwise) of N parts possible. That is to say, it is the ratio of (A8-2) to (A8-1):

$$P_{df} = \binom{M - D}{N} \Big/ \binom{M}{N} \tag{A8-3}$$

D can also be expressed as pM, where p is the fraction of the part population defective, and (A8-3) can be expressed as

$$P_{df} = \binom{M - PM}{N} \Big/ \binom{M}{N} = \frac{(M - PM)!\,(M - N)!}{(M - PM - N)!\,M!} \tag{A8-4}$$

A9   DEVICE FAILURE RATE/COMPLEXITY RELATIONSHIPS

Given the failure rate model for monolithic bipolar and MOS digital devices, the failure rate, λ_p, of an operating device is given by

$$\lambda_p = \Pi_q(C_1\Pi_t\Pi_v + C_2\Pi_e)\Pi_l$$

where:
   λ_p = device failure rate expressed in terms of failures/10^6 hours
   Π_l = device maturity factor
   Π_q = device quality factor
   Π_t = technology-based temperature acceleration factor
   Π_e = application environment factor
   C_1 = circuit complexity factor based on gate count
   C_2 = circuit complexity factor based on number of functioning pins and package type

Based on data in Mil-Hdbk-217E, C_1 will generally follow the approximate relationship

$$C_1 \sim k(g)^{1/2} \tag{A9-1}$$

where
   k = some constant
   g = number of gates making up the device

For monolithic bipolar and MOS digital devices, based on information in Mil-Hdbk-217E, for example, C_1 takes on the values shown below

Gate count	C_1
1–100	0.01
101–1000	0.02
1001–3000	0.04
3001–10 000	0.08
10 001–30 000	0.16

In this case a value of $k = 0.000\,85$ will provide a reasonable fit over most gate counts.

The behavior of C_1 with increased complexity is significantly less than linear. As seen above, C_1 increases generally with gate count level, but within given levels, gate count has little effect on C_1. The general relationship, (A9-1), provides for interpolation over and within each boundary set.

C_2 is a function of number of pins, (n_p), on a device package connected to some substrate location. For hermetic DIPS with solder or weld seal or leadless chip carriers, C_2 is computed in Mil-Hdbk-217E as

$$C_2 = (2.8 \times 10^{-4})(n_p)^{k_2}$$

Other package types have similar relationships for C_2. Generally,

$$C_2 \sim k_1(n_p)^{k_2} \qquad \text{(A9-2)}$$

where λ_p can then be expressed as a function of the failure rate contributions due to gate complexity and package complexity:

$$\lambda_p = \Pi_q \Pi_l (\lambda_{pg} + \lambda_{pc}) \qquad \text{(A9-3)}$$

where
$\lambda_{pg} = C_1 \Pi_t \Pi_v \sim k(g)^{1/2} \Pi_t \Pi_v = $ failure rate contribution due to gate complexity
$\lambda_{pc} = C_2 \Pi_e \sim k_1(n_p)^{k_2} \Pi_e = $ failure rate contribution due to package complexity

and

$p_{\lambda g} = \lambda_{pg}/(\lambda_{pg} + \lambda_{pc}) = $ proportion of failure rate contributed by gate count
$p_{\lambda c} = \lambda_{pc}/(\lambda_{pg} + \lambda_{pc}) = $ proportion of failure rate contributed by package complexity

The next question asked is how much does the failure rate of each device increase as a function of increase in complexity? The measure of complexity used will be defined as $z = $ the factor of increase in gate count. z can be thought of as the *maximum device equivalence factor*, the maximum number of devices replaceable by a new device with an increased gate count. For example, a situation where a single device containing 800 gates has the "conceptual" potential of replacing two 400-gate devices.

If the number of gates were to increase by a factor $z(z > 1)$, the value of C_1 would change to

$$C_1 \sim k(zg)^{1/2}$$

The behavior of C_2 with respect to increased complexity in terms of pins, n_p, is exponential as can be seen from examination of the relationship between C_2 and n_p discussed earlier. If the number of pins were to increase by a factor

$w(w > 1)$, the value of C_2 would change to

$$C_2 \sim k_1(wn_p)^{k_2}$$

Assume that an increase in gate count by a factor z will be incurred and a corresponding increase in pin count by a factor of w. Then the failure rate, λ_{pz}, for the complexity enhanced chip can be expressed as

$$\lambda_{pz} \sim \Pi_q \Pi_l [k(zg)^{1/2} \Pi_t \Pi_v + k_1(wn_p)^{k_2} \Pi_e]$$

But since we previously defined

$$\lambda_{pg} \sim k(g)^{1/2} \Pi_t \Pi_y, \qquad \lambda_{pc} \sim k_1(n_p)^{k_2}$$

λ_{pz} can be expressed as

$$\lambda_{pz} \sim \Pi_q \Pi_l (z^{1/2} \lambda_{pg} + w^{k_2} \lambda_{pc})$$

Then the *proportional increase in failure rate*, $P_{\lambda p}$, necessary to achieve a device equivalence factor of z and a pin count increase factor of w is

$$P_{\lambda p} \sim \frac{\lambda_{pz}}{\lambda_p} - 1 = \frac{(z^{1/2} \lambda_{pg} + w^{k_2} k \lambda_{pc})}{\lambda_{pg} + \lambda_{pc}} - 1$$

$$= (z^{1/2} P_{\lambda g} + w^{k_2} P_{\lambda c}) - 1 \tag{A9-4}$$

A10   THE VARIANCE ASSOCIATED WITH EQUIPMENT FAILURE-RATE PREDICTION VERSUS THE VARIANCE ASSOCIATED WITH PART FAILURE-RATE PREDICTION

In Chapter 8 we defined

$\lambda_{pi} =$ the *mean* or *predicted* failure rate for part type i

$\sigma_{pi} =$ the standard deviation of the failure rate of part i

$Q_i =$ the quantity of parts in the equipment of type i

$$\mu_{\lambda ae} = \lambda_e = Q_1 \lambda_{p1} + Q_2 \lambda_{p1} + \cdots + Q_m \lambda_{pm} \tag{A10-1}$$

= the mean or predicted failure rate of an equipment

$$\sigma_{\lambda ae} = [(Q_1 \sigma_{p1})^2 + (Q_2 \sigma_{p1})^2 + \cdots + (Q_m \sigma_{pm})^2]^{1/2} \tag{A10-2}$$

= the standard deviation of the predicted equipment failure rate

Further define

$$\sigma_{pi} = k_i \lambda_{pi}, \qquad \sigma_{\lambda ae} = k_0 \lambda_e, \qquad k_i, k_0 > 0$$

where $k_i \lambda_{pi}$ represents the magnitude of the standard deviation of failure rate for each part type in terms of multiples, k_i, of the mean value of failure rate (which is the predicted value) for the part type; $k_0 \lambda_e$ represents the magnitude of the standard deviation of failure rate for the equipment in terms of multiples, k_0, of the mean value (predicted value) of failure rate of the equipment.

In order to show that the degree of variation in an equipment level prediction (*relative to the value of the equipment prediction*) is *less* than the largest degree of variation present at any part level (*relative to the value of the part prediction*), it must be shown that k_0 is less than the largest value of k_i.

Since each part of the same type is assumed to have the same value σ_{pi} and since $\sigma_{pi} = k_i \lambda_{pi}$ and $\sigma_{\lambda ae} = k_0 \lambda_e$ (A10-2) can be expressed as

$$\sigma_{\lambda ae} = k_0 \lambda_e = \left[\sum_{}^{m} (Q_i k_i \lambda_i)^2 \right]^{1/2} \tag{A10-3}$$

where

Q_i = total number of parts of type i in the equipment
m = total number of part types in the equipment

First assume that all parts have the same value of k_i: (A10-3) can then be expressed as

$$k_0 \lambda_e = k_i \left[\sum_{}^{m} (Q_i \lambda_i)^2 \right]^{1/2}$$

which, after some simple manipulation, takes the form

$$\left(\frac{k_0}{k_i} \right)^2 = \sum_{}^{m} \left[Q_i \left(\frac{\lambda_i}{\lambda_e} \right) \right]^2 \tag{A10-4}$$

But

$$\sum_{}^{m} Q_i \left(\frac{\lambda_i}{\lambda_e} \right) = 1, \qquad \text{since } \sum_{}^{m} Q \lambda_i = \lambda_e \tag{A10-5}$$

and because of (A10-5), since each $Q_i \lambda_i / \lambda_e < 1$,

$$\sum_{}^{m} \left[Q_i \left(\frac{\lambda_i}{\lambda_e} \right) \right]^2 < 1$$

Hence,

$$\frac{k_0}{k_i} < 1, \qquad k_0 < k_i$$

This proves that, given the same relative variance in all part predictive models, their application will yield a smaller relative variance for equipment-level predictions.

Next assume that one part type exhibits a larger value of k_i than the rest and define this as k_z. (A10-3) can then be expressed as

$$(k_0 \lambda_e)^2 = \sum_{i=1}^{m-1} (Q_i k_i \lambda_i)^2 + (Q_z k_z \lambda_z)^2$$

or

$$(k_0)^2 = \sum_{i=1}^{m-1} (k_i)^2 \left(\frac{Q_i \lambda_i}{\lambda_e}\right)^2 + (k_z)^2 \left(\frac{Q_z \lambda_z}{\lambda_e}\right)^2 \tag{A10-6}$$

We want to show that

$$\left(\frac{k_0}{k_z}\right)^2 < 1$$

We know that

$$\left(\frac{k_0}{k_z}\right)^2 = \sum_{i=1}^{m-1} \left(\frac{k_i}{k_z}\right)^2 \left(\frac{Q_i \lambda_i}{\lambda_e}\right)^2 + \left(\frac{Q_z \lambda_z}{\lambda_e}\right)^2 \tag{A10-6}$$

And

$$\sum_{i=1}^{m-1} \frac{Q_i \lambda_i}{\lambda_e} + \frac{Q_z \lambda_z}{\lambda_e} = 1 \tag{A10-7}$$

But since $(Q_i \lambda_i / \lambda_e)$, $(Q_z \lambda_z / \lambda_e) < 1$, then

$$\sum_{i=1}^{m-1} \left(\frac{Q_i \lambda_i}{\lambda_e}\right)^2 + \left(\frac{Q_z \lambda_z}{\lambda_e}\right)^2 < 1 \tag{A10-8}$$

And since $(k_i/k_z) < 1$

$$\sum_{i=1}^{m-1} \left(\frac{k_i}{k_z}\right)^2 \left(\frac{Q_i \lambda_i}{\lambda_e}\right)^2 + \left(\frac{Q_z \lambda_z}{\lambda_e}\right)^2 < \sum_{i=1}^{m-1} \left(\frac{Q_i \lambda_i}{\lambda_e}\right)^2 + \left(\frac{Q_z \lambda_z}{\lambda_e}\right)^2 \qquad \text{(A10-9)}$$

Which shows that $(k_0/k_z)^2 < 1$ and hence $k_0 < k_z$.

A11   RELIABILITY EXPRESSIONS RESULTING AND FAILURE-RATE PREDICTION BASED ON AVERAGE HAZARD OVER AN ESTABLISHED PERIOD OF TIME

In Sections 2.1.5 and 7.3 the concept of reliability prediction based on a determination of average hazard rate over an established period of time was discussed. The approach facilitated reliability prediction for items which exhibit nonconstant hazard rates described in terms of its application to Weibull distributions. The approach summed hazard rate over a period of time T_p and yielded an item average hazard rate per hour, $\lambda(T_p)$, pertinent to item operation over its first T_p hours of use.

In accord with that concept, the average hazard (failure rate) of an item over T_p, $\lambda(T_p)$, and reliability expressed over T_p takes the form

$$R(T_p) = \exp(-[\lambda(T_p)] T_p) \qquad \text{(A11-1)}$$

which defines reliability over the first T_p hours of operation (a specific number of operating hours).

In some instances it may be desired to make predictions of failure rate or assessments of reliability with respect to other periods of operating time. The following provides a rationale for that purpose. As before, the Weibull distribution will be used to illustrate the procedure (see Section 2.1.2).

The Weibull distribution takes the form

$$f(t) = \frac{\beta}{\alpha} \left(\frac{t}{\alpha}\right)^{\beta-1} \exp\left[-\left(\frac{t}{\alpha}\right)^{\beta}\right] \qquad \text{(A11-2)}$$

where
β = a shape parameter
α = a scale parameter
t = operating time

The reliability (probability of no failure) over an interval of time $T = (T_2 - T_1)$ can be expressed as

$$R(T_2 - T_1) = \exp\left[-\left(\frac{1}{\alpha}\right)^{\beta}(T_2^{\beta} - T_1^{\beta})\right] \quad \text{(A11-3)}$$

where

T_2 = time at the end of the interval of time
T_1 = acquired operational history before interval of operation begins

[Note: (A11-3) will be exact if no failures have occurred prior to T_1 and provides a reasonable approximation if the number of failures which occurred is small with respect to the number of parts in the item. It was shown in Chapter 1 that only a small proportion of parts fail during service life. As a consequence, (A11-3) is treated as a satisfactory approximation to $R(T_2 - T_1)$ over service life.]

The average hazard (the failure rate) over any specific interval of time $T = (T_2 - T_1)$ is

$$\lambda(T_2 - T_1) = \lambda(T) = \frac{\left(\frac{1}{\alpha}\right)^{\beta}(T_2^{\beta} - T_1^{\beta})}{T_2 - T_1} \quad \text{(A11-4)}$$

Assume that the average hazard (the failure rate) used for prediction purposes was based on the first T_p hours of operating time. Then

$$\lambda(T_p) = \frac{\left(\frac{1}{\alpha}\right)^{\beta} T_p^{\beta}}{T_p} = \left(\frac{1}{\alpha}\right)^{\beta} T_p^{\beta-1} \quad \text{(A11-5)}$$

It is assumed that the shape parameter β associated with the failure distribution of the item is known or can be estimated from known characteristics of the parts or from empirical data.

Expressing (A11-5) as

$$\left(\frac{1}{\alpha}\right)^{\beta} = \frac{\lambda(T_p)}{T_p^{\beta-1}}$$

and substituting in (A11-4)

$$\lambda(T) = \lambda(T_2 - T_1) = \frac{\lambda(T_p)(T_2^{\beta} - T_1^{\beta})}{T_p^{\beta-1}(T_2 - T_1)} \quad \text{(A11-6)}$$

and

$$R(T_2 - T_1) = \exp(-[\lambda(T_p)/T_p^{\beta-1}][T_2^{\beta} - T_1^{\beta}]) \qquad \text{(A11-7)}$$

A12   CALCULATING THE EFFECTS OF NONOPERATING FAILURE RATES ON THE NUMBER OF FAILURES PREDICTED FOR OPERATING PERIODS

Let

λ_e = the predicted failure rate of an item in its operating state

λ_{en} = the predicted failure rate of an item in a nonoperating state

$$d = \frac{\lambda_{en}}{\lambda_e}$$

P = the proportion of time a unit is in an operating state

$(1 - P)$ = the proportion of time a unit is in a nonoperating state

T = total hours in a year $(365 \times 24) = 8760$

PT = total hours per year in an operating state

$(1 - P)T$ = total hours per year in a nonoperating state

Recall from the discussion at the end of Section 1.3, it was shown that in general λT_0 = expected number of failures which will occur over an interval of operating time T_0. Similarly,

$\lambda_e PT$ = the expected number of failures per year that can be anticipated due to failures while in operating states

$\lambda_{en}(1 - P)T$ = the expected number of failures per year that can be anticipated due to failures which occur during nonoperating states

Logically, then

$$\frac{\lambda_{en}(1 - P)T}{\lambda_e PT} = \Delta F$$

which is the proportional increase in failures over that expected to occur by virtue of the prediction of λ_e. Note that if

$$\lambda_{en} = 0$$

signifying that the nonoperating failure rate is zero, then the proportional increase in failures over that expected from the prediction λ_e, is equal to zero. And if

$$\lambda_{en} = d\lambda_e, \qquad \text{where } d < 1$$

then

$$\Delta F = \frac{\lambda_{en}(1 - PT)}{\lambda_e PT} = \frac{d\lambda_e(1 - P)T}{\lambda_e PT}$$

$$= \frac{d(1 - P)}{P} \tag{A12-1}$$

is the proportional increase in failures over that expected to occur by virtue of the prediction of λ_e.

Using Maintenance Checks to Blunt the Effects of Nonoperating Failures on Mission Reliability

Given that the following is known or assumed:

- The time between missions, T_m, or the range that T_m takes say, between T_{m1} and T_{m2}
- Failures in operating and nonoperating states follow exponential distributions
- Performance checks are performed T_C hours prior to a mission or every T_C hours*

then the proportion of total nonoperating failures which will occur between performance checks and mission starts, $P(T_C)$ (given a value of T_m and assuming instantaneous replacement of the failed items when detected) is

$$P(T_C) \leqslant \frac{\displaystyle\int_{T_m - T_C}^{T_m} f(t)\, dt}{\displaystyle\int_0^{T_m} f(t)\, dt} = \frac{\displaystyle\int_0^{T_C} f(t)\, dt}{\displaystyle\int_0^{T_m} f(t)\, dt} \tag{A12-2}$$

---

* If performance checks are performed T_C hours prior to a mission $P(T_C)$ is equal to the quantity shown. If checks are performed every T_C hours by the clock and call to mission is a random event, then checks are performed a *maximum* of T_C hours prior to mission start and $P(T_C)$ is less than the quantity shown.

where

$$\int_0^{T_m} f(t)\, dt = \text{proportion of population of nonoperating items which fail prior to } T_m$$

$$\int_{T_m - T_C}^{T_m} f(t)\, dt = \text{proportion of population of nonoperating items which fail between the time a performance check occurs and mission time}$$

Since an exponential failure distribution has been assumed (see Section 2.1.1.2):

$$\int_{T_m - T_C}^{T_m} f(t)\, dt = \int_0^{T_C} f(t)\, dt$$

where

$$f(t) = \lambda_{en} \exp(-\lambda_{en} t)$$

and

$$P(T_c) = \frac{1 - \exp(-\lambda_{en} T_C)}{1 - \exp(-\lambda_{en} T_m)}, \qquad 0 < T_C \leqslant T_m \qquad \text{(A12-3)}$$

Recalling that if x is small, say $x \leqslant 0.02$, e^{-x} can be approximated as

$$e^{-x} \sim 1 - x$$

Since, in general, both $\lambda_{en} T_m$ and $\lambda_{en} T_C$ will meet that criterion, we can write

$$\exp(-\lambda_{en} T_m) \sim 1 - \lambda_{en} T_m, \qquad \exp(-\lambda_{en} T_C) \sim 1 - \lambda_{en} T_C$$

and express $P(T_C)$ as

$$P(T_C) \sim \frac{T_C}{T_m}$$

Where just a possible range for T_m is known (i.e., T_m usually between T_{m1} and T_{m2}), two values for $P(T_C)$ can be calculated, generating a range for $P(T_C)$. As a consequence, the expected number of nonoperating failures per year that can impact mission performance (occur between a check and a mission start), can be expressed as

$$\lambda_{en}(1 - P)P(T_C)T \sim \lambda_{en}(1 - P)T\left(\frac{T_C}{T_m}\right) \tag{A12-4}$$

The proportion of increase in mission related failures over that expected to occur from the prediction of λ_e, ΔF, now takes the form

$$\Delta F_C = \frac{\lambda_{en}(1 - P)P(T_C)T}{\lambda_e PT} \sim \frac{d(1 - P)T_C}{T_m P} \tag{A12-5}$$

Therefore we blunt the effects of nonoperating failure rate on mission reliability by performing checks T_C hours prior to a mission or every T_C hours, so as to allow for replacement of the item, if it is in a failed state, prior to mission start. As a consequence, after such checks have been performed, either:

1. The item will be found to be in a failed state, it will be replaced and put back into its nonoperating status. The probability that it will fail prior to the mission start will be equal to or less than $[1 - R(T_C)]$. We say less than because the mission might well start at some time less than T_C hours after replacement.
2. The item will be found to be in a *nonfailed* state and nothing more will be done until mission start. Since the item follows an exponential failure distribution with its memoryless characteristic, the proportion of items will which check "good" but fail prior to mission start equals $[1 - R(T_C)]$, where

$$R(T_C) = \exp(-\lambda_{en} T_C)$$

Hence, we can make the general statement that if checks are performed T_C hours prior to a mission or every T_C hours, the proportion of items which will be in a failed state at mission start will be less than or equal to $[1 - R(T_C)]$.

A13   MISSION RELIABILITY FOR A SYSTEM MADE UP OF *n* COMPONENTS EACH HAVING AN EXPONENTIAL DISTRIBUTION OR WEIBULL DISTRIBUTION

As was indicated in Section 1.3, the most general form of $R(T)$ appropriate for any distribution of failure takes the form

$$R(T)_i = R(T_{ci} - T_{oi})_i = \frac{R(T_{ci})}{R(T_{oi})} = \exp\left(-\int_{T_{oi}}^{T_{ci}} h(t)\, dt\right) \tag{A13-1}$$

where

$R(T)_i$ denotes item reliability. An item can be thought of as a part, a printed circuit board, an assembly, or a modular unit anything subject to replacement upon failure.

$R(T)_i = R(T_{ci} - T_{oi})_i =$ probability that item i will perform satisfactorily over an interval (a mission) of time $T = (T_{ci} - T_{oi})$, given the item has acquired an operational history equal to T_{oi}, has experienced no failure prior to T_{oi} (is operable at T_{oi})

T_{oi} = the operating time *on item i* prior to a use interval (mission) start. Think of it as a reading on a time meter installed on item i, where a reading of $T_{oi} = 0$ denotes a new item

T_{ci} = total operating time *on item i* at the end of the mission

$h(t)$ = hazard rate associated with item i's failure distribution

T $= T_{ci} - T_{oi} =$ time associated with the mission

A system is made up of many items. It is similar to a car; it functions over many years and during that time replacements of various parts or components are made. As a consequence, the system might have acquired an operational history of, say, T_{oi} hours, and experienced a number of failures during that time such that it contains a quantity of items with shorter operational histories than that of the system itself. As a consequence, system reliability, $R(T)_{er}$, can be expressed as

$$R(T)_{er} = R(T_s - T_o)_{er} = \left[\prod_i^{n-r} R(T_s - T_o)_i \right]\left[\prod_j^r R(T_{cj} - T_{oj})_i \right] \qquad \text{(A13-2)}$$

where

$R(T)_{er} = R(T_s - T_o)_{er} =$ the probability that the system will perform satisfactorily over an interval of time $T = (T_s - T_o)$, given that it has acquired an operational history prior to the mission of T_o hours and *experienced failures in r different items* which were replaced by new items during T_o, and all system items are operable at the start of T

T_o = prior operating time (history) on the system at the time of mission start. Think of it as a reading on a time meter installed on the system. It *does not necessarily* denote failure free operation over that period

T_s = total operating time (history) on the *system* at the end of the mission

n = number of items making up the system

$n - r$ = the number of items which have not failed or been replaced by the time they acquire T_o hours of use

r = the number of different items which were replaced one or more times due to failure over T_o

T_{oj} = the operating time history associated with the current replacement for item j when the time meter *on the system* reads T_o

T_{cj} = the total operating time on the jth replacement for item j at the end of the mission

T = $(T_s - T_o) = (T_{cj} - T_{oj}) = (T_{ci} - T_{oi})$ = the mission time; the next T hours of operation

If *no* failures have occurred in T_o (T_o = prior *system operating history*), $T_{oi} = T_o$ for all items and system reliability, $R(T)_{er} = R(T)_{eo}$ (i.e., $r = 0$), and reliability over the next T hours of operation (i.e., between T_o and T_s), takes the form

$$R(T)_{eo} = R(T_s - T_o)_{eo} = \prod_i^n R(T_{ci} - T_{oi})_i = \prod_i^n R(T_s - T_o)_i \qquad \text{(A13-3)}$$

If the failure distribution for every item is exponential, the hazard, $h(t)_i$, is equal to a constant λ_i. Each item which had not as yet experienced failure since the system was put into service would have its $T_{oi} = T_o$ and $T_{ci} = T_s$. The reliability of each such item, $R(T_{ci} - T_{oi})_i$, over an interval T, could be expressed as

$$R(T_{ci} - T_{oi})_i = \frac{R(T_s)^*}{R(T_o)} = \exp\left(-\int_{T_o}^{T_s} \lambda_i \, dt\right) = \exp[-\lambda_i(T_s - T_o)]$$

$$= \exp(-\lambda_i T) \qquad \text{(A13-4)}$$

The reliability of each replacement item in the system over the same next interval, T, is signified by $R(T_{cj} - T_{oj})_j$. Each value of T_{oj} can be considered unique (after all, different replacements incorporated into the system to take the place of different failed items, at different times would have different operating histories) but, in every case, $T_{cj} - T_{oj} = T$. As a result, if each replacement item had an exponential distribution of time to failure with hazard λ_j, reliability over T would take the form

$$R(T_{cj} - T_{oj})_j = \frac{R(T_{cj})}{R(T_{oj})} = \exp\left(-\int_{T_{ij}}^{T_{oj}} \lambda_j \, dt\right) = \exp[-\lambda_j(T_{cj} - T_{oj})]$$

$$= \exp(-\lambda_j T) \qquad \text{(A13-5)}$$

---

* Recall one of the results of Appendix 0.

Note that for (A13-4) and (A13-5) the terms which represent prior operating hours T_{oi} and T_{oj} drop out, only the difference between the two values is relevant. This will occur only for an exponential distribution of time to failure.

Using the results of (A13-4) and (A13-5), both (A13-2) and (A13-3) could be expressed as

$$R(T)_{er} = R(T_s - T_o)_{er} = \exp\left[-T\left(\sum_i^{n-r} \lambda_i + \sum_j^r \lambda_j\right)\right] = \exp - (\lambda_e T) \quad \text{(A13-6)}$$

for any value of r, where $\lambda_e = \sum_i^n \lambda_i$, and all replacement items have the same hazard rates as the ones they replace. This means if the failure distribution of the items are exponential:

1. The prior operational history of the items or the system does not matter.
2. The number of items which have been replaced previously does not matter.

$R(T)_{er}$ will take on the same value for any interval of the same length T, no matter what value r, no matter what the value T_o has taken.

In addition, it is clear from (A13-3), (A13-4), and (A13-6) that $R(T)_{er}$, descriptive of the reliability of a system, takes the same general mathematical form as $R(T_{ci} - T_{oi})_i$, which is descriptive of the reliability of an item that exhibits an exponential distribution of time to failure. We can therefore conclude that a series-connected system made up of items which have exponential distributions of time to failure will exhibit an exponential distribution of time to failure.

If the distribution of time to failure for items is not exponential (with its associated constant failure hazard rate), a more involved, complex relationship will result. For example, if the distribution of failures for a component is Weibull (see Section 2.1.2), the hazard, $h(t)$, is

$$h(t) = \frac{\beta}{\alpha}\left(\frac{t}{\alpha}\right)^{\beta - 1} \quad \text{(A13-7)}$$

where
β, α = shape and scale parameter, respectively, associated with the distribution
t = a particular point in operating life history

$$R(T)_i = R(T_{ci} - T_{oi})_i = \exp\left(-\int_{T_{oi}}^{T_{ci}} h(t)\,dt\right) = \exp\left[-\left(\frac{1}{\alpha_i}\right)^{\beta_i}(T_{ci}^{\beta_i} - T_{oi}^{\beta_i})\right] \quad \text{(A13-8)}$$

Note that $R(T_{ci} - T_{oi})_i$ will take on different values for the same value of T. For example, $R(T_{ci} - T_{oi})_i$ will take on one value when $T_{ci} = 520$, $T_{oi} = 500$, and another when $T_{ci} = 1020$, $T_{oi} = 1000$, even though in both cases $T = 20$.

$R(T)_{e0}$ is then expressed as

$$R(T)_{e0} = R(T_s - T_o)_{s0} = \prod_i^n R(T_{ci} - T_{oi})_i = \prod_i^n R(T_s - T_o)_i$$

(if no failures have occurred $T_{ci} = T_s$, $T_{oi} = T_o$) as a result

$$R(T)_{e0} = \exp\left[-\sum_i^n \left(\frac{1}{\alpha_i}\right)^{\beta_i} (T_s^\beta - T_o^\beta) \right] \qquad \text{(A13-9)}$$

$R(T)_{er}$ is expressed as

$$R(T)_{er} = \exp\left[-\sum_{i=1}^{n-r} \left(\frac{1}{\alpha_i}\right)^{\beta_i} (T_s^{\beta_i} - T_o^{\beta_i}) - \sum_{j=1}^{r} \left(\frac{1}{\alpha_j}\right)^{\beta_j} (T_{cj}^{\beta_j} - T_{oj}^{\beta_j}) \right] \text{(A13-10)}$$

where

All parameters are defined as before

β_k = shape parameter associated with the failure distribution of the kth component

α_k = scale parameter associated with the failure distribution of the kth component

which means that if the failure distribution of each item is Weibull, the prior operational history of each component matters and knowledge of them is required in order to evaluate mission reliability, $R(T_s - T_o)_e$, over service life.

If we assume that all components comprising a system have the same value of the shape parameter, β, where $\beta_i = \beta$, then (A13-9) can be further simplified:

$$R(T_s - T_o)_{e0} = \exp\left[-(T_s^\beta - T_o^\beta) \sum_i^n \left(\frac{1}{\alpha_i}\right)^\beta \right] \qquad \text{(A13-11)}$$

If we equate $[\sum_i^n (1/\alpha_i)^\beta]^{1/\beta} = (1/\alpha_o)$, α_o becomes a scale parameter and (A13-11) represents a measure of mission reliability indicative of a Weibull distribution of time to failure. However, even with that assumption, $R(T)_{er}$, (A13-10), for values of $r \geq 1$ will never take on a form consistent with a Weibull distribution. As a result, technically speaking, a repairable system, even if it did originally exhibit a Weibull distribution to first failure, cannot exhibit one after that point.

Equation (A13-10) can be applied only when estimates of the operating hours on replacement items are available. (The age of *replacement* components *is not always zero when used.* That is to say, they are not always brand new. As service life continues and as items fail, they are usually repaired, generally replacing only a few new parts, and used as replacements for future failed items. As a consequence, a replacement item in the middle to late phases of service life is very likely not brand new.) The analysis required for that type of situation is far from trivial. For additional information, reference is made to Ref. 12 in Chapter 12.

However, if

1. There is a need to define $R(T_s - T_o)_{e0}$ when $T_o > 0$, and no failures have been experienced to T_o, (A13-9) will provide the exact answers.
2. If the system is comprised of a large number of components and each component failure is repaired by replacing a very, very small proportion of the parts making up the component, and the component is then reinserted in the system (or if it is made up of a very large number of components and only one or a very few are replaced at each system failure), (A13-9) will serve as a satisfactory approximation of (A13-10) over general service life.

(Recall from Section 1.1 that only a small fraction of the parts in an equipment fail over service life, so that $R(T_s - T_o)_{er}$ will be primarily driven by the characteristics of the original parts. This can be seen by examination of (A13-10), in particular the general magnitude of difference between the terms

$$\sum_{i=1}^{n-r}\left(\frac{1}{\alpha_i}\right)^{\beta_i}(T_s^{\beta_i} - T_o^{\beta_i}) \quad \text{and} \quad \sum_{j=1}^{r}\left(\frac{1}{\alpha_j}\right)^{\beta_j}(T_{cj}^{\beta_j} - T_{oj}^{\beta_j})$$

A14 EFFECT ON STATE AVAILABILITY OF THE STRATEGY OF SHUTTING DOWN THE SYSTEM AT THE FIRST COMPONENT FAILURE

If a system is shut down while repairs are made on a failed component (all components in a "shut down" system have $\lambda = 0$), state availability must be calculated differently from the case when the nonfailed components are allowed to operate while repair of failed units take place.

Assume that a system is made up of n components where the failure of any one will cause system failure, and where

$A_i = M_i/(R_i + M_i) =$ probability that component i is operable at a given time $=$ the proportion of the time component i is operable
$n\quad =$ number of components
$M_i =$ mean time to failure of component i
$R_i\quad =$ mean time to repair of component i

The probability (proportion of time) a system made up of a number of components is in a particular state is A_{sj} where, for example,

$A_{s0} = A_s = \prod_{i=1}^{n} A_i =$ probability (proportion of time) all components of the system are operable

$A_{s1} = (1 - A_1)\prod_{i=2}^{n} A_i =$ probability (proportion of time) all components of the system, with the exception of component 1 are operable

$A_{s2} = (1 - A_2)(A_1)\prod_{i=3}^{n} A_i =$ probability (proportion of the time) all components with the exception of component 2 are operable

$A_{s3} = (1 - A_1)(1 - A_2)\prod_{i=3}^{n} A_i =$ probability (proportion of the time) all components of the system with the exception of components 1 and 2 are operable

$\vdots$

$A_{sm} = \prod_{i=1}^{n} (1 - A_i)_i =$ probability (proportion of the time) that all components in the system will be inoperable

where $m + 1 =$ all possible system states.

If a failure occurred such that a system could no longer perform its function and the remaining functioning components were not shut off while repairs were made on the failed component(s), the operable units would continue to experience their operating failure rates and additional failures could accrue. It would be possible for the system to enter any of the above states. The probability of being in any one of the states at any time is certainty since the $(m + 1)$ *states include every possible state,* and the system *must* be in one of them. *Hence the probability of being in one of any set of "possible" states is always* $= 1$:

$$\sum_{j=0}^{m} A_{sj} = 1 \qquad\qquad (A14\text{-}1)$$

In accord with our strategy, however, if the system were shut down upon failure, it would be *impossible* for the system to enter most of the states above and the world of possible states shrinks. For example, for a simple series system where all components must operate in order to perform a function, if the system is shut down after one component fails, it would be impossible to reach a state where two or more components were failed.

If we denote

$$\Omega = \text{the set of "now possible" states}$$

$$\phi = \text{the set of "now impossible" states}$$

(A14-1) can now be expressed as

$$\sum_{\Omega} A_{sj} = \sum_{\phi} A_{sj} = 1 \qquad \text{(A14-2)}$$

and obviously then

$$\sum_{\Omega} A_{sj} < 1$$

but logically the sum of all "now possible" states must be certainty (i.e., equal to 1).

Since the world of possible states has shrunk, adjustments must be made to each member of $\sum_{\Omega} A_{sj}$ such that the summation of all "now possible" states equate to 1.

Equation (A14-2) can be expressed as:

$$\sum_{\Omega} A_{sj} = 1 - \sum_{\phi} A_{sj} \qquad \text{(A14-3)}$$

Dividing both sides by $1 - \sum_{\phi} A_{sj}$ results in

$$\frac{\sum_{\Omega} A_{sj}}{1 - \sum_{\phi} A_{sj}} = 1 \qquad \text{(A14-4)}$$

Hence, dividing each of the "old" values of A_{sj} *belonging to* Ω by $1 - \sum_{\phi} A_{sj}$ (or by $\sum_{\Omega} A_{sj}$ which equals $1 - \sum_{\phi} A_{sj}$) will yield a new probability (availability) value, A_{nsj}, for that state which is consistent with the strategy of shutting a system down after system failure:

$$A_{nsj} = \frac{A_{sj}}{\sum_\Omega A_{sj}}$$

(A14-5)

where

$$\sum_\Omega A_{nsj} = 1$$

Assume a series-connected system made up of n components. System availability, A_{s0}, assuming the system will not be shut down upon failure of one or more components, is defined as

$$A_{s0} = \prod_{i=1}^{n} A_i; \qquad A_i = \frac{M_i}{M_i + R_i}$$

The availability of the system, assuming a shut down of all operating units when one component fails (and the system is no longer capable of performing its function), A_{ns0}, is

$$A_{ns0} = \frac{A_{s0}}{\sum_\Omega A_{sj}}$$

(A14-6)

where Ω is made up of states 0 and 1, 2, 8, 16, 32, etc. (see Chart 9-1) the only states possible under the strategy. The latter are the only states which can be entered from state 0. They are the only states associated with a single failed component. The sum of the A_{sj} associated with all such "single component failed" states is denoted as A'_{s1} where

$$A_{s0} = \prod_{i}^{n} A_i$$

(A14-7)

$$A'_{s1} = \sum_{j=1}^{n} (1 - A_j)_j \prod_{i=j+1}^{n} A_1 \prod_{k=0}^{j-1} A_k$$

(A14-8)

Since

$$\sum_\Omega A_{sj} = A_{s0} + A'_{s1}$$

and

$$A_i = \frac{M_i}{R_i + M_i}$$

after some manipulation, A_{ns0}, (i.e., (A14-6)) the availability for *any* series system, reduces to

$$A_{ns0} = \frac{1}{1 + \sum_{i=1}^{n} (R_i/M_i)} \tag{A14-9}$$

A15 DERIVATION OF MEAN TIME TO FAILURE FOR FULL-ON REDUNDANT SYSTEMS

From Chapter 10,

$$\text{MTTF}_s = \text{MTFF}_s = \int_0^{\infty} [1 - (1 - e^{-\lambda t})^N]^L \, dt \tag{A15-1}$$

Let

$$p = e^{-\lambda t}$$

Then (A15-1) is transformed into

$$\frac{1}{\lambda} \int_0^1 \frac{1}{p} [1 - (1 - p)^N]^L \, dp \tag{A15-2}$$

Recalling the binomial expansion from any mathematical handbook,

$$(1 - x)^L = \sum_{k=0}^{L} \binom{L}{k} (-1)^k x^k$$

where

$$\binom{L}{k} = \frac{L!}{k!(L-k)!}$$

and equating $(1 - p)^N = x$, (A15-2) is transformed into

$$\frac{1}{\lambda} \int_0^1 \frac{1}{p} \sum_{k=0}^L \binom{L}{k} (-1)^k (1 - p)^{kN} \, dp$$

$$= \frac{1}{\lambda} \int_0^1 \frac{1}{p} \, dp + \frac{1}{\lambda} \sum_{k=1}^L \binom{L}{k} (-1)^k \int_0^1 \frac{(1 - p)^{kN}}{p} \, dp \qquad \text{(A15-3)}$$

Integrating by parts yields

$$\int_0^1 \frac{(1 - p)^{kN}}{p} \, dp = - \sum_{s=1}^{kN} \frac{1}{s} - \int_0^1 \frac{1}{p} \, dp$$

(A15-3) can therefore be expressed as

$$\text{MTTF}_s = \text{MTFF}_s = \frac{1}{\lambda} \sum_{k=1}^L \binom{L}{k} (-1)^{k+1} \sum_{s=1}^{kN} \frac{1}{s}$$

When $L = 1$,

$$\text{MTTF}_s = \text{MTFF}_s = \text{MTTF}_i = \text{MTFF}_i = \frac{1}{\lambda} \sum_{s=1}^N \frac{1}{s}$$

A16 AVAILABILITY EXPRESSIONS FOR MAINTAINED STANDBY REDUNDANT SUBSYSTEMS

A redundant subsystem is one where a total of n components are used. Only the minimum number of components, d, necessary to perform the subsystem function are energized and have a failure rate; the remaining components are initially in a standby mode of operation with a failure rate of 0. As each operating component fails, a standby component is energized and takes its place. The failed component is immediately subject to maintenance by a repair team and when repaired becomes a standby component. The subsystem fails when an operating component fails and no components are on standby (all are being repaired).

Assume a redundant standby subsystem made up of a total of n identical components such that:

1. They have a failure rate of λ when energized, a failure rate of 0 when on standby or in repair.
2. They have a repair rate of μ when under repair.
3. They have exponential distributions of time to failure and repair.
4. Sufficient maintenance resources are available such that, for all practical purposes, a repair can start on a component at the time it fails.
5. There is perfect failure detection and switching.
6. Only one operating component is necessary to perform the subsystem function (we will generalize to the case where more than one is required later).
7. Only one component can fail at a given instant. A fraction of a second may separate any two failures, but they cannot fail simultaneously.
8. If two or more components are under repair at the same time, only one will have its repair completed at a given instant.
9. One component cannot fail at the same instant that another has repair completed.
10. It is possible for nothing to happen at a given instant.
11. Upon subsystem failure the operable components are not shut down.

If the subsystem contains n individual components, then $(n + 1)$ different states are possible and:

1. All possible subsystem states are defined as in Chapter 10-2.
2. All possible ways the subsystem can directly transition to each state from another state are defined.
3. Expressions are developed describing the probability of the subsystem being in a given state at a given instant of time.

In accord with the rules and assumptions, following Chart 10-2 the only way for the subsystem to be in

State (0) *at time* $(t + \Delta t)$ is for it to have been in state (1) at time t and complete the repair of the failed component during Δt; or for it to have been in state (0) at time t and have no failures occur during Δt.

State (1) *at time* $(t + \Delta t)$ is for it to have been in state (0) at time t and have one of the operating components fail during Δt; or for it to have been in state (2) at time t and have a repair completed during Δt; or for it to have been in state (1) at time t and have no failures or repair occur during Δt.

⋮

State $(n - d)$ *at time* $(t + \Delta t)$ is for it to have been in state $(n - d - 1)$ at time t and have one of the operating components fail during Δt; or for it to have been in state $(n - d + 1)$, a subsystem failure state, at time t and have a repair completed during Δt; or for it to have been in state $(n - d)$ at time t and have no failures or repairs occur during Δt.

⋮

State $(n - d + 1)$ *at time* $(t + \Delta t)$ (a failed state) is for it to have been in state $(n - d)$ at time t and have one of the operating components fail during Δt; or for it to have been in state $(n - d + 2)$ (an extended failure state) at time t and have a repair completed during Δt; or for it to have been in state $(n - d + 1)$ at time t and have no failures or repairs during Δt.

⋮

State (n) *at time* $(t + \Delta t)$ (an extended failure) state is for it to have been in state $(n - 1)$ at time t and have the last operating component fail during Δt; or for it to have been state (n) at time t and have no repairs completed in Δt.

As before in Chapter 9, we define:

$\mu \Delta t$ = probability that any single failed component will be repaired in Δt. If there were m components under repair, $m\mu \Delta t$ = the probability of one repair during Δt

$(1 - \mu \Delta t)$ = probability that any single failed component will not be repaired in Δt. If there were m components being repaired, $(1 - m\mu \Delta t)$ = the probability that no components will be repaired during Δt

$d\lambda \Delta t$ = probability that one of the d operating components will fail during Δt

$(1 - d\lambda \Delta t)$ = probability that none of the d operating components will fail during Δt

$[1 - (Z_s \lambda + Q_s \mu) \Delta t]$ = probability that no transitions occur from a given state s during Δt

where
Z_s = number of operating components in state s
Q_s = number of components under repair in state s

The probability of being in each state can be placed in a form suitable to the definition of differential equations. The expressions for the $(n + 1)$ states take the form:

$$P_0(t + \Delta t) = P_0(t)(1 - Z_0 d\lambda \, \Delta t) + P_1(t)\mu \, \Delta t$$

$$P_1(t + \Delta t) = P_0(t)Z_0 d\lambda \, \Delta t + P_2(t)2\mu \, \Delta t + P_1(t)[1 - (Z_1 d\lambda + \mu) \Delta t]$$

$$\vdots$$

$$P_{(n-d)}(t + \Delta t) = P_{(n-d-1)}(t)Z_{(n-d-1)}d\lambda \, \Delta t + P_{(n-d+1)}(t)[(n - d + 1)\mu \, \Delta t]$$
$$+ P_{(n-d)}(t)[1 - (Z_{(n-d)}d\lambda + (n - d)\mu) \Delta t]$$

$$\vdots$$

$$P_n(t + \Delta t) = P_{(n-1)}(t)Z_{(n-1)}d\lambda \, \Delta t + P_n(t)(1 - n\mu) \Delta t$$

where

$$Z_i = \frac{(n - i)}{d} \text{ and}$$

If $Z_i = (n - i)/d < 1$, $Z_i = (n - i)/d$
If $Z_i = (n - i)/d \geqslant 1$, $Z_i = 1$

Transformation to a set of differential equations gives

$$\dot{P}_0(t) = -Z_0 d\lambda P_0(t) + \mu P_1(t)$$

$$\dot{P}_1(t) = Z_0 d\lambda P_0(t) - (Z_1 d\lambda + \mu)P_1(t) + 2\mu P_2(t)$$

$$\vdots$$

$$\dot{P}_{(n-d)}(t) = Z_{(n-d-1)}d\lambda P_{(n-d-1)}(t) + (n - d + 1)\mu P_{(n-d+1)}(t)$$
$$- [Z_{(n-d)}d\lambda + (n - d)\mu]P_{(n-d)}(t)$$

$$\vdots$$

$$\dot{P}_n(t) = Z_{(n-1)}d\lambda P_{(n-1)}(t) - n\mu P_n(t)$$

If a steady-state solution to the set of equations is required (i.e., as $t \to \infty$), note that, as $t \to \infty$,

$$\dot{P}_i(t) \to 0, \qquad P_i(t) \to P_i$$

As a result, the above equations may be expressed as

$$0 = \mu P_1 - Z_0 d\lambda P_0$$

$$0 = -(Z_1 d\lambda + \mu)P_1 + Z_0 d\lambda P_0 + 2\mu P_2$$

$$\vdots$$

$$0 = Z_{(n-d-1)} d\lambda P_{(n-d-1)} + (n-d+1)\mu P_{(n-d+1)}$$

$$\quad - [Z_{(n-d)} d\lambda + (n-d)\mu] P_{(n-d)}$$

$$\vdots$$

$$0 = Z_{(n-1)} d\lambda P_{(n-1)} - n\mu P_n$$

where

$$\sum_{i=0}^{n} P_i = 1$$

where P_i = the probability (or the proportion of the time) that the subsystem will at any random point in time be in a state where i components will be in repair and a total of $(n-i)$ components will be operable (either operating or standby).

Using the above equations, we solve for each value of P_i in terms of P_n, i.e.,

$$P_i = C_i P_n$$

In general, it will be found that

$$P_i = \frac{n!}{i!}\left(\frac{\mu}{d\lambda}\right)^{n-i}\left(\frac{1}{\prod\limits_{j=Z_i}^{Z_{(n-1)}} j}\right)P_n, \qquad \text{for } i \leqslant n - 1 \qquad (A16\text{-}1)$$

where
   n = total number of redundant components in the subsystem
   d = number of operating components necessary for the subsystem to perform its function

Since, as indicated before, the summation of all possible state probabilities must equal 1,

$$\sum_{i=0}^{n} P_i = P_n + P_n\left[\sum_{i=0}^{n-1}\frac{n!}{i!}\left(\frac{\mu}{d\lambda}\right)^{n-i}\left(\frac{1}{\prod\limits_{j=Z_i}^{Z_{(n-1)}} j}\right)\right] = 1 \qquad (A16\text{-}2)$$

and P_n can be evaluated as

$$P_n = 1 \left/ \left[1 + \sum_{i=0}^{n-1} \frac{n!}{i!} \left(\frac{\mu}{d\lambda}\right)^{n-i} \left(\frac{1}{\prod\limits_{j=Z_i}^{Z_{(n-1)}} j}\right) \right] \right. \qquad \text{(A16-3)}$$

If a subsystem requires d operating components in order for it to perform its function, its availability, A_s, can be expressed as

$$A_s = P_n \sum_{i=0}^{n-d} \frac{n!}{i!} \left(\frac{\mu}{d\lambda}\right)^{n-i} \left(\frac{1}{\prod\limits_{j=Z_i}^{Z_{(n-1)}} j}\right) \qquad \text{(A16-4)}$$

A17 STEADY-STATE MTBF AND MEAN TIME TO FAILURE OF MAINTAINED FULL-ON REDUNDANT SUBSYSTEMS

We start the development of such measures by defining the various subsystem states. A subsystem state is described in terms of how many units are operating and how many are in a failed state.

The following is a list of the possible states for a parallel system comprised of n units.

State	Units operating	Units failed
0	n	0
1	$(n-1)$	1
2	$(n-2)$	2
3	$(n-3)$	3
⋮	⋮	
$(n-d)$	d	$(n-d)$
⋮	⋮	
n	0	n

As can be seen from the above and from Fig. 10.9, transitions can take place only between adjacent states. It is assumed that:

1. More than one failure occurring at exactly the same instant has a probability of 0.
2. More than one repair being completed at the same instant has a probability of 0.
3. A failure and repair manifesting themselves simultaneously has a probability of 0.

4. If the system is in state i ($i > 0$), it can go either to state ($i + 1$) or to state ($i - 1$).

Further, assuming that the system is in state (0), at $t = 0$, in order for the system to enter a failed state, say, (n), it must at some time go from state (0) to state (1), from state (1) to state (2) from state (2) to state (3), etc., and reach state ($n - 1$) before entering n.

Each state has associated with it a unique failure rate, λ_i, and repair rate, μ_i, computed as

$$\lambda_i = (n - i)\lambda, \qquad i \leqslant n$$

$$\mu_i = i\mu, \qquad i > 0 \text{ (assumes that as soon as a unit fails, repair is begun).}$$

Since we are dealing with exponential density functions for repair and failure:

$$\mu_i + \lambda_i = \text{total rate of transition from state } i$$

That is to say, given the fact the subsystem is in state i, it can transition either to state ($i + 1$), denoting a failure in one of the operating units, or to state ($i - 1$), denoting a repair in one of the units in maintenance. The sum ($\mu_i + \lambda_i$) represents the total *rate of transition* from state i.

Expected time in state i, $E(i)$, before transition is

$$E(i) = \frac{1}{\lambda_i + \mu_i} \tag{A17-1}$$

The probability of going to state ($i + 1$) on the next transition, given that the subsystem is presently in state i, can be represented as

$$P[(i + 1)/i] = \frac{\lambda_i}{\lambda_i + \mu_i} \tag{A17-2}$$

(This, of course, signifies a component failure.)

The probability of going to state ($i - 1$) on the next transition, given that the subsystem is presently in state i, can be represented as

$$P[(i - 1)/i] = \frac{\mu_i}{\lambda_i + \mu_i} \tag{A17-3}$$

(this, of course, signifies a repair) where

$$P[(i - 1)/i] + P[(i + 1)/i] = 1$$

Given the characteristics above and using expectation, the expected length of time to go from state i to state $(i + 1)$, $E[(i + 1)/i]$, can be formulated:

$$E[(i + 1)/i] = P[(i + 1)/i]E(i) + P[(i - 1)/i][E(i) + E[i/(i - 1)] + E[(i + 1)/i]]^*$$

(A17-4)

Rearranging and grouping terms,

$$E[(i + 1)/i] = \frac{P[(i + 1)/i]E(i) + P[(i - 1)/i][E(i) + E[(i + 1)/i]}{1 - P[(i - 1)/i}$$ (A17-5)

Substituting (A17-1)–(A17-3) into the above and expanding:

$$E[(i + 1)/i] = \frac{1}{\lambda_i} + \left(\frac{\mu_i}{\lambda_i}\right)E[i/(i - 1)]$$

$$= \frac{1}{\lambda_i} + \left[\frac{\mu_i}{\lambda_i}\right]\left[\frac{1}{\lambda_{i-1}} + \left[\frac{\mu_{(i-1)}}{\lambda_{(i-1)}}\right]E[(i - 1)/(i - 2)]\right]$$ (A17-6)

and finally

$$\frac{1}{\lambda_i} + \sum_{k=0}^{i-1}\left(\prod_{j=k+1}^{i}\mu_j \Bigg/ \prod_{j=k}^{i}\lambda_j\right)$$

Since by definition steady-state MTBF, M_{ss}, is

$$M_{ss} = E[(n - d + 1)/(n - d)]$$

And since as indicated before

$$\lambda_i = (n - i)\lambda$$

$$\mu_i = i\mu$$

(A17-6) can be expressed as

$$M_{ss} = \frac{1}{d\lambda} + \sum_{k=0}^{n-d-1}\left[\prod_{j=1+k}^{n-d} ju \Bigg/ \prod_{j=k}^{n-d}(n - j)\lambda\right]$$ (A17-7)

---

*For more information on the application of the expectation procedure to the evaluation of redundant subsystems, see R. Dick (1963) "The MTBF and Availability of Compound Redundant Systems," Proceedings, Fourth Annual New York Conference on Electronic Reliability, sponsored by the IEEE.

Equation (A17-7) can be put in another form expanding the summation term by term

$$\sum_{k=0}^{n-d-1} \left[\prod_{j=k+1}^{n-d} ju \bigg/ \prod_{j=k}^{n-d} (n-j)\lambda \right] = (d-1)!(n-d)! \sum_{k=0}^{n-d-1} \frac{\mu^{n-d-k}}{\lambda^{n-d+1-k}(n-k)!\,k!}$$

And after some manipulation the above can be put in the form

$$= \frac{d!(n-d)!}{d\lambda^{n-d+1}\mu^d} \sum_{k=0}^{n-d-1} \frac{\mu^{n-k}\lambda^k}{(n-k)!\,k!}$$

(A17-7) can now be expressed as

$$M_{ss} = \frac{1}{d\lambda} + \frac{d!(n-d)!}{d\lambda^{n-d+1}\mu^d} \sum_{k=0}^{n-d-1} \frac{\mu^{n-k}\lambda^k}{(n-k)!\,k!}$$

$$= \frac{d!(n-d)!}{d\lambda^{n-d+1}\mu^d} \sum_{k=0}^{n-d} \frac{\mu^{n-k}\lambda^k}{(n-k)!\,k!} \qquad (A17\text{-}8)$$

which is the same as the expression for M_{ss} derived in Chapter 10 using an alternate approach.

Relationship of M_{ss} to Mean Time to Failure or Mean Time to First Failure

It is obvious that in order to fail (assuming the units of the subsystem were all operating at $t = 0$), the subsystem must gracefully degrade. That is to say, from state (0) it must eventually go to state (1), from state (1) it must eventually go to state (2), etc., and the average time for each graceful degradation form state (i) to $(i + 1)$ (taking into account transitions from (i) to $(i - 1)$ is accounted for by (A17-4).
    Therefore,

$$\sum_{i=0}^{n-d} E[(i + 1)/i] = \text{MTTF, MTFF} \qquad (A17\text{-}9)$$

That is, the sum of the expected times to transition from state (i) to $(i + 1)$ to $(i + 2)$, etc., is the expected time to go from state (0) to state $n - d + 1$, which is the mean time to system failure. Repeated use of (A17-4) in (A17-6) to

$$\text{MTTF} = \text{MTFF} = \sum_{i=0}^{n-d} E[(i + 1)/i]$$

$$= \sum_{i=d}^{n} [i!(n-i)!/i\lambda^{n-i+1}\mu^i] \sum_{k=0}^{n-i-1} [\mu^{n-k}\lambda^k]/[(n-k)!\,k!]$$

or its equivalent

$$= \sum_{i=d}^{n} \left([1/i\lambda] + \sum_{k=0}^{n-i-1} \left[\prod_{j=k+1}^{n-i} ju \Big/ \prod_{j=k}^{n-i} (n-j)\lambda \right] \right) \qquad \text{(A17-10)}$$

A18   STEADY-STATE MTBF AND MEAN TIME TO FAILURE OF MAINTAINED STANDBY REDUNDANT SUBSYSTEMS

The concepts of the previous section can be used to develop a relationship for steady-state MTBF and MTTF for a subsystem employing standby redundancy.

Assume a subsystem comprised of n redundant identical units (each with identical values of λ and μ). In addition, all failure and repair times of units take on exponential density functions as defined previously. Assume that d units, $d < n$, must operate at all times in order for the system to operate. All other units (both in standby and in repair) are not energized and have $\lambda = 0$. They remain unenergized until a failure occurs in one of the operating units, and only then is one energized. A repair action is immediately started on the failed unit. When repaired, the unit enters the standby pool. The subsystem fails when one of the d operating units fails and no standby unit is available to take its place (i.e., all $(n-d)$ units under repair when a failure occurs). The system has a number of possible states:

State	Units operating	Units in standby	Units under repair
0	d	$(n-d)$	0
1	d	$(n-d-1)$	1
2	d	$(n-d-2)$	2
$\vdots$	$\vdots$	$\vdots$	$\vdots$
$(n-d)$	d	0	$(n-d)$
n	0	0	n

Again transition can take place only among adjacent states. That is, given that the system is in state (i) $(i \geqslant 0)$, the system can next go to either state $(i+1)$ or $(i-1)$, and no others.

Naturally, for example, when two of n units are undergoing repair (are failed), the next transition must be either to one unit undergoing repair (signifying one unit repaired) or three units undergoing repair (signifying another operating unit failed while repair of the two failed units continued).

It is also clear that immediately before the system fails, (i.e., reaches state $(n-d+1)$), the system must be in state $(n-d)$.

Each state (i) has associated with it transition rates: a state failure rate of λ_i, which will indicate the state's propensity to transition to state $(i + 1)$; and a system repair rate μ_i, which will indicate its propensity to transition to state $(i - 1)$.

For standby redundancy,

$\lambda_i = d\lambda, \quad i \leqslant n$

$\mu_i = i\mu, \quad i > 0$ (assumes that as soon as a unit fails, repair is begun)

Since we are dealing with exponential density functions for repairs and failure, expected time in state (i) before *any* transition is

$$E(i) = \frac{1}{d\lambda + i\mu} \tag{A18-1}$$

The probability of going to state $(i + 1)$ on the next transition, given that the system is in state (i) at present (signifies an additional failure) is

$$P[(i + 1)/i] = \frac{d\lambda}{d\lambda + i\mu} \tag{A18-2}$$

The probability of going to state $(i - 1)$ on the next transition, given that the system is in state (i) at present (signifies a repair completed) is

$$P[(i - 1)/i] = \frac{i\mu}{d\lambda + i\mu} \tag{A18-3}$$

$$P[(i - 1)/i] + P[(i + 1)/i] = 1$$

The expected length of time to go from state (i) to state $(i + 1)$, $E[(i + 1)/i]$, can be formulated as

$$E[(i + 1)/i] = P[(i + 1)/i] \cdot E(i) + P[(i - 1)/i]$$
$$\times \{E(i) + E[i/(i - 1)] + E[(i + 1)/i]\}$$

Substituting (A18-1)–(A18-3) into the above, regrouping, and simplifying

$$E[(i + 1)/i] = \frac{1}{d\lambda} + \frac{i\mu}{d\lambda} E[i/(i - 1)] \tag{A18-4}$$

where $i > 0$ (if $I = 0$, $E[i/(i - 1)] = 0$).

Expansion of (A18-4) leads to the general result for any i,

$$E[(i + 1)/i] = \sum_{k=0}^{i} \frac{i!(\mu)^k}{(i - k)!(d\lambda)^{k+1}} \tag{A18-5}$$

If $E[(i + 1)/i]$ denotes the average time to transition from state (i) to state $(i + 1)$, then steady-state MTBF, M_{ss}, is

$$M_{ss} = E[(n - d + 1)/(n - d)]$$

Assuming the units of the system are all in operable condition at $t = 0$, in order to fail the system must go from state (0) to state (1), from state (1) to state (2), from (2) to state (3), etc., until the failure state $(n - d + 1)$ is reached. The average time for such a transition, the MTTF, is

$$\text{MTTF} = \sum_{i=0}^{n-d} E[(i + 1)/i] \tag{A18-6}$$

From (A18-5),

$$\text{MTTF} = \sum_{i=0}^{n-d} E[(i + 1)/i] = \sum_{i=0}^{n-d} \sum_{k=0}^{i} \frac{i!\,\mu^k}{(i - k)!\,(d\lambda)^{k+1}}$$

A19   APPROXIMATE DISTRIBUTION OF A PROPORTION FOR LARGE SAMPLE SIZES

This approximation is based on the fact that when sample size is large and when $(1 - p)n$ and pn are both >5 the binomial distribution may be approximated by a normal distribution having a mean $\mu = np$ and a standard deviation $\sigma = [np(1 - p)]^{1/2}$. In accord with the discussion of the standard normal distribution in Section 2.2.3, the quantity

$$K_m = \frac{n\bar{p} - np}{\sqrt{np(1 - p)}} \tag{A19-1}$$

is distributed as a standard normal variable where

$\bar{p}$ $= (n - r_0)/n =$ any observed value of proportion of "good" items
n $=$ the number of items in the sample
r_0 $=$ the number of items that failed to perform successfully—a random value
$n\bar{p}$ $=$ an observed sample estimate based on $\bar{p}$
p $=$ the actual proportion of "good" items in the population from which the sample was chosen
np $=$ the expected number (mean) of "good" items in the sample
K_m $=$ the normal distribution coefficient indicating the probability that a given magnitude of difference between p and $\bar{p}$ will occur

Canceling out like terms (A19-1) can be expressed in terms which allow its use to develop confidence bounds for proportions

$$K_m = \frac{\bar{p} - p}{\sqrt{\dfrac{p(1 - p)}{n}}} \tag{A19-2}$$

We rearrange the above such that if p = the true value of the mean p of the time a sample value of $\bar{p}$ would result such that

$$\bar{p} \leqslant p + K_m \left[\frac{p(1 - p)}{n} \right]^{1/2} \tag{A19-3}$$

and

$$\bar{p} \geqslant p - K_m \left[\frac{p(1 - p)}{n} \right]^{1/2}$$

where
$P \quad = \int_{-\infty}^{K_m} \phi(u)\, du$
$\phi(u)$ = the normal distribution density function

A20   TRANSFORMATION OF THE WEIBULL DISTRIBUTION TO AN EXPONENTIAL FORM TO FACILITATE RELIABILITY MEASUREMENT AND DEMONSTRATION

The reliability function and the mean of the Weibull distribution (see Section 2.1.2 for more detail) may be expressed as

$$R(T) = \exp\left[-\left(\frac{T}{\alpha} \right)^{\beta} \right] \tag{A20-1}$$

$$\text{MTTF} = \phi = \alpha\Gamma\left(1 + \frac{1}{\beta} \right) \tag{A20-2}$$

or we can express

$$\alpha = \frac{\phi}{\Gamma[1 + (1/\beta)]} \tag{A20-3}$$

where

T = interval of operation measured from the time the item was first energized

α = a scale parameter

β = a shape parameter

It is assumed that β is known or can be closely approximated. Since β is considered known, α and T can be transformed such that

$$\left(\frac{1}{\alpha}\right)^{\beta} = \frac{1}{\alpha_x} \tag{A20-4}$$

or

$$\alpha_x = \alpha^{\beta} = \left(\frac{\phi}{\Gamma[1 + (1/\beta)]}\right)^{\beta} \tag{A20-5}$$

$$T^{\beta} = T_x \tag{A20-6}$$

As a consequence of the transformation, (A20-1) can be expressed as

$$R(T^{\beta}) = R(T_x) = \exp(-T_x/\alpha_x) \tag{A20-7}$$

An exponential distribution in transformed time T_x, where T, and β are defined as before.

For reliability measurement, the objective is to define upper and lower confidence bounds (and intervals) for MTTF = ϕ.

Since β is known and (A20-2) shows a direct proportionality between ϕ and α, any confidence bound on α can be directly related to a confidence bound on ϕ. Further, from (A20-4) it is clear that α_x can be directly related to α. As a result any confidence bound on α_x can be related to a confidence bound on α.

Equation (A20-7) indicates that α_x can be treated as the exponential parameter θ is, after the time transformation described in (A20-6) is applied, and when β is considered known.

Since, from (A20-5),

$$\phi = \alpha_x^{1/\beta}\Gamma\left(1 + \frac{1}{\beta}\right) \tag{A20-8}$$

A lower confidence bound on α_x, treating it as though it were the MTBF parameter, would correspond to a lower bound on α which could be transformed into a lower bound on ϕ. A similar logic pertains to the estimation of an upper confidence bound.

Index